Surveying for Construction
Fourth Edition

Surveying for Construction

Fourth Edition

William Irvine, FRICS

Lecturer in Land Surveying
Glasgow College of Building and Printing

McGRAW-HILL BOOK COMPANY

London · New York · St Louis · San Francisco · Auckland · Bogotá · Caracas · Lisbon · Madrid
Mexico · Milan · Montreal · New Delhi · Panama · Paris · San Juan · São Paulo
Singapore · Sydney · Tokyo · Toronto

Published by

McGRAW-HILL Book Company Europe

SHOPPENHANGERS ROAD · MAIDENHEAD · BERKSHIRE · SL6 2QL · ENGLAND

TEL: 01628 23432; FAX: 01628 770224

British Library Cataloguing in Publication Data

Irvine, William
  Surveying for Construction. — 4Rev.ed
  I. Title
  526.9024624

    ISBN 0-07-707998-1

Library of Congress Cataloging-in-publication Data

Irvine, William Hyslop.
  Surveying for construction / William Irvine. — 4th ed.
    p. cm.
  Includes index.
  ISBN 0-07-707998-1 :
  1. Surveying. 2. Building sites. I. Title.
  TA625.I78 1995
  526.9′2—dc20

                                94-48598
                                      CIP

1234 CUP 98765

Typeset by Paston Press Ltd · Loddon · Norfolk
Printed and bound in Great Britain at the University Press · Cambridge
Printed on permanent paper in compliance with ISO Standard 9706

Contents

Preface

This is the fourth edition of *Surveying for Construction*. It has been necessitated by five factors: (a) the continuing developments in surveying instrumentation and computing techniques, (b) the resulting demise of some traditional methods of surveying, (c) the introduction of the competence-based method of learning and assessment in colleges of education and universities, (d) the widely held belief among academics and employers that the standard of education of school leavers is in decline and (e) the return of adult learners to education, many of whom are lacking in, or have forgotten, the basics of mathematics and measurement fundamentals. To reconcile these five factors in any textbook is a difficult task, since the first two place a greater degree of emphasis on the higher end of surveying education, while the latter three indicate a need to return to the basics.

The problem is further exacerbated by the fact that students are nowadays computer literate and are accustomed to communicating and learning interactively, through computer programs and games, which encourage learning via examples and self-assessments.

This book attempts to guide students towards certain objectives by (a) introducing theories and principles in a condensed but, hopefully explicit manner, (b) exemplifying these theories in a series of well-structured examples, with which the student can work interactively, (c) providing self-assessment exercises (with answers) at frequent intervals throughout every subject area and (d) including a project (Chapter 18), which, from my experience in teaching all levels of surveying students, adequately reinforces the lessons.

This fourth edition has been completely revised and rewritten wherever necessary. In response to numerous requests, the section on setting out has been greatly amended and enlarged and a completely new chapter on radial positioning and use of mapping systems has been added.

With the advance in surveying technology, many traditional surveying methods and techniques have been rendered obsolete. All have been removed, leaving a thoroughly modern textbook which should adequately satisfy the needs of students who are studying surveying to first year degree level.

Acknowledgements

My sincere thanks are due to the following individuals and organizations:

The British Standards Institution, for permission to reproduce and adapt drawings from *BS 4484: Measuring Instruments for Construction Works, Part 1 1969; Metric Graduation and Figuring of Instruments for Linear Measurement*.

D. T. F. Munsey FRICS and the Technical Press Ltd for allowing the reproduction of tacheometric tables from the book *Tacheometric Tables for the Metric User*.

E. Mason (Editor) and Virtue and Co. Ltd for permission to adapt material from *Surveying* by J. L. Holland, K. Wardell and A. G. Webster.

Sokkia UK Ltd; Geotronics Ltd; Leica UK Ltd; Hall Watts Ltd; Pentax Ltd; Laser Alignment Inc.; Microsoft Corporation; Projectina Co. Ltd; Rabone Chesterman Ltd; Topcon Corporation of America; for permission to use and adapt illustrations, etc., from their technical literature.

The City and Guilds of London Institute; The Institute of Building; The Royal Institution of Chartered Surveyors; The Scottish Vocational Education Council; for permission to use questions from their examination papers.

The responsibility for the drawing and accuracy of all diagrams, the compilation and solutions of all questions and the compilation of all computer programs and spreadsheets is entirely my own. If I have inadvertently used material without permission or acknowledgement, I sincerely apologise and hope that any oversight will be excused.

CHAPTER 1

Surveying fundamentals

Objective

After studying this chapter, the reader should be able to demonstrate an understanding of the units of measurement and scaling of plans, by producing a drawing of a small surveyed area of ground, to scale.

Every day of the year, in many walks of life, maps and plans are in common use. These maps and plans include street maps, charts of lakes and rivers, underground railway maps, construction site plans and architectural plans and sections.

All of these plans are drawn to scale by cartographers, engineers, architectural draughtsmen or surveyors, from measurements of distances, heights and angles. The measurements are made by surveyors or engineers, using surveying instruments such as tapes, levels and theodolites. They are employed in the field of land surveying, which, put in its simplest form, is the science and art of measuring, recording and drawing to scale, the size and shape of the natural and man-made features on the surface of the earth.

All measurements of height and distance and all scaled drawings use the metric system of measurement, in common with most countries of the world, but angles are measured in sexagesimal units. This study of surveying therefore begins with a revision of the most commonly used units of measurement.

1. Units of measurement

(a) Linear

The metric system was officially introduced in France in the year 1799 and was standardized in 1964. The Système Internationale (SI) sets out the basic and derived units which were agreed internationally at that time.

The following units are most important to the surveyor:

Quantity	Unit	Symbol
Length	metre	m
Area	square metre	m^2
Volume	cubic metre	m^3
Mass	kilogramme	kg
Capacity	litre	l

Taking any one of the above-listed quantities as the basic unit, a table of multiples and sub-multiples is derived by prefixing the basic unit as shown in Table 1.1.

Prefix	Multiplication factor	Derived unit	SI recommended unit
kilo	1000	kilometre	kilometre (km)
hecto	100	hectometre	
deca	10	decametre	
		metre	metre (m)
deci	0.10	decimetre	
centi	0.01	centimetre	
milli	0.001	millimetre	millimetre (mm)

Table 1.1

In Table 1.1, it should be noticed that only three units are recommended for general use. This is true for other quantities also and Table 1.2 shows the small selection of the units included in the SI system that are now in common use.

Quantity	Recommended SI unit	Other units that may be used
Length	kilometre (km) metre (m) millimetre (mm)	centimetre (cm)
Area	square metre (m^2) square millimetre (mm^2)	square centimetre (cm^2) hectare (100 m × 100 m) (ha)
Volume	cubic metre (m^3) cubic millimetre (mm^3)	cubic decimetre (dm^3) cubic centimetre (cm^3)
Mass	kilogramme (kg) gramme (g) milligramme (mg)	
Capacity	cubic metre (m^3) cubic millimetre (mm^3)	litre (l) millilitre (ml)

Table 1.2

Finally, it is necessary to see exactly how mass, volume and capacity are related. Table 1.3 shows the basic relationship from which others may be deduced.

Volume	Mass	Capacity
1 cubic metre	1000 kilogrammes	1000 litres
1 cubic decimetre	1 kilogramme	1 litre
1 cubic centimetre	1 gramme	1 millilitre

Table 1.3

(b) Angular

In many countries of the world, angles are measured in degrees, which is a sexagesimal unit. Degrees are subdivided into minutes and seconds, in exactly the same manner as time. Thus there are sixty minutes in one degree and sixty seconds in one minute.

It is worth noting that on the Continent of Europe, angles are measured in grades. There are four hundred grades in one complete revolution, whereas there are three hundred and sixty degrees in one revolution.

Examples

Linear measurements should be written to three decimal places to avoid confusion, unless required otherwise.

1 Find the sum of the following measurements:
(a) 1 metre and 560 millimetres

Answer: 1.000
   + 0.560
   = 1.560 m

(b) 15 metres and 31 centimetres

Answer: 15.000
   + 0.310
   = 15.310 m

(c) 25 cm and 9 cm and 8 mm

Answer: 25.000
   + 0.090
   + 0.008
   = 25.098 m

2 Calculate the area (in hectares) of a rectangular plot of ground measuring 85 m long by 160 m wide.

Answer: $85.000 \times 160.000 = 13\,600 \text{ m}^2$
   = 1.360 ha

Angular measurements are written in sexagesimal form as follows:

10 degrees, 23 minutes and 18 seconds =
   $10°\ 23'\ 18''$

---

3 In surveying, angles have to be added or subtracted frequently. Find the value of:
(a) $23°\ 24'\ 30'' - 10°\ 18'\ 15''$

Answer: $23°\ 24'\ 30''$
   $- 10°\ 18'\ 15''$
   $= 13°\ 06'\ 15''$

(b) $56°\ 35'\ 20'' - 15°\ 19'\ 45''$

Answer: $56°\ 35'\ 20''$
   $- 15°\ 19'\ 45''$

Since 1 min = 60 s, the calculation becomes

   $56°\ 34'\ 80''$
   $- 15°\ 19'\ 45''$
   $= 41°\ 15'\ 35''$

(c) $14°\ 35'\ 52'' + 12°\ 39'\ 23''$

Answer: $14°\ 35'\ 52''$
   $+ 12°\ 39'\ 23''$
   $= 26°\ 74'\ 75''$

Since 1 min = 60 s and 1 deg = 60 min, the answer becomes

   $26°\ 74'\ 75''$
   $= 26°\ 75'\ 15''$
   $= 27°\ 15'\ 15''$

Adding the figure 60 to either minutes or seconds is, of course, done mentally.

The calculations are facilitated, on a pocket calculator, by the key marked DMS or [° ' "]. All calculators work differently and users should consult their calculator manual for full instructions.

Using a Casio calculator, the problem in Example 3(c) is solved as follows:

Operation	Calculator instruction
1 Enter 14 deg	Press 14, [° ' "]
2 Enter 35 min	Press 35, [° ' "]
3 Enter 52 s	Press 52, [° ' "]
4 Add	Press [+]
5 Enter 12 deg	Press 12, [° ' "]
6 Enter 39 min	Press 39, [° ' "]
7 Enter 23 s	Press 23, [° ' "]
8 Answer (decimal)	Press [=]
9 Answer (sexagesimal)	Press INV, [° ' "]

Exercise 1.1

1 Write the following measurements in metres, to three decimal places:
(a) 4 metres and 500 millimetres
(b) 3 m and 17 mm
(c) 3 m, 40 cm and 67 mm
(d) 100 m and 9 mm
(e) 10 cm

2 Calculate the following, giving answers to three decimal places:
(a) 3 m + 4 cm + 250 mm + 35 cm
(b) 10 m − 82 cm + 140 mm + 120 cm
(c) 435 mm + 965 mm + 8 mm
(d) 10.326 m + 9 mm − 126 mm + 17.826 m
(e) 10 m + 10 cm + 100 mm − 15 cm − 50 mm
3 Calculate the area (in hectares) of the following rectangular plots of land:
(a) 100 m long by 150 m wide
(b) 90 m long by 53 m wide
(c) 90.326 m long by 265.112 m wide
4 (a) Find the sum of the three angles of a triangle, measured using a theodolite (a theodolite is a surveying instrument used for the accurate measurement of angles).

Angle ABC: 58° 17′ 40″, angle BCA: 67° 23′ 20″ and angle CAB: 54° 18′ 20″.
(b) Find the error of the survey.

2. Understanding scale

The end product of a survey is usually the production of a scaled drawing and throughout the various chapters of this book scaled drawings will have to be made. A scale is a ratio between the drawing of an object and the actual object itself.

Example

4 Figure 1.1 is the drawing of a two pence piece. The diameter of the coin on the drawing is 13 mm. An actual two pence piece has a diameter of 26 mm; hence the scale of the drawing is

$$\frac{\text{Plan size}}{\text{Actual size}} = \frac{13\,\text{mm}}{26\,\text{mm}} = \frac{1}{2}$$

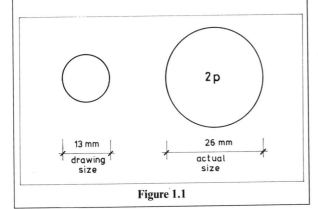

2 p

| 13 mm | | 26 mm |
| drawing size | | actual size |

Figure 1.1

(a) Methods of showing scale

The scale of a map or plan can be shown in three ways:

1. *It may simply be expressed in words*
For example, 1 centimetre represents 1 metre. By definition of scale, this simply means that one centimetre on the plan represents 1 metre on the ground.

2. *By a drawn scale*
A line is drawn on the plan and is divided into convenient intervals such that distances on the map can be easily obtained from it. If the scale of 1 centimetre represents 1 metre is used, the scale drawn in Fig. 1.2 would be obtained.

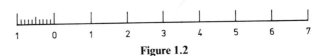

Figure 1.2

Figure 1.2 is an example of an open divided scale in which the primary divisions (1.0 metre) are shown on the right of the zero. The zero is positioned one unit from the left of the scale and this unit is subdivided into secondary divisions. An alternative method of showing a drawn scale is to fill in the divisions, thus making a filled line scale, an example of which is shown in Fig 1.3.

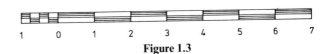

Figure 1.3

3. *By a representative fraction*
With this method of showing scale, a fraction is used in which the numerator represents the number of units on the map (always 1) and the denominator represents the number of the same units on the ground. With a scale of 1 centimetre represents 1 metre, the representative fraction will be 1/100, shown as 1:100, since there are 100 centimetres in 1 metre.

A representative fraction (RF) is the international way of showing scale. Any person looking at the RF on a map thinks of the scale in the units to which he is accustomed. A scale of 1:129 600 would mean to an American that 1 inch equals 129 600 inches or 2 miles and to a European that 1 millimetre equals 129 600 millimetres or 129.6 metres.

Examples

5 Calculate the scale of a plan where 1 mm represents 0.5 m.

$$\text{Scale} = \frac{\text{plan size}}{\text{actual size}} = \frac{1\,\text{mm}}{0.5\,\text{m}} = \frac{1\,\text{mm}}{500\,\text{mm}} = \frac{1}{500}$$

6 Figure 1.4 is the scale drawing of a badminton court drawn to a scale of 1:250. Using the open divided scale provided, measure:
(a) the overall length and breadth of the court,
(b) the distance between the tram lines,
(c) the size of the service courts.

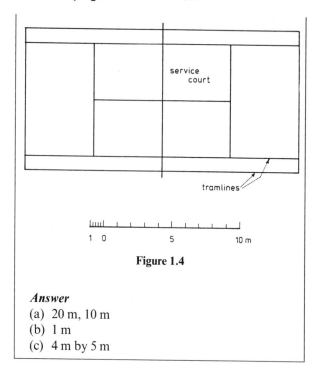

Figure 1.4

Answer
(a) 20 m, 10 m
(b) 1 m
(c) 4 m by 5 m

(b) Conversion of area by representative fractions

Frequently, in survey work, the area of a parcel of land is measured on a scaled plan, using an instrument called a planimeter (Chapter 14). A planimeter measures areas in square centimetres (cm^2) and the actual ground area has to be calculated. If the RF on a plan is very large, say 1:4, one unit on the plan will represent four units on the ground. A square of 1 unit on the plan will therefore represent a ground area of 4 units × 4 units. From these facts emerges a simple formula:

$$\text{Plan scale} = 1{:}4$$
$$\text{Plan area} = 1 \times 1 \text{ sq. units}$$
$$\text{Therefore ground area} = (1 \times 4) \times (1 \times 4) \text{ sq. units}$$
$$= 1 \times (4 \times 4) \text{ sq. units}$$
$$= 1 \times 4^2 \text{ sq. units}$$
$$\text{i.e. ground area} = \text{plan area} \times 4^2$$
$$= \text{plan area} \times (\text{scale factor})^2$$

$$\text{Therefore plan area} = \frac{\text{ground area}}{4}$$
$$= \text{ground area} \times (\text{RF})^2$$

Examples

7 An area of 250 cm^2 was measured on a plan, using a planimeter. Given that the plan scale is 1:500, calculate the ground area in m^2.

Answer
$$\text{Plan area} = 250 \text{ cm}^2$$
$$\text{RF (scale)} = 1{:}500$$
$$\text{Ground area} = \text{Plan area} \times (\text{scale factor})^2$$
$$= (250 \times 500^2) \text{ cm}^2$$
$$= 250 \times \frac{500^2}{100^2} \text{ m}^2$$
$$= (250 \times 25) \text{ m}^2$$
$$= 6250 \text{ m}^2$$

8 An area was measured on a plan by a rule as 175 × 250 mm. Calculate the ground area in square metres if the scale is:
(a) 1:2000
(b) 1:500

Solution
(a)
$$\text{RF} = 1{:}2000$$
$$\text{i.e. 1 mm (plan)} = 2000 \text{ mm (ground)}$$
$$\text{Therefore 1 mm}^2 \text{ (plan)} = 2000 \times 2000 \text{ mm}^2$$
$$\text{(ground)}$$
$$\text{Plan area} = 175 \times 250 \text{ mm}^2$$
$$\text{Therefore ground area} = 175 \times 250 \times 2000$$
$$\times 2000 \text{ mm}^2$$
$$= 175 \times 250$$
$$\times \frac{2000}{1000} \times \frac{2000}{1000} \text{ m}^2$$
$$= 175\,000 \text{ m}^2$$
(b) By formula:
$$\text{Plan area} = \text{ground area} \times (\text{RF})^2$$
$$\text{Therefore ground area} = (175 \times 250)$$
$$\times 500^2 \text{ mm}^2$$
$$= 175 \times 250$$
$$\times \frac{500}{1000} \times \frac{500}{1000} \text{ m}^2$$
$$= 10\,937.5 \text{ m}^2$$

Check on solution Scale (a) is four times smaller than (b). Therefore there will be sixteen times the area for the same size plan:

$$10\,937.5 \times 16 = 175\,000 \text{ m}^2$$

Exercise 1.2

1 Calculate the scale of a plan where 1 cm represents 20 m.

2 Figure 1.5 is the plan view of a house drawn to scale 1:100. Beginning at point A and moving in a clockwise direction, measure the lengths of the various walls, using a scale rule.

3 A parcel of ground was measured on a 1:250 scale map, using a planimeter, and found to be 51.25 cm^2. Calculate the ground area in hectares.

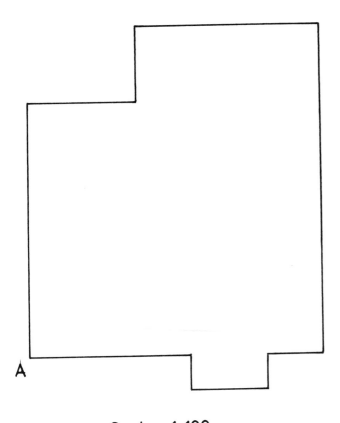

Scale 1:100

Figure 1.5

4 The dimensions of a room on a 1:50 scale plan are 60 mm × 85 mm. Calculate the area of the room in m^2.

3. Drawing to scale

A surveyor's main objective is to achieve accuracy in field operations. However, unless the results can be depicted accurately, legibly and pleasingly on paper, proficiency in the field is robbed of much of its value.

Equipment required for plotting:

1. *Paper*
The survey when plotted may have to be referred to frequently, over a number of years, and it is essential therefore that the material on which it is plotted should be stable. Modern drawing materials are excellent in that respect and most show little or no shrinkage over a long period.

2. *Scale rule*
A scale rule should conform to British Standard 1347. Most scale rules are made of plastic. They are double sided and are usually manufactured with eight scales:

1:1, 1:5, 1:20, 1:50, 1:100, 1:200, 1:1250 and 1:2500

Other common plotting scales, 1:500 and 1:1000, can be derived simply by multiplying the scale units of the 1:50 and 1:100 scales by the factor 10.

3. Other equipment includes two set squares, 45° and 60°, varying grades of pencils, 4H to H, sharpened to a fine point, paper weights, curves, inking pens, pricker pencil, erasers, steel straight edge and spring-bow compasses.

(a) Procedure in plotting

Example

9 Figure 1.6 is an example of a surveyor's field measurements of a small ornamental garden. These field measurements are to be plotted to scale 1:100 on an A4 size drawing sheet, to produce a finished drawing.

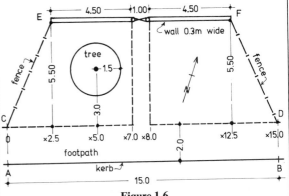

Figure 1.6

Solution
1. *Orientate the survey*
The north direction on any plan should point towards the top of the paper and the subject of the plan should lie in the centre of the sheet.

2. *Centre the drawing on the paper*
The overall size of the garden, as determined from the field measurements, is 15 m long and 8 m wide. At scale 1:100, the overall size of an A4 sheet excluding the border is 28 m by 19 m. The starting point A should therefore lie 6.5 m from the left edge and 5.5 m from the bottom of the paper (Fig. 1.7).

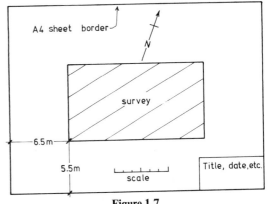

Figure 1.7

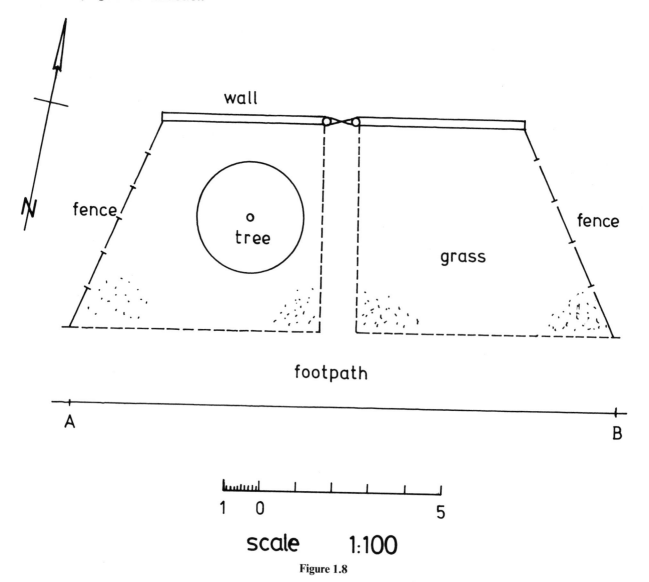

Figure 1.8

3. *Plot a base line* (Fig. 1.8)
Line AB, being the longest line, is chosen as the base line. Beginning at point A, the line is drawn parallel to the bottom of the sheet and scaled accurately at 15.0 m to represent the toe of the kerb.

4. *Complete the survey*
Using two set squares, line CD is drawn parallel to AB, at a distance of 2 m to the north of AB. Between points C and D, the various running dimensions from 2.5 to 12.5 m are accurately scaled and, using a set square, the remaining dimensions to E, F and the tree are scaled at right angles to line CD.

5. *Make a finished drawing*
The surveyed details are drawn in ink, using a drawing pen, set square and compasses, to produce a finished drawing in accordance with British Standard BS 1192, Part IV. Suitable symbols are shown in Fig. 2.2 in Chapter 2. A scale is drawn parallel to the bottom edge of the sheet and a title, date, etc., are added in a box in the bottom right-hand corner of the sheet.

Exercise 1.3

1 Plot the field notes of Fig. 1.6 to scale 1:100 and make a finished drawing of the result. Compare the drawing with Fig. 1.8.

4. Answers

Exercise 1.1

1 (a) 4.500 m
   (b) 3.017 m
   (c) 3.467 m

(d) 100.009 m
(e) 0.100 m
2 (a) 3.640 m
(b) 10.520 m
(c) 1.408 m
(d) 28.035 m
(e) 10.000 m
3 (a) 1.500 ha
(b) 0.477 ha
(c) 2.395 ha

4 (a) 179° 59′ 20″
(b) 00° 00′ 40″

Exercise 1.2

1 1:2000
2 6.75 m, 3.00 m, 2.00 m, 5.00 m, 8.75 m, 1.50 m, 0.90 m, 2.100 m, 0.90 m, 4.40 m
3 0.032 ha
4 12.75 m^2

Understanding maps and plans

Maps and plans are representations, on paper, of physical features on the ground. It is important to note the difference between a map and a plan. A plan will accurately define widths of roads, sizes of buildings, etc.; in other words, every feature is exactly true to scale. A map, on the other hand, is a representation, no matter how accurately it may be shown.

As an example, a winding country road about the width of a car measures almost 1 millimetre on a 1:50 000 map. This represents 50 metres, far in excess of the actual width of the road. Of the Ordnance Survey productions only the 1:1250 and 1:2500 can be considered to be plans in the strict sense of the word.

Site plans, on the other hand, are large-scale productions and are essential to the planning and development of a construction project.

1: Site plans

Figure 2.1 is part of a plan of a proposed development, drawn to scale 1:200.

(a) Code of symbols

In order to read a site plan, it is necessary to understand the code of signs and symbols used to depict ground features. Figure 2.2 shows some of the symbols of the Code of Signs, BS 1172, Part IV, together with alternatives commonly used by architects and engineers.

The signs are, as far as possible, plan views of the actual objects being portrayed. Thus, a tree would appear as a circle, the diameter of which represents the spread of the branches. Other features require some annotation; thus a small circle with the letters LP represents a lamp post and a small square with the letters MH represents a manhole.

Examples

1 Using the BS code of symbols of Fig. 2.2, list the features forming the boundaries of the site in Fig. 2.1.

Answer

South boundary .... a hedge
West boundary .... partly a hedge, partly a stob and wire fence
North boundary .... a timber fence
East boundary .... partly a brick wall, partly a stob and wire fence

2 State the number of trees on the site.

Answer

Six

3 State the number and type of manholes on Hillside Road.

Answer

Two foul manholes (sewerage) and two storm water manholes (rainfall)

4 Describe the proposed development works on site.

Answer

The building is a nursing home, serviced by a driveway from Hillside Road

(b) Scale

A full explanation of scales and scaling was given in Chapter 1 and should be revised, if required.

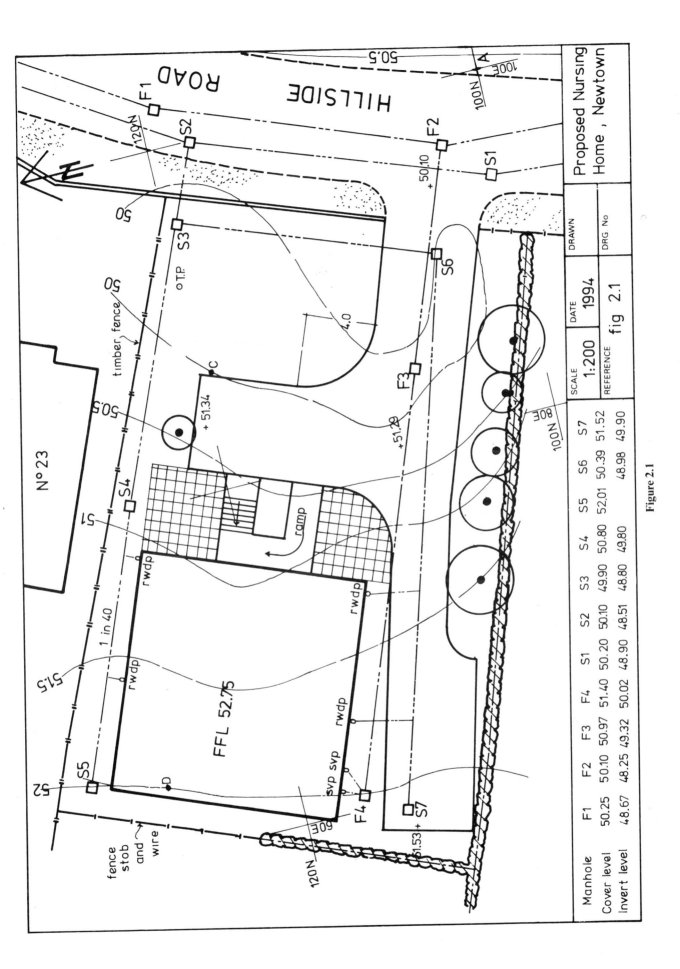

Figure 2.1

Manhole	F1	F2	F3	F4	S1	S2	S3	S4	S5	S6	S7
Cover level	50.25	50.10	50.97	51.40	50.20	50.10	49.90	50.80	52.01	50.39	51.52
Invert level	48.67	48.25	49.32	50.02	48.90	48.51	48.80	49.80		48.98	49.90

9

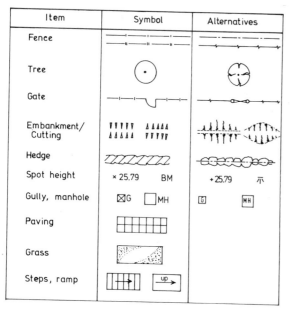

Item	Symbol	Alternatives
Fence		
Tree		
Gate		
Embankment/ Cutting		
Hedge		
Spot height	× 25.79 BM	+25.79
Gully, manhole	⊠G ▢MH	⌷G ▣MH
Paving		
Grass		
Steps, ramp		up

Figure 2.2

Example

5 Using the scale stated on the plan, measure the overall size of the building and the various widths of the proposed driveway.

Answer

Building	14 m × 13.5 m
Driveway	5.5 m, 3.5 m, 5.5 m, 5.5 m

(c) Grid lines

On almost every site plan there is, or should be, a north point indicating the direction of north chosen for that particular plan. There is also a family of lines, drawn parallel to, and perpendicular to, the north direction at some convenient interval. These lines form a grid on which it is possible to position features by coordinates referenced to some origin.

On Fig. 2.1 point A is the origin of the grid, but instead of having coordinates of zero metres east and zero metres north, the point has been given coordinates of 100 m east and 100 m north. The grid lines are drawn at 20 m intervals. Any point may be coordinated by scaling the distance to the point from the grid lines. Thus manhole F1 has coordinates of 102 m east, 119 m north, stated to the nearest metre.

Examples

6 State the coordinates, to one metre, of the following features:
(a) storm manhole S4,
(b) the south-west corner of the proposed building,
(c) the single tree to the east of the building.

Answer
(a) 81 m; 126 m
(b) 60 m; 118 m
(c) 84 m; 122 m

7 State the features that have the following coordinates:
(a) 84, 108
(b) 71, 107
(c) 90, 99

Answer
(a) manhole F3
(b) tree
(c) junction of fences

(d) Surface relief

Undulations in the ground surface are shown on a plan by contour lines and spot heights. Spot heights are called levels. A full study of contours and levels is made in succeeding chapters of this book. Using contour lines and levels, it is possible to deduce the slope of the ground

Examples

8 From a study of Fig. 2.1, state the value of:
(a) the highest contour line,
(b) the lowest contour line,
(c) the height interval between successive contour lines.

Answer
(a) 52.0
(b) 50.0
(c) 0.5

9 Describe the general slope of the ground from east to west.

Answer
The ground falls from a height of 50.5 m at Hillside Road to a height of 50 m and then rises to 52 m at the western end of the site.

10 State the level of the proposed driveway at:
(a) the short side road,
(b) the western parking area.

Answer
(a) 51.29
(b) 51.53

(e) Gradients

The gradient between any two points on the ground, a road or sewer can be calculated by (a) computing the difference in height between the points and (b) scaling the horizontal distance between them. From these dimensions, the gradient between the points is expressed as the rise or fall over the horizontal length.

Example

11 Calculate the average gradient of the ground between points C and D on the plan.

Answer
Gradient = rise (C to D) over horizontal length
 CD
  = (52 − 50) m over 24 m
  = 2 m over 24 m
  = 1 in 12 or 1/12

The level of the inside of the bottom of a sewer pipe is called the invert level. In order to function properly a sewer pipe must be laid on a gradient, between two invert levels.

Example

12 Calculate the gradient of the sewer pipe between foul manholes 3 and 4.

Answer
 Invert level manhole 3 = 49.32
 Invert level manhole 4 = 50.02
 Rise (manhole 4 to 3) = 0.70
 Length (manhole 4 to 3) = 24.6 m
   Gradient = 0.70 in 24.6
    = 1 in 35 or 1/35

It is commonplace on the continent of Europe to express gradients as percentages. In order to change any vulgar fraction into a percentage, the fraction is simply multiplied by 100.

Examples

13 Express the following gradients as percentages:
(a) 1 in 10
(b) 1 in 50
(c) 1 in 16.67

Answer
(a) 1 in 10 = (1/10 × 100)% = 10%
(b) 1 in 50 = (1/50 × 100)% = 2%
(c) 1 in 16.6 = (1/16.67 × 100)% = 6%

---

14 Express the following percentage gradients as fractions:
(a) 25% (b) 1.667% (c) 6.535%

Answer
(a) 25% = 1 in (100/25) = 1 in 4
(b) 1.667% = 1 in (100/1.667) = 1 in 60
(c) 6.535% = 1 in (100/6.535) = 1 in 15.3%

Exercise 2.1

Using site plan of Fig. 2.1:

1 Measure the length of the fence on the western boundary, using a scale rule.
2 Estimate or interpolate, using the contour lines, the existing ground level at the four corners of the proposed building.
3 State the floor level of the proposed building.
4 Calculate the invert level of the storm water manhole 5.
5 Calculate the gradient of the storm water drain between storm water manholes 6 and 3.
6 State the coordinates of:
(a) storm water manhole 3
(b) foul water manhole 4
(c) the telephone pole in the north-east corner of the site
7 State the features that have coordinates of:
(a) 88.6, 113.2
(b) 78.0, 120.0
(c) 65.0, 115.8
8 Count the number of soil ventilation pipes and rain water downpipes that serve the proposed building.

Exercise 2.2

Using site plan of Fig. 2.3:

1 State the scale of the plan.
2 Using the scale, measure the overall size of the lecture room and of the proposed building.
3 State the coordinates (to the nearest metre) of the centre of the lecture room.
4 State the coordinates of manholes S1 and S2.
5 State the features that have the following coordinates:
(a) 53E; 103N (b) 89E; 112N (c) 39E; 117N
6 State the number of trees on the site.
7 Describe the general slope of the ground from north to south across the site.
8 Calculate the gradient of the pipe connecting manhole S2 to manhole S3.
9 Calculate the gradient of the slope of the ground between points A and B and between points C and D.
10 Calculate the invert level of the pipe entering the proposed building from manhole F2.

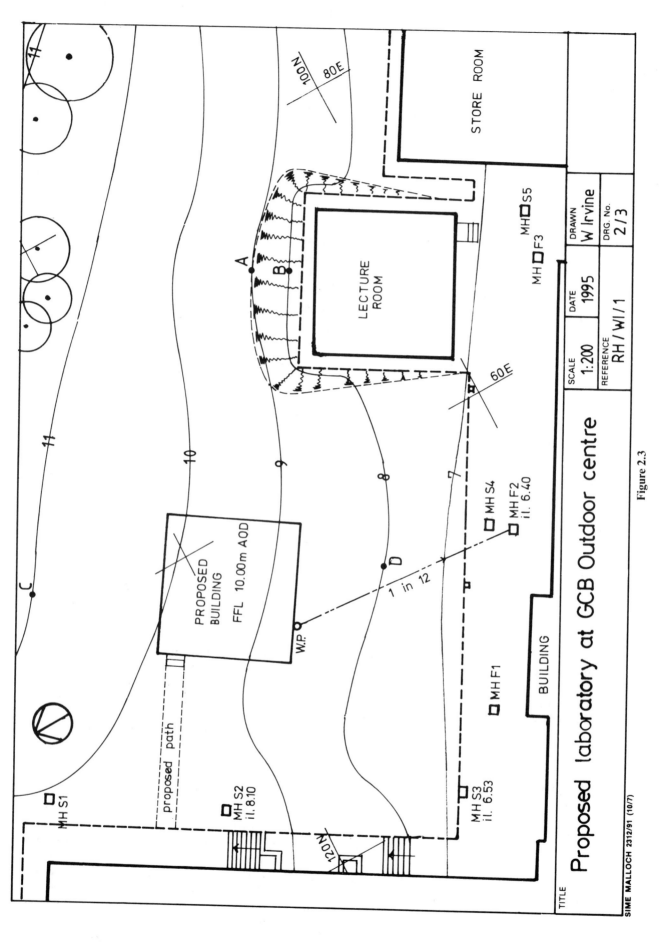

STORE ROOM

100N

80E

A
B

LECTURE ROOM

MH☐S5

MH☐F3

60E

C

10

9

8

7

11

11

PROPOSED BUILDING

FFL 10.00m AOD

W.P.

1 in 12

D

☐MH S4

MH F2
il. 6.40

☐MH F1

BUILDING

proposed path

☐
MH S2
il. 8.10

☐MH S1

☐
MH S3
il. 6.53

120N

SIME MALLOCH 2312/91 (10/7)

TITLE	Proposed laboratory at GCB Outdoor centre		
SCALE 1:200	DATE 1995	DRAWN W Irvine	
REFERENCE RH / WI / 1		DRG. No. 2/3	

Figure 2.3

12

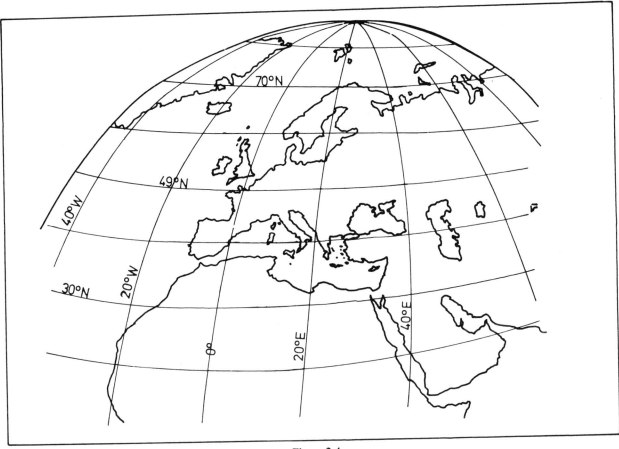

Figure 2.4

2. Ordnance Survey maps and plans

Figure 2.4 shows the Northern Hemisphere of the Earth. The small circles parallel to the Equator are lines of latitude. Latitude increases northwards from 0° at the Equator to 90° at the North Pole. Great Britain lies to the north of the 49° line of latitude.

The lines at right angles to the Equator are meridians of longitude, which are great circles passing around the Earth and meeting at the North and South Poles. Longitude is measured east or west of the Greenwich meridian which passes through London and has a value of zero degrees (0°).

(a) National Grid

Figure 2.5 is an enlarged version of part of Fig. 2.4 and shows Great Britain lying northwards of the 49° north line of latitude and practically bisected by the 2° west line of longitude. This line is called the Central Meridian. The two lines cross at right angles and their intersection forms the true origin of a rectangular grid covering the whole of Great Britain, called the National Grid (Fig. 2.6).

The north–south lines of the grid are parallel to the Central Meridian and cannot therefore point to true north. The direction in which they do point is called Grid North.

Since all east–west lines of the National Grid are parallel to the 49° north line of latitude they point truly east–west.

The 49° north line of latitude was chosen as the axis for the National Grid because all points in Great Britain lie to the north of it, while the 2° west line of longitude was chosen because it runs roughly centrally through the country. The use of these two lines as axes means that grid coordinates of points west of the Central Meridian would be negative while points on the mainland of the extreme north of Scotland would have north coordinates in excess of 1000 kilometres.

In order to keep all east–west coordinates positive and all north coordinates less than 1000 kilometres, the origin of the National Grid was moved northwards by 100 km and westwards by 400 km to a point south-west of the Scilly Isles. This point is called the false origin (Fig. 2.6). Any position in Great Britain is therefore known by its eastings followed by its northing, which are respectively the distances east and north of the false origin.

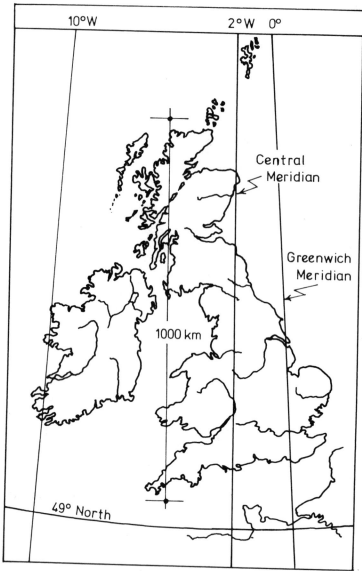

Figure 2.5

(b) Grid references

Commencing at the false origin, the National Grid of 100 km squares covers the country as shown in Fig. 2.6. Each 100 km square is given a separate letter of the alphabet (I being excepted); therefore there are 25 squares of 100 km side in any 500 km block. In order to differentiate between 500 km blocks, each is given a prefix letter H, J, N, O, S or T (anagram ST JOHN or JOHN ST).

Figure 2.7 shows how the National Grid reference is given for any point in the British Isles. The 100 km easting figure is followed by the 100 km northing figure and translated into letters of the alphabet. The letters are followed by the remaining easting figures and northing figures, quoted to any degree of accuracy, ranging from 10 kilometres to 1 metre.

Example

15 The National Grid coordinates of Ben Nevis are 216745E, 771270N while those of Cardiff Castle are 318100E, 176610N. Quote (a) the NG 1 metre and (b) the NG 100 metre reference of each point.

Solution

Ben Nevis
(a) NN1674571270
(b) NN167712
Cardiff Castle
(a) ST1810076610
(b) ST181766

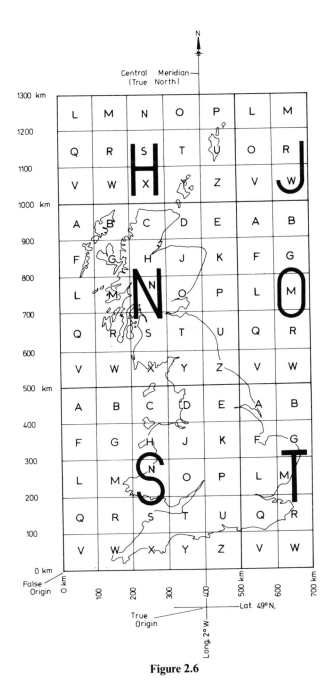

Figure 2.6

(c) Scales of OS maps and plans

Ordnance Survey maps can be conveniently divided into two classes:

1. *Small-scale maps*
1:625 000
1:250 000
1:50 000
1:25 000

2. *Large-scale plans*
1:10 000
1:2500
1:1250

The 1:25 000 map can be considered as being the smallest map of value to the builder. It is a convenient map to use as a base and is shown in Fig. 2.8(b).

Of the Ordnance Survey production only the 1:1250 and 1:2500 can be considered to be plans in the strict sense of the word.

(d) Reference numbers of maps and plans

It has already been shown in Fig. 2.6 that the major blocks of the National Grid have sides of 100 km. Thus block NS is bounded on the north and south by the 700 km and 600 km northing lines and on the west and east by the 200 km and 300 km easting lines respectively.

Figure 2.8 shows this block subdivided into one hundred blocks each measuring 10 km square.

1:25 000 scale maps

Taking a 10 km square block bounded by Grid lines 670 km north, 660 km north, 250 km east and 260 km east (Fig. 2.8(b)) and drawing it to scale 1:25 000 produces a square of 400 mm side. This is the typical format for the small-scale 1:25 000 maps. Points of detail are shown by conventional signs and variation in height by contours at 5 m vertical

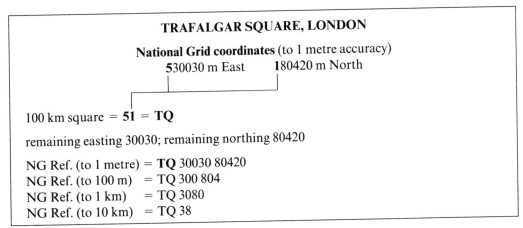

Figure 2.7

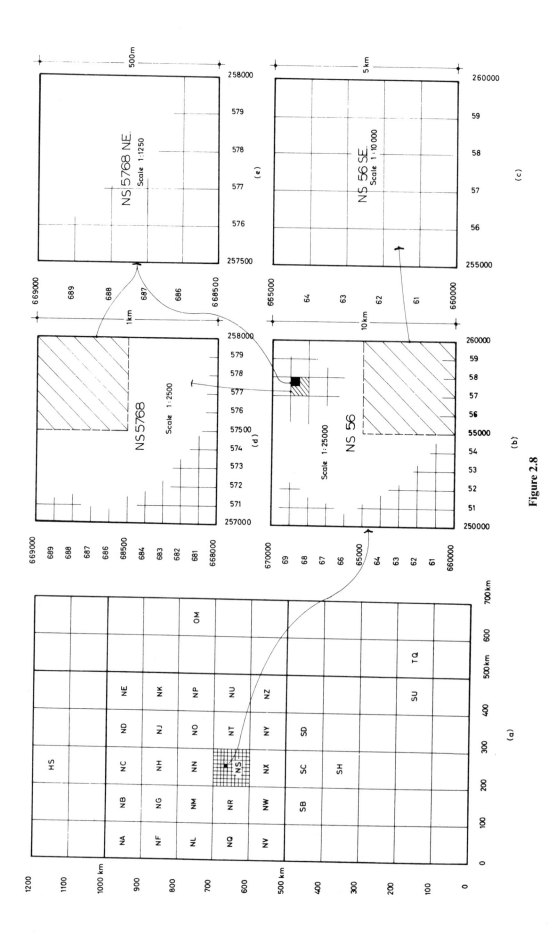

Figure 2.8

intervals, with the National Grid superimposed at 1 km intervals.

Each map has a unique reference referred to its south-west corner. Since the side of the map is 10 km long, the reference must be given to the 10 km figure. Thus the reference of the map shown in Fig. 2.8(b) is derived as follows:

Coordinates of south-west corner
250 000 m east: **660** 000 m north
= **250** km **east 660** km north
100 km Grid reference = **26** = **NS**

Therefore 10 km Grid reference = **NS 56** = NS 56

The 1:25 000 map is the base map on which the various series of larger scale maps and plans are built.

1:10 000 scale maps

Figure 2.8(c) shows a ground square of 5 km side drawn to a scale of 1:10 000 to produce the format for the 1:10 000 scale series of maps. Certain Town and Country Planning matters and Development proposals are shown on this scale. The map really represents one-quarter of the area of the 1:25 000 scale map shown in Fig. 2.8(b).

Details on these maps are shown true to scale and only the widths of narrow streets are exaggerated. Surface relief is shown by contour lines at 10 metre vertical intervals in mountainous areas and 5 metre vertical intervals in the rest of the country. The National Grid is superimposed at 1 km intervals.

Example

16 Figure 2.9 shows a 1:25 000 scale OS map hatched in black. The south-west corner has a 10 km easting figure of **44** and a 10 km northing figure of **26**. Its 10 km reference is therefore **SP 46**. List the reference numbers of the maps immediately adjoining it.

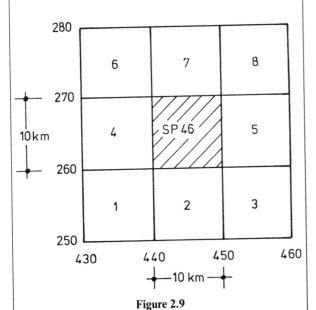

Figure 2.9

Answer

Map	10 km easting	10 km northing	Reference
1	43	25	SP 35
2	44	25	SP 45
3	45	25	SP 55
4	43	26	SP 36
5	45	26	SP 56
6	43	27	SP 37
7	44	27	SP 47
8	45	27	SP 57

Example

17 Figure 2.10 shows a 1:10 000 scale OS map hatched in black. It is the north-east quarter of the corresponding 1:25 000 sheet **SP 46**. Its reference number is therefore **SP 46 NE**. List the 1:10 000 scale reference numbers of the maps numbered 1 to 8.

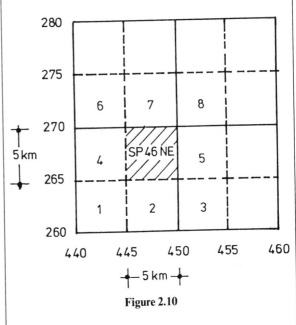

Figure 2.10

Answer

Map	5 km easting	5 km northing	Map reference
1	440	260	SP 46 SW
2	445	260	SP 46 SE
3	450	260	SP 56 SW
4	440	265	SP 46 NW
5	450	265	SP 56 NW
6	440	270	SP 47 SW
7	445	270	SP 47 SE
8	450	270	SP 57 SW

The map is referenced as a quadrant of the 1:25 000 map of which it is part. Thus the 1:10 000 scale map of Fig. 2.8(c), being the south-east quadrant of the 1:25 000 map **NS 56**, is given the reference **NS 56 SE**.

1:2500 scale plans

Figure 2.8(d) shows a 1 km square taken from the 1:25 000 map NS 56 and enlarged to scale 1:2500 to produce the format for the 1:2500 National Grid series of plans.

The whole of Great Britain, except moorland and mountain area, is covered by this series. All details are true to scale, and areas of parcels of land are given in acres and hectares. Surface relief is shown by means of bench marks and spot heights, and the National Grid is superimposed at 100 metre intervals.

This plan is probably the most commonly used plan in the construction industry. Most site and location plans are shown on this scale. The plan reference is again made to the south-west corner and is given to 1 km. Thus the reference of the map shown in Fig. 2.8(d) is derived as follows:

Coordinates of south-west corner
257 000 m east: **668** 000 m north
= **257** km east: **668** km north
100 km Grid reference **26** = **NS**

Therefore 1 km Grid reference = **NS 5768**

Example

18 Figure 2.11 shows a 1:2500 scale OS plan hatched in black. The south-west corner has a 1 km easting figure of **445** and a 1 km northing figure of **255**. Its 1 km reference is therefore **SP 4555**. List the reference numbers of plans 1 to 8 immediately adjoining it.

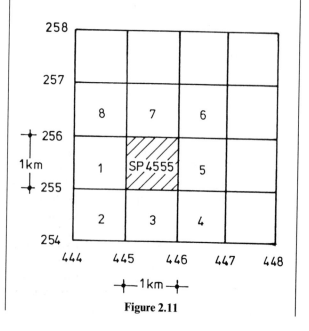

Figure 2.11

Answer

Map	1 km easting	1 km northing	Plan reference
1	444	255	SP 4455
2	444	254	SP 4454
3	445	254	SP 4554
4	446	254	SP 4654
5	446	255	SP 4655
6	446	256	SP 4656
7	445	256	SP 4556
8	444	256	SP 4456

It has been found that it is more convenient and economical to produce the 1:2500 plans in pairs. The large-scale 1:2500 plan covers an area of 2 kilometres east–west by one kilometre north–south. The grid line forming the western edge of the sheet is always an even number. Thus plan **NS 5768** (Fig. 2.8(d)) would be the eastern half of sheet **NS 5668–5768**. This reference may be shortened to read **NS 56/5768**.

1:1250 scale plans

Figure 2.8(e) shows a ground square of 500 m side drawn to a scale of 1:1250 to produce the format for the 1:1250 National Grid series of plans. The plan represents one-quarter of the area of the 1:2500 plan of Fig. 2.8(d).

These maps are the largest scale published by the Ordnance Survey and cover only urban areas. All details are true to scale. Surface relief is shown by bench marks and spot heights and the National Grid is carried at 100 m intervals.

The plan is referenced as a quadrant of the 1:2500 plan of which it is a part. Thus the 1:1250 scale plan of Fig. 2.8(e), being the north-east quadrant of the 1:2500 plan **NS 5768**, is given the reference **NS 5768 NE**.

Example

19 Figure 2.12 shows a 1:1250 OS plan, hatched in black. It is the NE quarter of the corresponding 1:2500 sheet **SP 4555**. Its reference number is therefore **SP 4555 NE**. List the 1:1250 scale reference numbers of the plans numbered 1 to 8 immediately adjoining it.

Answer

Map	0.5 km easting	0.5 km northing	Plan reference
1	445.0	255.5	SP 4555 NW
2	445.0	255.0	SP 4555 SW
3	445.5	255.0	SP 4555 SE
4	446.0	255.0	SP 4655 SW
5	446.0	255.5	SP 4655 NW
6	446.0	256.0	SP 4656 SW
7	445.5	256.0	SP 4556 SE
8	445.0	256.0	SP 4556 SW

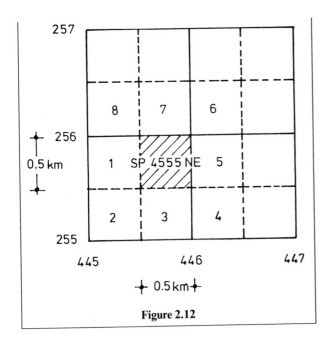

Figure 2.12

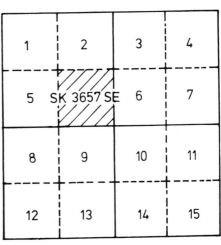

Figure 2.14

Exercise 2.3

1 Figure 2.13 shows diagrammatically the 1:2500 scale Ordnance Survey plan **SK 5265**. List the reference numbers of the eight plans immediately adjacent to it.

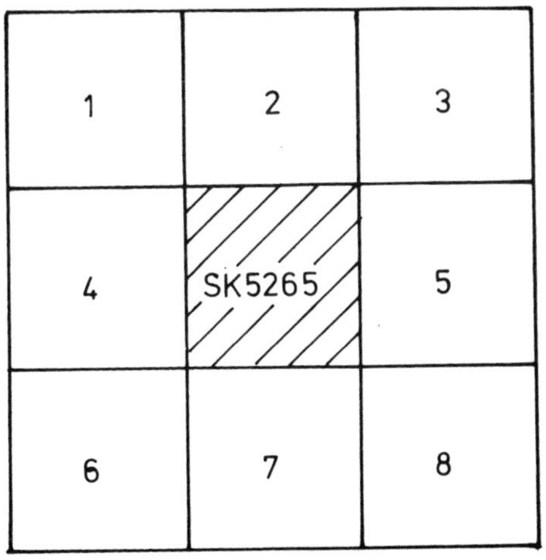

Figure 2.13

2 Give the reference number of the 1:1250 plan containing the point 533545E and 180202N (Tower Bridge, London).

3 Make a diagram to show the relationship between the true north and Grid north meridians at a point whose NG reference is SN 5461.

4 Figure 2.14 shows diagrammatically the 1:1250 scale Ordnance Survey plan **SK 3657 SE**. List the reference numbers of the fifteen plans adjacent to it.

(e) Types of OS maps

The maps produced by the Ordnance Survey Department are best described as topographical maps. As the name implies, they show the topography of the ground, i.e. the hills and valleys, rivers, roads and man-made features.

The large-scale plans are often referred to as Cadastral maps, though this is not strictly true. The word Cadastral means that the plans show property boundaries, when in fact they show the physical boundaries of property, e.g. fence lines, boundary walls, etc. The physical boundary is not necessarily the legal boundary, and to this extent the name is misapplied.

In order to read and understand Ordnance Survey maps, it is necessary to understand scales, grid references and directions — all of which have been mentioned. The methods of showing the rise and fall of the ground (or surface relief) and the conventional signs for features on the map must also be mastered.

(f) Surface relief

On the large-scale and small-scale maps, there are three methods of showing surface relief.

By contour lines

A contour line is a line joining all points of equal height. The 1:10 000 scale map is the largest scale that shows contours, the interval being 10 m, 5 m or the metric equivalent of 50 ft, depending upon the nature of the terrain and available survey information.

On other small-scale maps the interval varies with the scale and series of the map, e.g. on some of the second series 1:50 000 scale maps the contour interval is 10 m, while on others the interval is the metric equivalent of 50 ft.

By spot levels

As the name would imply, a spot level is simply a dot or spot on the map, the level of which is printed alongside, and this is the method of showing variations in elevation on the large-scale maps. These levels are shown to the nearest 100 mm, i.e. one decimal place of metres.

By bench marks

Bench marks (Fig. 2.15) are permanent marks chiselled on buildings showing the height of that particular point above Ordnance Datum. Sometimes the bench mark is a bronze plaque let into the wall on which an arrow records the height.

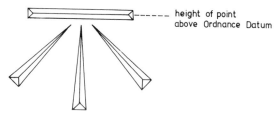
height of point
above Ordnance Datum

Figure 2.15

Ordnance Datum is the point of reference to which all bench marks, spot levels and contour lines are referred, and is the mean level of the sea as recorded over many years at Newlyn Harbour, Cornwall.

The values of these bench marks are shown to the nearest centimetre on the 1:2500 and 1:1250 sheets. The latest series of these sheets show only the positions and not the values of the bench marks. In order to find the value of a particular bench mark, it is necessary to scale its coordinates from a plan and consult a bench mark list published by the OS where the location and value of the bench mark may be read off against its coordinates.

(g) Conventional signs

Since it is impossible truly to represent each individual feature on any map, a code of signs has been devised for use with each scale of OS map. As far as possible, the signs are facsimiles of the actual object, e.g. conifers and deciduous trees, cliffs and quarries, but very often recourse is made to simply printing the initials of the feature, e.g. T for telephone kiosk and PH for public house.

A selection of signs and abbreviations used on the 1:2500 and 1:1250 sheets is shown in Fig. 2.16.

Exercise 2.4

Using Fig. 2.17, which is a facsimile Ordnance Survey 1:2500 scale plan, and the list of conventional signs and abbreviations in Fig. 2.16, answer the following questions:

1 State the values of the two spot levels along the main roadway and calculate the gradient of the roadway between these points.
2 State the value of the bench mark on the railway bridge.
3 State the features that have the following coordinates:
(a) 4730, 6767 (b) 4729, 6781
(c) 4719, 6778 (d) 4704, 6777
4 State the coordinates (to the nearest 10 metres) of:
(a) the centre of the orchard
(b) the telephone call box
(c) the public house
(d) the electricity pylon
5 State the direction in which the electricity transmission line runs.

(h) Parcel numbers and areas on 1:2500 sheets

The Ordnance Survey has a legal obligation to supply the areas of enclosures for all districts of England, Scotland and Wales. The areas are shown, as far as is practicable, on the 1:2500 plans.

As has already been stated, these plans are not Cadastral plans and do not show any details of ownership of property. They show only physical boundaries. The boundaries enclose areas of land known as parcels. A parcel may be a single feature, e.g. a field, or it may consist of several small areas grouped together, e.g. some small islands in a river or a combination of a field, ditch and wood.

Each parcel is given a four-figure reference number representing the National Grid 10 metre reference of the centre of the parcel. Thus in Fig. 2.18, the wooded area has a reference number of 0914 derived as follows:

Full National Grid coordinates of the centre of the area:

354 090 m east
639 140 m north

The 354 km east figure and the 639 km north figure are not required when giving a reference for any point on any 1:2500 scale plan. The plan is only 1 km square and the kilometre figures therefore cannot change. References can be given completely by using only the metre figures, in which case the centre of the wooded area has a reference of 090 m east and 140 m north. The reference to the nearest 10 m is therefore 09 east and 14 north. Thus the parcel number is 0914.

Whenever a parcel is composed of two or more separate areas of land, they are joined on the sheet by an elongated symbol called a brace. In Fig. 2.18, the bracing symbol indicated by ① simply joins together

Conventional Signs — Scales 1:1250 and 1:2500

ꭤ .... Bracken	⊞ .... Electricity Pylon	+ .... Surface Level
⚵ .... Coniferous Tree (Surveyed)	_ E T L _ .... Electricity Transmission Line	△ .... Triangulation Station
⚵ ⚵ ....Coniferous Trees (Not Surveyed)	⚯ ⚮ .... Marsh, Saltings	∫ .... Area Brace (1:2500 scale only)
⚶ ⚶ .... Coppice, Osier	♧ ....Orchard Tree	Perimeter of built-up area with single acreage (1:2500 scale only)
⚘ ....Non-coniferous Tree (Surveyed)	⚯ .... Reeds	� ....
⚘ ⚘ ....Non-coniferous Trees (Not Surveyed)	⁘ ....Rough Grassland	
⚭ .... Antiquity (site of)	Ᏼ .... Scrub	**Roofed Building** **Slopes**
►►► ....Direction of water flow	Ᏻ ᗅ	
↑ B M ....Bench Mark (Normal)	⌃ ᨆ ....Heath	

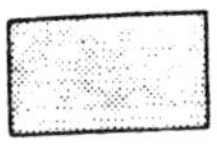

Boundaries

England, Wales & Scotland

• • • • • • • • •	——— Civil Parish Boundary
Boro (or Burgh) Const . Co Const	——— Parly & Ward Boundaries based on civil parish
Boro (or Burgh) Const & Ward Bdy	——— Parly & Ward Boundaries not based on civil parish
Co Const Bdy	

Examples of Boundary Mereings

o—⦻—o ...	........Symbol marking point where boundary mereing changes
• • • Und • • •	.............Undefined boundary
• • • Def • • •	....Original boundary feature destroyed or defaced
C B ....Centre of Bank	E K ....Edge of Kerb
C C ....Centre of Canal, etc.	F F ....Face of Fence
C D ....Centre of Ditch, etc.	F W ....Face of Wall
C R ....Centre of Road, etc.	S R ....Side of River, etc.
C S ....Centre of Stream, etc.	T B ....Top of Bank
C O C S............Centre of Old Course of Stream	Tk H ....Track of Hedge
C C S............Centre of Covered Stream	Tk S ....Track of Stream
4ft R H ....4 feet from Root of Hedge	

Abbreviations

B H ....Beer House	P ....Pillar, Pole or Post
B P, B S ....Boundary Post, Boundary Stone	P C ....Public Convenience
Cn, C...........................Capstan, Crane	P C B ....Police Call Box
Chy ....Chimney	P T P .... Police Telephone Pillar
D Fn...........................Drinking Fountain	P O.......................................Post Office
El P.........................Electricity Pillar or Post	P H.......................................Public House
E T L.......................Electricity Transmission Line	Pp ....Pump
F A P............................Fire Alarm Pillar	S B, S Br...................Signal Box, Signal Bridge
F S ....Flagstaff	S P, S L...................Signal Post, Signal Light
F B.................................Foot Bridge	Spr..Spring
G P..................................Guide Post	S, S D............................Stone, Sundial
H.........................Hydrant or Hydraulic	Tk..Tank or Track
L B...................................Letter Box	T C B.............................Telephone Call Box
L C..............................Level Crossing	T C P.............................Telephone Call Post
L Twr..............................Lighting Tower	Tr..Trough
L G..............................Loading Gauge	Wr Pt, Wr T...............Water Point, Water Tap
Meml...................................Memorial	W B...Weighbridge
M P U..............................Mail Pick-up	W...Well
M H.....................................Manhole	Wd Pp.......................................Wind Pump
M P...................Mile Post or Mooring Post	M H or L W...............Mean High or Low Water
M S...................................Mile Stone	(England and Wales)
N T...................................National Trust	M H or L W S............Mean High or Low Water
N T L...............................Normal Tidal Limit	Springs (Scotland)

Figure 2.16

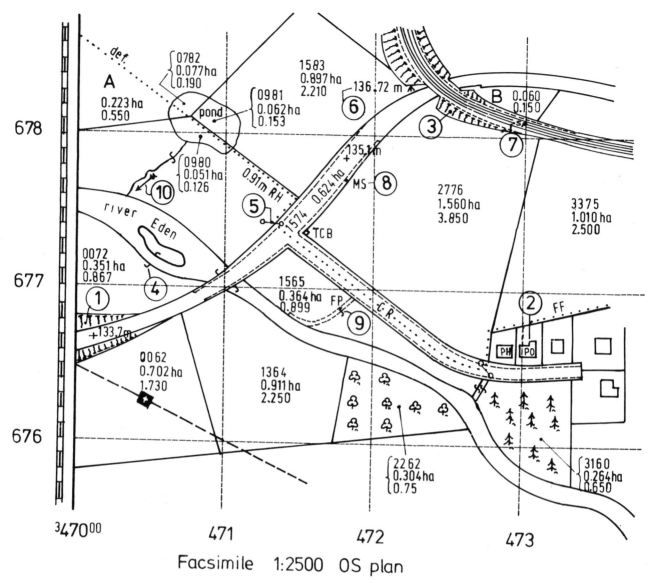

Facsimile   1:2500   OS plan

Figure 2.17

two parcels for the purpose of area measurement, i.e. 0.545 hectares.

When a parcel is bounded by double features of equal importance, e.g. a double fence or hedge indicated by ②, centre braces are used to show that the area has been measured to the centre of the space between the two features.

An open brace indicated by ③ is used to link parts of a parcel which are separated, e.g. the island in the River Leven is separated from the south bank of the river, but is joined by the open brace for the purpose of area measurement, i.e. 0.574 hectares.

In towns and villages containing many parcels in built-up areas, the parcels are simply banded together and treated as a whole. The perimeter of this type of parcel is denoted by a series of regularly spaced symbols resembling belisha beacons.

Example

20 (a) Using Fig. 2.18 quote the 10 m reference of the centre of field at 354320E, 639130N.
(b) Calculate the area of the field in hectares and in acres.

Answer
(a) Full NG coordinates
$$= 354320 \text{ east}   639130 \text{ north}$$
Omit figures   354..0 and 639..1
Therefore 10 metre reference
$$= ...32. \text{ and } ...13.$$
$$= 3213$$

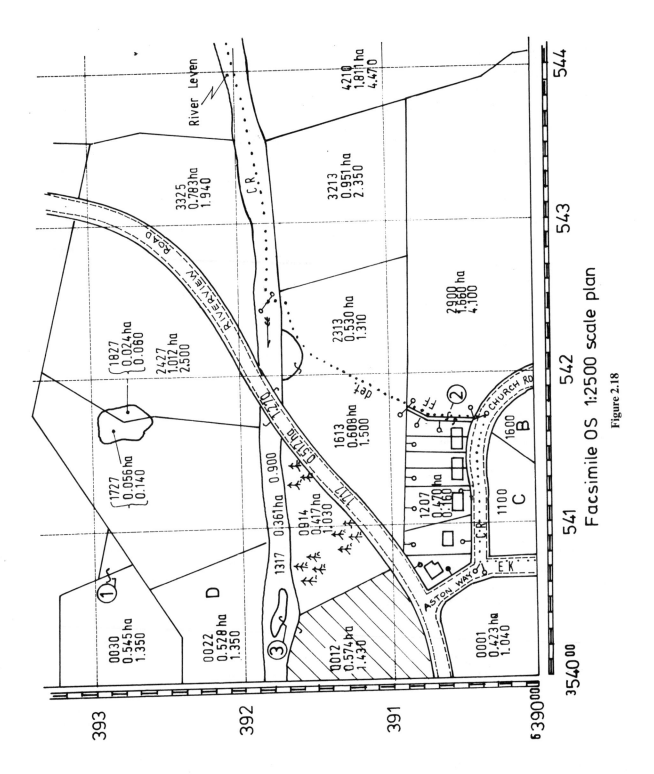

Facsimile OS 1:2500 scale plan

Figure 2.18

23

(b) The field is trapezoidally shaped.

$$\text{Therefore area} = \tfrac{1}{2}(90 + 100) \times 100 \text{ m}^2$$
$$= 95 \times 100 \text{ m}^2$$
$$= 9500 \text{ m}^2$$
$$= 0.95 \text{ ha}$$

To convert hectares to acres:

$$1 \text{ hectare} = 10\,000 \text{ m}^2$$
$$= 10\,000 \times 3.2808^2 \text{ ft}^2$$
$$= 10\,000 \times (3.2808^2/9) \text{ yd}^2$$

Since there are 4840 yd^2 in 1 acre

$$1 \text{ ha} = \frac{10\,000 \times (3.2808^2/9)}{4840}$$
$$= 2.471 \text{ acres}$$
$$\text{Field area (acres)} = 0.95 \times 2.471$$
$$= 2.347 \text{ acres}$$

Where a sheet edge cuts across any particular parcel, part of the reference number is the mid-point of that sheet edge intercepted by the parcel. Thus in Fig. 2.18, the area shown cross-hatched has a reference number 0012 derived from coordinates 5400E, 3912N.

Any parcel lying in the corner of a sheet is given an arbitrary reference number from 0001 to 0006 depending upon the position of the map. The system is so devised that a number allotted to a corner parcel will not be repeated elsewhere on any of the plans surrounding that corner.

On the latest series of 1:2500 sheets the area of each parcel is given in hectares and in acres. Both of these figures are printed immediately below the reference number. The wooded area is therefore parcel number 0914 of area 0.417 hectares or 1.03 acres.

Whenever possible, parcel boundaries are natural features such as hedges or streams. Thus the wooded area 0914 is bounded by the river, hedge and road. However, wherever an administrative boundary, e.g. parish or district boundary, divides an enclosure, each part is treated as a separate parcel. Parcel numbers 1613 and 2313 are parts of the same field, divided by a parish boundary into areas of 0.608 hectares and 0.530 hectares respectively.

Example

21 In Fig. 2.18 determine the parcel numbers of fields B and C and D.

Answer
Fields B and C are cut by the east/west sheet edge and the parcels presumably extend on to the next sheet below. The northing 10 m refer-

ence figure is therefore 00. The easting 10 m reference for field C is 11 and the parcel number is 1100.

Field B   Easting 10 m reference = 16
Northing 10 m reference = 00
Parcel number = 1600

Field D   Easting 10 m reference = 00
Northing 10 m reference = 22
Parcel number = 0022

(i) Boundaries

The Ordnance Survey is required by law to show administrative boundaries on its maps. The minimum scales of the maps vary, depending upon the type of boundary being depicted, but, in practice, all boundaries are shown at scales of 1:1250 and 1:2500 by the range of symbols shown in Fig. 2.16, under the heading 'Boundaries'.

Boundaries may coincide with certain physical features e.g, hedges and fences, or may run along the centre of a road, river or railway. In other cases boundaries may be offset from certain physical features and a description or mereing of the location of the boundary is printed on the map alongside the feature.

When the type of mereing changes (as from the face of a fence to the edge of a kerb) a special symbol resembling barbells is placed on the map at the point of change. Sometimes the symbol is bent to enable it to be positioned on the map without interfering with any map detail.

Example

22 Using Fig. 2.18 describe the parish boundary that runs approximately south-westwards across the plan.

Answer
The boundary runs westwards along the centre (CR) of the River Leven. It then changes to run southwards across a field. The boundary has been defaced (def). It changes at the south end of the field to run along the face of a fence (FF) and again changes to run across Church Road and continues westwards along the centre of this roadway (CR) to the junction with Aston Way, where it turns southwards and runs along the edge of a kerb (EK).

Exercise 2.5

Using Fig. 2.17:

1 State the parcel number containing the island in the River Eden.

2 Determine the total area of the pond in hectares and acres.

3 State the parcel numbers of fields marked A and B.

4 Describe the parish boundary running diagonally across the plan.

5 State:

(a) the parcel number of the roadway,

(b) the area in hectares and

(c) convert this area into acres.

3. Answers

Exercise 2.1

1 12 m

2 SE 51.2 m, SW 52.2 m, NW 52.0 m, NE 51.1 m

3 FL 52.75 m

4 Invert level S5 is 50.20 m

5 Gradient 1 in 86

6 (a) 96 119   (b) 61 116   (c) 92 120

7 (a) Centre of curve   (b) Stairs   (c) RWDP

8 SVPs 2, RWDPs 4

Exercise 2.2

1 1:200

2 Length 8.4 m, breadth 8.4 m: length 8.0 m, breadth 8.0 m

3 68 east, 102 north

4 S1 manhole, 52E, 134N; S2 manhole 46E, 125N

5 (a) Manhole S4   (b) tree   (c) steps

6 Five

7 Ground falls from 11 m on the north to 7 m on the south

8 Gradient 1 in 8.9

9 Between A and B gradient — 1 in 2 and between C and D gradient — 1 in 7

10 Invert level, 7.57 m

Exercise 2.3

1 1, SK 5166   2, SK 5266   3, SK 5366
4, SK 5165   5, SK 5365   6, SK 5164
7, SK 5264   8, SK 5364

2 TQ 3380 SE

3 See Fig. 2.19

4 1, SK 3657NW   2, SK 3657NE   3, SK 3757NW
4, SK 3757NE   5, SK 3657SW   6, SK 3757SW
7, SK 3757SE   8, SK 3656NW   9, SK 3656NE
10, SK 3756NW   11, SK 3756NE   12, SK 3656SW
13, SK 3656SE   14, SK 3756SW   15, SK 3756SE

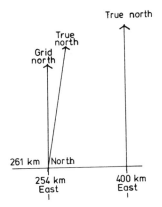

Figure 2.19

Exercise 2.4

1 133.7 m, 135.1 m, 1 in 153

2 136.72 m

3 (a) Post office   (b) signal post   (c) milestone
(d) direction of flow of stream

4 (a) 4722,6762   (b) 4715,6773
(c) 4729,6766   (d) 4705,6763

5 ESE

Exercise 2.5

1 0072

2 0.190 ha, 0.47 ac

3 0083, 2982

4 Parish boundary—running from north-west to south-east—boundary is defaced across a field and through a pond: a line running parallel to a hedge and lying 0.91 m from the roots of a hedge; a line running along the centre line of a road: a line running along the face of a fence passing behind the post office

5 (a) 1574   (b) 0.624 ha   (c) 1.54 ac

4. Project

Chapter 18 is a project covering all chapters of this textbook. It is intended that the project build into a complete portfolio of the surveying work required in the survey and the setting out of a building or engineering development.

If the reader wishes to participate in the project, he or she should now turn to page 388 and attempt Section 1.

CHAPTER 3

Linear surveying

Objective

After studying this chapter, the reader should be able to make a linear survey of a parcel of ground, apply the necessary corrections to the survey data and produce a scaled drawing of the results.

Every surveying technician will, at some stage of his or her career, have to make a survey of a small parcel of ground, using a tape and a few ancillary pieces of equipment. This method of surveying is known as linear surveying, because the survey is carried out by measuring only the lengths of lines. No angular measurements are made at all.

Linear surveys were fashionable for all sizes of ground parcels before the advent of electromagnetic distance measurement, but nowadays only small sites, probably less than a hectare in extent, are surveyed using this method.

Consider Fig. 3.1. It shows a site adjacent to a main road, which is to be developed as a small building estate. An accurate site plan is required for the development. Note that the figure is actually a photograph of Blairvadach Residential Education Centre at Rhu, Scotland. The centre is used by surveying students of Glasgow College of Building and frequent reference will be made to the site throughout this book. It will be referred to simply as the GCB Outdoor Centre.

In order to make a scaled drawing of the site it is necessary to master:

Figure 3.1

(a) the geometric principles of making the survey,
(b) the technique of measuring lengths using a tape,
(c) the method of carrying out the survey and recording the measurements and
(d) the method of plotting these measurements to scale.

1. Principles of linear surveying

As already stated, the objective of any survey is the production of the scaled drawing of a parcel of ground. A survey makes use of the principles of geometry. The simplest possible geometric figure is a triangle, which of course is a three-sided figure, the sides of which join three points. If the distance between two of the points is measured the points are fixed. The problem, then, is to fix the third point in relation to the two known points.

Figure 3.2(a) illustrates the situation. A large tree, Z (position unknown), is to be fixed by survey to a straight fence XY. Two methods are available.

(a) Trilateration

The word means measurement of three sides. When the principle is applied to Fig. 3.2(a), the lengths XY, YZ and XZ are all measured, using a tape. Length XY is then drawn to scale on paper, point Z is fixed to XY, using a pair of compasses and drawing arcs XZ and YZ to intersect in the point Z (Fig. 3.2(b)).

(b) Lines at right angles (offsets)

In the field lines XO and OY are measured along the fence XY. Offset OZ is measured at right angles to line XY. Using a set square and scale rule, point Z is plotted at right angles to line XY, thus establishing the correct relationship of the tree to the fence (Fig. 3.2(c)).

Example

1 Plot the position of the tree in relation to the fence (Fig. 3.2(a)), to a scale of 1:200, from the following surveyed data:

(a) Trilateration (b) Offsets
length XY = 25.00 length XO = 15.50
length XZ = 17.00 length OY = 9.50
length YZ = 11.80 Offset OZ = 7.00

Tree: 3.0 m, radius–branch spread

Answer (see Fig. 3.3, page 28)

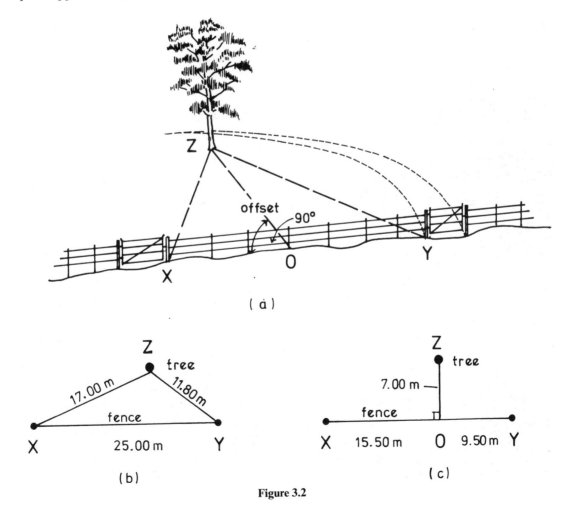

(a)

(b)   (c)

Figure 3.2

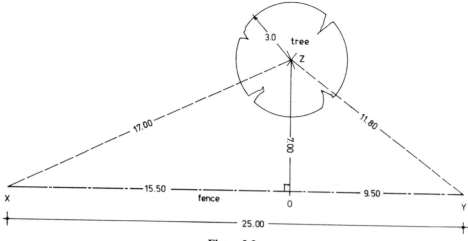

Figure 3.3

(c) Framework and detail surveys

The principles of trilateration and offsetting are applied to all linear surveys, which are carried out in two parts:

1. Framework survey

A framework is established over the whole site to form a sound geometrical figure, which can be readily plotted (Fig. 3.4). The detail survey is then added to the framework; thus the fundamental surveying rule of 'working from the whole to the part' is complied with.

2. Detail survey

The fences, wall, stream, trees, buildings, etc., are the details that are added to the framework by offsetting, as shown.

Figure 3.4 is the actual survey layout of the GCB Outdoor Centre site (Fig. 3.1). The procedure used to carry out the survey is fully explained in Sec. 3.

2. Linear measuring techniques

(a) Equipment

All linear measuring equipment should conform to the British Standards Specification 4484:1969. Such instruments will then have clear legible graduations and unambiguous figuring. The instruments required on a linear survey vary from the simple folding 1 metre rule to the long 50 metre steel tape.

Table 3.1 shows the lengths, graduations and method of figuring of the principal linear measuring instruments. Land chains are included in the table but, being obsolescent, they are not used in construction surveying.

Folding boxwood rule

The 1 metre folding rule is made from high-quality boxwood with one central brass joint and two folding hinges. The rule is made in such a way that the fine graduations are always protected (Fig. 3.5).

Measuring instrument	Length (m)	Graduations			Method of figuring
		Major	Inter	Fine	
Folding rule	1	10 mm	5 mm	1 mm	10 mm intervals using 3-digit numbers
Folding and multi-folding rods	1, 1.5, 2	10 mm	5 mm	1 mm	10 mm intervals using 3-digit numbers
Steel pocket rule	2	10 mm	5 mm	1 mm	10 mm intervals using 3-digit numbers
Steel tapes	10, 20, 30	100 mm	10 mm	5 mm	100 mm intervals in decimals of a metre
		First and last metres further subdivided into divisions of			
Synthetic tapes	10, 20, 30	100 mm	50 mm	10 mm	50 mm marks are denoted by arrows
Land chains	20	1 m	—	200 mm	Yellow tallies at 1 metre intervals; 5 and 10 m shown by red tallies

Table 3.1

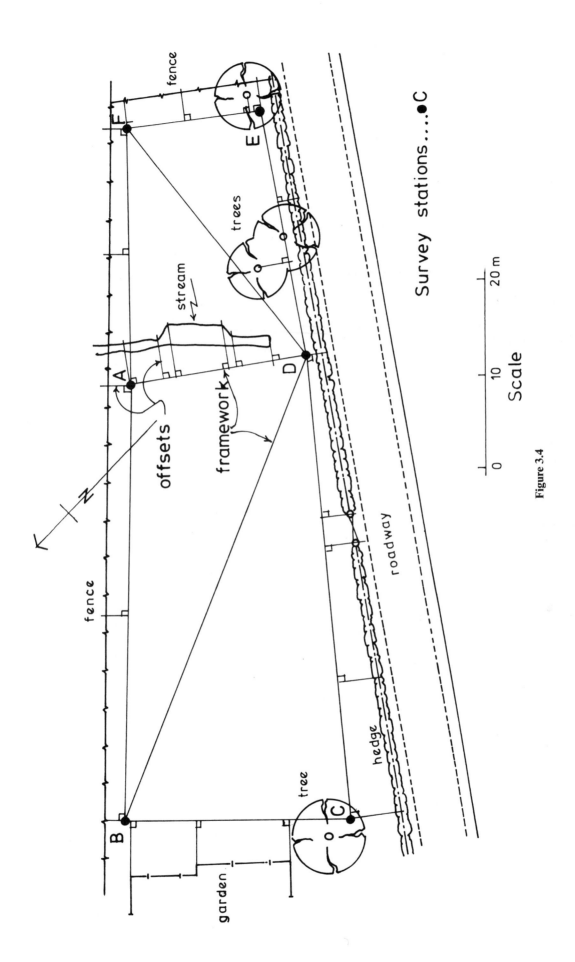

offsets

framework

stream

trees

fence

fence

garden

hedge

tree

roadway

Survey stations....●C

Scale

0   10   20 m

Figure 3.4

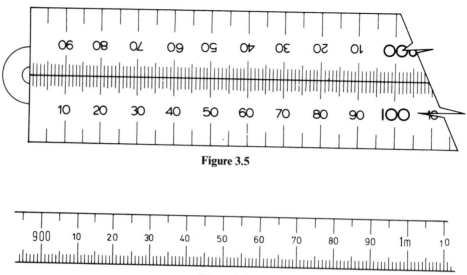

Figure 3.5

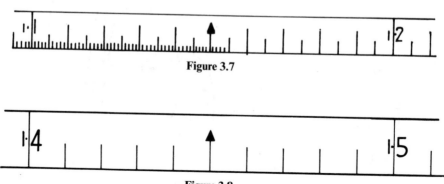

Figure 3.6

Steel pocket rule

The rule is contained in a strong lightweight case, which is 50 mm wide at the base to enable inside measurements to be taken easily. The rule is graduated on one side only with fine graduations on the bottom edge. The top edge shows 5 mm coarse graduations. The 1 metre graduation mark is shown in a contrasting colour on the ruler and a quick reading repeater figure, 1 m, occurs at every 100 mm to facilitate reading, without scanning the instrument (Fig. 3.6).

Steel tapes

These tapes are made from hardened tempered steel sheathed in durable white plastic. The tapes are 10 mm wide with black and red figures and black graduations. They are very tough and the graduations almost indestructible. The tapes are housed in open-frame plastic winders fitted with a quick rewind handle. They are subdivided in millimetres throughout, figured at every 10 mm with quick-reading metre figures in red at every 100 mm and whole metre figures are in red (Fig. 3.7).

Synthetic tapes

These tapes are manufactured from multiple strands of fibreglass and coated with PVC. Fibron is impervious to water and can be wound back into the case without damage, even when wet. The tapes are graduated throughout in metres and decimals, the finest graduation being 10 mm (Fig. 3.8).

(b) Use of the tape

Reading the tape

On every survey, there will inevitably be a variety of long and short, flat and inclined, lines to be measured accurately. Figure 3.9(a) shows a short survey line AB marked on the ground by two pegs. The distance AB is shorter than one length of tape. The measurement of the line AB is obtained by unreeling the tape and straightening it along the line between the pegs. The zero point of the tape (usually the end of the handle) is held against station A by the rear tape person (called the follower). The forward end of the tape is read against station B by the forward tape person (called the leader) after it has been carefully tightened.

Figure 3.7

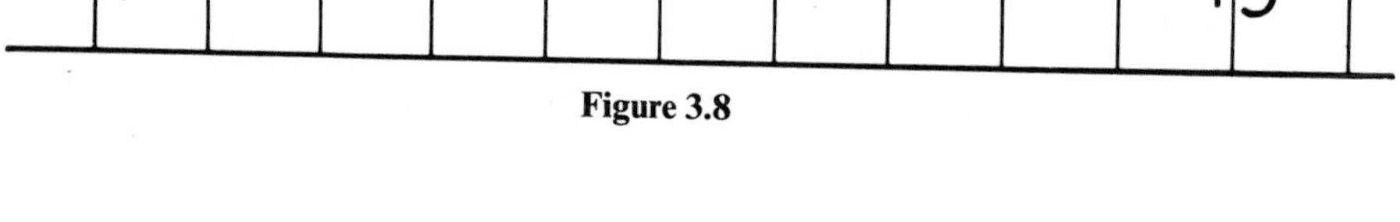

Figure 3.8

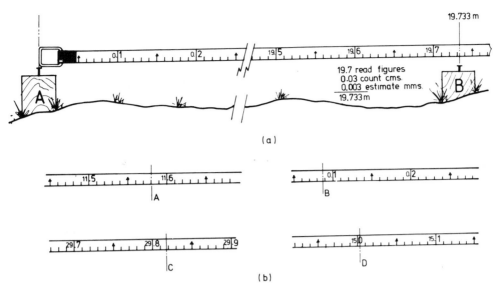

Figure 3.9

Example

2 Figure 3.9(b) shows tape readings A, B, C and D. Read the metres and decimals from the tape, count the centimetres and estimate the millimetres to give the correct reading at each point.

Answer

A, 11.580 m   B, 0.088 m   C, 29.818 m
D, 15.003 m

On surveys most of the lines will be considerably longer than one tape length and a sound operational technique is required. Two ancillary pieces of equipment are necessary, namely ranging rods and marking arrows.

Ranging rods are 2 metre long, round, wooden poles, graduated into 500 mm divisions and painted alternately red and white. They have a pointed metal shoe for penetration into the earth.

Marking arrows are made from steel wire, 375 mm long, pointed at one end, a 30 mm loop at the other and painted in fluorescent paint. They are made up in sets of ten. Both instruments are shown in Fig. 3.10.

Two surveyors are required to measure a long line. The leader's job is to pull the tape in the required direction and mark each tape length. A known number of arrows and a ranging rod are carried by the leader. The follower's job is to align the tape and count the tape lengths.

In Fig. 3.11, a line AB is to be measured across a gently sloping grassy field. The follower holds the zero end of the tape against station A and the leader pulls the tape towards station B. When the tape has been laid out the leader holds the ranging rod vertically approximately on the line. The follower signals

to the leader to move it until it is exactly on the line AB. The tape is tightened between the newly erected rod and station A and an arrow is pushed into the ground at the 20 mark of the tape.

The follower moves forward to this new point and the whole procedure is repeated for the remainder of the line until station B is reached. The follower gathers the marking arrows and the number of tape lengths measured is the number of arrows carried by the follower. The portion of tape between the last arrow and station B is then measured and added to the number of complete tape lengths to produce the total length of the line.

Inclined measurements

When any measured distance is to be shown on a plan, the horizontal distance is required and any inclined distance must be converted to its horizontal equivalent before plotting.

Figure 3.12(a) shows a survey line measured between two stations A and B. The line is not horizontal. Trigonometrically, the inclined distance is the hypotenuse of a right-angled triangle ABC.

In $\triangle$ABC

$$AC/AB = \cos x$$
$$\text{Therefore } AC = AB \times \cos x$$
i.e. plan length = slope length $\times$ cos inclination

The angle of inclination is measured in the field using some form of clinometer, the most common instrument being the Abney level.

Abney level

This instrument consists of a square-section sighting tube, 127 mm long, fitted with a draw tube extension, which extends to 178 mm. The draw tube is fitted with a pin-hole sight, and a horizontal cross wire, at the

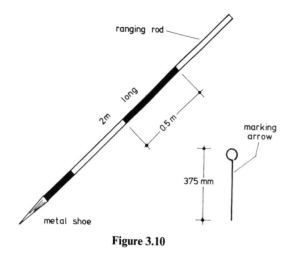

Figure 3.10

viewing end, completes the sighting arrangements (Fig. 3.12(b)).

Screwed to the rectangular sighting tube is a semi-circular arc graduated in degrees and read by a vernier. A spindle running through the arc carries a bracket holding a spirit level. A highly polished mirror is fitted within the sighting tube at an angle of 45° to the line of sight, allowing the observer to view simultaneously the spirit level via the mirror and some distant target against the cross-hair.

To measure an angle of inclination the Abney level is placed to the eye so that the bubble is apparent in the mirror. The sighting tube is tilted to observe the forward station and the slow motion screw controlling the spirit level is activated until the view shown in Fig. 3.12(c) is obtained. The spirit level will now be in the centre of its run and the vernier will have been moved by the slow motion screw over the graduations giving the angle of inclination to half of a degree.

Step taping

An alternative method of obtaining horizontal measurements, without using angle-measuring instru-

ments, is that known as step taping. This is a field method where the horizontal distances are obtained directly.

Three men are required, one leader, one follower and one observer. The follower's duties are exactly as before, namely aligning the leader, holding the end of the tape on the marks left by the leader and collecting the arrows.

Once aligned by the follower, the leader holds the tape horizontally. Considerable tension is required to straighten the tape and avoid sagging. The horizontal position is estimated by the observer, who signals to the follower and leader when this position has been attained. On receiving the signal, the leader drops a drop-arrow (a marker arrow with a lead weight attached) from the handle of the chain, thus transferring the horizontal distance to the ground (Fig. 3.13). Alternatively, a plumb-bob or plumbed ranging rod may be used for the transference.

The length of steps that can be adopted is limited by the gradient. At no time should the tape be above the leader's eye level, because plumbing becomes very difficult. As the gradient increases the length of step must therefore decrease. The maximum length of unsupported tape should not exceed 10 metres.

When the observer has noted the length of the first step in a book, the second and third steps are measured and the procedure repeated until the whole line has been measured. The summation of the steps will produce the required horizontal distance.

Step taping can also be performed when measuring uphill. In this case, the follower has to hold the handle of the tape above the mark left by the leader. For this purpose a plumb-bob is used. Simultaneously, therefore, the follower is applying tension to the tape horizontally and plumbing the handle above a ground mark. The observer may assist in the plumbing, but generally it is more difficult to step-tape uphill than down.

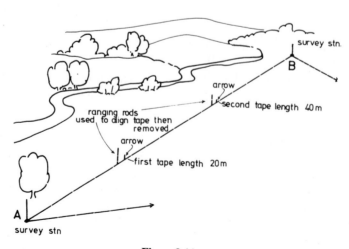

Figure 3.11

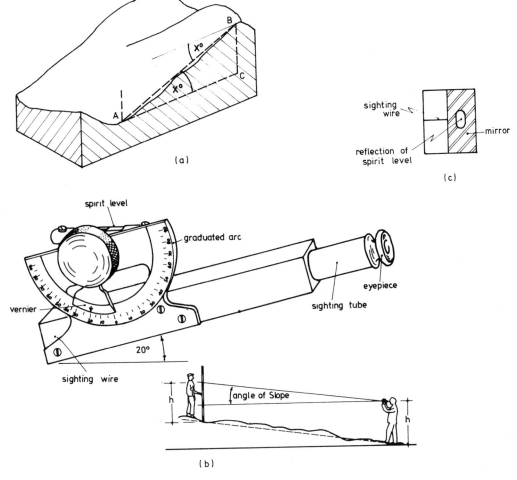

Figure 3.12

(c) Errors in measurement

All measurements made with tapes are subject to some form of error, no matter how carefully any line is measured. The error may be due to the surveyor's carelessness or inexperience, it may be due to physical site conditions or it may be inherent in the instrument being used.

Errors are divided into three classes.

Gross errors

Gross errors are blunders. They arise from inexperience, carelessness or lack of concentration on the part of the observer. Examples of gross errors are:

1. *Misreading the tape graduations*

This is the most common error in survey operations, e.g. six metres and forty millimetres is 6.040 m not 6.400 m.

2. *Miscounting the number of tape lengths*

During the measurement of long lines the surveyor may lose count of the number of tape lengths measured. A careful count of the number of arrows used on a long line should be made before and after the measurement of the line in an attempt to minimize the occurrence of this particular error.

3. *Booking errors*

The booker may simply write down the wrong measurement, particularly in windy, wet weather when it is difficult to hear and write.

All of these errors may be detected by measuring every line twice.

Constant errors

Constant errors are those that occur no matter how often a line is measured and checked. The error will always be of the same sign for any one tape or for any given set of circumstances. Examples of constant errors are:

1. *Misalignment of the tape*

This error is perhaps the simplest which demonstrates this class of error. If any line is to be measured between two points it must be measured on a straight line. Suppose, however, that a tree stump is in the direct line of measurement, and the distance is simply measured around it, producing a deviation, e, from the line, at the obstruction. The length of line measured will be too long no matter how often the line is measured and regardless of the number of chains or tapes used.

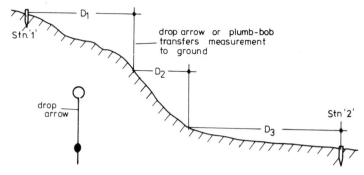

Figure 3.13

By using the theorem of Pythagoras and a binomial extension the error can be shown to be

$$e^2/L$$

where e is the amount of deviation of the tape and L is the total measured length.

Example

3 A survey line PQ is measured round a large tree stump causing a deviation (e) from the straight line of 1.0 m. The measured length PQ is 110.65 m. Calculate the correct length of the straight line PQ.

Solution

$$\text{Measured length} = 110.65 \text{ m}$$
$$\text{Correction} - e^2/L = -1.0^2/110.65$$
$$= -0.01 \text{ m}$$
$$\text{Therefore correct length} = 110.64 \text{ m}$$

From the example it is evident that the error is not serious even for such a large deviation. In a linear survey, however, offsets have to be measured from the line and complications arise when the tape is not straight.

2. Standardization
It is very important, before commencing a survey, that the measuring instrument being used is exactly the right length. It must be compared with some standard length, probably a new tape kept solely for that purpose. If the tape is not the standard length it will give a wrong measurement.

Suppose a tape is used to measure a room and the width is found to be 10 m exactly. If a knot is tied in the tape, obviously making it shorter, and the room is again measured, the width will appear to be greater than 10 m. This shows that a short tape, when used to measure distances, will produce lengths that are greater than the correct length, and conversely a long tape will produce lengths that are less than the correct length. The correction to any measured length is found from the formula:

$$c = (L - l) \text{ per tape/chain length}$$

where c = correction
         L = actual length of tape
         l = nominal length of tape

Examples

4 A line AB is measured using a tape of nominal length 20 m and is found to be 65.32 m long. When checked against a standard, the tape was found to be 50 mm too long. Calculate:
(a) the correction to the length AB,
(b) the correct length of AB.

Solution

Nominal length (l) of tape
$$= 20.00 \text{ m}$$
Number of tape lengths in line AB
$$= \frac{65.32}{20.00} = 3.266$$
Actual length (L) of tape
$$= 20.05 \text{ m}$$
$$\text{Correction } c = (20.05 - 20.00) \times 3.266$$
$$= +0.16 \text{ m}$$
$$\text{Correct length AB} = (65.32 + 0.16) \text{ m}$$
$$= 65.48 \text{ m}$$

5 A survey line XY is measured using a 30 m tape which owing to previous damage is actually 40 mm short. The line measured 147.36 m. Calculate:
(a) the correction to the measured length XY,
(b) the corrected length of line XY.

Solution

Nominal length (l) of tape = 30.00 m
Actual length (L) of tape = 29.96 m
Number of tape lengths
$$= \frac{147.36}{30.0} = 4.912$$
$$\text{Correction } c = (29.96 - 30.00) \times 4.912$$
$$= -0.20 \text{ m}$$
Therefore

Correct length XY = 147.36 − 0.20 = 147.16 m

3. Slope

All distances measured on a survey are slope lengths and must be converted into plan lengths before plotting. Slope lengths are longer than plan lengths; hence a constant error will be made by ignoring the inclination.

The formula (derived previously) for converting slope to plan lengths is as follows:

Plan length = slope length × cos angle of inclination

Example

6 A survey line AB was measured along a 6° gradient. The slope distance was 49.75 m. Calculate the plan length of the line.

Solution

$$\text{Plan AB} = \text{slope AB} \times \cos 6°$$
$$= 49.75 \times \cos 6°$$
$$= 49.48 \text{ m}$$

Some survey lines will have several changes of gradient along their lengths. In such cases each inclined section is treated separately and its plan length is calculated as in Example 7. All of the plan lengths are then added to produce the total horizontal length of the line.

Examples

7 Figure 3.14 shows a straight line ABCD which has three distinct changes of gradient along its length. Each gradient was obtained by Abney level. Calculate the plan length of the line AD given the following field results:

Line	Section	Slope length (m)	Inclination (deg)
AD	AB	84.40	−5.0
	BC	47.21	+2.0
	CD	39.47	+6.5

Solution

Plan length   AB = 84.40 × cos 5° = 84.08
                 BC = 47.21 × cos 2° = 47.18
                 CD = 39.47 × cos 6.5° = 39.22

Total length AD = 170.48 m

In practice it is likely that a combination of errors will be present and separate corrections will have to be made for standardization and slope.

8 A survey line XY is measured along a steep gradient of 8° by two survey teams A and B. Team A used a 30 m Fibron tape which when checked was found to have shrunk by 50 mm. The measured length was 85.24 m. Team B used a 20 m steel tape which was actually 20 mm too long. The measured length was 84.94 m.

Calculate the most probable horizontal corrected length XY.

Solution

Team A
  Actual tape length (L) = 29.95
  Nominal tape length (l) = 30.00
                 L − l = −0.05
  Number of tape lengths = 85.24/30 = 2.841
                 Correction = −0.05 × 2.841
                            = −0.14 m
       Corrected length = 85.24 − 0.14
                            = 85.10 m

Team B
  Actual tape length (L) = 20.10
  Nominal length (l) = 20.00
                 L − l = +0.10
  Number of tape lengths = 84.94/20 = 4.247
                 Correction = +0.02 × 4.247
                            = +0.08
       Corrected length = 84.94 + 0.08
                            = 85.02 m
  Average slope length = ½(85.10 + 85.02)
                            = 85.06 m
     Plan length XY = 85.06 × cos 8°
                            = 84.23 m

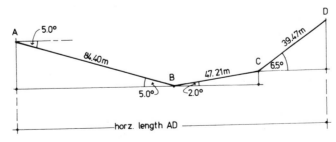

Figure 3.14

Human errors

This class of error arises from defects of human sight and touch, when marking the various tape or chain lengths and when estimating the readings on the tape when they do not quite coincide with a graduation mark. The chances are that two people will mark or estimate slightly differently. It is also reasonable to assume that no one person will overestimate or underestimate every reading, nor mark every tape length too far forward. These small errors tend to be compensatory and have relatively little significance at this level of surveying.

Conclusions

In every measured length there will be errors that fall into one of three classes.

1 Gross errors

These are mistakes that should not occur in practice provided each line is measured at least twice.

2 Constant errors

All of these errors are cumulative and have a very great effect on the accuracy of any length. Every tape should be checked to ensure that it is the correct length, and each tape length must be accurately aligned. The gradient must be determined along each measured length. In linear surveying, temperature and tension corrections are *not* necessary. They are applied in theodolite traversing (Chapter 8) where accuracy is essential.

3. Human errors

These are small residual errors of human sight and touch and tend to be compensating.

Exercise 3.1

1 Calculate the correct length of the line in each of the following:

Line	Measured length	Nominal length of tape	Actual length of tape
AB	32.51	10.00 m	10.02 m
XY	172.34	30.00 m	29.93 m
MN	63.85	20.00 m	19.96 m

2 Team A measured line FG using a 20 m tape which had stretched by 100 mm. The measured length was 93.52 m. Team B measured the line using a 50 m tape which had shrunk by 20 mm. The measured length was 94.05 m. Calculate the average length of line FG.

3 A line AB known to be 65.35 m long was checked using a 20 m tape. It was found to measure 65.10 m. Calculate the actual length of the tape.

4 Calculate the corrected plan length of line AB measured in 3 sections as follows:

Line AB	Section 1	Section 2	Section 3
Measured length	36.50	19.26	52.77
Angle of slope	2°	3.5°	5°

5 Calculate the corrected plan length of line RS given the following data:

Measured length	83.19 m
Angle of slope	4°30′
Nominal length of tape	20.00 m
Tape shrinkage	50 mm

3. Procedure in linear surveying

Where comparatively small areas have to be surveyed a linear survey might be used. As already mentioned, the principle of linear surveying is to divide the area into a number of triangles, all the sides of which are measured. The errors that can arise when measuring have already been discussed. It is obvious therefore that great care must be exercised for every measurement.

(a) Reconnaissance survey

On arrival at the site (Fig. 3.1), the survey team's first task is to make a reconnaissance survey of the area, i.e. the team simply walks over the area with a view to establishing the best sites for survey stations. The sites must be chosen with care and are in fact governed by a considerable number of factors.

Working from the whole to the part

This is the fundamental rule of all survey operations. It can be illustrated by a simple analogy. When a rectangular jigsaw puzzle is being manufactured, the whole picture is painted and then broken down into several hundred pieces. The alternative would be to manufacture one piece and then add the others to form the picture. Clearly the latter method would give a very inaccurate picture and certainly would not make a rectangle when fitted together.

The area to be surveyed is treated as a whole and is then broken down into several triangles rather than the reverse.

Formation of well-conditioned triangles

The triangles into which the area is broken should be well-conditioned, i.e. they should have no angle less then 30° nor greater than 120°. These are minimum conditions. The ideal figure is an equilateral triangle and every effort should be made to have triangles whose angles are all around 45° to 75°. The reason is simply that it is much easier and much more accurate to plot such a triangle as opposed to plotting a badly conditioned triangle.

Good measuring conditions

All of the lines of the survey must be accurately measured and it is sound practice to select lines that are going to be physically easy to measure. Roads and paths are usually constructed along even gradients and present good measuring conditions. Lines that change gradient frequently are best avoided. Heavy undergrowth, for example, or railway embankments and cuttings, present difficulties and may make an otherwise good line completely immeasurable.

Permanency of the stations

The survey stations may have to be used at some future date when setting-out operations take place. They may, therefore, have to be of a permanent nature. Examples of permanent marks are shown in Fig. 3.15.

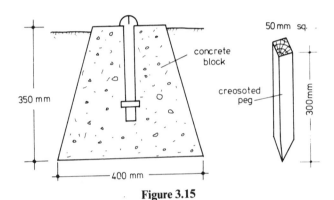

Figure 3.15

The marks must be sited in places that do not inconvenience anyone: for example, concrete blocks can do considerable damage to ploughs, etc., and cannot be placed in the middle of a field.

Referencing the stations

When the stations have to be used again it is necessary to be able to find them easily. If possible they should be placed near some permanent objects, fence posts, gates, bus stops, lamp standards, etc., from which measurements may be taken to the survey stations. This is known as 'referencing the station'. Each station should be referenced in case any check measurements are to be taken in the event of survey errors. A typical situation for a station is shown referenced in Fig. 3.16.

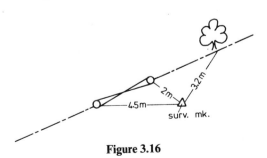

Figure 3.16

Obstructions to measuring

When siting any station with consideration for the factors outlined above it may be found that the survey line will cross a pond, river or railway cutting, which will present a considerable problem to measurement. While such problems can be overcome they should be avoided if at all possible.

Intervisibility of stations

Strictly speaking it is necessary to be able to see only from any one station to the other two stations of any triangle. Check measurements have to be made, however, and wherever possible an attempt should be made to see as many stations as possible from any one station.

Check measurements or tie lines

Before the survey is complete, check measurements must be made. A check line is a dimension that will prove the accuracy of part, of all, of a survey. In Fig. 3.17(a), line CF has been measured to check triangles ABD, BCD and ADF.

On completion of the plotting, the scaled distance of this line must agree with its actual measured length. If it does, the survey is satisfactory, but if not, there is an error in one or more lines and each must therefore be remeasured until the error is found—hence the necessity for referencing each station. This procedure could take a considerable time and a better method is to check each triangle in turn by providing it with its own check line. When plotted triangle by triangle, each must check; if it does not, it is then necessary to measure only the sides of the faulty triangle. Figure 3.17(b) shows the two methods of checking a triangle.

(b) Conducting a survey

Surveying the framework

Consider Fig. 3.1, where a plot of land at the GCB Outdoor Centre is to be developed as a housing site. Figure 3.17 shows the survey layout, consisting of a trilateration framework and a series of offsets.

Once the trilateration stations have been selected, the various lines are measured, using the methods previously described. It must be borne in mind continuously that the plan length is required and all gradients must be carefully observed and measured, or step taping must be employed.

The three sides and check line of each triangle should be measured before another triangle is attempted. The survey may begin on any station. In Fig. 3.17(a) the order of measuring the sides is AB, BD, DA and check line AG. In measuring line BD a ranging pole should be left at station G. It will then serve as a check point for triangle BCD, producing a check line CG. Only the lines BC and CD remain to be measured to complete the second triangle. There is therefore economy of movement of the survey party

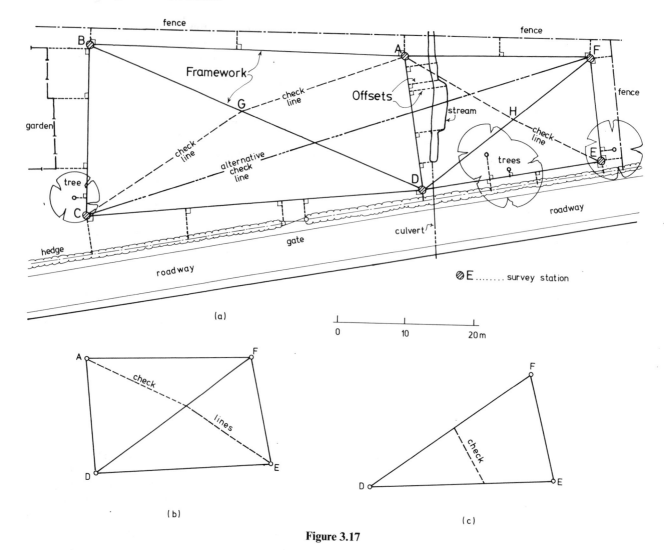

(a)

(b)

(c)

Figure 3.17

in using this technique, resulting in a considerable saving of time.

In Fig. 3.17(a) the survey lines are sited as close as possible to the details that have to be surveyed. These details include the hedges, trees, fences, building and stream.

Offsetting

The principle of offsetting has already been explained in Sec. 1(b). Offsets are short lengths measured to all points of detail from points along the main framework lines. These latter points are called chainages. Thus, any point of detail must have at least two measurements to fix its position, namely a chainage and an offset. Figure 3.18 shows surveyors measuring an offset to a tree which lies close to a main framework line.

Wherever possible, the offsets are measured at right angles to the survey lines, the right angle being judged by eye. The maximum length of offset is determined by the scale to which the plan is to be plotted.

On average, the naked eye can detect a distance of 0.25 mm, on paper. When a survey is plotted to a

scale of 1:1000, the eye will be able to detect (0.25×1000) mm actual ground size. Thus any error of over 0.25 metre will be detectable in the plotting. The average person is able to judge a right angle to $\pm 3°$.

Therefore, in Fig. 3.19 the detail point A will be fixed to within $\pm x$ metres. If x exceeds 0.25 metre, it will be plottable; thus, the maximum offset at which

Figure 3.18

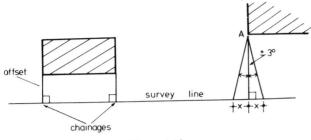

Figure 3.19

framework, complete with check lines, to survey the area.

2 Figure 3.22 shows a survey framework A–F. Add a series of suitable check lines to the survey.

3 The area of ground in Fig. 3.23 is to be converted into tennis courts. Devise a linear survey framework suitable for surveying the site.

4 Figure 3.24 shows a survey line XY, from which some ground detail is to be surveyed by offsetting. Show on the plan the required offsets.

this error will be indiscernible will be

$$(0.25 \text{ cotangent } 3°) \text{ metres} = 0.25 \times 19.08 \text{ metres}$$
$$= 5.00 \text{ metres approx.}$$

At this scale (1:1000) no offset should be allowed to exceed 5 metres and as the plotting scale decreases so the maximum length of offset increases.

It may be necessary or desirable to fix certain points of detail by more than one offset, whereupon the estimation of right angles ceases. In Fig. 3.20(a) certain details are fixed from two chainage points by 'oblique' offsets. This is a more accurate method and is used to fix important details, like the corners of a house. It is necessary to use this method when the maximum allowable right-angled offset is to be exceeded.

The principle of well-conditioned triangles should again be adhered to, and the distance between chainage points should be approximately equal to the oblique offsets.

When a building or wall, etc., lies at an angle to a survey line, it may be desirable to use 'in line' offsets. Such offsets are very similar to 'oblique' offsets but have the advantage that they are measured on the line of the detail feature, and when plotting is being done a 'bonus' is thereby obtained.

Exercise 3.2

1 Figure 3.21 shows a parcel of ground that is to be developed as a shopping centre. Devise a linear survey

(c) Recording the survey

One of the most important and probably most under-rated points in a linear survey is recording the information. There is little point in measuring accurately if the survey notes are not clear and intelligible. At the outset, a clear legible style of 'booking' must be adopted, if confusion is to be avoided at a later date.

The survey field book on which the notes are recorded measures 200 mm × 110 mm. It is normally rainproof and consists of about 100 leaves, ruled with two red centre lines, about 15 mm apart, running up and down both sides of each leaf. The double red centre lines represent the chain or tape as it lies on the ground.

Referencing the survey

The first task in booking is to make a reference sketch of the survey as a whole. The sketch is drawn to show the main survey stations in their correct relationship. Figure 3.25 is the survey reference sketch of the GCB Outdoor Centre.

As each line is measured, its length is written alongside the line, together with any gradient values. The measurements are always written in the direction of travel and arrows, indicating gradients, always point downhill. From this reference sketch, the basic framework can be plotted. Very often, the survey will not be plotted by the surveyor who made the survey, but by a draughtsman. Consequently, the sketch must be absolutely clear.

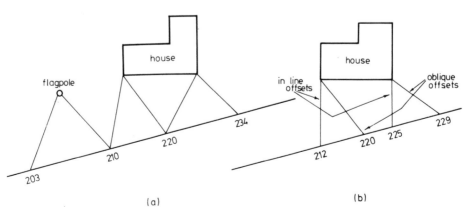

Figure 3.20

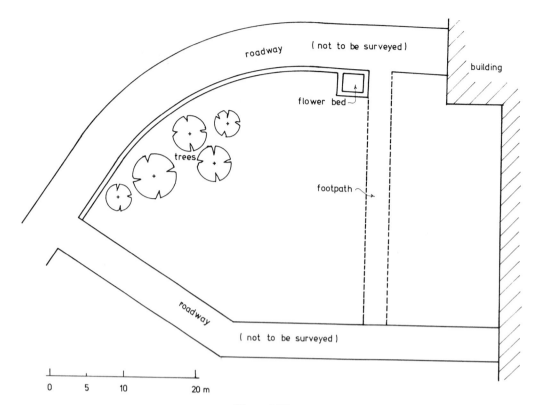

Figure 3.21

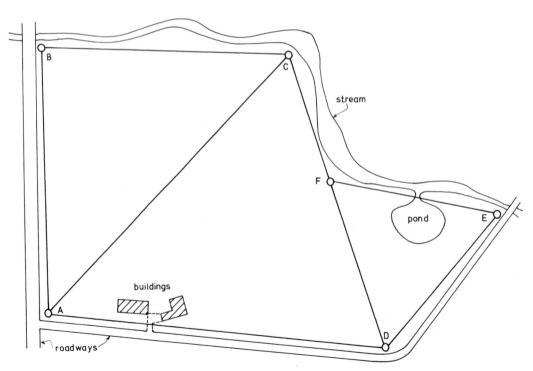

Figure 3.22

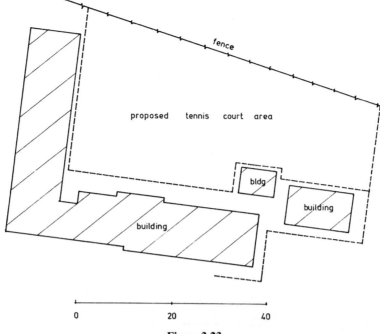

Figure 3.23

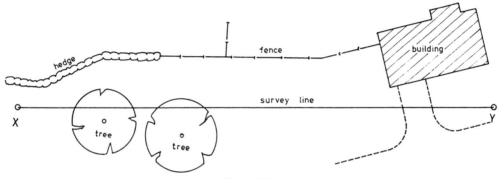

Figure 3.24

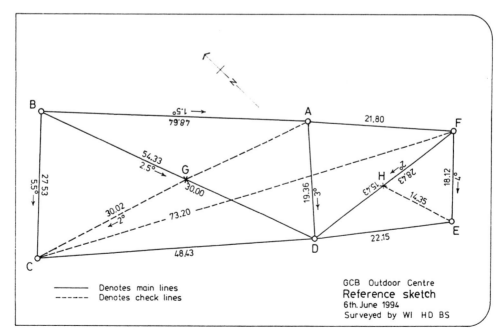

Figure 3.25

Booking the details

In Figure 3.17, the details to be surveyed from the main survey lines include the road, hedges, fences, building and stream. These are surveyed by taking offsets, at selected chainages, along the main survey lines. Figure 3.26 shows line BC being measured in the direction of B towards C. The field book is opened at the back and station B is positioned on the chain line (double red centre lines) at the bottom of the page. The booking proceeds up the page. Should it be necessary to use more than one page for any line, the page is simply flicked over and the booking continues on the reverse side. In this way, continuity of the line is preserved in the field book.

This method of booking is called the 'double line' method and all chainages are noted within the red lines. The various features to which offsets are taken are drawn in their relative positions and the offsets to right or left of the chain line are noted graphically. Virtually no attempt is made to draw to scale. The emphasis is placed rather on relative positioning of the features and on clear legible figuring.

A second method of booking, the 'single line' method, finds favour among a large number of surveyors. A blank field book is used, whereon a single pencil line is drawn to represent the chain line. Chainages are noted alongside the chain line and offsets shown graphically to right and left, as before. Figure 3.27 shows the same information as in Fig. 3.26 but as booked in 'single line' style.

The choice of method is largely personal, there being no overwhelming advantages in either. The

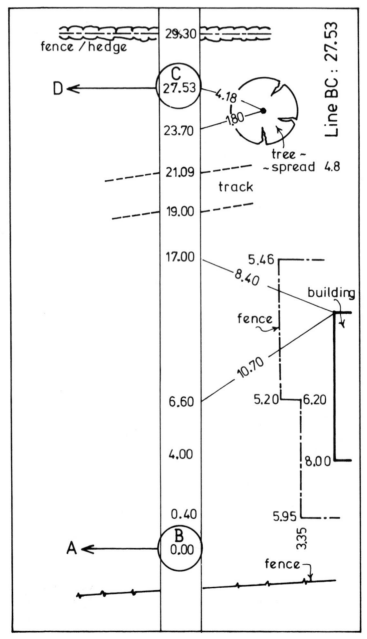

Figure 3.26

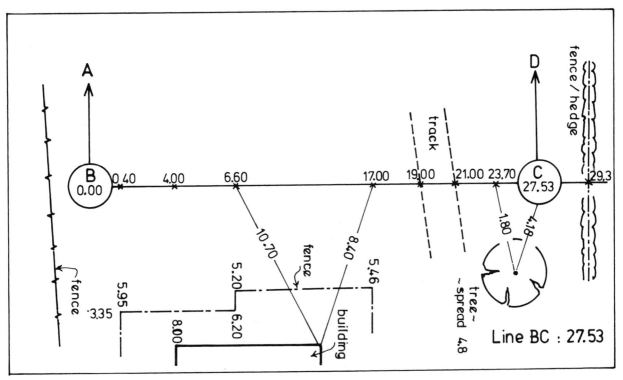

Figure 3.27

double line method does have the advantage that the chainages are contained within the red lines and there is therefore no possibility of their being confused with offsets. On the other hand, where features of any width, such as streams, roadways, etc., cross the chain line, their shape is distorted. In Fig. 3.26 a track crosses the line at an angle, at chainages of 19 and 21 metres. It appears to have a distinct change of direction. In Fig. 3.27 the same track crosses the chain line and is undistorted because the line has no width.

Example

9 Figure 3.28 shows line AD of the GCB Outdoor Centre survey in its correct relationship to the stream and wildlife pond. Using a 1:200 scale rule as the equivalent of a tape, make an offset survey of the stream and pond and show how the dimensions would be booked in double line form.

Answer (Fig. 3.29)

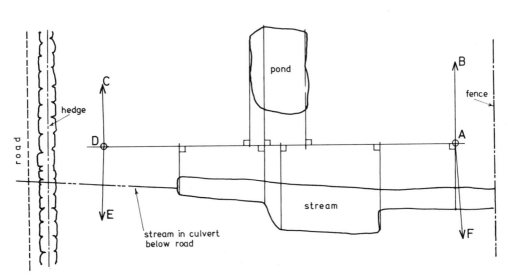

Figure 3.28

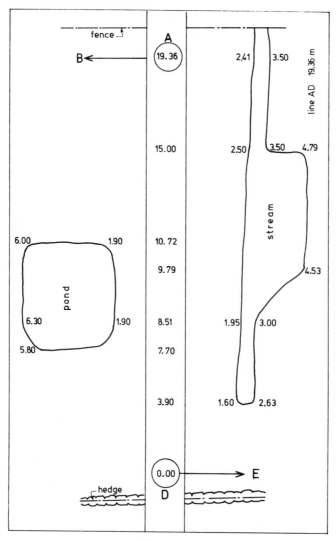

Figure 3.29

4. Plotting the survey

Undoubtedly the surveyor's main objective is to achieve accuracy in the field operations. However, unless results can be depicted accurately, legibly and pleasingly on paper, proficiency in the field is robbed of much of its value.

While it can be said that a natural flair for drawing can greatly assist the surveyor in the presentation of results, it is none the less true that a systematic approach to plotting will produce a perfectly acceptable plan of the results of the fieldwork.

(a) Results of the fieldwork

Figures 3.30 and 3.31 show the results of the survey of the GCB Outdoor Centre. They are to be plotted to a suitable scale on paper.

(b) Plotting equipment

1. Paper. The survey when plotted may have to be referred to frequently, over a number of years, and it is essential therefore that the material on which it is plotted should be stable. Modern 'plastic' drawing materials such as Permatrace are excellent in this respect.

2. The introduction of the metric system has simplified the problem of buying suitable scales, since almost all scales used nowadays are direct multiples of each other. The British Standard 1347 scale rule contains no fewer than eight scales—1:1, 1:5, 1:20, 1:50, 1:100, 1:200, 1:1250 and 1:2500. Other common plotting scales, 1:500 and 1:1000, can be derived simply by multiplying the scale units of the 1:50 and 1:100 scales by the factor 10. Modern plastic scale rules are ideal.

Linear survey   GCB Outdoor Centre

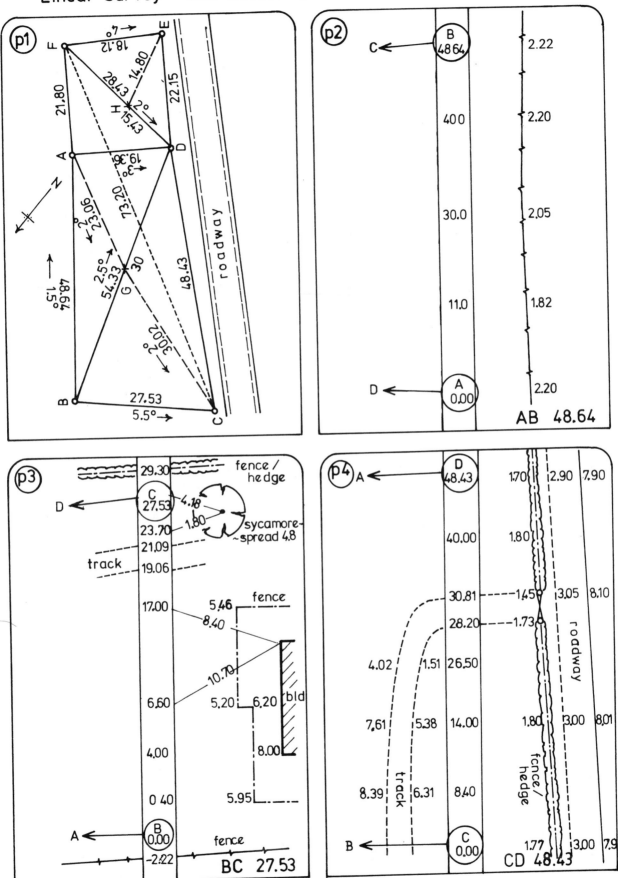

Figure 3.30

Linear survey (cont.)   GCB Outdoor Centre

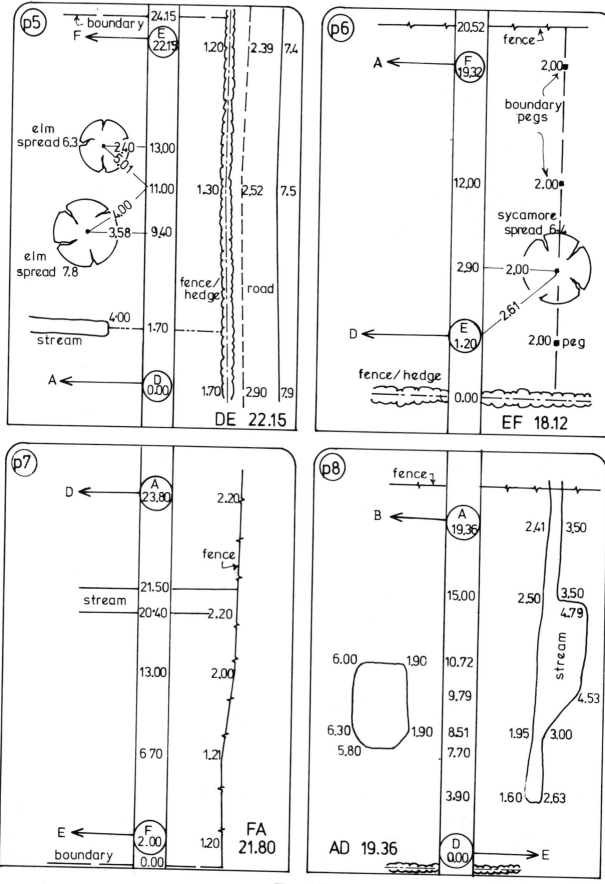

Figure 3.31

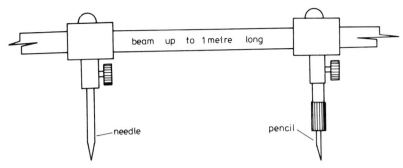

beam up to 1 metre long

needle

pencil

Figure 3.32

3. Beam compasses. Since the lines of chain surveys can be very long, large radius compasses are necessary. Ordinary compasses are unwieldy at large radius and beam compasses are used in plotting work (Fig. 3.32).

4. Other equipment includes two set squares, 45° and 60°, varying grades of pencils, 4H to H, sharpened to a fine point, paper weights, curves, inking pens, pricker pencil, eraser, steel straight edge and spring-bow compasses.

(c) Procedure in plotting

Orientation

Most maps and plans are drawn and interpreted looking north towards the top of the paper. It is customary on linear surveys to find north by some means, e.g. by compass or more roughly by the sun and a watch. The plotting material should always be oriented to north such that the top and bottom of the paper are respectively north and south.

Rough sketch

If a sketch of the survey is roughly drawn to scale it will greatly facilitate centralizing the survey on the drawing paper and will result in a much more balanced appearance.

Scale

A line scale, filled or open divided, is next drawn along the bottom. This scale drawn on the paper is necessary to detect possible shrinkage or expansion of the drawing material.

Calculation of plan lengths

Before any line of any survey can be plotted, the horizontal (plan) length must be calculated. In Sec. 2(b) it was shown that

Plan length = slope length × cos inclination

In the GCB Outdoor Centre survey reference sketch (Fig. 3.30), the slope length of every line of the survey is shown, together with its angle of inclination. The plan length of each line is calculated as follows.

Example

10		Angle of	
	Slope	inclination	Plan length (m)
Line	length	(deg)	(slope × cos angle)
AB	48.64	1.5	48.62
BC	27.53	5.5	27.40

Exercise 3.3

1 Calculate the plan lengths of the following lines of the GCB Outdoor Centre survey (Fig. 3.30).

		Angle of inclination
Line	Slope length	(deg)
BD	54.33	2.5
BG	30.00	2.5
CG	30.02	2.0
GA	23.06	2.0
AD	19.36	3.0
FE	18.12	4.0
FH	15.43	2.0

Plotting the framework

The details in Figs 3.30 and 3.31 are to be plotted to a scale of 1:200 on an A3 size sheet of drawing paper. The reference sketch (Fig. 3.30) indicates that the line DA of the survey points northwards. Line DA should, therefore, point towards the top of the sheet. From the survey sketches, the overall size of the survey is approximately 70 m east–west and 38 m north–south. The A3 plotting sheet measures (at scale 1:200) 84 m by 60 m. In order that the plotted survey finishes in the middle of the drawing sheet, point B will have to be placed 11 m from the left edge and 16 m from the top edge of the sheet (Fig. 3.33), to allow for offsets.

Station B is now fixed as the starting point of the plotting. From B, the line BA is drawn parallel to the top edge of the sheet and the horizontal length of 48.62 m is accurately scaled. An arc of a circle of length BD (54.28 m) is drawn using the beam compasses and a second arc of length DA (19.33 m) is drawn to intersect the first at point D. The check point G, along the line BD, is marked and the distance GA is scaled. The scaled length should agree closely

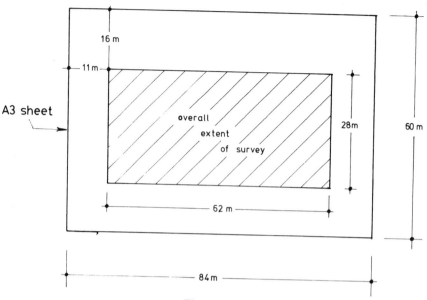

Figure 3.33

with the actual plan length of GA (23.05 m). Should it fail to do so, it indicates that either the plotting of one or more lines of the triangle ABD is in error or the field measurement of one or more of the lines is wrong. This is precisely the purpose of the tie line GA.

Assuming that the scaled length GA agrees with the field length, the triangle BCD is then plotted from the line BD in the same way as was the triangle ABD, and the length of its check line is compared with the actual plan length of 30.00 m.

The triangles AFD and DFE are plotted in like manner, thereby completing the construction of the framework. The various survey lines are then lightly drawn in ink and each survey station marked by a small circle.

Plotting the details

It now remains to plot the details along each chain line. On the line AB, for example, the various chainages of 0, 11, 30 and 40 m are marked off. Short right-angled lines, lying to the right of the survey line, are marked off and the various offsets are scaled along them. At 11 m, an offset of 1.82 m is scaled to the right and at 30 m, a 2.05 m offset is scaled. When these points are themselves joined by a straight line, they form the line of the fence, lying to the right of the survey line AB.

Along line BC, there are two sets of oblique offsets, which fix the positions of a tree and a corner of a house. These offsets are arced using a pair of compasses, the intersection of the 10.70 m arc from chainage 6.60 m and the 8.40 m arc from chainage 17.00 m forming a corner of the house.

Along each chain line the plotting of the detail is carried out with great care until the whole survey has been plotted. The various details are then drawn in

ink and a suitable title, north point, etc., are added. Figure 3.34 shows the survey, completely plotted to a scale of 1:250 but reduced to fit the book format.

Exercise 3.4

1 Using Figs 3.30 and 3.31 plot a GCB Outdoor Centre survey to scale 1:200.
2 Figure 3.35 shows the station points and lines relating to the survey of a plot of land. Utilizing the measured data shown in the figure, plot to a scale of 1:500 the main survey lines including all the detailed information on lines BC and CD.
(City and Guilds of London Institute)
3 (a) Describe the principle of trilateration.
 (b) Define (i) offsets, (ii) tie lines, (iii) station points.
 (c) List four possible causes of error in taking linear measurements.
 (d) Describe how to measure a line up a steep slope.
(City and Guilds of London Institute)

5. Negotiating obstructions

Despite the precautions taken during reconnaissance to site the survey stations in suitable positions, it might be impossible to avoid some obstacles to measuring. Such obstacles might well be in the form of a wood, small hill, change of gradient, river, railway cutting or even a building. They can generally be overcome by using 'field geometry' methods.

In some of the methods it is essential to be able to construct a right angle accurately either by linear means or by using some form of small hand instrument.

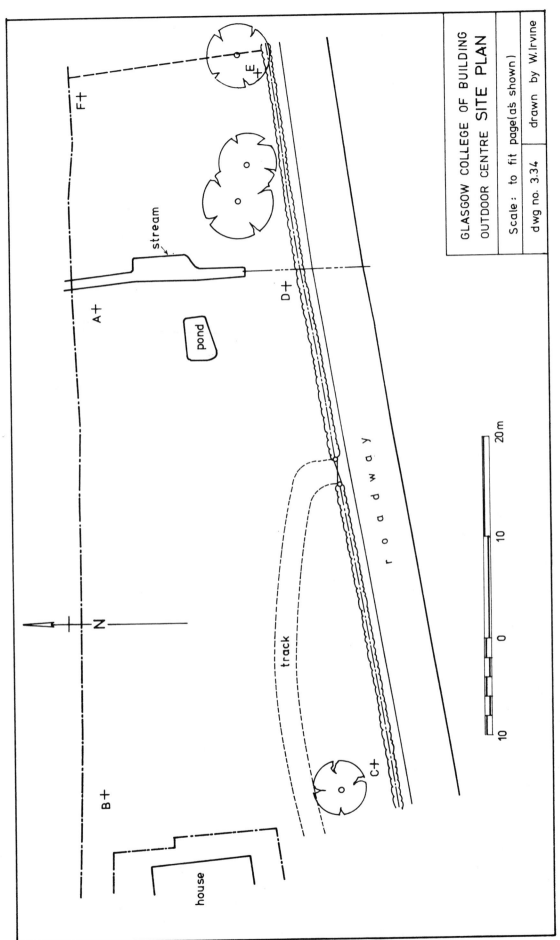

GLASGOW COLLEGE OF BUILDING
OUTDOOR CENTRE SITE PLAN

Scale : to fit page (as shown)

dwg no. 3.34   drawn by W.Irvine

Figure 3.34

49

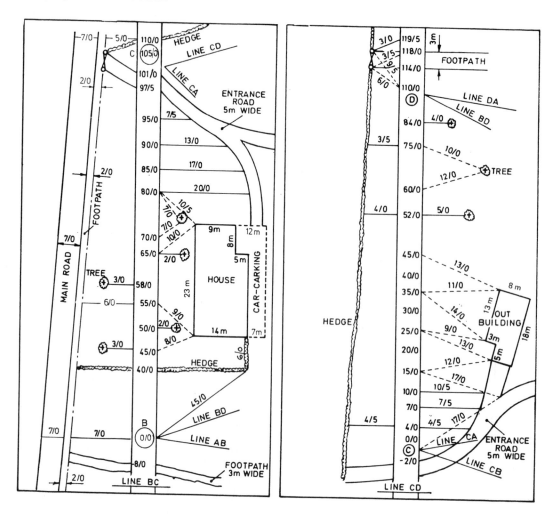

Distances (m)	
AB	82.000
BC	105.000
CD	110.000
DA	181.000
BD	186.000
AC	151.500

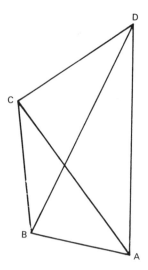

Figure 3.35

(a) Constructing a right angle

By linear means

In Fig. 3.36, B is a point on the mean survey line AC from which a right angle is to be erected. Equal distances BX and BY are laid off to right and left on the point B. From X and Y equal arcs XZ and YZ are described and the point Z defined. BZ will then be

at right angles to the survey line AC. A right-angled triangle can also be established using the principle of Pythagoras. The basic relationship of the three sides is

$$(2n + 1):2n(n + 1):2n(n + 1) + 1$$

If $n = 1$ then the relationship becomes 3:4:5.

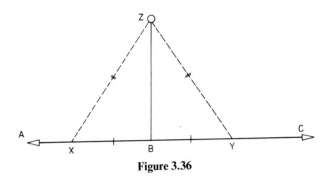

Figure 3.36

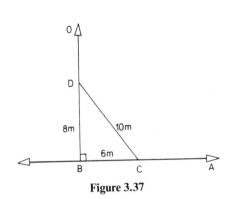

Figure 3.37

In Fig. 3.37, AB is the survey line and B the point at which the right angle is to be established. BC is measured out for a distance of 6 metres and C is accurately aligned on the survey line. The tape handle is held at B and the 18 metre marker held at C. The 8 metre marker is then pulled laterally from B forming a triangle BCD. The length BD is 8 metres and CD is 10 metres, while BC has already been measured as 6 metres. Thus, a triangle has been formed whose sides are in the ratio 3:4:5 and a right angle has thereby been established at B.

Sometimes, a right angle has to be established from a point to a survey line. In Fig. 3.38, X is an external point and AB is a survey line. If X is less than one tape length from the line, a right angle can be established quite easily by holding the tape handle at X and swinging the tape in an arc, cutting the survey line at points C and D. The required angle to X will occur at E, the mid-point of CD.

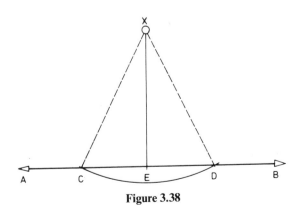

Figure 3.38

In Fig. 3.39, X is more than one tape from the survey line. The point C on the survey line, at which the right angle will occur, is roughly estimated and the distance CX paced or estimated. The same distance is then stepped out to point D. Length XD is accurately measured and bisected at O. If an arc of a circle is described with radius OX, it will cut the survey line at points D and C. The angle XCD is a right angle since it stands on a diameter of XD of a circle.

Use of hand instruments
There are several small hand instruments to assist the surveyor in setting out right angles:

1. *The cross staff*
Probably the simplest instrument is the cross staff which is essentially a metal cylinder in which four slots are cut at 90° intervals around the cylinder. The instrument is held on top of a ranging pole, or a special rod, and a sight taken along the survey line. A right angle can then be set out by sighting through the cross slits (Fig. 3.40).

2. *The prism square*
In this simple instrument, use is made of a pentagonal glass prism PQRST (Fig. 3.41). The faces PT and RS, if produced, contain 45°. The observer holds the prism at eye level and looks over the instrument to view directly target 1. The incident rays carrying the image of target 2 are reflected to the observer's eye. When target 1 and the reflected image of target 2 coincide a right angle will have been formed.

(b) Obstructions

An obstacle or obstruction is any object that obstructs the direct measurement or ranging of any line. They can be divided into several groups.

Obstructions to ranging only
Where a survey line has been established and both ends are not intervisible, a straight line must still be established between the points before the measurements can be conducted. In Fig. 3.42, survey stations A and D are not intervisible because of the intervening high ground.

Interpoles B_1 and C_1 are placed between A and D, with no attempt being made to align them. The only

Figure 3.39

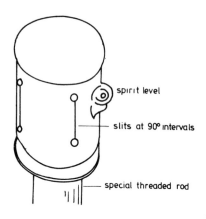

spirit level

slits at 90° intervals

special threaded rod

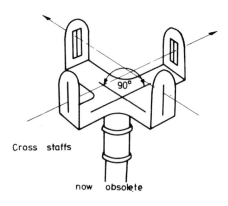

Cross staffs

now   obsolete

Figure 3.40

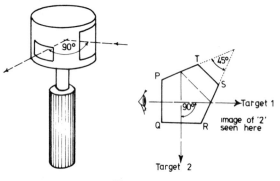

Figure 3.41

Obstructions to measuring only

1. *Obstructions that can be measured around*

Figure 3.43 shows a pond which lies directly on the line of survey stations X and Y, making measurement of that part of the survey line over the pond impossible. At a point A near the pond, a right angle is set out to point B by any of the previously described methods, and the distance AB is measured. The distance from point B to point C on the survey line at the far end of the pond is measured. By the theorem of Pythagoras, the inaccessible distance AC is calculated:

$$AC = \sqrt{(BC^2 - AB^2)}$$

condition governing their position is that C_1 must be visible from A and B_1 visible from D.

On the line C_1A, pole B_1 is ranged to its new position B_2, forming a straight line AB_2C_1. On the line B_2D, pole C_1 is ranged to C_2. B_2 is then ranged into the new position B_3, on the line AC_2, and the procedure is continued until A, B, C and D form a straight line.

Figure 3.44 shows an alternative method of obtaining the distance across the pond. At points A, B, C and D, right angles are set out and distances AB, BC and CD measured. If all setting out is accurate AB will equal CD and the inaccessible distance AD will be equal to CB.

Many other methods are available for obtaining this distance but one, in particular (Fig. 3.45), is

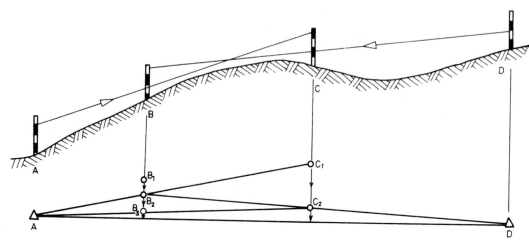

Figure 3.42

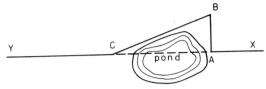

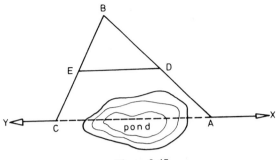

Figure 3.45

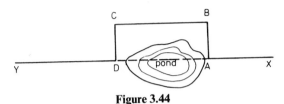

Figure 3.43

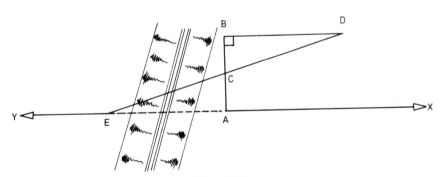

Figure 3.44

worthy of closer study, because it is a method used for setting out parallel lines.

At point A, any random line is set out to point B. The distances AB and BC are measured accurately and bisected at D and E respectively. Triangles BED and BCA are similar and DE is parallel to CA. The sides of the triangles have been laid out in proportion of 1 to 2 and CA is therefore equal to twice ED:

$$\frac{DE}{CA} = \frac{BD}{BA} = \frac{1}{2}$$

Therefore CA = 2DE.

2. *Obstructions that cannot be measured around*
The obstructions envisaged under this heading are rivers and railway cuttings, exceeding one tape in width. Again, many variations are possible. Figure 3.46 shows a survey line XY crossing a railway cutting; at point A, a right angle is set out to B and the distance AB measured and bisected at C. At B another right angle is set out towards D and the point established where this line intersects the line EC produced. Lines BD and AE are parallel and triangles AEC and BCD are congruent. The inaccessible distance AE therefore equals BD.

An alternative method of determining the unknown distance is shown in Fig. 3.47. A right angle is again set out at point A, any point B established and the distance AB measured. At B an optical square is used to sight C, across the railway cutting. A right angle is set out from B and the point D established on the line XX such that CBD is the right angle. Lines BD and AD are measured.

In the figure, triangles BAD and CBD are similar, since both have a right angle and the angle D is

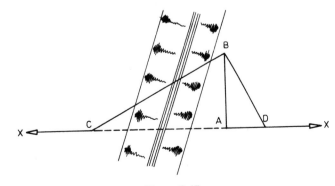

Figure 3.46

Figure 3.47

common to both. All three angles in both triangles are therefore equal: a condition for similarity.

Therefore
$$\frac{CD}{BD} = \frac{BD}{AD}$$

and
$$CD = \frac{BD^2}{AD}$$

However
$$CD = CA + AD$$

Therefore $(CA + AD) = \dfrac{BD^2}{AD}$

and
$$CA = \frac{BD^2}{AD} - AD$$

Obstructions to measuring and ranging

Despite taking every precaution to avoid obstacles, the occasion arises where a building or wood lies on the survey line and the line can neither be measured nor ranged. Figure 3.48 shows a typical situation, where a building has been erected on the line XY some time after the points X and Y were chosen.

At point X, a line XA is laid out to pass close to the building. Point A is selected such that angle YAX is

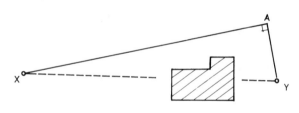

building erected on line XY

Figure 3.48

right-angled, the right angle being made by using a prism square or cross staff.

Using the theorem of Pythagoras

$$\text{Length } XY = \sqrt{XA^2 + AY^2}$$

Example

11 A survey line XY is interrupted by a wide river and the distance across the river is required. On the near bank a right angle is set out at point A for a distance of 20 metres and a point B established. At point C, 40 metres back from A on the line AX, a second right angle is set out to D, on the same side of AX as point B. The point D is aligned with Y on the far bank and with B, already established. The line CD is found to measure 31 metres. Calculate the length of the inaccessible portion AY.

Solution (Fig. 3.49)

Triangles YAB and YCD are similar.

Therefore
$$\frac{YA}{AB} = \frac{YA + AC}{CD}$$

i.e.
$$\frac{YA}{20} = \frac{YA + 40}{31}$$

Therefore
$$31\,YA = 20\,YA + 800$$

and
$$YA = \frac{800}{11}$$
$$= 72.72 \text{ metres}$$

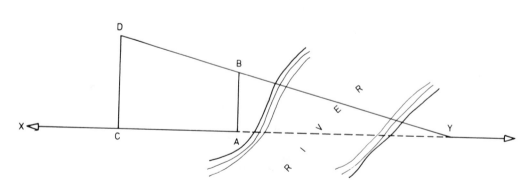

Figure 3.49

6. Answers

Exercise 3.1

1 32.58 m, 171.94 m, 63.72 m
2 94.00 m
3 20.077 m
4 108.27 m
5 82.726 m

Exercise 3.2

1 See Fig. 3.50.
2 See Fig. 3.51.
3 See Fig. 3.52.
4 See Fig. 3.53.

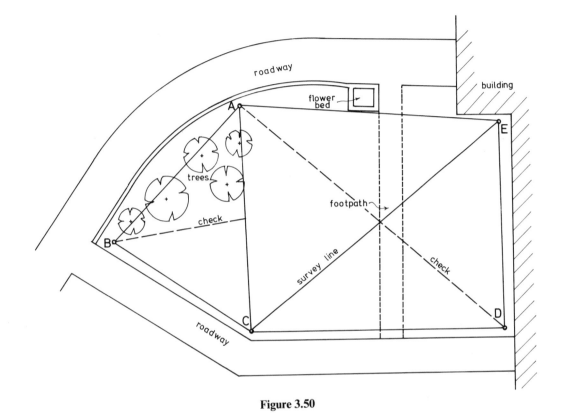

Figure 3.50

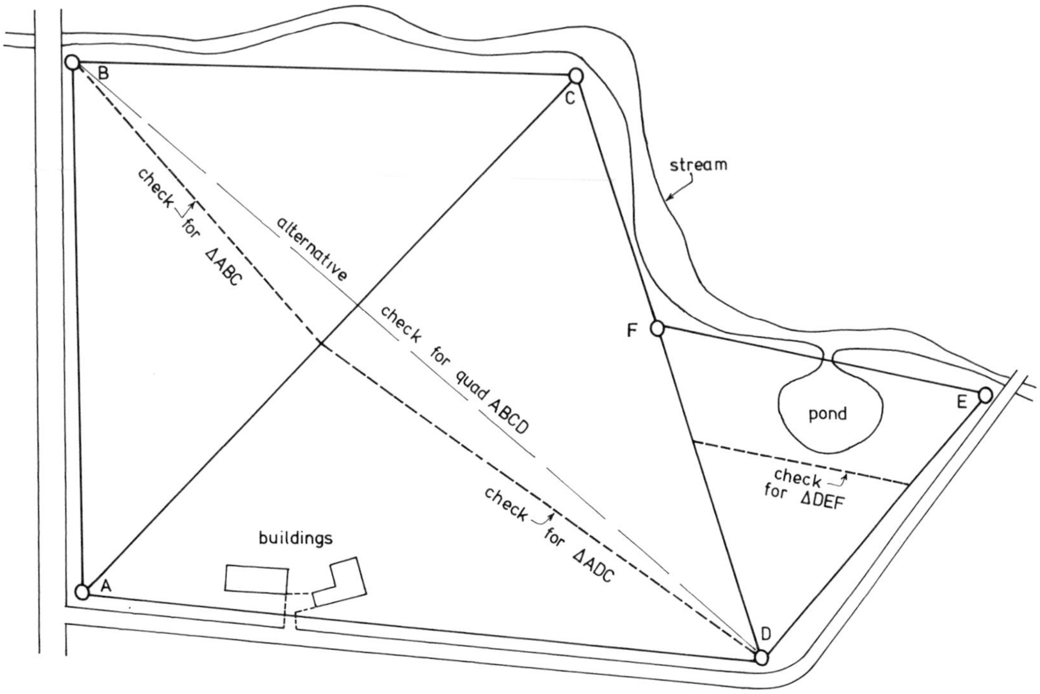

Figure 3.51

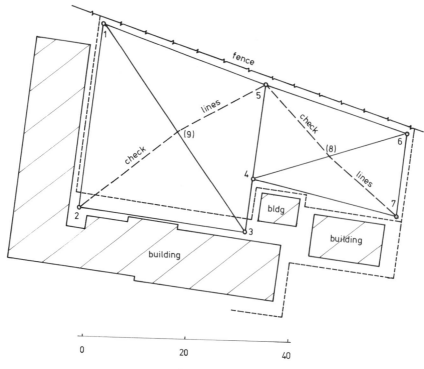

Figure 3.52

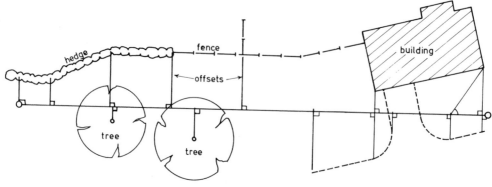

Figure 3.53

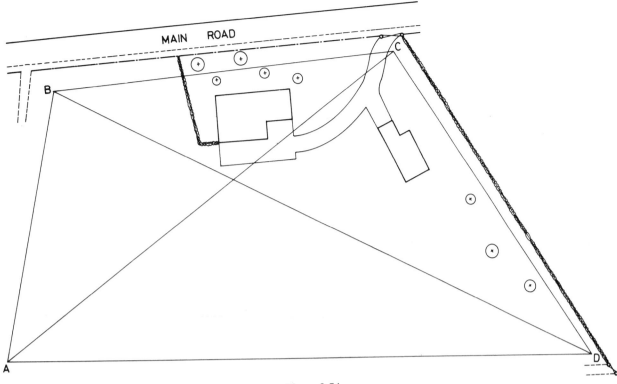

Figure 3.54

Exercise 3.3

1 BD 54.28 m
BG 29.97 m
CG 30.00 m
GA 23.05 m
AD 19.33 m
FE 18.08 m
FH 15.42 m

Note: The differences between the slope and horizontal distances are in this case unplottable. These differences can be significant, however, in other circumstances.

Exercise 3.4

1 See Fig. 3.34.

2 See Fig. 3.54.
3 Descriptive question:
(a) See page 27.
(b) See pages 27, and 37.
(c) See page 33.
(d) See page 32.

7. Project

Chapter 18 is a project covering all chapters of this textbook. It is intended that the project build into a complete portfolio of the surveying work required in the survey and the setting out of a building or engineering development.

If the reader wishes to continue work on the project or begin work at this stage, he or she should now turn to page 390 and attempt Section 2.

CHAPTER 4

Levelling

Objective

After studying this chapter, the reader should be able to make a levelling survey and calculate the results relative to some chosen datum.

Chapter 3 dealt with the simple principles of representing the earth's features in two dimensions on a plan. To be of any practical value, however, the third dimension, namely the height of the feature, must be shown by some means on the plan. In surveying, these heights are found by levelling. The heights of the points are referred to some horizontal plane of reference called the datum.

Figure 4.1 shows a table and chair standing on a level floor. The difference in height between them could be found using a rule in a variety of ways:

(a) by measuring upwards from a horizontal plane, namely the floor;
(b) by measuring downwards from a horizontal plane, namely the ceiling;
(c) by measuring downwards from an imaginary horizontal plane established by a spirit level.

If the spirit were held at eye level, say 1.500 m, and the horizontal plane extended over the table and chair, the plane would cut the rule held on the table at 0.850 m and on the chair at 1.050 m.

In practical levelling, the simple rule is replaced by a levelling staff while the spirit level is replaced by a surveying instrument called a 'level'. The level is, in fact, only a spirit level attached to a telescope which is mounted on a tripod. If the table and chair are replaced by two points on the surface of the earth, the simple illustration in Fig. 4.1 becomes an actual levelling exercise in Fig. 4.2.

It will be seen that the height of point A above datum is $1.500 - 0.850 = 0.650$ m while the height of C is $1.500 - 1.050 = 0.450$ m above datum. The datum, in this case, is an imaginary horizontal plane through the top of peg B.

If any Ordnance Survey map is examined, the heights of several points on the map will be seen to be shown by spot levels. These heights are measured above a datum called Ordnance Datum (OD), which is actually the mean level of the sea recorded at Newlyn Harbour, Cornwall, over the period 1915 to 1921. From this datum, levellings have been conducted throughout the country and the levels of numerous points permanently established by bench marks (Fig. 2.15), described in Sec. 2(f) in Chapter 2.

Any levelling done on a building site could therefore be referred to the OD by simply taking a reading to a bench mark instead of the peg B. On many sites, however, this will not be necessary and a peg concreted into the ground would serve as a bench mark.

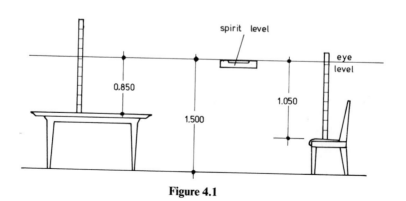

Figure 4.1

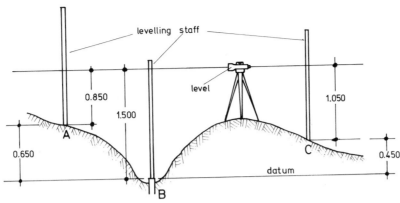

Figure 4.2

1. Levelling instruments

(a) The surveying telescope

An essential part of every levelling instrument is the telescope (Fig. 4.3) which is used to sight a levelling staff (a long rule described in detail in part (f)). Figure 4.3(b) shows the optical arrangements of the telescope of a typical surveying instrument.

In order to use the instrument, the observer places an eye behind the eyepiece and manually aligns the

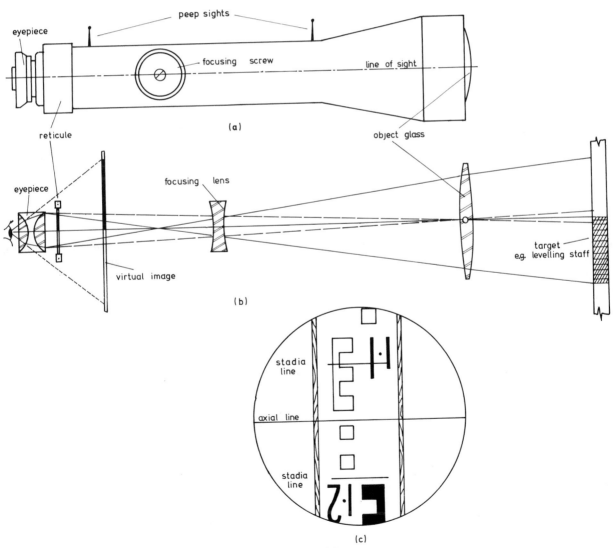

Figure 4.3

telescope on to the distant levelling staff. The light rays, originating at the staff, pass into the telescope through the object glass and in the process are inverted. The rays continue through the telescope and are captured as an image on a glass disc, called the reticule.

A series of lines is etched on the reticule. The observer sees these lines and the inverted image of the staff when an eye is placed to the eyepiece. The eyepiece is really an arrangement of lenses which magnifies the image and allows the observer to read the graduations on the staff (Fig. 4.3(c)).

It is almost certain that the reticule lines, or the image of the staff, or both, will be out of focus, and in order to read the staff graduations clearly the observer must make two adjustments to the focusing arrangements:

1. The eyepiece must be focused on to the reticule until the lines of the reticule are seen clearly and sharply. This is done by slowly rotating the eyepiece in either a clockwise or anticlockwise direction. This process is known as eliminating parallax.
2. The image of the staff must be focused on to the reticule by means of the internal focusing lens. This lens is activated by the focusing screw which is situated on the right-hand side of the telescope (Fig. 4.3(a)).

When focusing has been completed the observer sees clearly the view shown in Fig. 4.3(c).

In most modern instruments optical alterations are made to re-invert the image and thus present an upright view of the staff.

(b) Categories of levelling instrument

Levels are categorized into three groups: (a) dumpy levels, which are obsolescent, (b) tilting levels, which, though in common use, are being rapidly superseded by (c) automatic levels. A selection of these levels is shown in Fig. 4.4.

(c) Dumpy level

Dumpy levels, though obsolescent, are used in a few educational establishments (where cash for replacement is in short supply) and by a few small building firms; hence some students may be obliged to use them. They are not used on major construction or civil engineering projects. Nevertheless, the dumpy level is worthy of study since the levelling arrangements and the telescope assembly are common to a variety of surveying instruments.

Figure 4.5 shows a dumpy level reduced to its simplest form. It comprises the following parts:

1. *Trivet stage*
The open threaded flat base plate which attaches the instrument to the tripod.

2. *Levelling arrangement*
The conventional three-screw type where the feet of the screws fit into the trivet stage.

3. *Tribrach*
The main 'platform' which sits on top of the levelling screws and carries the remainder of the instrument.

It should be noted at this juncture that the trivet stage remains fixed in position, its sole function being to hold the instrument firmly on the tripod. The tribrach, however, can be tilted by movement of the levelling screws. The three parts are collectively known as the levelling head.

4. *Telescope*
Mounted on a vertical spindle which is free to rotate within the tribrach. The optical arrangement of the telescope has already been discussed. The principal axis is known as the line of sight or line of collimation.

5. *Spirit level*
Mounted on the telescope in such a fashion that it can be adjusted relative to the instrument if required. Figure 4.5 shows the spirit level attached to the telescope with one end free to pivot while the other end can be raised or lowered by means of the capstan screws.

Setting up the dumpy level (temporary adjustments)
The temporary adjustments are performed every time the instrument is set up. Three distinct operations are involved:

1. *Setting the tripod*
This is probably the most underrated aspect in setting up the surveying instrument, yet a few seconds spent at this point will save much time and effort later on.

Two legs of the tripod should be pushed firmly into the ground. If the tripod is being set on sloping ground, the two legs should be downhill. The third leg is manoeuvred until the tripod head is approximately level and is then pushed firmly into the ground.

2. *Levelling the instrument*
The levelling screws are positioned as in Fig. 4.6. The telescope is turned until it lies along the line of any two screws, say B and C, and the position of the spirit level bubble is observed. The levelling screws are held by the thumb and forefinger of each hand and the screws turned in opposite directions. The bubble will be seen to move along the bubble tube in the direction of movement of the left thumb—hence the 'left thumb rule'. The movement is continued until the bubble is centralized. If the telescope is now turned through 90° it will lie over screw A. Using screw A only and

(a) Wild No 1 Dumpy level

(b) Topcon Auto level ATG-7

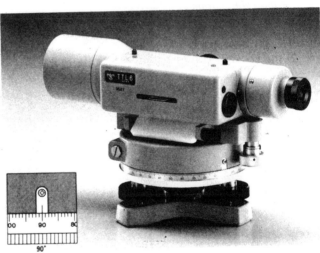

(c) Sokkia TTL6 tilting level showing coincidence bubble reader and three footscrews levelling arrangement

Figure 4.4

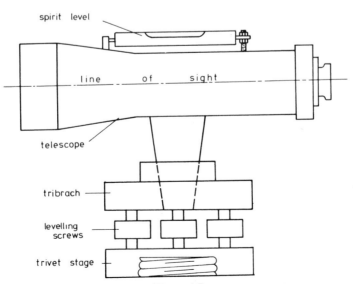

Figure 4.5

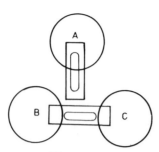

Figure 4.6

moving it by the left thumb, the bubble is centralized once more. Theoretically a horizontal plane through the telescope is established by these two levelling operations but, in practice, both operations will have to be repeated two or three times before the bubble remains central for both positions.

3. Parallax elimination

Parallax must be removed before any observations can be made to a levelling staff. The method of removing parallax has been described in part (a) dealing with optical principles.

The dumpy level is now ready to observe and a sight can be taken to a staff held in any position, whereupon the bubble of the spirit level should be central. In practice this might well not be the case for a variety of reasons, namely:

(a) imperfect adjustment of the instrument,
(b) wind pressure,
(c) the observer's movements around the tripod,
(d) soft ground causing the instrument to sink,
(e) unequal expansion of the various parts of the instrument by the sun.

Before the staff is read, however, the bubble must be perfectly centred. Therefore continual small adjustments of the spirit level have to be made by the only means available, namely the footscrews. Each movement of the screws, however, will alter the height of the line of sight, causing errors. These errors will be small and of practically no significance by the whole operation of relevelling several times is very time consuming and annoying.

This disadvantage is overcome by using a tilting level.

(d) Tilting level

A typical tilting level is shown in Fig. 4.7. It comprises the following parts:

1. The levelling head

The levelling head is made up of the same three components as the dumpy level, namely the trivet stage, the levelling arrangement and the tribrach.

The levelling arrangement is either a conventional three-screw system as on a dumpy level or some form

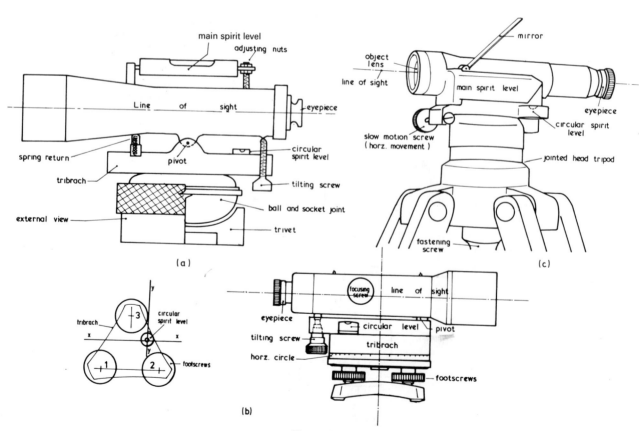

Figure 4.7

of ball and socket joint (Fig. 4.7(a)) which allows much faster levelling of the instrument to take place.

It is used in conjunction with a small relatively insensitive circular spirit level mounted on the tribrach. The tribrach can thus be levelled completely independently of the telescope and main spirit level. The instrument can be rotated about the vertical axis and the bubble of the circular spirit level will remain in the centre, indicating that the tribrach is approximately level.

2. *The telescope*
On a tilting level, the telescope is not rigidly attached to the tribrach but is supported by a central pivot. It is therefore capable of a small amount of vertical movement. This movement gives the instrument a tremendous advantage over a dumpy level. The vertical movement is imparted to the telescope by a tilting screw passing through the tribrach at the eyepiece end of the telescope. A spring loaded return, mounted on the tribrach at the objective end of the telescope, works in sympathy with the tilting screw to elevate or depress the telescope (Fig. 4.7(a)).

3. *Spirit level*
The main spirit level is mounted on top or on the side of the telescope.

Setting up the tilting level (temporary adjustments)
1. The tripod is set up with the levelling head approximately level and secured to the instrument by the ring clamp (Fig. 4.7(a)) or fastening screw (Fig. 4.7(c)).
2. The bubble of the circular screw level is centralized by one of the three methods (a), (b) or (c), below, thus making the vertical axis of the instrument approximately vertical and the line of sight approximately horizontal.

(a) If the instrument has three footscrews (Fig. 4.7(b)) the bubble is moved in the x direction using footscrews 1 and 2 in the same manner as with a dumpy level. Without rotating the instrument, the bubble is moved in the y direction using footscrew 3 until it is centred.

(b) Using a ball and socket joint (Fig. 4.7(a)), the ring clamp is released and the bubble is centralized by hand and the clamp tightened.

(c) Using the jointed head principle (Fig. 4.7(c)), the fastening screw is released and the instrument is shifted over the spherical surface of the tripod head by hand until the bubble is centralized. The fastening screw is then tightened.
3. Parallax is now eliminated by focusing the eyepiece on to the diaphragm until the cross lines appear sharp and clearly defined.
4. The staff is sighted using the slow motion screw if necessary and brought into sharp focus using the focusing screw on the side of the telescope.
5. The main spirit level is accurately centralized using the tilting screw and a reading taken on the staff.

When the staff is removed to another station the bubble of the main spirit level will move off-centre. However, a small rotation of the tilting screw will quickly bring it back to centre and another sight can be taken. In contrast to the dumpy level this re-levelling will not alter the height of the plane of collimation since the telescope is pivoted centrally.

Example

1 What is the *essential* difference between a dumpy and tilting level?

Answer
The essential difference between a dumpy and a tilting level is that the vertical axis of the dumpy level is made perfectly vertical by the levelling operation. The telescope, being fixed at right angles to the vertical axis, will revolve around it, sweeping out a horizontal plane. The telescope of the tilting level possesses a limited amount of vertical movement and can be made horizontal even though the vertical axis of the instrument is not truly vertical.

Coincidence bubble reader
In both the dumpy and tilting levels, the bubble of the spirit level is set by eye, to the centre of the bubble tube graduations. Many modern tilting levels, however, incorporate an optical system whereby the images of both ends of the bubble are viewed side by side in the same field of view using a coincidence bubble reader.

A system of 45° prisms reflects the images of the ends of the bubble as in Fig. 4.8. The observer sees both ends of the bubble through the viewing eyepiece of a level or theodolite. As the spirit level is made to tilt in these surveying instruments, the ends of the bubble are seen to move relative to each other until they coincide in the field of view, whereupon the bubble is accurately centred. A large increase in accuracy in setting the bubble is achieved by this means.

Figure 4.9 shows a WILD tilting level utilizing such a system.

(e) Automatic level

In conventional tilting levels the line of sight is, or should be, parallel to the axis of the telescope. It is only horizontal when the bubble of the properly adjusted spirit level is central. In the automatic levels the line of sight is levelled automatically (within certain limits) by means of an optical compensator inserted into the path of the rays through the telescope.

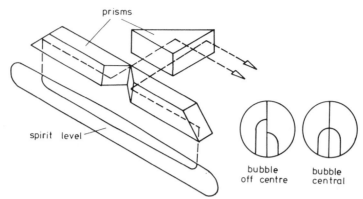

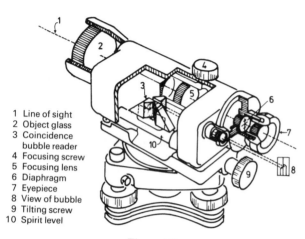

Figure 4.8

1 Line of sight
2 Object glass
3 Coincidence
   bubble reader
4 Focusing screw
5 Focusing lens
6 Diaphragm
7 Eyepiece
8 View of bubble
9 Tilting screw
10 Spirit level

Figure 4.9

There are currently over fifty different automatic levels produced by various manufacturers and most of these instruments utilize a different compensator. Fundamentally, the compensator is a form of prism suspended like a pendulum and acting as a mirror to direct the incoming horizontal ray from the staff through the centre of the diaphragm.

Basic principle of the compensator

Figure 4.10(a) shows a telescope in which two mirrors have been placed at 45° to the telescope axis. The horizontal ray of light entering the objective glass through the optical centre will be reflected at 90° from mirror A on to mirror B, where it will once again be reflected at 90° to pass through the centre of the diaphragm C.

In Fig. 4.10(b) the telescope has been tilted through a small angle of 1°. Relative to the horizontal plane the mirrors A and B therefore lie at an angle of 44°. The horizontal ray of light (solid line) entering through the optical centre of the object glass strikes mirror A, is reflected to strike mirror B and is reflected from it at an angle of 44°, i.e. diverging from the original ray (shown dashed) by 1°. It no longer passes through the centre of the diaphragm.

If mirror A could be maintained in a position at 45° to the horizontal, however, the incoming horizontal ray would be reflected vertically from the surface towards B. It will strike mirror B at 46° and be reflected from it at the same angle, thereby converging on the original ray at 1° to pass through the centre of the diaphragm C (Fig. 4.10(c)).

Using this system the compensator (mirror A) must be placed exactly mid-way between the object glass and diaphragm.

It must be emphasized at this point that the deviation of 1° in the example could not be tolerated in practice. This figure was used for illustrative purposes only. The maximum deviation is of the order of ±15 minutes of arc.

The system, as previously described, is applied almost exactly in the Nikon AP level. It has an automatic compensation prism suspended by a special metal plate mounted on ball bearings to maintain the horizontal line of sight automatically (Fig. 4.11).

There are several variations of this system but in all of them there is fundamentally some form of prism suspended like a pendulum which directs the horizontal ray through the centre of the diaphragm. Figure 4.12 shows a WILD automatic engineer's level. The compensator consists of two fixed prisms and one suspended prism. The pendulum and damping devices have been omitted in the figure for clarity.

Damping systems

Most pendulum devices are suspended on wires (Sokkisha B2, Wild NAK). A few are suspended in the force field of a permanent magnet (Kern GKIA). Naturally the pendulum device oscillates and these oscillations must be speedily damped before a reading can be taken on a staff. The damping is achieved by attaching the compensator to a piston which is free to move backwards and forwards within a short cylinder (Fig. 4.13). The compensator's oscillations move the piston, causing air to be expelled from the cylinder. This creates a vacuum which opposes the motion of

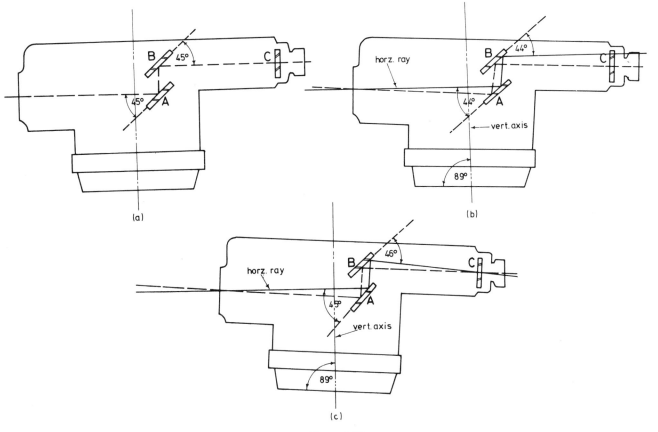

Figure 4.10

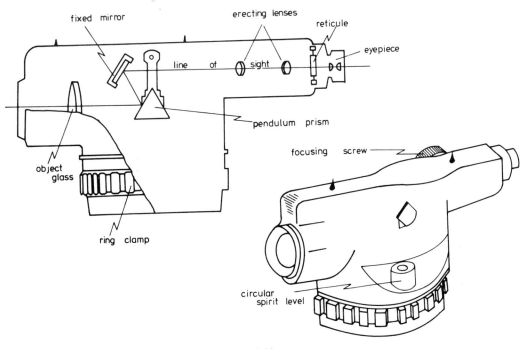

Figure 4.11

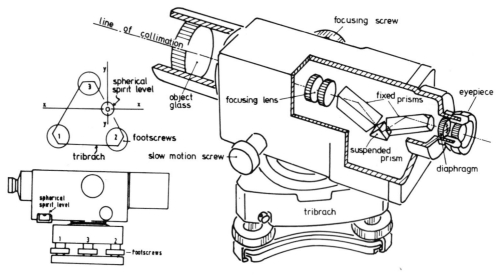

Figure 4.12

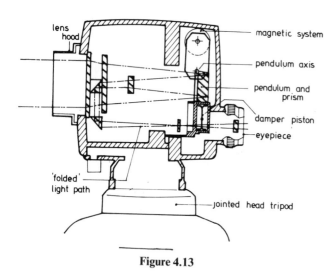

Figure 4.13

the piston and brings the compensator to rest in a fraction of a second.

Alternatively, the damping of the compensator is achieved magnetically. A small magnet is built into the telescope housing. It produces a north–south magnetic field. The pendulum compensator is positioned within this magnetic field and the oscillations of the pendulum produce a braking force which is directly proportional to the movement of the pendulum.

Setting up the automatic level

1. The instrument is set up with the levelling head approximately level and the instrument securely attached using the fastening screw.

2. On all automatic levels there is a small circular spirit level which is centred in exactly the same manner as a tilting level via a three-screw arrangement, a ball and socket joint or a jointed head system.

3. When the spirit level has been centred the vertical axis of rotation of the instrument is approximately

vertical. The compensator automatically levels the line of sight for every subsequent pointing of the instrument.

4. Parallax is eliminated as before, the staff is sighted and brought into focus and the staff reading noted.

(f) Levelling staff

The levelling staff should conform with British Standard Specification 4484 (1969). A portion of such a staff is shown in Fig. 4.14. The length of the instrument is 3, 4 or 5 m, while the width of reading face must not be less than 38 mm. Different colours must be used to show the graduation marks in alternate metres, the most common colours being black and red on a white ground. Major graduations occur at

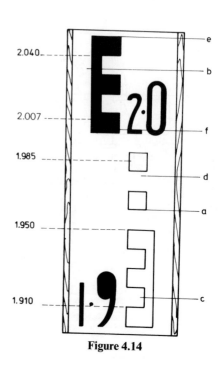

Figure 4.14

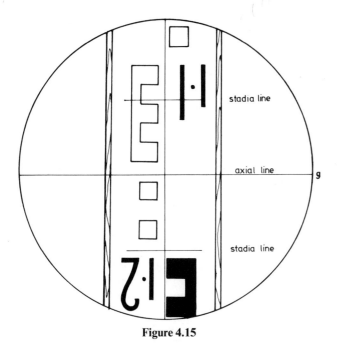

stadia line

axial line g

stadia line

Figure 4.15

on the side of the telescope brings the staff clearly into focus. If parallax has been properly eliminated, the diaphragm should be sharply outlined against the staff and there should be no apparent movement of the cross-hairs over the staff as the head is moved up and down.

If a tilting level is being used, the main spirit level is now accurately centred. The dumpy and automatic levels will already be levelled accurately. The observer must now remove his or her hands from the tripod and instrument. If the tripod is leant on at this stage the line of collimation will be tilted upwards and a wrong reading will be taken.

Of the two sets of figures on either side of the axial line, the lower is noted. The 10 mm graduations are counted and the final millimetre placing is estimated. The complete reading is then booked, the spirit level checked, the staff reading taken once more and checked against the booked figures.

The staffholder is then directed to the next staff station where the reading procedure is repeated.

2. Levelling procedure (using the rise and fall method)

(a) Levelling between two points

Figure 4.16(a) shows a peg A situated beside a manhole cover at the bottom of a fairly steep slope. The height of the peg A is known to be 95.400 m above Ordnance Datum (AOD), having already been surveyed from a nearby Ordnance Survey bench mark (OSBM). The known height of peg A is called the reduced level (RL) of A; i.e.

$$RL\ A = 95.400\ m\ AOD$$

The peg A need not have a known height relative to Ordnance Datum, in which case it would be assigned an arbitrary level of, say, 10.000 m or 100.000 m above an assumed datum (AD). The reduced level of A would therefore be written as

$$RL\ A = 10.000\ m\ AD$$

The reduced level of peg B, at the top of the slope, is required.

100 mm intervals, the figures denoting metres and decimal parts. Minor graduations are at 10 mm intervals, the lower three graduation marks of each 100 mm division being connected by a vertical band to form a letter E. Thus, the 'E' band covers 50 mm and its distinctive shape is a valuable aid in reading the staff. Minor graduations of 1 mm may be estimated.

In Fig. 4.14 various staff readings are shown to illustrate the method of reading. Owing to the optical principles of the telescope, staff readings will be inverted with many instruments. Figure 4.15 shows such an inverted view through the telescope of a levelling staff set up about 8 metres away.

Example

2 Make a list of the staff readings 'a' to 'f' in Fig. 4.14 and reading 'g' in Fig. 4.15.

Solution

a, 1.960 m	b, 2.033 m	c, 1.915 m
d, 1.978 m	e, 2050 m	f, 2.002 m
g, 1.156 m		

(g) Taking a reading

The method of taking a reading differs slightly with the type of levelling instrument being used. In general, however, the instrument is set up and levelled accurately; then parallax is eliminated.

The staffholder holds the staff on the mark and ensures that it is held perfectly vertically facing towards the instrument. The observer directs the telescope towards the staff and using the focusing screw

Procedure

The level is set up approximately midway between the points, accurately levelled, and a first reading is taken to a staff held vertically on A. Let the reading be 2.500 m. The staff is then transferred to B, held vertically on the station and a second reading taken. Let this reading be 0.500 m.

From the sketch it is obvious that point B is higher than A by $2.500 - 0.500 = 2.000$ m. In other words, the ground rises from A to B by 2.000 m (Fig. 4.16(b)).

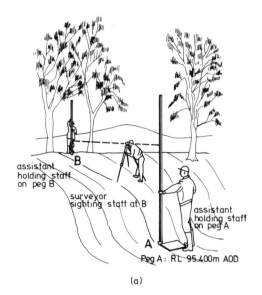

assistant
holding staff
on peg B

surveyor
sighting staff at B

B

assistant
holding staff
on peg A

A

Peg A: RL 95.400m AOD

(a)

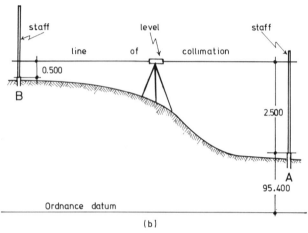

staff   level   staff

line   of   collimation

0.500

B

2.500

A
95.400

Ordnance datum

(b)

Figure 4.16

In this example, if the height of the ground at A is 95.400 m above datum, the height of the ground at B above datum can be readily deduced. Since the ground rises by 2.000 from A to B, point B must lie 95.400 + 2.000 = 97.400 m above datum.

This is the basis for all levelling work no matter how long or complex the particular levelling may be. It is essential therefore to understand fully the above principle.

In general terms, where the height above datum of any point is required a staff reading is taken to it and compared with a staff reading taken to a point of known height above datum. A comparison of the readings indicates a rise or fall of the ground between them. The unknown height is then determined by adding the rise to, or subtracting the fall from, the known height.

Booking and reducing the readings

As in all aspects of surveying, the fieldwork must be properly recorded and calculated. In levelling, two methods are available, viz (a) the rise and fall method (RF method) and (b) the height of the plane of collimation method (HPC method).

The RF method requires a slightly longer time than the HPC method to calculate the results but has the advantage that the results are more easily checked. This section is based on the rise and fall system. All levellings must be properly recorded in a field book ruled as in Table 4.1.

At any instrument setting the first sight taken is called a 'backsight' (BS). In Fig. 4.16 the sight to A is therefore a backsight and the reading of 2.500 is entered in the BS column. The description of the observed point is entered in the 'remarks' column.

The final sight taken is called the 'foresight' (FS). In the example the foresight is the sight to B and the reading 0.500 is entered in the FS column.

The rise or fall of the ground is calculated by *always* subtracting the second reading from the first. If a positive result is obtained there is a rise between the points. Similarly, a negative result indicates a fall of the ground. In the example:

$$BS\ A = 2.500$$
$$FS\ B = 0.500$$
$$Difference\ (A - B) = +2.000\ m\ (rise)$$

In levelling the term 'reduced level' is used to denote the height of any point above datum. Since A lies 95.400 m above datum, this figure is entered in the reduced level column alongside A.

The reduced level of B is the algebraic sum of the reduced level A and the rise or fall from A to B:

$$Reduced\ level\ A = 95.400$$
$$Rise\ A\ to\ B = +2.000$$
$$Therefore\ reduced\ level\ B = +97.400\ m$$

Example

3 The data shown in Table 4.2 relate to three different levellings. Calculate the reduced level of the second point in each case by completing the table.

BS	IS	FS	Rise	Fall	Reduced level	Distance	Remarks
2.500					95.400		A. Peg level
		0.500	2.000		97.400		B. Peg level

Table 4.1

	BS	IS	FS	Rise	Fall	Reduced level	Distance	Remarks
(a)	3.250		1.130			135.260		Bench mark A Kerb B
(b)	0.752		2.896			73.270		Temperature bench mark Peg A
(c)	2.111		2.397			55.210		OS bench mark Ground level K

Table 4.2

Solution

(a) $3.250 - 1.130 = (\text{rise } 2.120) + 135.260$
$= 137.380$ m (reduced level B)

(b) $0.752 - 2.896 = (\text{fall } 2.144) + 73.270$
$= 71.126$ m
(reduced level peg A)

(c) $2.111 - 2.397 = (\text{fall } 0.286) + 55.210$
$= 54.924$ m (ground level K)

(b) Flying levelling

Procedure

Where two points, A and B, lie a considerable distance apart or are at largely different elevations, more than one instrument setting is necessary. In Fig. 4.17, points A and B lie about 250 metres apart. The reduced level of A is 23.900 m and the reduced level B is required.

The instrument is set about 40 m from A (position 1) and a backsight of 4.200 m taken. The staff is removed to a point about 40 m beyond the instrument and a foresight (0.700 m) taken. The reduced level of X is therefore calculated, as before:

$$\begin{aligned}
\text{Backsight to A} &= 4.200 \\
\text{Foresight to X} &= \underline{0.700} \\
\text{Difference (A} - \text{X)} &= \underline{+3.500} \text{ m (rise)} \\
\\
\text{Reduced level A} &= 23.900 \\
\text{Rise A to X} &= \underline{+3.500} \\
\text{Reduced level X} &= \underline{27.400} \text{ m}
\end{aligned}$$

Table 4.3 shows the field booking and reduction as it would be done on lines 1 and 2 of the table.

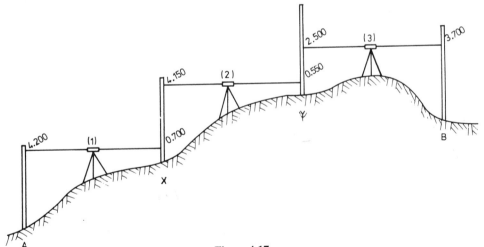

Figure 4.17

	BS	IS	FS	Rise	Fall	Reduced level	Distance	Remarks
Line 1 Line 2 Line 3 Line 4	4.200		0.700	3.500		23.900 27.400		A. Ground level X. Ground level CP

Table 4.3

	BS	IS	FS	Rise	Fall	Reduced level	Distance	Remarks
Line 1	4.200					23.900		A. Ground level
Line 2	4.150		0.700	3.500		27.400		X. Ground level CP
Line 3			0.550	3.600		31.000		Y. Ground level CP

Table 4.4

No further staff readings can be taken beyond X because the line of sight will run into the ground. The level is therefore removed to position (2), the staff again held on X facing towards the instrument and a sight is taken. Since this is the first sight at the new setting, it is by definition a backsight and the reading 4.150 is entered in the BS column. The reading must be entered on line 2 because this is the line referring to point X (Table 4.4). The staff is then removed to Y and a foresight taken. The reading 0.550 is entered in the field book on line 3 in the FS column. The reduced level of Y is calculated thus:

$$\begin{aligned} \text{Backsight to X} &= \quad 4.150 \\ \text{Foresight to Y} &= \quad 0.550 \\ \hline \text{Difference (X − Y)} &= +3.600 \text{ m (rise)} \end{aligned}$$

$$\begin{aligned} \text{Reduced level of X} &= 27.400 \\ \text{Rise X to Y} &= \quad 3.600 \\ \hline \text{Reduced level of Y} &= 31.000 \text{ m} \end{aligned}$$

It should be noticed that the booking and reduction of the second instrument setting is exactly the same as for the first setting and indeed all levelling reductions are performed in exactly this manner.

Since the line of sight would again run into the ground if continued beyond Y, the level is removed to position (3). The staff is held on Y and turned to face towards the new instrument position. A backsight is taken, resulting in a reading of 2.500. A foresight is then taken to B where the reading is 3.700. Table 4.5 shows the readings entered in their appropriate columns on lines 3 and 4 respectively. The reduced level of B is calculated thus:

$$\begin{aligned} \text{Backsight to Y} &= \quad 2.500 \\ \text{Foresight to B} &= \quad 3.700 \\ \hline \text{Difference (Y − B)} &= -1.200 \text{ m (fall)} \end{aligned}$$

$$\begin{aligned} \text{Reduced level Y} &= \quad 31.000 \\ \text{Fall Y to B} &= \; -1.200 \\ \hline \text{Reduced level B} &= \quad 29.800 \text{ m} \end{aligned}$$

The points X and Y are points where both foresights and backsights are taken as already seen, the instrument position is changed between the foresight and backsight and the points are called 'change points'. The letters CP are often inserted in the remarks column but are not absolutely necessary. The complete reduction is shown in Table 4.5.

Arithmetic check

As in all surveying operations, a check should be provided on the arithmetic. Lines 5, 6 and 7 provide such a check.

A moment's thought will show that the last reduced level is calculated as follows:

$$\begin{aligned} \text{Last reduced level} &= \text{first reduced level} \\ &\quad + \text{all rises} - \text{all falls} \\ \text{Therefore last reduced level} &= \text{first reduced level} \\ &\quad + \text{sum rises} \\ &\quad - \text{sum falls} \end{aligned}$$

Each rise or fall, however, is the difference between its respective backsight and foresight. Therefore, the sum of the rises minus the sum of the falls must equal the sum of the backsights minus the sum of the foresights.

	BS	IS	FS	Rise	Fall	Reduced level	Distance	Remarks
Line 1	4.200					23.900		A. Ground level
Line 2	4.150		0.700	3.500		27.400		X. Ground level CP
Line 3	2.500		0.550	3.600		31.000		Y. Ground level CP
Line 4			3.700		1.200	29.800		B. Ground level
Line 5	10.850		4.950	7.100	1.200	29.800		
Line 6	−4.950			−1.200		−23.900		
Line 7	=5.900			=5.900		=5.900		

Table 4.5

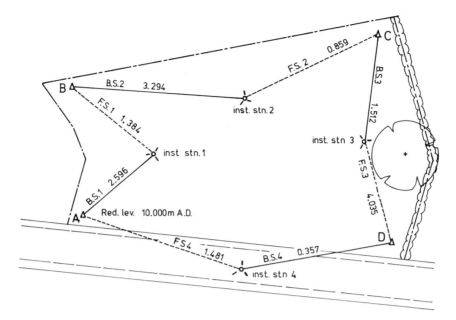

Figure 4.18

The complete check is therefore:

(Last reduced level − first reduced level)
$$= \text{(sum rises − sum falls)}$$
$$= \text{(sum BS − sum FS)}$$

i.e.

$$(29.800 − 23.900) = (7.100 − 1.200)$$
$$= (10.850 − 4.950)$$
$$= 5.900$$

This particular levelling is an example of flying levelling. The shortest route between the points A and B is chosen and as few instrument settings as possible are used.

Examples

4 In Fig. 4.18, four pegs are spaced around a construction site. These pegs are to be used as temporary bench marks for the duration of the site. A flying levelling was made around the pegs as shown, in order to establish the reduced levels of the pegs. Book the readings and calculate the reduced levels of the pegs, assuming that peg A has an assumed level of 10.000 m AD.

Answer (Table 4.6)

5 The data shown in Table 4.7 relate to a flying levelling from an Ordnance Survey bench mark (OSBM) to a proposed temporary bench mark (TBM). Calculate the reduced level of the TBM.

Solution
$$2.57 − 2.16 = \text{(rise 0.41)} + 85.36$$
$$= 85.77 \text{ (reduced level CP 1)}$$
$$1.03 − 1.52 = \text{(fall 0.49)} + 85.77$$
$$= 85.28 \text{ (reduced level CP 2)}$$
$$2.03 − 2.16 = \text{(fall 0.13)} + 85.28$$
$$= 85.15 \text{ (reduced level TBM)}$$

Check (BS column − FS column)
$$= \text{(rise column − fall column)}$$
$$= \text{(reduced level TBM − reduced level OSBM)}$$
$$= −0.21$$

BS	IS	FS	Rise	Fall	Reduced level	Remarks
2.596					10.000	A
3.294		1.384	1.212		11.212	B
1.512		0.859	2.435		13.647	C
0.357		4.035		2.523	11.124	D
		1.481		1.124	10.000	E
7.759		7.759	3.647	3.647	10.000	
−7.759			−3.647		−10.000	
0.000			0.000		0.000	

Table 4.6

BS	IS	FS	Rise	Fall	Reduced level	Remarks
2.57					85.36	OSBM
1.03		2.16				CP1
2.03		1.52				CP2
		2.16				TBM

Table 4.7

Closed circuit of levelling

It is good practice, when making any levelling survey, to finish the survey on the point where it commenced. A closed circuit is thus formed.

In the previous subsection, the arithmetic check on any levelling survey was shown to be:

(Last reduced level − first reduced level)
  = (sum of rise column − sum of fall column)
  = (sum of BS column − sum of FS column)

Since the last point is actually also the first point in a closed circuit, then the last reduced level minus the first reduced level should be zero. The sum of the BS column should therefore equal the sum of the FS column.

In the field, the levelling can be very quickly verified by simply checking the sum of the BS column against that of the FS column.

While in theory the difference between the sums of the columns should be zero, in practice this is unlikely to happen, due to the minor errors that inevitably arise during a levelling. These errors are considered fully in Sec. 5.

1. Top

2. Mid

3. Lower

Figure 4.19

Exercise 4.1

1 Figure 4.19 shows a section of a levelling staff viewed through the telescope of a levelling instrument. State the readings of the three cross lines.
2 Figure 4.20 illustrates the linear framework of the

GCB Outdoor Centre and Table 4.8 shows the readings of a flying levelling, conducted from a nearby Ordnance Survey bench mark, around the pegs of the framework survey. Calculate the reduced levels of the survey pegs.

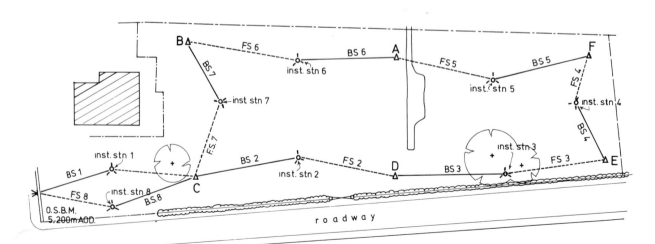

Figure 4.20

BS	IS	FS	Rise	Fall	Reduced level	Remarks
1.955					5.200	OSBM
1.315		2.030				Station C
1.243		0.885				Station D
2.071		1.485				Station E
1.570		0.880				Station F
1.835		1.590				Station A
0.631		0.540				Station B
1.200		3.289				Station C
		1.130				OSBM

Table 4.8

(c) Series levelling

When the reduced levels of many points are required, the method known as series levelling is used.

Points observed from single instrument station

In Fig. 4.21, the reduced levels of six points A to F are required. The instrument is set up accurately and a sight taken to A. Since this is the first sight it is of course a backsight. Its value is entered in the BS column.

The points B, C, D and E are sighted in turn and finally point F is observed. Point F is the last sight taken and is therefore, by definition, a foresight. The readings taken at B, C, D and E are intermediate between the first and last sights and are in fact called intermediate sights. The readings are entered in the IS column as in Table 4.9.

The rise or fall is required between successive points A to B, B to C, C to D, etc.

$$
\begin{array}{rl}
\text{Backsight A} = & 0.510 \\
\text{Intermediate sight B} = & 3.720 \\
\hline
\text{Difference} & -3.210 \text{ fall}
\end{array}
$$

$$
\begin{array}{rl}
\text{Intermediate sight B} = & 3.720 \\
\text{Intermediate sight C} = & 0.920 \\
\hline
\text{Difference} & +2.800 \text{ rise}
\end{array}
$$

$$
\begin{array}{rl}
\text{Intermediate sight C} = & 0.920 \\
\text{Intermediate sight D} = & 0.920 \\
\hline
\text{Difference} & 0.000
\end{array}
$$

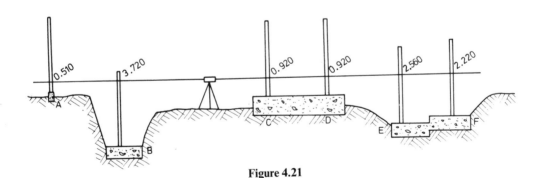

Figure 4.21

BS	IS	FS	Rise	Fall	Reduced level	Distance	Remarks
0.510					107.520		A. (TBM)
	3.720			3.210	104.310		B. Foundation level 1
	0.920		2.800		107.110		C. Foundation level 2
	0.920				107.110		D. Foundation level 2
	2.560			1.640	105.470		E. Foundation level 3
		2.220	0.340		105.810		F. Foundation level 4
0.510		2.220	3.140	4.850	105.810		
2.220			−4.850		−107.520		
−1.710			−1.710		−1.710		

Table 4.9

Intermediate sight D =   0.920
Intermediate sight E =   2.560
Difference   − 1.640

Intermediate sight E =   2.560
Foresight F =   2.220
Difference   + 0.340 rise

Check

Sum of backsights (one only) =   0.510
Sum of foresights (one only) =   2.220
Difference   − 1.710 fall

Sum of rises (2.8000 + 340) =   3.140
Sum of falls (3.210 + 1.640) =   4.850
Difference   − 1.710 fall

The arithmetic checks at this stage and it is therefore safe to proceed to the reduced level column. As in the previous example, the reduced levels are obtained by algebraically adding the rises or falls to the first reduced level in succession.

Reduced level A =   107.520
− fall (3.210) A to B =   104.310
+ rise (2.800) B to C =   107.110
No rise or fall C to D =   107.110
− fall (1.640) D to E =   105.470
+ rise (0.340) E to F =   105.810
Check   Last reduced level =   105.810
− first reduced level =   − 107.520
Difference =   − 1.710 fall

The arithmetical check provided in all levellings checks solely that the observed readings entered in the field book have been correctly computed. It does not prove that the reduced level of any point is correct. If verification of each level is required, the fieldwork must be repeated from a different instrument set-up point as in Example 6.

Example

6 The levelling shown in Fig. 4.21 was re-observed from a second instrument station. The observations were as follows:

A, 0.240   B, 3.450   C, 0.655
D, 0.650   E, 2.290   F, 1.955

Enter these results in a levelling table, then calculate and check the reduced levels of the stations from the temporary bench mark A (reduced level 107.520 m).

The re-levelling shows that there is a 5 mm difference in the reduced levels of points C and F, which is acceptable.

Answer (Table 4.10)

Exercise 4.2

1 Figure 4.22 shows the station points of the GCB Outdoor Centre linear survey. Table 4.11 shows the results of a levelling of these stations from a single instrument set-up. Calculate the reduced levels of the stations.

Note. This method of levelling is not recommended for the establishment of temporary bench marks, since any of the readings might be wrong and would not be detected. Furthermore, the lengths of the sights are not balanced and instrumental errors would affect the results.

Multiple instrument settings (change points)
Figure 4.23 shows a small site where reduced levels are required around the perimeter at various identifiable features. One benchmark is available.

Whenever possible the levelling instrument should be set in the centre of the site and observations made to all points sequentially as in Example 6. Frequently this will not be possible for the following reasons:

BS	IS	FS	Rise	Fall	Reduced level	Remarks
0.240					107.520	A (TBM)
	3.450			3.210	104.310	B
	0.655		2.795		107.105	C
	0.650		0.005		107.110	D
	2.290			1.640	105.470	E
		1.955	0.335		105.805	F
0.240		1.955	3.135	4.850	105.805	
− 1.955			− 4.850		− 107.520	
− 1.715			− 1.715		− 1.715	

Table 4.10

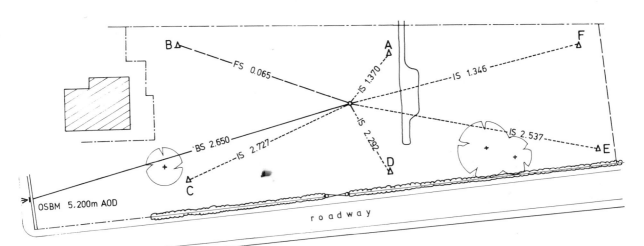

Figure 4.22

BS	IS	FS	Rise	Fall	Reduced level	Remarks
2.650					5.200	OSBM
	2.727					C
	2.292					D
	2.537					E
	1.346					F
	1.370					A
		0.065				B

Table 4.11

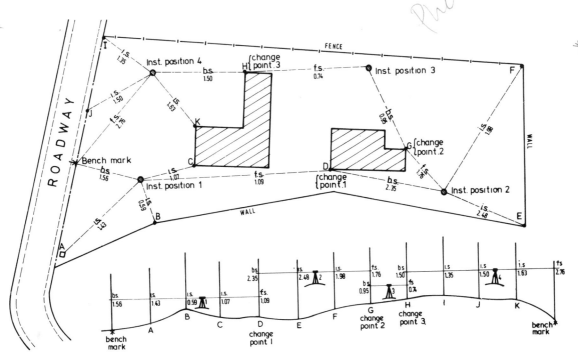

Figure 4.23

1. Buildings obstruct the lines of sight.
2. The lengths of some of the sights are too great.
3. The point being observed may be at a higher level than the instrument. The horizontal line of sight through the instrument therefore strikes the ground before reaching the point.
4. The point being observed may be at a much lower level than the instrument in which case the line of sight through the instrument passes over the staff held on the point.

In Fig. 4.23 some of these difficulties are present, so a set-up position has to be chosen from which the bench mark is visible. The instrument is set up at position 1 and a backsight is taken to the BM (RL 35.27 m), followed by sights to as many other points as possible or desirable. In this case only the points A, B, C and D are visible from the instrument.

Intermediate sights are taken to points A, B and C, and a foresight taken to the final point D. The readings are entered in the field book (Table 4.12, section ①) and point D is noted as a change point.

It is now possible, *though not usual practice*, to reduce the levels in exactly the same manner as before. The relevant calculations are detailed below and entered in Table 4.12, section ②:

(a) Calculation of rise or fall between successive points BM to A, A to B, B to C, and C to D.

$$
\begin{array}{lr}
\text{Backsight to BM} & 1.56 \\
-\text{Intermediate sight to A} & -1.43 \\
\hline
= +0.13 \text{ rise}
\end{array}
$$

$$
\begin{array}{lr}
\text{Intermediate sight to A} & 1.43 \\
-\text{Intermediate sight to B} & -0.59 \\
\hline
= +0.84 \text{ rise}
\end{array}
$$

$$
\begin{array}{lr}
\text{Intermediate sight to B} & 0.59 \\
-\text{Intermediate sight to C} & -1.07 \\
\hline
= -0.48 \text{ fall}
\end{array}
$$

$$
\begin{array}{lr}
\text{Intermediate sight to C} & 1.07 \\
-\text{Intermediate sight to D} & -1.09 \\
\hline
= -0.02 \text{ fall}
\end{array}
$$

(b) Calculation of reduced levels. These are obtained by algebraically adding the rises or falls to the first reduced level in succession.

$$
\begin{array}{lll}
\text{Reduced level BM} & = 35.27 \\
+ \text{ rise (0.13)} & \text{BM to A} & = 35.40 \text{ (RL A)} \\
+ \text{ rise (0.84)} & \text{A to B} & = 36.24 \text{ (RL B)} \\
- \text{ fall (0.48)} & \text{B to C} & = 35.76 \text{ (RL C)} \\
- \text{ fall (0.02)} & \text{C to D} & = 35.74 \text{ (RL D)}
\end{array}
$$

The fieldwork and associated calculations finished on point D, which is a change point. The second series of readings must therefore begin on point D which in effect has become a temporary bench mark.

The instrument is removed to instrument set-up point 2 and a backsight taken to the point D. It is entered in the field book (Table 4.12, section ③) in the BS column on line D. Thus there are two field readings on the same line since they were taken to the same station, i.e. point D.

The levelling is continued with another series of readings taken to any further accessible points. These are intermediate sights to E and F and a foresight to G, which becomes the second change point. These results are calculated and entered in the field book (Table 4.12, section ④):

(a)
$$
\begin{array}{lr}
\text{Backsight to D} & 2.35 \\
-\text{Intermediate sight to E} & -2.48 \\
\hline
= -0.13 \text{ fall}
\end{array}
$$

BS	IS	FS	Rise	Fall	Reduced level	Distance	Remarks
1.56					35.27	Not	Bench mark
①	1.43		0.13	②	35.40	required	A. Manhole
	0.59		0.84		36.24		B. Fence
	1.07			0.48	35.76		C. Corner of building
2.35		1.09		0.02	35.74		D. Corner of building change point 1
③	2.48			0.13	35.61		E. Fence
	1.98		0.50	④	36.11		F. Fence
0.95	⑤	1.76	0.22		36.33		G. Corner of building change point 2
1.50		0.74	0.21	⑥	36.54		H. Corner of building change point 3
	1.35		0.15		36.69		I. Fence
⑦	1.50		⑧	0.15	36.54		J. Fence
	1.63			0.13	36.41		K. Corner of building
		2.76		1.13	35.28		Bench mark
6.36		6.35	2.05	2.04	35.28		
−6.35			−2.04		−35.27		
=0.01			=0.01		=0.01		

Table 4.12

Intermediate sight to E   2.48
− Intermediate sight to F   − 1.98
$$= \underline{+0.50} \text{ rise}$$

Intermediate sight to F   1.98
− Foresight to G   − 1.76
$$= \underline{+0.22} \text{ rise}$$

(b) The reduced levels E, F and G are calculated from change point D.

Reduced level = 35.74
− fall (0.13) to E = 35.61 (RL E)
+ rise (0.50) to F = 36.11 (RL F)
+ rise (0.22) to G = 36.33 (RL G)

The instrument is removed to set up point 3 and the third series of levels is taken. From this position only the points G and H can be seen. Point G is a backsight. The reading is entered in the field book (Table 4.12, section ⑤) in the BS column, on line G. Point H is a foresight and is the next change point.

The relevant calculations to obtain the reduced level of H, shown in Table 4.12, section ⑥, are as follows:

(a)   Backsight to G   0.95
     − Foresight to H   − 0.74
$$= \underline{+0.21} \text{ rise}$$

(b)   Reduced level G = 36.33
     + rise (0.21) to H = 36.54 (RL H)

From instrument position 4, the fourth series of levels is observed, beginning with a backsight to station H and finishing with a foresight to the BM. They are calculated in the manner of the foregoing paragraphs. The observations and relevant results are shown in Table 4.12, sections ⑦ and ⑧:

(a)   Backsight H   1.50
     − Intermediate sight I   − 1.35
$$= \underline{+0.15} \text{ rise}$$

Intermediate sight I   1.35
− Intermediate sight J   − 1.50
$$= \underline{-0.15} \text{ fall}$$

Intermediate sight J   1.50
− Intermediate sight K   − 1.63
$$= \underline{-0.13} \text{ fall}$$

Intermediate sight K   1.63
− Foresight L   − 2.76
$$= \underline{-1.13} \text{ fall}$$

(b)   Reduced level H = 36.54
     + rise (0.15) to I   = 36.69 (RL I)
     − fall (0.15) to J   = 36.54 (RL J)
     − fall (0.13) to K   = 36.41 (RL K)
     − fall (1.13) to L   = 35.28 (RL)

The arithmetical check shows that the levels have been correctly calculated and also shows that an error of 10 mm has been made in the field work. The normal acceptable limit of error is ± 20 mm for this levelling.

It was stated earlier that it is unusual to calculate the results in parallel with the field work for the simple reason that if any observation is wrongly read the calculated results must also be wrong. It is therefore wise to complete the field work and verify that the levelling closes before calculating any results.

The logical order in any levelling calculation is therefore:

1. Complete the fieldwork.
2. Check that the sum BS column = sum FS column within the acceptable limit of error. If the error exceeds these limits there is no point in continuing the calculation.
3. Calculate rises and falls.
4. Check that (sum rise column − sum fall column) = (sum BS column − sum FS column).
5. Calculate reduced levels.
6. Check that (last RL − first RL) = (sum rise column − sum fall column).

Example

7 Figure 4.24 shows the positions of a level and staff set up on the line of a proposed drainage installation. Draw up a page of a level book and

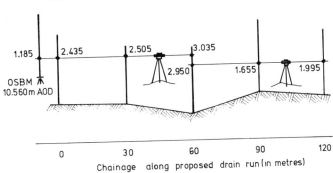

1.185   2.435   2.505   3.035

2.950   1.655   1.995

OSBM
10.560 m AOD

0   30   60   90   120

Chainage along proposed drain run (in metres)

Figure 4.24

BS	IS	FS	Rise	Fall	Reduced level	Distance	Remarks
1.185					10.560		OBM
	2.435			1.250	9.310	0	
	2.505			0.070	9.240	30	
2.950		3.035		0.530	8.710	60	
	1.655		1.295		10.005	90	
		1.995		0.340	9.665	120	
4.135		5.030	1.295	2.190	9.665		
−5.030			−2.190		−10.560		
−0.895			−0.895		−0.895		

Table 4.13

reduce the readings, applying the necessary checks.

Answer (Table 4.13)

Exercise 4.3

1 Figure 4.25 shows the station points of the GCB Outdoor Centre linear survey and Table 4.14 shows the results of a levelling of these stations from multiple set-up points. Calculate the reduced levels of the stations.

3. Levelling procedure (using the HPC method)

A second method of calculating the reduced levels of survey points is that known as the height of the plane of collimation method (HPC method). The fieldwork for the HPC method and for the rise and fall method is exactly the same. The calculation of the results is different.

The line joining the centre of the object glass to the centre of the reticule (Fig. 4.3) is the line of collimation of the telescope. When the telescope is rotated it will sweep out a horizontal plane known as the plane of collimation (Fig. 4.26).

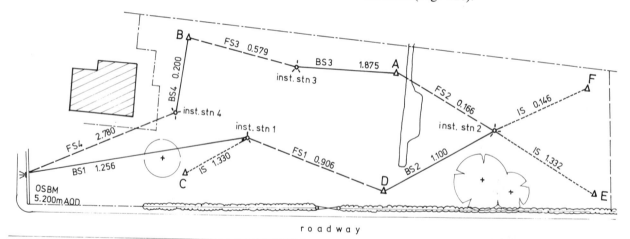

Figure 4.25

BS	IS	FS	Rise	Fall	Reduced level	Remarks
1.256					5.200	OSBM
	1.330					C
1.100		0.906				D
	1.332					E
	0.146					F
1.875		0.166				A
0.200		0.579				B
		2.780				OSBM

Table 4.14

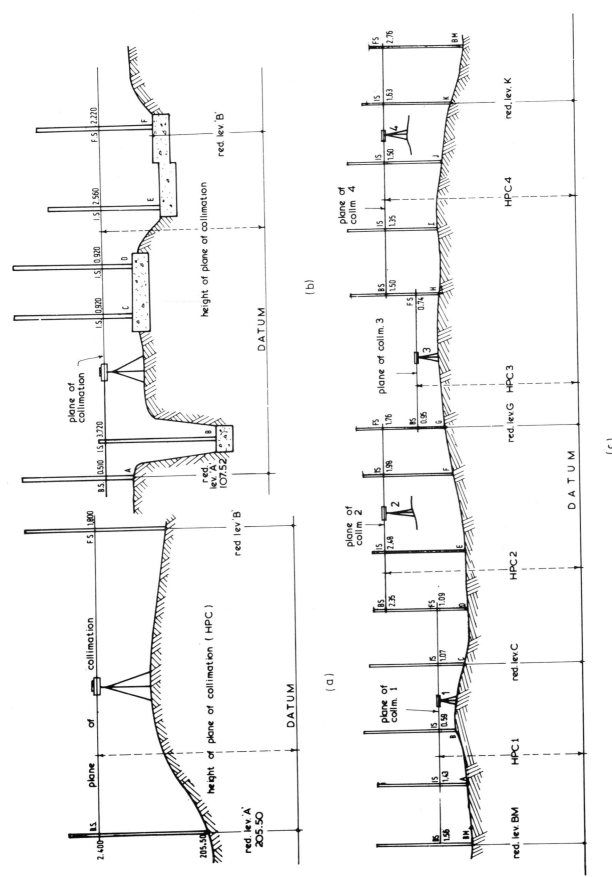

Figure 4.26

(a) Levelling between two points

In Fig. 4.26(a) the plane of collimation cuts a levelling staff held on a point A where the reduced level is 205.500 m. The staff reading is 2.400 m.

In this method of reduction the height of the plane of collimation above datum is required for every instrument setting. Clearly, in the figure the height of the plane of collimation (HPC) above datum is the reduced level of A (205.500 m) plus the staff reading 2.400 m at that point.

i.e. HPC = reduced level A + staff reading
          = 205.500 + 2.400
          = 207.900 m

If another reading of 1.800 m is observed at B, the reduced level of the point B can be readily found. The plane of collimation is still at a height of 207.900 when pointing to B and since ground level B lies 1.800 m below the HPC the reduced level of B is HPC minus staff reading B.

i.e. reduced level B = HPC − staff reading
                    = 207.900 − 1.800
                    = 206.100 m

Generally, then, the height of the plane of collimation is the reduced level of any point plus the staff reading *at that point* and the reduced level of any other point is the HPC minus the staff reading at the point.

The field book is ruled differently, of course, and Table 4.15 shows the book ruled for the collimation system of reduction. Since the actual observations are in no way altered, the sighting to A is a backsight while B is a foresight. The HPC 207.900 is entered on line 1. The HPC is written only once, opposite the

backsight, and is understood to refer to the whole of that instrument setting. The reduced level of B, 206.100, is entered on line 2 in its appropriate column.

(b) Series levelling

Points observed from single instrument station

In Fig. 4.26(b) the reduced levels of six points are required. (The figure and the readings used are those of Fig. 4.21 where the levels were calculated by the rise and fall method.)

(a) The height of collimation is found as before.

  HPC = reduced level A + staff reading (BS)
      = 107.52 + 0.51
      = 108.03 m

The HPC is entered in the field book in line 1 (Table 4.16). It is then held to apply to the whole of that instrument setting and need not be repeated down the whole HPC column as it is in the table.

(b) The reduced level of every other point is found thus:

RL (any point) = HPC − staff reading at that point

Therefore

  RLB = 108.03 − 3.72 = 104.31
  RLC = 108.03 − 0.92 = 107.11
  RLD = 108.03 − 0.92 = 107.11
  RLE = 108.03 − 2.56 = 105.47
  RLF = 108.03 − 2.22 = 105.81

The reduced levels are entered in the RL column (Table 4.16) opposite their respective staff readings. They agree with the levels calculated in Table 4.9 by the rise and fall method.

	BS	IS	FS	HPC	Reduced level	Distance	Remarks
Line 1 Line 2	2.400		 1.800	207.900	205.500 206.100		A. Ground level B. Ground level

Table 4.15

BS	IS	FS	HPC	Reduced level	Distance	Remarks
0.51	 3.72 0.92 0.92 2.56	 2.22	108.03 108.03 108.03 108.03 108.03 108.03	107.52 104.31 107.11 107.11 105.47 105.81		A. (TBM) B. Foundation level 1 C. Foundation level 2 D. Foundation level 2 E. Foundation level 3 F. Foundation level 4
0.51 −2.22 −1.71		2.22		105.81 −107.52 − 1.71		Check

Table 4.16

BS	IS	FS	HPC	Reduced level	Distance	Remarks
0.240			107.760	107.520		A. (TBM)
	3.450		107.760	104.310		B. Foundation level 1
	0.655		107.760	107.105		C. Foundation level 2
	0.650		107.760	107.110		D. Foundation level 2
	2.290		107.760	105.470		E. Foundation level 3
		1.955	107.760	105.805		F. Foundation level 4
0.240		1.955		105.805		
−1.955				−107.520		Check
−1.715				− 1.715		

Table 4.17

Multiple instrument settings

Figure 4.26(c) shows a whole series of points where the reduced levels are required. (The figure and the readings are those of Fig. 4.23 where the levels were reduced by the rise and fall method.)

(a) The height of collimation of the first instrument setting is found by adding the reduced level of the bench mark and the backsight reading observed to it.

$$\text{HPC (1)} = \text{reduced level BM} + \text{staff reading (BS)}$$
$$= 35.27 + 1.56$$
$$= 36.83 \text{ m}$$

The HPC is entered in the field book (Table 4.18, line 1).

(b) The reduced levels of stations A, B, C and D, observed from instrument position 1, are obtained by subtracting their staff readings from the HPC. Thus:

$$\text{Reduced level (any point)} = \text{HPC} - \text{staff reading}$$

Therefore

Reduced level A = 36.83 − 1.43 = 35.40 m
Reduced level B = 36.83 − 0.59 = 36.24 m
Reduced level C = 36.83 − 1.07 = 35.76 m
Reduced level D = 36.83 − 1.09 = 35.74 m

These levels are entered in the field book (Table 4.18, lines 2–5) opposite their respective staff readings.

(c) The fieldwork and calculations finished with a foresight to point D which is a change point. The instrument is removed to set-up point 2 and a backsight taken to station D. The instrument therefore has a new height of collimation which is found as before:

$$\text{HPC (2)} = \text{reduced level (change point)}$$
$$+ \text{staff reading (BS)}$$
$$= 35.74 + 2.35$$
$$= 38.09 \text{ m}$$

The HPC (2) is entered in the field book (Table 4.18, line 5) opposite the BS, to which it refers.

(d) The reduced levels of the second set of points are obtained as before by subtracting their staff readings from the HPC. Thus:

$$\text{Reduced level (any point)} = \text{HPC(2)} - \text{staff reading}$$

Therefore

Reduced level E = 38.09 − 2.48 = 35.61 m
Reduced level F = 38.09 − 1.98 = 36.11 m
Reduced level G = 38.09 − 1.76 = 36.33 m

These levels are entered in the field book (Table 4.18, lines 6–8) opposite their respective staff readings.

(e) Point G is a change point. The instrument is removed to set-up point 3 and a backsight observed to change point G immediately followed by a foresight to H.

$$\text{HPC (3)} = \text{reduced level G} + \text{BS}$$
$$= 36.33 + 0.95$$
$$= 37.28 \text{ m (Table 4.18, line 8)}$$

Line	BS	IS	FS	HPC	Reduced level	Remarks
1	1.56			36.83	35.27	BM
2		1.43			35.40	A
3		0.59			36.24	B
4		1.07			35.76	C
5	2.35		1.09	38.09	35.74	D
6		2.48			35.61	E
7		1.98			36.11	F
8	0.95		1.76	37.28	36.33	G
9	1.50		0.74	38.04	36.54	H
10		1.35			36.69	I
11		1.50			36.54	J
12		1.63			36.41	K
13			2.76		35.28	BM
14	6.36	12.03	6.35		35.28	
15	−6.35				−35.27	
16	0.01				0.01	

Table 4.18

Reduced level H = HPC (3) − foresight
$$= 37.28 - 0.74$$
$$= 36.54 \text{ m (Table 4.18, line 9)}$$

(f) The instrument is set up at station 4 and the levelling is closed from change point H to the bench mark.

HPC (4) = reduced level H + backsight
$$= 36.54 + 1.50$$
$$= 38.04 \text{ m (Table 4.18, line 9)}$$
Reduced level (any point) = HPC (4) − staff readings
Reduced level I = 38.04 − 1.35 = 36.69 m
Reduced level J = 38.04 − 1.50 = 36.54 m
Reduced level K = 38.04 − 1.63 = 36.41 m
Reduced level BM = 38.04 − 2.76 = 35.28 m

The levels are entered in the field book (Table 4.18, lines 10–13).

Arithmetic check   The commonly applied check is shown in lines 14, 15 and 16. It is the same as the check in the rise and fall system in that the difference between the first and last reduced levels equals the difference between the sum of the BS column and the sum of the FS column, namely 0.01.

While this check is very widely used it does *not* completely check the levelling and many serious errors have arisen on site because this fact is not appreciated. Suppose, for the moment, that the RL of point B had been found to be 37.24 m instead of 36.24 m. The check shown above would still work and indeed if the reduced level of *every* intermediate sight were wrong, the check would continue to be successful. The reason for this is simply that a reduced level does not depend on the value of the preceding level as it does in the rise and fall system.

The complete arithmetical check shown below is complex and is almost certainly the reason for the adoption of the simple check shown above.

Complete check:

Sum of reduced levels (except first) = sum (each height of collimation × number of IS and FS observed from each) − sum (IS column + FS column)

In Table 4.18,

Sum of RLs except first = 432.65 m
Sum of HPC × number of IS and FS
$$= \quad 36.83 \times 4 = \quad 147.32$$
$$+ 38.09 \times 3 = +114.27$$
$$+ 37.28 \times 1 = + \ 37.28$$
$$+ 38.04 \times 4 = +152.16$$
$$\overline{\qquad\qquad\quad 451.03}$$

Sum of IS column = 12.03
Sum of FS column = 6.35
$$\overline{\qquad\qquad 18.38}$$

(451.03 − 18.38) = 432.65 m

Example

9 The level booking sheet (Table 4.19) shows the ground levels through which it is proposed to run a drain, rising from manhole 1 to manhole 5. Calculate the reduced levels of each staff position and apply a full arithmetic check.

(SCOTVEC—Ordinary National Diploma in Building)

Backsight	Intermediate sight	Foresight	Height of collimation	Reduced level	Distance	Remarks
1.579				100.000	0.0	BM 1 on cover plate; manhole 1
	1.295				20.0	
	1.873				40.0	
	2.018				60.0	
	1.884				80.0	Manhole 2; cover level
	1.625				100.0	
2.441		1.000			105.0	Start of steps
	1.807				118.5	Top of steps
	1.495				122.3	Manhole 3; cover level
		1.020		102.000	135.0	BM 2 on cover plate; manhole 5

Table 4.19

Answer (Table 4.20)

Check:

Sum of RL (except first) = 1005.73
$$= (101.579 \times 6)$$
$$+ (103.02 \times 4)$$
$$- (13.804 + 2.020)$$
$$= 1005.73$$

Undoubtedly, the collimation system is best for setting out levels and since a great deal of setting out has to be done on building sites, this is probably the reason for the popularity of the method among building engineers.

(c) Comparison of reduction methods

The rise and fall system provides a complete check on the total working with ease, whereas the collimation system of checking is very tedious.

The rise and fall system takes longer to work out but takes much less time to check than does the collimation system. The overall time is much the same for both systems.

The rise and fall system should be used where the levelling involves a great number of intermediate sights.

Exercise 4.4

1 Figure 4.27 shows levelling information observed as part of a proposed road. The readings 1.727 and 0.573 were on a change point while the remainder of the readings at points A, B, C, D, E, F and G were taken at 30 m intervals along the line of the proposed road. Given that the reduced level of point A is 13.273 m above datum, calculate the reduced levels of all points along the road.
(City and Guilds of London Institute Surveying and Levelling Examination)

Backsight	Intermediate sight	Foresight	Height of collimation	Reduced level	Distance	Remarks
1.579			101.579	100.000	0.0	BM 1 on cover plate; manhole 1
	1.295			100.284	20.0	
	1.873			99.706	40.0	
	2.018			99.561	60.0	
	1.884			99.695	80.0	Manhole 2; cover level
	1.625			99.954	100.0	
2.441		1.000	103.020	100.579	105.0	Start of steps
	1.807			101.213	118.5	Top of steps
	1.495			101.525	122.3	Manhole 3; cover level
	1.807			101.213	129.6	Manhole 4; cover level
		1.020		102.000	135.0	BM 2 on cover plate; manhole 5
4.020 −2.020 =2.000	13.804	2.020		102.000 − 100.000 = 2.000		

Table 4.20

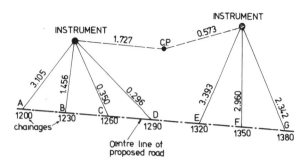

Figure 4.27

4. Inverted staff readings

In all of the previous examples on levelling, the points observed all lay below the line of sight.

Frequently on building sites, the reduced levels of points above the height of the instrument are required, e.g. the soffit level of a bridge or underpass, the underside of a canopy, the level of roofs, eaves, etc., of buildings. Figure 4.28 illustrates a typical case.

The reduced levels of points A, B, C and D on the frame of a multi-storey building require checking. The staff is simply held upside down on the points A and C and booked with a *negative sign* in front of the reading, e.g. −1.520. Alternatively, the reading may be put in brackets, e.g. (1.520), or an asterisk may be put alongside the figure, e.g. 1.520*. Such staff readings are called inverted staff readings.

Table 4.21 shows the readings observed to the points A, B, C and D on the multi-storey building of Fig. 4.28. The levels may be reduced by either (a) the rise and fall method or (b) the HPC method.

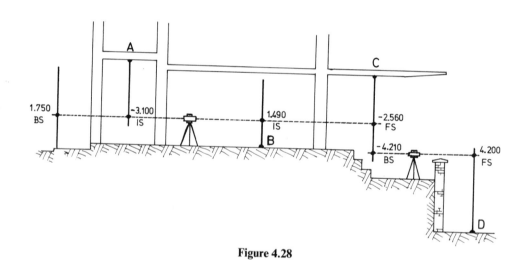

Figure 4.28

BS	IS	FS	Rise	Fall	Reduced level	Distance	Remarks
1.750					72.300		Bench mark
	−3.100		4.850		77.150		A. Frame (lift shaft)
	1.490			4.590	72.560		B. Floor level
−4.210		−2.560	4.050		76.610		C. Canopy
		4.200		8.410	68.200		D. Kerb
−2.460		+1.640	8.900	13.000	68.200		
−1.640			−13.000		−72.300		
−4.100			−4.100		−4.100		

Table 4.21

(a) Reduction by the rise and fall method

The rise or fall is required between successive points as before. The second reading is subtracted from the first as follows:

BS to bench mark =	(1.750)
IS to frame A = −	(−3.100)
Difference BM to A =	+4.850 rise
IS to frame A =	(−3.100)
IS to floor level B =	−(1.490)
Difference A to B =	−4.590 fall
IS to floor level B =	1.490
FS to canopy C = −	(−2.560)
Difference B to C =	+4.050 rise
BS to canopy C =	(−4.210)
FS to kerb D =	−(4.200)
Difference C to D =	−8.410 fall

The reduced levels are obtained by the algebraic addition of the rises and falls as before (Table 4.21).

The arithmetical check is applied in the usual manner. The BS and FS columns are added algebraically, the inverted staff readings being regarded as negative.

$$\begin{aligned}
\text{(Last RL − first RL)} &= \text{(sum rises − sum falls)} \\
&= \text{(sum BS − sum FS)} \\
&= 68.200 - 72.300 \\
&= 8.900 - 13.000 \\
&= -2.460 - 1.640 \\
&= -4.100 \text{ m}
\end{aligned}$$

(b) Reduction by the HPC method

The usual HPC rules are followed, viz.

$$\text{HPC} = \text{reduced level of point} + \text{staff reading at point}$$

RL point = HPC − staff reading at point

$$\begin{aligned}
\text{HPC (1)} &= \text{reduced level BM + BS staff reading} \\
&= 72.300 + 1.750 \\
&= 74.050
\end{aligned}$$

$$\begin{aligned}
\text{RL A} &= \text{HPC(1) − IS staff reading A} \\
&= 74.050 - (-3.100) \\
&= 77.150 \text{ m}
\end{aligned}$$

$$\begin{aligned}
\text{RL B} &= \text{HPC(1) − IS staff reading B} \\
&= 74.050 - 1.490 \\
&= 72.560 \text{ m}
\end{aligned}$$

$$\begin{aligned}
\text{RL C} &= \text{HPC(1) − FS staff reading C} \\
&= 74.050 - (-2.560) \\
&= 76.610 \text{ m}
\end{aligned}$$

$$\begin{aligned}
\text{HPC (2)} &= \text{reduced level change point C} \\
&\quad + \text{BS staff reading C} \\
&= 76.610 + (-4.210) \\
&= 72.400 \text{ m}
\end{aligned}$$

$$\begin{aligned}
\text{RL D} &= \text{HPC(2) − FS staff reading D} \\
&= 72.400 - 4.200 \\
&= 68.200 \text{ m}
\end{aligned}$$

The complete reduction is shown in Table 4.22.

Exercise 4.5

1 A tunnel, being driven as part of a water supply scheme, is to be checked for possible subsidence of the roof line. Levels were taken as shown in Fig. 4.29 and Table 4.23. Reduce the levels by (a) the rise and fall method and (b) the height of collimation method.

5. Errors in levelling

As in all surveying operations, the sources and effects of errors must be recognized and steps taken to eliminate or minimize them. Errors in levelling can be classified under several headings.

BS	IS	FS	HPC	Reduced level	Remarks
1.750			74.050	72.300	Bench mark
	−3.100			77.150	A. Frame liftshaft
	1.490			72.560	B. Floor level
−4.210		−2.560	72.400	76.610	C. Canopy
		4.200		68.200	D. Kerb
−2.460	−1.610	1.640		68.200	
−1.640				−72.300	
−4.100				−4.100	

Table 4.22

Check:

$$\begin{aligned}
\text{Sum of RLs except first} &= 294.52 \\
&= (74.050 \times 3) + 72.400 - (-1.61 + 1.64) \\
&= 222.15 + 72.40 - 0.03 \\
&= 294.52
\end{aligned}$$

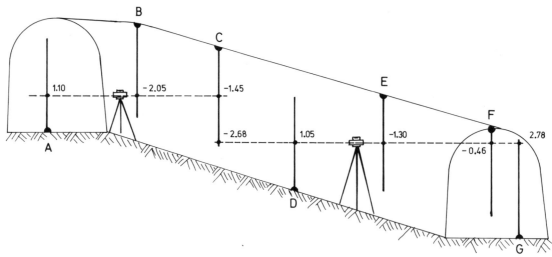

Figure 4.29

BS	IS	FS	Reduced level	Remarks
1.10			10.000	A. (TBM) floor level
	−2.05			B. Roof level
−2.68		−1.45		C. Roof level
	1.05			D. Floor level
	−1.30			E. Roof level
	−0.46			F. Roof level
		2.78		G. Floor level

Table 4.23

(a) Gross errors

Mistakes arising in the mind of the observer may be due to carelessness, inexperience or fatigue.

1. *Wrong staff readings*
This is probably the most common error of all in levelling. Examples of wrong staff readings are: misplacing the decimal point, reading the wrong metre value and reading the staff wrong way up.

2. *Using the wrong cross-hair*
Instead of reading the staff against the axial line, the observer reads against one of the stadia lines. This error is common in poor visibility.

3. *Wrong booking*
The reading is noted with the figures interchanged, e.g. 3.020 instead of 3.002.

4. *Omission or wrong entry*
A staff reading can easily be written in the wrong column or even omitted entirely.

5. *Spirit level not centred*
The staff is read without centring the bubble.

All of these errors can be small or very large and every effort must be made to eliminate them. The only way to eliminate gross errors is to make a double levelling, i.e. to level from A to B then back from B to A. Theoretically the levelling should close without any error but this will very seldom happen. However, the error should be within acceptable limits, certainly not more than 1 mm per 50 m for levellings less than 1 kilometre.

(b) Constant errors

These errors are due to instrumental defects and will always be of the same sign.

1. *Non-verticality of the staff*
This is a serious source of error. Instead of being held vertically the staff may be leaning forward or backward. In Fig. 4.30, the staff is 3° out of vertical. If a reading of 4.000 m is observed, it will be in error by

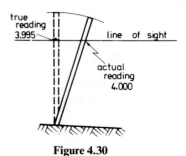

Figure 4.30

5 mm. The correct reading is

$$4 \times \cos 3° = 3.995 \, \text{m}$$

The error can be eliminated by fitting the staff with a circular spirit level. The staffholder must ensure that the bubble is centred when the staff is being read. A second method of eliminating the error is for the staffholder to swing the staff slowly backwards and forwards across the vertical position during the observation. The observer then reads the lowest reading.

2. *Collimation error in the instrument*

In a properly adjusted level the line of sight must be perfectly horizontal when the bubble of the spirit level is central. If this condition does not prevail there will be an error in the staff reading. In Fig. 4.31 the line of sight is inclined and the resultant error is *e*, increasing with length of sight. The error can be entirely eliminated by making backsights and foresights equal in length. The error *e* will be the same for each sight and the true difference in level will be the difference in the readings. It is fair to say that collimation errors on instruments are fairly common on building sites. Since the resulting errors can be completely eliminated by taking backsights and foresights of equal length, this practice should always be adhered to.

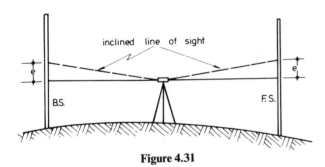

Figure 4.31

3. *Staff graduation error*

This error is not common nowadays. It was common at the transition stage when Imperial staffs were being converted to metric by sticking metric graduation strips to the faces of old staffs. Care had to be taken to ensure that the zero of the strips coincided with the bottom of the staff and that the various staff sections were properly graduated.

(c) **Random errors**

These errors are due to physical and climatic conditions. The resulting errors are small and are likely to be compensatory.

1. *Effect of wind and temperature*

The stability of the instrument may be affected, causing the height of collimation to change slightly.

2. *Soft and hard ground*

When the instrument is set on soft ground it is likely to sink slightly as the observer moves around it. When set in frosty earth, the instrument tends to rise out of the ground. Again the height of collimation changes slightly.

3. *Change points*

At any change point the staff must be held on exactly the same spot for both foresight and backsight. A firm spot must be chosen and marked by chalk. If the ground is soft a change plate (Fig. 4.32) must be used. The plate is simply a triangular piece of metal bent at the apices to form spikes. A dome of metal is welded to the plate and the staff is held on the dome at each sighting.

Figure 4.32

4. *Human deficiencies*

Errors arise in estimating the millimetre readings, particularly when visibility is bad or sights are long.

All of the errors in this class tend to be small and compensating and are of minor importance only in building surveying.

6. Permanent adjustments

In the preceding section on errors, it was pointed out that the line of collimation might not be parallel to the axis of the bubble tube. The levelling is still correct provided the sights are of equal length. This is not always possible, however, particularly when many intermediate sights are being taken and the only way in which these errors can be eliminated is to ensure that the instrument is in good adjustment.

Prior to the commencement of any contract the levelling instrument must be tested and, if necessary, adjusted. The adjustments of the various instruments are slightly different, because of their differing constructions.

(a) **Dumpy level**

1. The vertical axis must be truly vertical when the bubble of the spirit level is central.

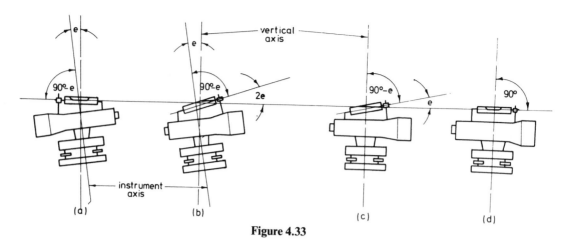

Figure 4.33

Test

(a) Set up the level and level the spirit level over two footscrews. If the instrument is not in good adjustment the relationship of the vertical axis to the spirit level will be as shown in Fig. 4.33(a), i.e. there is an inclination error, *e*.

(b) Turn the telescope through 90° and recentralize the bubble.

(c) Repeat these operations until the bubble remains central for test positions (a) and (b) above.

(d) Turn the telescope until it is 180° from test position (a) above. The vertical axis will still be inclined with an error *e* and the bubble of the spirit level will no longer be central. It will, in fact, be inclined to the horizontal at an angle of 2*e* (Fig. 4.33(b)). The number of divisions, *n*, by which the bubble is off-centre is noted.

Adjustment

(e) Turn the footscrews until the bubble moves back towards the centre by *n*/2 divisions, i.e. by half the error. The vertical axis is now truly vertical (Fig. 4.33(c)).

(f) Adjust the spirit level by releasing the capstan screws and raising or lowering one end of the spirit level until the bubble is exactly central. The other half *n*/2 of the error is thereby eliminated (Fig. 4.33(d)), and the spirit level axis is at right angles to the vertical axis.

2. The line of collimation must be parallel to the newly adjusted spirit level; in other words, the line of collimation must be horizontal.

The question is often asked 'How can the line of sight be in error in a dumpy level?' The answer is that the instrument can receive a knock, thereby displacing the reticule vertically. Even without any knocks the displacement can take place through natural wear and tear, effects of temperature, etc.

Test

The test called the 'two-peg' test is as follows:

(a) Select two points A and B 60 m apart and hammer two pegs firmly into the ground.

(b) Set up the instrument exactly midway between the pegs and level carefully.

(c) Sight staff A and note the reading.

(d) Remove the staff to B. Read the staff and adjust the height of peg B until exactly the same reading as at A is obtained.

(e) The two pegs A and B will be exactly the same height irrespective of whether the line of collimation is horizontal or not. In Fig. 4.34 the line of collimation is inclined upwards by an angle (α), causing the staff reading at A to be in error by an amount *e*. Since peg B is exactly the same distance from the instrument as A there will also be an error *e*.

(f) Remove the instrument to B and, using a plumb-bob, set it up so that the eyepiece is exactly over peg B. Level carefully (Fig. 4.34).

(g) Hold a staff on B and read it against the eyepiece of the instrument.

(h) Remove the staff to A and read it. The reading should be identical to the staff reading at B in test position (g) above. If not, there is a collimation error caused by the reticule being displaced vertically.

Adjustment

(i) Release the antagonistic adjusting screws and move the reticule up, in this case (Fig. 4.35) until the required reading is obtained at A.

(j) Since pegs A and B are at exactly the same height, the line of sight is now horizontal, i.e. it is parallel to line AB and the instrument is in adjustment.

(b) Tilting level

1. The tilting level differs from the dumpy in that the vertical axis need only be approximately vertical for the instrument to be in good adjustment. This condition is fulfilled satisfactorily by centring the small circular spirit level at each set-up.

2. The line of collimation must be horizontal when the bubble of the main spirit level is central.

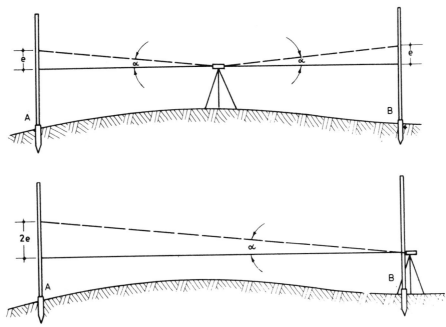

Figure 4.34

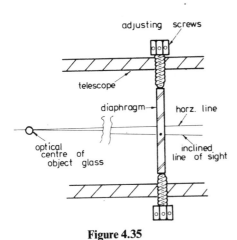

Figure 4.35

Test

The test is identical to that for the dumpy level, namely the 'two-peg' test.

Suppose that the test has been carried out and it is discovered that the line of collimation is not horizontal, i.e. staff reading A does not equal staff height B when the instrument is set up over B.

Adjustment

The line of sight can be tilted using the tilting screw until the reading at A agrees with staff height B. The line of sight is now horizontal, since pegs A and B are at the same height. However, since the tilting screw has been moved, the mains spirit level will no longer be central.

It is adjusted by means of the spirit level capstan screws until the bubble is exactly central. The main spirit level is therefore parallel to the now horizontal line of sight and the condition is fulfilled.

(c) Automatic level

The vertical axis need only be approximately vertical, as with the tilting level. Once more this condition is adequately fulfilled using the small circular spirit level.

The line of collimation must be horizontal when the small circular spirit level is central. The 'two-peg' test is carried out as before and if the line of collimation is found to be in error the reticule can usually be adjusted, as with the dumpy level. If this is not possible the compensator unit can be adjusted, but this is definitely not a job for the surveyor and the instrument should be returned to the manufacturer.

In the two-peg test the pegs A and B need not be level. It is also common practice to set the instrument beyond peg B instead of over it at the second set-up.

Example

10 In a two-peg test a levelling instrument was set up exactly midway between pegs A and B, which are 80 m apart. The pegs were adjusted until the same reading was obtained over each.

The instrument was then set over peg B where its height was measured as 1.350 m and a reading of 1.450 m was made to a staff held on peg A.
(a) Calculate the collimation error of the instrument.
(b) Express the error as (i) a percentage and (ii) an angle of elevation or depression.

Solution

Since the instrument was set up midway between the pegs and the same reading was obtained at each peg, the pegs must be level.

(a) Instrument at peg B:

Apparent difference in level = 1.450 − 1.350
$$= 0.10 \text{ m}$$

i.e. the line of collimation is elevated by 0.10 m over a length of 80 m.

(b) (i)   Percentage error = $\dfrac{0.10}{80} \times 100$

$$= 0.125 \text{ per cent}$$

(ii) Tan elevation angle = $\dfrac{0.10}{80}$

$$= 0.00125$$
$$\text{Angle} = +00° \, 04'$$

In the two-peg test, the pegs A and B need not be set level, as is evidenced in the following example.

Example

11 A levelling instrument was set up exactly midway between two pegs P and Q lying 70 m apart, and the following readings were obtained to a staff held vertically on the pegs in turn (Fig. 4.36):

Reading to P, 0.765 m; reading to Q, 1.395 m

When the instrument was set over peg P, the following information was obtained:

Instrument height P, 1.305 m;
staff reading Q, 1.855 m

Calculate the collimation error of the instrument and express it as an angle of elevation or depression.

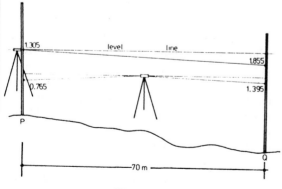

Figure 4.36

Solution

Instrument midway between pegs P and Q:

True difference in level QP = 1.395 − 0.765
$$= 0.630 \text{ m}$$

Instrument set over peg P:

Apparent difference in level QP = 1.855 − 1.305
$$= 0.550 \text{ m}$$

Since the apparent difference is less than the true difference in level, the line of collimation is depressed by

$$0.630 - 0.550 \text{ m} = 0.080 \text{ m over 70 m}$$

Angle of depression = $\tan^{-1} \dfrac{0.080}{75.0}$

$$= \tan^{-1} 0.001\,07$$
$$= 00° \, 03' \, 56''$$

Exercise 4.6

1 Calculate the error in a staff reading of 2.950 m where the staff is inclined to the vertical by an angle of 5°.

2 A tilting level has a proven collimation error of + 00° 08'. The following readings were taken at the stated distances using the level:

2.530 at 10.5 m (A)
1.350 at 25.0 m (B)
−3.005 at 38.5 m (C) (inverted staff)

Correct the staff readings for collimation error.

3 A tilting level was checked by the two-peg test method and the following staff readings were obtained:

Staff position	Staff reading	Distance to staff (m)	Instrument position
Peg X	0.975	40	Mid-way between
Peg Y	1.235	40	peg X and peg Y
Peg Y	1.480	80	Over peg X
Peg X	1.300		

Calculate the amount of instrumental error as a percentage of the distance measured.

7. Curvature and refraction

Throughout this chapter, reference has been made to a horizontal line of sight as distinct from a level line. If the earth is considered to be a perfect sphere (Fig. 4.37) a level line would be, at all points, equidistant from the centre. However, the line of sight through a levelling instrument is a horizontal line tangential to the level line. If a staff were held on B, the staff reading observed from A would therefore be too

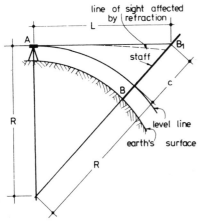

Figure 4.37

great, by BB_1. This is the curvature correction, c, and it can be calculated as follows.

In the triangle L is the length of sight in kilometres and R is the mean radius of the earth (6370 km). By Pythagoras's theorem:

$$(R + c)^2 = R^2 + L^2$$

i.e.

$$R^2 + c^2 + 2Rc = R^2 + L^2$$

Therefore

$$c(c + 2R) = L^2$$
$$c = L^2/(c + 2R)$$

Since c is so small compared with R, it can be ignored. Therefore,

$$c = (L^2/2R) \text{ kilometres}$$

i.e.

$$c = \left(\frac{L^2}{12\,740}\right) \text{ km}$$

However, c is required in metres while L remains in kilometres:

$$c = \left(\frac{L^2 \times 1000}{12\,740}\right) \text{ metres}$$
$$c = 0.0785L^2 \text{ metres (where } L \text{ is in kilometres)}$$

The line of sight is not really horizontal. It is affected by refraction in such a manner that the line of sight is bent downwards towards the earth. Refraction is affected by pressure, temperature, latitude, humidity, etc., and its value is not constant. Its value is taken as $\frac{1}{7}$ curvature and is opposite in effect to that of curvature. Thus,

$$\text{Combined correction} = 0.0785L^2 - \tfrac{1}{7}(0.0785L^2)$$
$$= \tfrac{6}{7}(0.0785L^2)$$
$$= 0.0673L^2 \text{ metres}$$

(where L is in kilometres).

Examples

12 Calculate the corrected staff reading for a sight of 1500 metres if the observed reading is 3.250.

Answer

$$\text{Length of sight} = 1.5 \text{ km}$$
$$\text{Correction for curvature} = (0.0673 \times 1.5^2) \text{ m}$$
$$\text{and refraction} = (0.0673 \times 2.25) \text{ m}$$
$$= 0.151 \text{ m}$$
$$\text{Observed staff reading} = 3.250$$
$$\text{Correction} = -0.151$$
$$\text{Correct staff reading} = 3.099 \text{ m}$$

13 Calculate the correction due to curvature and refraction over a length of sight of 120 metres.

Answer

$$c = (0.0673 \times 0.12^2) \text{ m}$$
$$= (0.0673 \times 0.0144) \text{ m}$$
$$= 0.001 \text{ m}$$

Since 0.001 m is negligible, the correction can be neglected for lengths of sight less then 120 metres. It is good practice, when levelling, to restrict the length of sight to about 50 metres. Furthermore, at this distance the staff reading should never be allowed to fall below 0.5 m because of the variation in refraction caused by fluctuations in the density of the air close to the ground.

8. Reciprocal levelling

The importance of having sights of equal length has already been stressed. Briefly, collimation errors are eliminated by this technique. If the length of sight does not exceed 120 metres curvature and refraction errors are negligible. However, there are occasions when a long sight must be taken and collimation and curvature errors are present. For example, in Fig. 4.38, the difference in level between two points A and B on opposite banks of a wide river is required.

The instrument is set at A and the instrument height, h_1, is measured. The staff is held on B and the staff reading s_1 recorded. In Fig. 4.38, r, the refraction error, and c, the curvature error, are clearly shown. Since AA_1 is the level surface through A the difference in level between A and B is the distance A_1B. Now

$$A_1B = h_1 + c - r - s_1$$

The instrument is then removed to B, the height h_2 measured and the staff reading s_2 recorded: r and c are the refraction and curvature errors as before. The

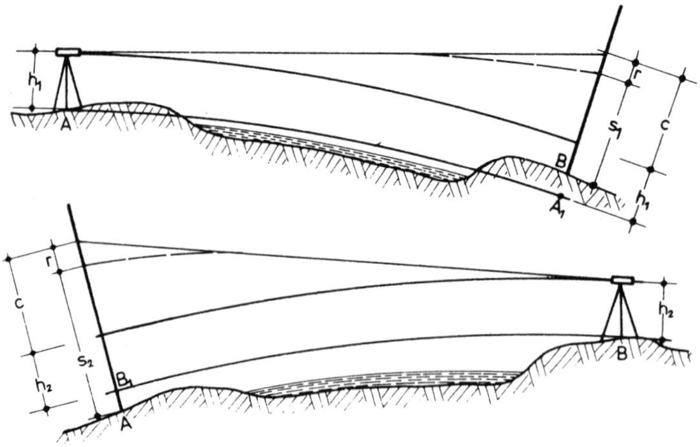

Figure 4.38

difference in level between B and A is the distance B₁A:

$$B_1A = s_2 + r - c - h_2$$

Note that since the sights are of equal length, the collimation error is the same and is therefore cancelled.

The difference in level is the mean of $A_1B + B_1A$

$$= \tfrac{1}{2}(h_1 + c - r - s_1 + s_2 + r - c - h_2)$$
$$= \tfrac{1}{2}(h_1 - h_2 + s_2 - s_1)$$

Strictly, this is not the true difference of level since the value of r will be slightly different in the two equations. However, it will be sufficiently close for building surveys provided the observations are made as soon after each other as possible.

Example

14 Observations were made between points X and Y on opposite sides of a wide water-filled quarry as follows:

Level at X   Instrument height = 1.350
                  Staff reading Y = 1.725

Level at Y   Instrument height = 1.410
                  Staff reading X = 1.055

Calculate the true difference in level between the stations and the reduced level of Y if X is 352.710 AOD.

Solution
True difference in level $= \tfrac{1}{2}(h_1 - h_2 + s_2 - s_1)$
$$= \tfrac{1}{2}(1.350 - 1.725$$
$$+ 1.055 - 1.410) \, \text{m}$$
$$= \tfrac{1}{2}(2.405 - 3.135) \, \text{m}$$
$$= \tfrac{1}{2}(-0.730) \, \text{m}$$
$$= -0.365 \, \text{m}$$

Reduced level X = 352.710 m AOD
Therefore fall X − Y = −0.365 m
and reduced level Y = 352.345 m AOD

Exercise 4.7

1 The following list of readings was taken in sequence during a levelling survey.

Reading (m)	Remarks
1.250	BM, 1.435 m AOD
1.285	Peg A
1.125	Peg B
0.810	Change point
1.555	
−1.400	Inverted staff reading taken to the underside of a bridge
1.235	Peg C
0.665	Change point
1.905	
0.070	BM, 4.600 m AOD

Adopt a standard form of booking and reduce the levels to Ordnance Datum.

(OND, Building)

2 The table below shows the results of a levelling along the centreline of a roadway where settlement has taken place. The road was initially constructed at a uniform gradient rising at 1 in 75 from A to B. Assuming no settlement has taken place at station A:
(a) Reduce the levels.
(b) Calculate the maximum settlement along the line AB.

BS	IS	FS	Remarks	Chainage (m)
3.540			OBM 78.675	
0.410		3.665		
0.525		2.245		
	2.840		Station A	0
	2.440			30
	2.045			60
2.475		1.655		90
	2.090			120
	1.700			150
	1.315			180
	0.900			210
2.465		0.485		240
	2.055			270
	1.645		Station B	300
		2.040	TBM 78.000	

(IOB, Site Surveying)

3 Table 4.24 shows the readings taken to determine the clearance between the river level and the soffit of a road bridge. Reduce the levels and determine the clearance between the river level and the soffit of the bridge.

BS	IS	FS	RL	Remarks
0.872			21.460	OBM
0.665		3.980		
	2.920			River level at A
	−1.332			Soffit of bridge at A
	−1.312			Soffit of bridge at B
	−1.294			Soffit of bridge at C
	−1.280			Soffit of bridge at D
	2.920			River level at D
4.216		0.597		
		1.155		OBM

Table 4.24

(City and Guilds of London Institute)

4 Table 4.25 shows a page of a level book in which certain entries are missing. Complete the missing entries and carry out the normal checks.

BS	IS	FS	Rise	Fall	Reduced levels
3.786					36.642
	—		2.474	0.648	38.468
	1.960			—	36.868
	—	3.560		—	34.042
	3.698			—	—
	0.670		—		35.560
	—	2.180	1.186	—	—
2.874	1.052			—	34.992
	—		1.158	—	—
	0.950		0.766		36.916
	1.412			—	—

(City and Guilds of London Institute)

Table 4.25

5 Table 4.26 shows the staff readings obtained from a survey along the line of a proposed roadway from A to B.
(a) On the table, reduce the levels from A to B applying all the appropriate checks.
(b) Given that the finished roadway has to be evenly graded from A to B, calculate the depths of cut or fill at 20 metre intervals between A and B.

9. Answers

Exercise 4.1

1 0.855(top), 0.900(mid), 0.945(bottom)
2 Table 4.27

BS	IS	FS	Rise Fall or HPC	Surface reduced level	Grade reduced level	Fill	Cut	Remarks
0.824				39.220				TBM 1
	1.628							A
	0.790							20 m from A
	0.383							40 m from A
2.154		1.224						60 m from A
	2.336							80 m from A
	2.757							100 m from A
2.555		0.461						Change point
	2.275							120 m from A
	0.436							140 m from A
	0.227							160 m from A
	0.716							180 m from A
	0.652							B, 200 m from A
		0.233						TBM 2

(SCOTVEC, Ordinary National Diploma in Building)

Table 4.26

BS	IS	FS	Rise	Fall	Reduced level	Remarks
1.955					5.200	OSBM
1.315		2.030		0.075	5.125	C
1.243		0.885	0.430		5.555	D
2.071		1.485		0.242	5.313	E
1.570		0.880	1.191		6.504	F
1.835		1.590		0.020	6.484	A
0.631		0.540	1.295		7.779	B
1.200		3.289		2.658	5.121	C
		1.130	0.070		5.191	OSBM
11.820		11.829	2.986	2.995	5.191	
−11.829			−2.995		−5.200	
−0.009			−0.009		−0.009	

Table 4.27

Checks indicate that (a) there is a survey error of 9 mm and (b) there is no arithmetical error.

Exercise 4.2

1 Table 4.28

BS	IS	FS	Rise	Fall	Reduced level	Remarks
2.650					5.200	OSBM
	2.727			0.077	5.123	C
	2.292		0.435		5.558	D
	2.537			0.245	5.313	E
	1.346		1.191		6.504	F
	1.370			0.024	6.480	A
		0.065	1.305		7.785	B
2.650			2.931	0.346	7.785	
−0.065			−0.346		−5.200	
2.585			2.585		2.585	

Table 4.28

Exercise 4.3

1 Table 4.29

BS	IS	FS	Rise	Fall	Reduced level	Remarks
1.256					5.200	BM
	1.330			0.074	5.126	C
1.100		0.906	0.424		5.550	D
	1.332			0.232	5.318	E
	0.146		1.186		6.504	F
1.875		0.166		0.020	6.484	A
0.200		0.579	1.296		7.780	B
		2.780		2.580	5.200	BM
4.431		4.431	2.906	2.906		
−4.431			−2.906			
0.000			0.000			

Table 4.29

Exercise 4.4

1 Table 4.30

BS	IS	FS	Rise	Fall	Reduced level	Distance	Remarks
3.105					13.273	1200	TBM A
	1.456		1.649		14.922	1230	B
	0.350		1.106		16.028	1260	C
	0.296		0.054		16.082	1290	D
0.573		1.727		1.431	14.651		Change point
	3.393			2.820	11.831	1320	E
	2.960		0.433		12.264	1350	F
		2.342	0.618		12.882	1380	G
3.678		4.069	3.860	4.251	12.882		
−4.069			−4.251		−13.273		
−0.391			−0.391		−0.391		

Table 4.30

Exercise 4.5

1 (a) Using the rise and fall method (Table 4.31)

BS	IS	FS	Rise	Fall	Reduced level	Remarks
1.10					10.00	A. (TBM) floor level
	−2.05		3.15		13.15	B. Roof level
−2.68		−1.45		0.60	12.55	C. Roof level
	1.05			3.73	8.82	D. Floor level
	−1.30		2.35		11.17	E. Roof level
	−0.46			0.84	10.33	F. Roof level
		2.78		3.24	7.09	G. Floor level
−1.58		1.33	5.50	8.41	7.09	
−1.33			−8.41		−10.00	
−2.91			−2.91		−2.91	

Table 4.31

(b) Using the HPC method (Table 4.32)

BS	IS	FS	HPC	Reduced level	Remarks
1.10			11.10	10.00	A. (TBM) floor level
	−2.05			13.15	B. Roof level
−2.68		−1.45	9.87	12.55	C. Roof level
	1.05			8.82	D. Floor level
	−1.30			11.17	E. Roof level
	−0.46			10.33	F. Roof level
		2.78		7.09	G. Floor level
	−2.76	1.33			

Sum RLs (except first) = 63.11
(11.10 × 2) + (9.87 × 4) = 61.68
61.88 − (−2.76) − 1.33 = 63.11 check

Table 4.32

Exercise 4.6

1 $2.950 \times \cos 5° = 2.939$ m

2 At 10.5 m, correct reading is

$$2.530 - 0.024 = 2.506 \text{ m}$$

At 25.0 m, correct reading is

$$1.350 - 0.058 = 1.292 \text{ m}$$

At 38.5 m, correct reading is

$$-3.005 - 0.090 = -3.095 \text{ m}$$

3 (a) When instrument is midway between X and Y, the fall from X to Y is $(0.975 - 1.235)$ m $= 0.260$ m.

(b) When instrument is over X at height 1.300, the reading on Y should be $1.300 + 0.260 = 1.560$ m.

(c) Actual reading on Y is 1.480 m. Therefore line of sight is depressed by $(1.560 - 1.480) = 0.080$ m and

$$\begin{aligned}\text{Error} &= 0.080 \text{ m over } 80.00 \text{ m} \\ &= (0.080/80.000) \times 100 \text{ per cent} \\ &= 0.10 \text{ per cent}\end{aligned}$$

Exercise 4.7

1 Table 4.33

BS	IS	FS	HPC	RL	Remarks
1.250			2.685	1.435	BM
	1.285			1.400	Peg A
	1.125			1.560	Peg B
1.555		0.810	3.430	1.875	CP
	−1.400			4.830	Inverted staff
	1.235			2.195	Peg C
1.905		0.665	4.670	2.765	CP
		0.070		4.600	BM
4.710	2.245	1.545		4.600	
−1.545				−1.435	
3.165				3.165	Simple partial check

Sum RLs (except first) $= 19.225$ m

$$\begin{aligned}2.685 \times 3 &= 8.055 \\ +\ 3.430 \times 3 &= 10.290 \\ +\ 4.670 \times 1 &= \underline{4.670} \\ &\ 23.015\end{aligned}$$

Sum IS column 2.245

Sum FS column $\underline{1.545}$

3.790

$23.015 - 3.790 = \overline{19.225}$ m

Table 4.33

2 Table 4.34

BS	IS	FS	Rise	Fall	Reduced level	Remarks	Chainage	Formation level	Settlement
3.540					78.675	OBM			
0.410		3.665		0.125	78.550				
0.525		2.245		1.835	76.715				
	2.840			2.315	74.400	Station A	0	74.400	0.000
	2.440		0.400		74.800		30	74.800	0.000
	2.045		0.395		75.195		60	75.200	0.005
2.475		1.655	0.390		75.585		90	75.600	0.015
	2.090		0.385		75.970		120	76.000	0.030
	1.700		0.390		76.360		150	76.400	0.040
	1.315		0.385		76.745		180	76.800	0.055
	0.900		0.415		77.160		210	77.200	0.040
2.465		0.485	0.415		77.575		240	77.600	0.025
	2.055		0.410		77.985		270	78.000	0.015
	1.645		0.410		78.395	Station B	300	78.400	0.005
		2.040		0.395	78.000	TBM			
9.415		10.090	3.995	4.670	78.000				
−10.090			−4.670		−78.675				
−0.675			−0.675		−0.675				

Table 4.34

Original level station A = 74.400. Since the construction gradient of the roadway is 1 in 75 rising from A, the formation levels of the chainage points must increase uniformly by (30/75) = 0.400 m, giving the formation levels in the table above. The maximum settlement occurs at chainage 180 m.

3 Table 4.35

BS	IS	FS	HPC	Reduced level	Remarks
0.872			22.332	21.460	OBM
0.665		3.980	19.017	18.352	
	2.920			16.097	River level at A
	−1.332			20.349	Soffit of bridge at A
	−1.312			20.329	Soffit of bridge at B
	−1.294			20.311	Soffit of bridge at C
	−1.280			20.297	Soffit of bridge at D
	2.920			16.097	River level at D
4.216		0.597	22.636	18.420	
		1.155		21.481	OBM
5.753	0.622	5.732		21.481	
−5.732				−21.460	
0.021				0.021	

Sum RLs (except first) = 171.733 = [22.332 + (19.017 × 7) + 22.636] − [0.622 + 5.732] = 171.733

Table 4.35

4 Table 4.36

BS	IS	FS	Rise	Fall	Reduced levels
3.786					36.642
	1.312		2.474		39.116
	1.960			0.648	38.468
0.872		3.560		1.600	36.868
	3.698			2.826	34.042
	0.670		3.028		37.070
2.238		2.180		1.510	35.560
	1.052		1.186		36.746
2.874		2.806		1.754	34.992
	1.716		1.158		36.150
	0.950		0.766		36.916
		1.412		0.462	36.454
9.770		9.958	8.612	8.800	36.454
−9.958			−8.800		−36.642
−0.188			−0.188		−0.188

Table 4.36

5 Table 4.37

BS	IS	FS	Height of collimation	Surface reduced level	Grade reduced level	Fill	Cut	Remarks
0.824			40.044	39.220				TBM 1
	1.628			38.416	38.416	—	—	A
	0.790			39.254	38.816		0.438	20 m from A
	0.383			39.661	39.216		0.445	40 m from A
2.154		1.224	40.974	38.820	39.616	0.796		60 m from A
	2.336			38.638	40.016	1.378		80 m from A
	2.757			38.217	40.416	2.199		100 m from A
2.555		0.461	43.068	40.513				Change point
	2.275			40.793	40.816	0.023		120 m from A
	0.436			42.632	41.216		1.416	140 m from A
	0.227			42.841	41.616		1.225	160 m from A
	0.716			42.352	42.016		0.336	180 m from A
	0.652			42.416	42.416		—	B, 200 m from A
		0.233		42.835				TBM 2
5.533	12.200	1.918		42.835				
−1.918				−39.220				
3.615				3.615				

Sum of RLs (except first) = 527.388 = [(40.044 × 4) + (40.974 × 3) + (43.068 × 6)] − [12.200 + 1.918]
= 527.388

Gradient A to B = (42.416 − 38.416) in 200
= 1 in 50

Rise per 20 m = (1/50) × 20
= 0.400 m

Table 4.37

10. Project

Chapter 18 is a project covering all chapters of this textbook. It is intended that the project build into a complete portfolio of the surveying work required in the survey and the setting out of a building or engineering development.

If the reader wishes to continue work on the project or begin work at this stage, he or she should now turn to page 390 and attempt Section 3.

CHAPTER 5

Contouring

Objective

After studying this chapter, the reader should be able to (a) make a levelling survey of a parcel of ground, process the data and construct a contoured plan of the parcel from the results; (b) add construction works to the plan and determine the limits of the resultant earthworks.

Chapter 3 dealt with the principles of making a linear survey and producing a two-dimensional plan from the results. Figure 3.34 is the end product and shows clearly the surveyed parcel of ground, bordered by hedges, fences, road, etc. The plan would be greatly enhanced and much more useful if the third dimension, namely height, could be represented in some way.

On large-scale plans of civil engineering and building projects, contouring is the commonly used method of showing height.

1. Contour characteristics

(a) Contour line

A contour line is a line drawn on a plan joining all points of the same height above or below some datum. The concept of a contour line can readily be grasped if a reservoir is imagined. If the water is perfectly calm, the edge of the water will be at the same level all the way round the reservoir forming a contour line. If the water level is lowered by, say, five metres the water's edge will form a second contour.

Further lowering of the water will result in the formation of more contour lines (Fig. 5.1). Contour lines are continuous lines and cannot meet or cross any other contour line, nor can any one line split or join any other line, except in the case of a cliff or overhang.

Figure 5.2 shows the contour plan and section of an island. The tide mark left by the sea is the contour line of zero metres value. If it were possible to pass a series of equidistant horizontal planes 10 metres apart through the island, their points of contact with the island would form contours with values of 10, 20, 30 and 40 m.

(b) Gradients

The height between successive contours is called the vertical interval or contour interval and is always constant over a map or plan. On the section the vertical interval is represented by AB. The horizontal distance between the same two contours is the distance BC. This is called the horizontal equivalent.

The gradient of the ground between the points A and C is found from

$$\text{Gradient} = \frac{AB}{BC} = \frac{\text{vertical interval}}{\text{horizontal equivalent}}$$

Since the vertical interval is constant throughout any plan the gradient varies with the horizontal equivalent, for example

$$\text{Gradient along AC} = \frac{10}{100} = \frac{1}{10} = 1 \text{ in } 10$$

$$\text{Gradient along DE} = \frac{10}{30} = \frac{1}{3} = 1 \text{ in } 3$$

(c) Reading contours

It should be clear from the above examples that the gradient is steep where the contours are close together and conversely flat where the contours are far apart.

In Fig. 5.3 three different slopes are shown. The contours are equally spaced in Fig. 5.3(a), indicating that the slope has a regular gradient. In Fig. 5.3(b) the contours are closer at the top of the slope than at the bottom. The slope is therefore steeper at the top

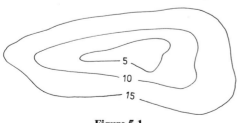

Figure 5.1

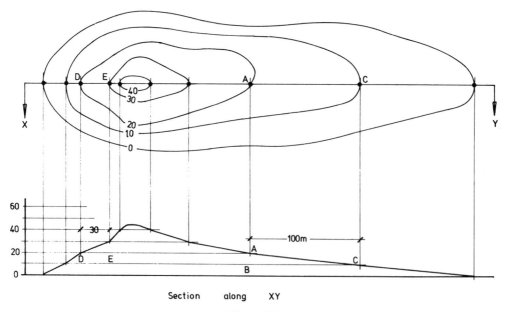

Figure 5.2

than at the bottom and such a slope is called a concave slope. Conversely, Fig. 5.3(c) portrays a convex slope.

A river valley has a characteristic 'V-shape' formed by the contours (Fig. 5.3(d)). The V always points towards the source of the river, i.e. uphill. In contrast, the contours of Fig. 5.3(e) also form a V-shape but point downhill forming a 'nose' or spur.

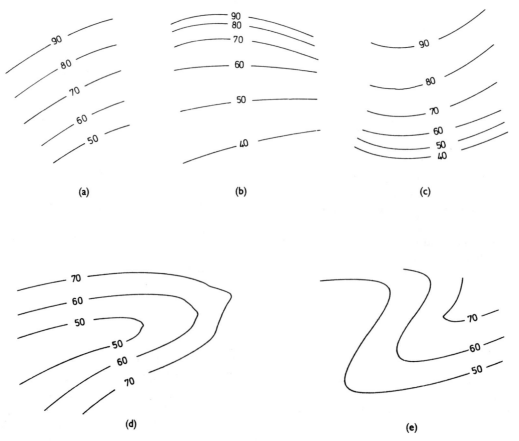

Figure 5.3

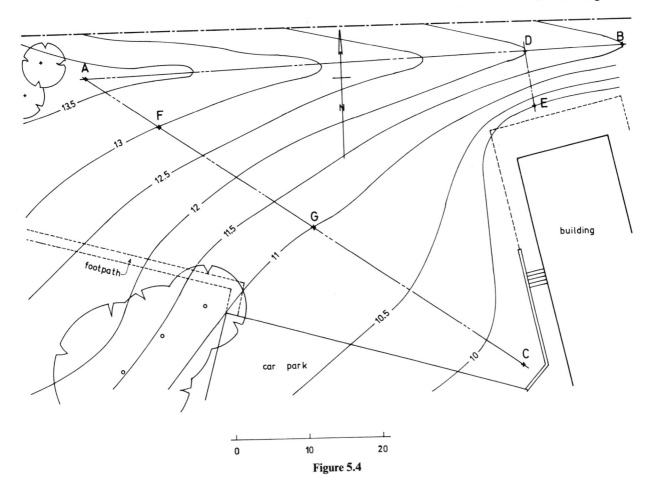

Figure 5.4

Example

1 Figure 5.4 shows contours at vertical intervals of 0.5 m over a building site, drawn at scale 1:500.
(a) Using a scale rule, measure the length and breadth of the site.
(b) Describe briefly the terrain along the lines AB and AC.
(c) Calculate the gradient between:
  (i) points D and E,
  (ii) points F and G.

Answer
(a) 80 m × 51 m
(b) From point A, the ground falls on a fairly regular slope, towards B. Line AB is, in fact, the line of a ridge, where the ground falls, both northwards and southwards, from the ridge line. On the line AC, the ground falls in the form of a concave slope, the slope being steeper towards the north end of the line.

  The whole area is the side of a hill, probably formed by bulldozing the area, during the construction of the building and car park.

(c) Line DE
    Fall = 2 m, horizontal distance = 8.0 m
    Therefore gradient = 1 in 4
   Line FG
    Fall = 2 m, horizontal distance = 25 m
    Therefore gradient = 1 in 12.5

Exercise 5.1

1 Figure 5.5 shows a contoured plan of two car parks, divided by a high retaining wall and bounded by various other walls and fences.
(a) Calculate the average gradient across the lines AB, CD and EF.
(b) Estimate, from the values of the contour lines, the height of the retaining wall dividing the car parks.

2. Methods of contouring

Once an accurate survey has been completed the surveyor knows the planimetric position of all points on the site relative to each other. The second task is to make a levelling to enable accurate positions of the contour lines over the site to be drawn.

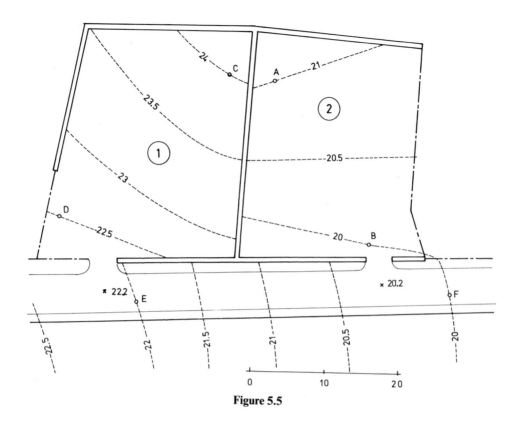

Figure 5.5

(a) Choice of vertical interval

The vertical interval of the contour lines on any plan depends on several factors, namely:

1. The purpose and extent of the survey
Where the plan is required for estimating earthwork quantities or for detailed design of works a small vertical interval will be required. The interval may be as small as 0.5 metre over a small site but 1 to 2 metres is more common, particularly where the site is fairly large.

2. The scale of the map or plan
Generally, on small-scale maps the vertical interval has to be fairly large. If not, some essential details might be obscured by the large number of contour lines produced by a small vertical interval.

3. The nature of the terrain
In surveys of small sites, this is probably the deciding factor. A close vertical interval is required to portray small undulations on relatively flat ground. Where the terrain is steep, however, a wider interval would be chosen.

The methods of contouring can be divided into (a) direct and (b) indirect methods. Before studying the direct method, it is necessary to understand how a point is physically set out, on the ground, at a predetermined height.

(b) Setting out a point of known level

The basic principle of setting out a point on the ground at a predetermined level is illustrated in Fig. 5.6. Point A is a temporary bench mark (RL 8.55 m AD). Markers, in the form of pegs or arrows, are to be placed in the ground, in positions where the ground has, firstly, a level of 9.00 metres and, secondly, a level of 8.00 metres.

Procedure

1. The observer sets up the level at a height, convenient for observing the TBM (RL 8.55 m) and takes a backsight staff reading. The reading is 1.25 m.

$$HPC = 8.55 + 1.25$$
$$= 9.80 \text{ m AD}$$

2. The staffman walks to an area where it is estimated that the ground has an approximate level of 9.00 m and turns the staff towards the instrument. The staff is then moved slowly up or down the ground slope until the base of the staff is at a height of 9.00 m exactly. This will occur when the observer reads 0.80 m on the staff, since

$$HPC = 9.80 \text{ m}$$
$$\text{Required level} = 9.00 \text{ m}$$
$$\text{Therefore staff reading} = 0.80 \text{ m (to achieve the required level)}$$

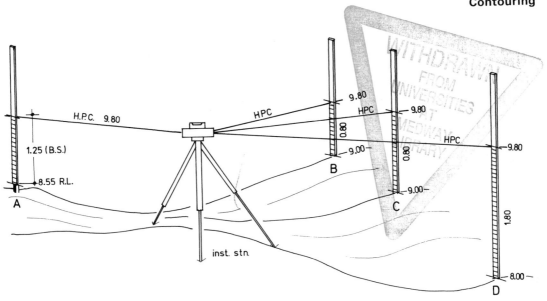

Figure 5.6

3. The ground position (B), is marked with an arrow or peg and the staff is removed to another location. The operation is repeated and a second peg is placed at C, when the observer again reads 0.80 m on the staff. This operation can be repeated any number of times.

4. When a different ground level, say 8.00 m, is to be marked, a new calculation is made, as follows:

$$
\begin{aligned}
\text{HPC} &= 9.80 \text{ m} \\
\text{Required level} &= 8.00 \text{ m} \\
\text{Therefore staff reading} &= 1.80 \text{ m (to achieve the}
\end{aligned}
$$
required level)

5. The observer directs the staffman downhill and directs him or her to move the staff until the reading becomes 1.80 m, whereupon the staffman marks the position (D) with a peg of a different colour.

(c) Direct method of contouring

Using this method the contour lines are physically followed on the ground. The work is really the reverse of ordinary levelling for whereas, by the latter operation, the levels of known positions are found, in contouring it is necessary to establish the positions of known levels, using the techniques described in part (b).

Figure 5.7 shows the plan of the site at the GCB Outdoor Centre. Contour lines are required at one metre vertical intervals. Temporary bench marks have been established on the linear survey stations A to F by flying levelling (Sec. 2(b), exercise 4.2, in Chapter 4).

In order to fix the contour positions two distinct operations are necessary.

Levelling

A level is set on site at some position from which a bench mark can be observed comfortably. In Fig. 5.7,

the level is set at position X and a backsight of 0.92 m observed on the BM (peg B, RL 7.78 m). The height of collimation = 7.78 + 0.92 = 8.70 m. From this instrument position, the contours at 7.00 m and 6.00 m can be observed. The staffman holds the staff facing the instrument and backs slowly downhill. When the observer reads 1.700 m, the bottom of the staff is at 7.000 m, since

$$
\begin{aligned}
\text{Height of collimation} &= 8.70 \text{ m} \\
7 \text{ metre contour} &= 7.00 \text{ m} \\
\text{Staff reading} &= 1.70 \text{ m}
\end{aligned}
$$

The staffman marks the staff position by knocking a peg into the ground. He or she then proceeds roughly along a level course, stopping at frequent intervals, while being directed to move the staff slowly up or downhill until the 1.700 m reading is observed from the instrument. A peg is inserted at each correct staff position. In Fig. 5.7, pegs 1 to 4 have been established on the 7.00 m contour. The 6.00 m contour is similarly established at pegs 5 to 9 using a staff reading of 2.70 m.

In order to set out the 5.00 m contour, a staff reading of 3.70 m is required. The contour line is denoted by pegs 10 to 14. Table 5.1 shows the appropriate booking. In order to check the levelling, an FS reading of 3.14 m has been taken to TBM peg D.

The eastern end of the site is too far removed from instrument station X to allow accurate contouring to take place. Consequently, a further instrument set-up is established at point Y and the levelling operation is continued from there.

Survey of the pegs

Figure 5.7 shows the various peg positions denoting contour lines 7.0, 6.0 and 5.0 m, on both sides of the small stream. The pegs are surveyed by offsets from

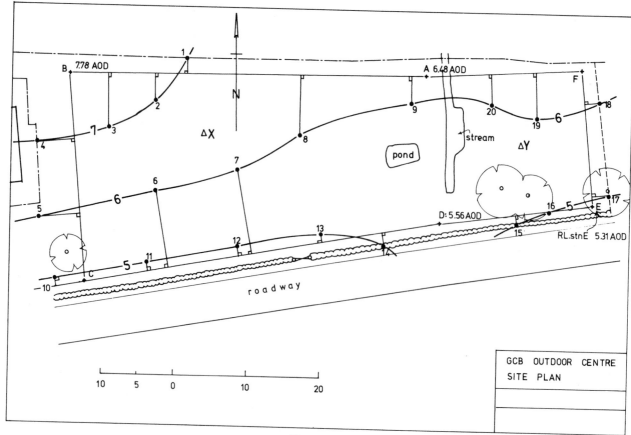

Figure 5.7

BS	IS	FS	HPC	Reduced level	Distance	Remarks
0.920			8.700	7.780		TBM peg B
	1.700		8.700	7.000		7 m contour line
	2.700		8.700	6.000		6 m contour line
	3.700		8.700	5.000		5 m contour line
		3.140	8.700	5.560		TBM peg D (RL 5.56 m)

Table 5.1

the existing linear survey framework, all of which are clearly shown on the figure.

On a larger site, instrumental methods would be employed, e.g. tacheometric and electromagnetic radial positioning. The plan positions of the contours are plotted directly on to the site plan and smooth curves drawn through them.

Example

2 The contour survey of the GCB Outdoor Centre site was completed on the east side of the stream by removing the instrument to set up point Y. Table 5.2 shows the relevant readings.

BS	IS	FS	HPC	Reduced level	Remarks
1.05				6.48	TBM
	?			6.00	6 m contour line
	?			5.00	5 m contour line
		2.23		?	TBM E (RL 5.31 m)

Table 5.2

Complete the table by calculating the staff readings required to set out the 6.0 and 5.0 m contour lines.

Answer

$$HPC = 6.48 + 1.05 = 7.53$$
Therefore staff reading $= 7.53 - 6.00$
$$= 1.53 \text{ (for 6 m contour)}$$
and $= 7.53 - 5.00$
$$= 2.53 \text{ (for 5 m contour)}$$
$$RL\,E = 7.53 - 2.23$$
$$= 5.30 \text{ (which checks to 1 cm)}$$

The 5 and 6 m contour lines are set out on the ground and marked by pegs 15 to 20 (Fig. 5.7).

(d) Indirect method of contouring

When using this method, no attempt is made to follow the contour lines. Instead a series of spot levels is taken at readily identifiable locations, e.g. at trees, gateposts, manholes and at intersections of walls and fences. Contour positions are then interpolated between them.

On open areas, where there are no easily identifiable features, a grid of squares or rectangles is set out on the ground and spot levels are taken to each point of the grid.

In theory, when contouring by this method, spot levels are required only at the tops and bottoms of all slopes. The ground is then treated as a plane surface between them. In practice, many more levels are taken. Mistakes are usually easy to identify.

Three distinct operations are involved in the indirect method of contouring.

Setting out a grid

1. On any site, a long side is chosen as the baseline and ranging rods or small wooden pegs are set out along the line at regular intervals. The interval depends on the contour vertical interval; the factors affecting this interval have been previously discussed. Briefly, the baseline intervals should be 5 or 10 metres where the vertical interval is to be close (say 0.5 or 1 m) and should be a maximum of 20 metres in all other cases.

2. At 30 metre intervals, or more closely if desired, lines are set out at right angles to the baseline, using a prism square or a tape. (The methods of setting out right angles are described in Section 5(a) in Chapter 3.) Further pegs are inserted at 10 metre intervals along these lines. Gaps are filled very simply, using a tape.

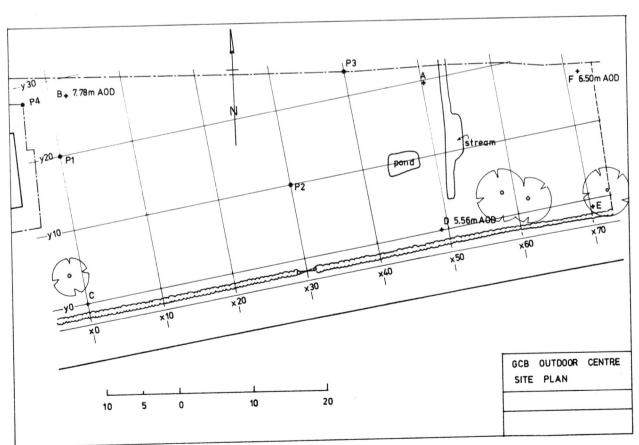

Figure 5.8

The numerous pegs must be easily identified at a later stage so each is given a grid reference. There are many reference systems in common use, each with its own advantages and disadvantages. Though slightly cumbersome, an excellent system is formed by giving all distances along and parallel to the baseline an x chainage and all distances at right angles to the baseline a y chainage.

On the GCB Outdoor Centre site (Fig. 5.8, page 105), line CD of the linear survey is chosen as the baseline. In order to cover the complete site, the line is extended eastwards, Pegs are then established at 10 metre intervals along this line and at right angles to it. Peg P1, therefore, has a grid reference of $(x0, y20)$, while the reference of P2 is $(x30, y10)$.

The grid values of any point on the boundary can be obtained by measuring along the appropriate line to the boundary. The point P3 would have the identification of $(x40, y23)$. This is the main advantage of the identification system.

Table for use with Exercise 5.2

BS	IS	FS	Rise or HPC	Fall	Reduced level	Distance	Remarks
0.560					7.780		Peg B
	3.480						0, 0
	2.110						0, 10
	1.190						0, 20
	0.340						0, 30
	3.530						10, 0
	1.910						10, 10
	1.200						10, 20
	1.010						10, 28
	3.540						20, 0
	2.230						20, 10
	1.530						20, 20
	1.360						20, 26
	3.570						30, 0
	2.310						30, 10
	1.880						30, 20
	1.700						30, 24
	3.110						40, 0
	2.450						40, 10
	1.870						40, 20
	1.730						40, 23
	3.750						0, −2
	3.740						10, −2
	3.750						20, −2
	3.940						30, −2
	3.240						40, −2
	3.540						−5, 0
	2.090						−5, 11
	1.160						−4, 21
	0.740						−5, 28
2.280		2.780					Peg D
	2.440						50, 0
	2.230						50, 10
	1.520						50, 20
	1.440						50, 22
	2.610						60, 0
	1.920						60, 10
	1.320						60, 20
	2.710						70, 0
	2.000						70, 10
	1.290						70, 18
	2.890						60, −2
	2.940						70, −2
	2.940						72, 0
	2.040						72, 10
	1.410						72, 20
		1.330					Peg F

Table 5.3

Levelling

A series levelling is conducted over the site and the reduced level of every point on the grid is obtained, as in the following Exercise 5.2. The calculated reduced levels are then added to the plan in their correct locations.

Exercise 5.2

1 Table 5.3 (opposite) shows the readings obtained during the grid levelling of the GCB Outdoor Centre site. Reduce the levels, using either the rise/fall or HPC method of reduction and plot the results on Fig. 5.8.

Interpolating the contours

Figure 5.9(a) shows a part of the grid, with its appropriate reduced levels. Contour lines may be interpolated on the grid either mathematically or graphically.

1. Mathematically

The positions of the contours are interpolated mathematically from the reduced levels by simple propor-

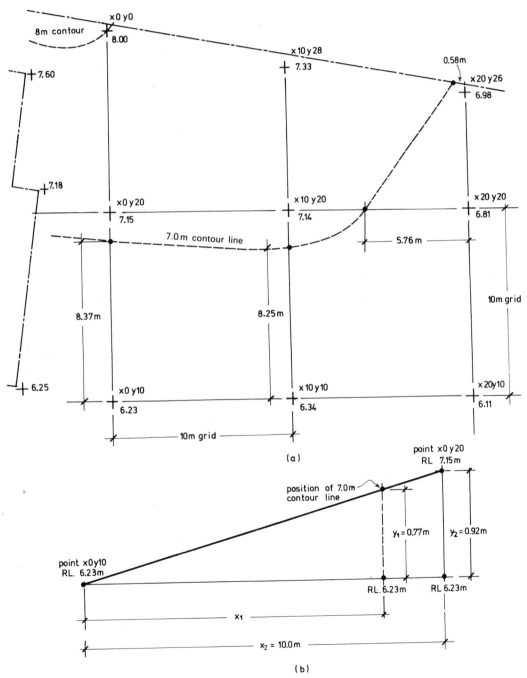

Figure 5.9

1	2	3	4	5	6	7
Lower grid point 1	Higher grid point 2	Contour line value	Horizontal distance from point 1 to 2	Difference in level (high point 2 − low point 1) (column 2 − column 1)	Difference in level (contour − low point 1) (column 3 − column 1)	Horizontal distance from low point to contour line $\left(\dfrac{\text{column 6}}{\text{column 5}}\right) \times$ column 4
$x0, y10$	$x0, y20$	7.00	10.00	0.92	0.77	8.37
$x10, y10$	$x10, y20$	7.00	10.00	0.80	0.66	8.25
$x20, y20$	$x10, y20$	7.00	10.00	0.33	0.19	5.76
$x20, y26$	$x10, y28$	7.00	10.20	0.35	0.02	0.57

Table 5.4

tion; e.g. the 7 m contour passes somewhere between the point $(x0, y10)$ and $(x0, y20)$ (Fig. 5.9(a)).

Referring to Fig. 5.9(b), the plan position of the contour line is calculated as follows:

Horizontal distance between $(x0, y10)$
$$\text{and } (x0, y20) = x_2 = 10\text{ m}$$

Difference in level between $(x0, y10)$
$$\text{and } (x0, y20) = y_2 = (7.15 - 6.23) = 0.92\text{ m}$$

Difference in level between $(x0, y10)$ and the
$$7.0\text{ m contour line} = y_1 = (7.0 - 6.23) = 0.77\text{ m}$$

Using the geometric principles of similar triangles,

$$\frac{x_1}{x_2} = \frac{y_1}{y_2}$$

Therefore $x_1 = \dfrac{y_1 \times x_2}{y_2}$

$$= \frac{0.77 \times 10}{0.92}$$

$$= 8.37\text{ m}$$

In other words, the horizontal distance from point $(x0, y10)$ to the 7.0 m contour line $= 8.37$ m

When a great many points are to be interpolated, a hand calculator is sometimes used, particularly when levels are taken to two decimal places. Use is made of a calculation table (Table 5.4). The interpolation of the 7 m contour, shown above, appears in the first line of the table while the remainder of the table is devoted to the calculations necessary for plotting the 7 m contour line completely.

The contour positions are plotted on the plan (Fig. 5.9(a)) and joined by a smooth curve. Similar calculations are carried out for other contour lines, which are then added to the plan.

2. Graphically

Figure 5.10 is a portion of the grid of a site survey, on which the position of the 94 m contour is to be plotted between the points $(x20, y20)$ and $(x20, y40)$. The respective levels of the points are 94.4 m and 92.5 m.

Using any suitable scale on a scale rule, the 2.5 graduation mark is placed against the 92.5 level and the scale positioned as shown in Fig. 5.10. The 4.0

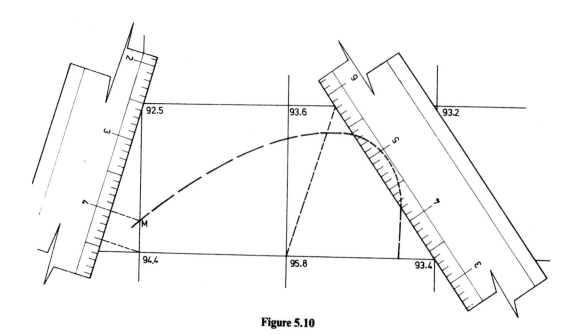

Figure 5.10

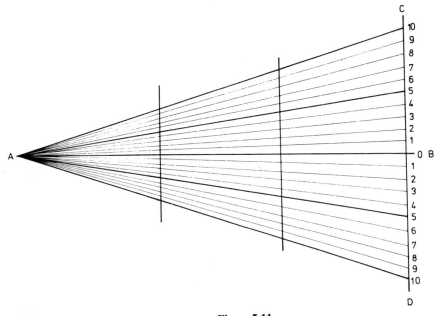

Figure 5.11

graduation representing the 94.0 contour line and the 4.4 graduation representing the 94.4 m level are marked on the plan. The latter mark is joined to the 94.4 m level and a parallel line drawn through the 4.0 m graduation to meet the grid line at M. From the similar triangles, formed by the scale, grid line and construction lines, point M is the true position of the 94 m contour on the grid line.

An alternative graphical device is the radial interpolation graph. The graph is drawn on tracing paper and is constructed as in Fig. 5.11. Two mutually perpendicular lines AB and CD are drawn. The shorter line CD is divided into twenty equal parts and radial lines are drawn from A to each of the divisions.

In Fig. 5.12, the graph is shown in use. The 92 m contour position is required between the stations $(x60, y40)$ and $(x60, y60)$ where the reduced levels are

93.2 and 91.6 m respectively. The difference in level between the points is therefore 1.6 m. The overlay is laid on the grid with AB and CD parallel to the grid lines, until 16 divisions are intercepted between the two grid stations.

Each division represents 0.1 m in this case and the 92 m contour will therefore be four divisions from station $(x60, y60)$. A pin-hole is made through the overlay at this point and the mark denoted by a small circle on the site plan. It should be noted that the overlay can be manoeuvred until any convenient number of divisions is intercepted by the grid stations. If, for example, eight divisions were chosen, each would represent 0.2 metre.

Since the overlay is not drawn to any scale, it can be kept and used for any contour interpolation. It is probably the fastest method of interpolating and it is

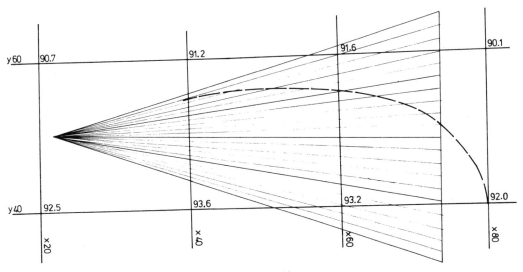

Figure 5.12

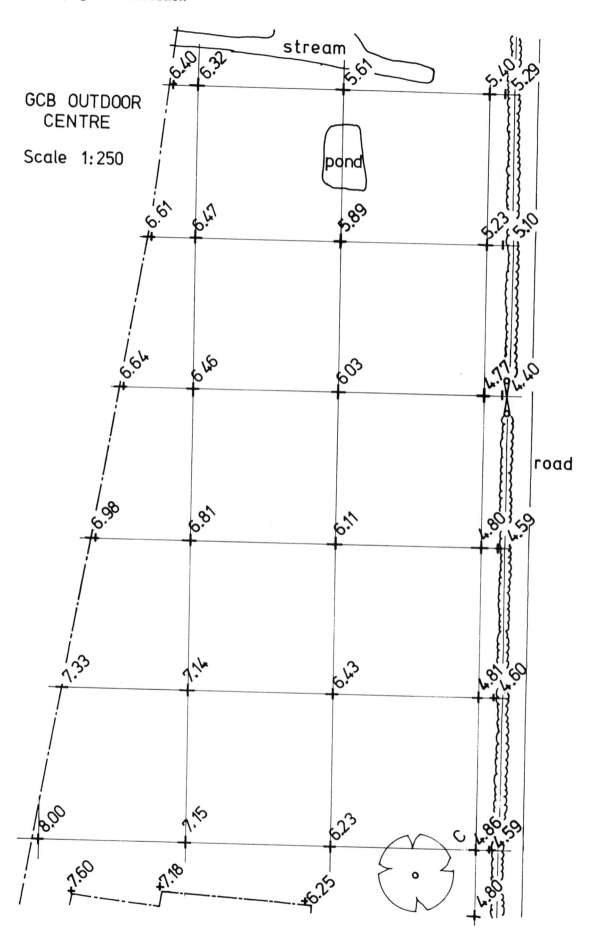

Figure 5.13

therefore worth while spending a few minutes' time in preparing a really good overlay.

Exercise 5.3

1 Figure 5.13 (opposite) shows the network of spot levels of the western part of the GCB Outdoor Centre site, plotted to a scale of 1:500. (The levels are, in fact, the answers to the problem set in Exercise 5.2.) Using any interpolation method, draw the 7, 6 and 5 m contour lines on the plan.

3. Uses of contour plans

Of the many uses to which contour maps are put, the following are the most relevant to building surveying.

(a) Vertical sections

A vertical section is really the profile of the ground that would be obtained by cutting the ground surface along any chosen line. Figure 5.14 shows part of the contoured plan of the GCB Outdoor Centre. A vertical section is to be drawn along the line XX.

The section is drawn as follows:

1. A datum line is drawn parallel to the line of the section, clear of the plan, and a height is selected for the datum. In this case, the chosen value is zero (0) metres.
2. The contour positions along the section line (points a, b, c, d) are projected at right angles, to reach the datum line (points e, f, g, h).
3. The contour values of 5, 6, 7 and 8 m are scaled at right angles to the datum, to form the ground surface points.
4. The ground surface points are joined to form the ground profile.

(b) Intersection of surfaces

Whenever earth is deposited it will adopt a natural angle of repose. The angle will vary according to

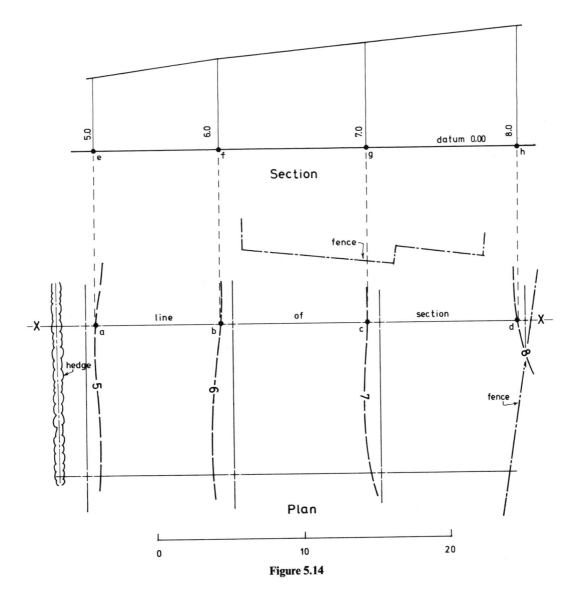

Figure 5.14

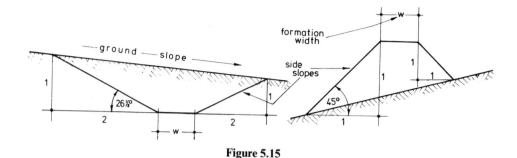

Figure 5.15

whether the material is clayey or sandy, wet or dry, but will almost certainly be between 45° and $26\frac{1}{2}°$; i.e. the material will form side slopes of 1 in 1 to 1 in 2. In Fig. 5.15 the cutting has side slopes that batter a 1 in 2 while the side slopes of the embankment batter at 1 in 1.

New construction work is often shown on plans by a system of contour lines. The portion of the embankment shown in Fig. 5.16 has a formation width of 10 metres, is 10 metres high and is to accommodate a level roadway 100 metres long. The sides of the embankment slope at 1 unit vertically to $1\frac{1}{2}$ units horizontally and the embankment is being constructed on absolutely flat ground.

Contour lines at 1 metre vertical intervals are drawn parallel to the roadway at equivalent $1\frac{1}{2}$ metre horizontal intervals. Since the embankment is formed on level ground the outer limits, denoting the bottom of the embankment, will be perfectly straight lines parallel to the top of the embankment. These lines will be 15 metres on either side of the formation width.

Very seldom will the ground surface be perfectly level. Figure 5.17 shows a series of contour lines portraying an escarpment, the southern scarp slope of which rises steeply from 0 to 9 m, while the dip slope slopes gently northwards from 9 to 6 m. The embankment already described is to be laid along the line AB. The perspective view in Fig. 5.17 illustrates the situation.

Approximately halfway along the embankment, the ground surface is at a height of 9 m, i.e. 1 m below the formation level of the embankment. It follows therefore that the width of the embankment will be very much narrower than at the commencement of the embankment, i.e. at point A. The plan position of the outer limits will no longer be a straight line. The actual position is determined by superimposing the embankment contours on the surface contours.

The intersection of similar values forms a point on the tail of the embankment (Fig. 5.18). When all of the intersection points are joined a plan position of the outer limits is obtained.

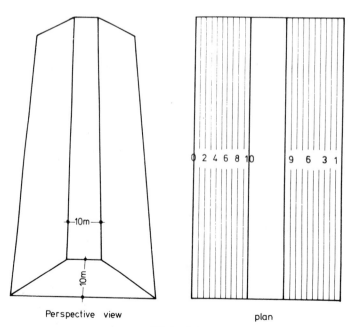

Perspective view   plan

Figure 5.16

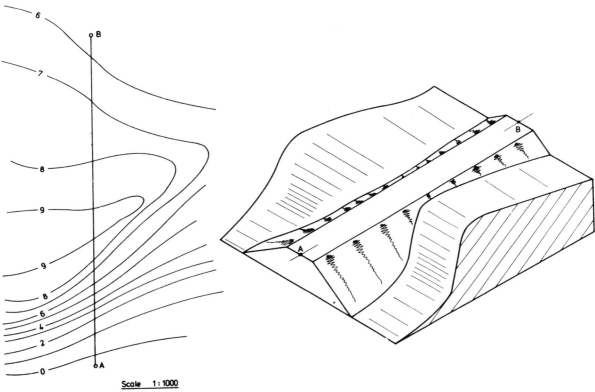

Figure 5.17

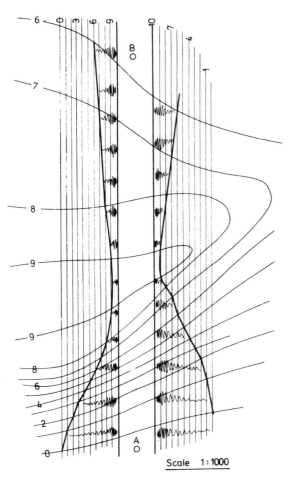

Figure 5.18

When cuttings are to be constructed, the procedure for drawing the outline is identical to that described above. However, when the contours along the cutting slopes are drawn, values increase with distance from the formation, whereas the contour values decrease on embankments.

Example

3 Figure 5.19 shows the position of a proposed factory building, together with ground contours at 1 metre vertical intervals. The formation level over the whole site is to be 0.5 m below the finished level, i.e. 23.000 m AOD, and any cuttings or embankments are to have side slopes of 1 unit vertically to 1 unit horizontally.

Draw on the plan the outline of the required earth works.

Answer (Fig. 5.20)
(a) There is no excavation or filling at point A, since the ground level and formation level are the same.
(b) All earthworks southwards are in cutting. Contours are drawn parallel to the building at 1 metre horizontal intervals and numbered from 23 to 28 m.
(c) The intersections of ground contours and earthwork contours of similar value form the edge of the embankments or cuttings.

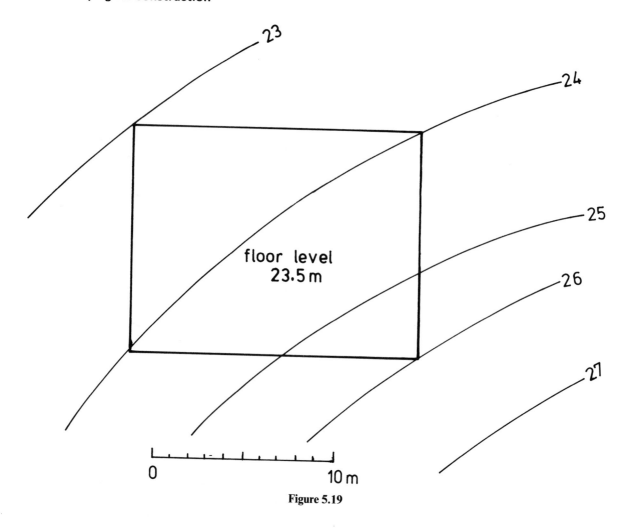

Figure 5.19

Sloping earthworks

When earthworks are to be formed to some specific gradient, the drawing of the contours is not so straightforward as with level works. Figure 5.21 shows an embankment being formed on flat ground, the level of which is 67.000 m. The gradient of the embankment is 1 in 20 rising from a formation level of 70.000 m at A towards B. The length of the embankment is 60 m and the sides slope at 1 in 2.

The embankment contours are formed by joining two points of the same level on the embankment sides.

The reduced level of point A is 70.000 m; therefore the reduced level of point B is

$$70.000 + (60/20) \text{ m} = 73.000 \text{ m}$$

A point of 70.000 m level will therefore be 3 metres vertically below point B. The horizontal equivalent of 3 m vertical at a gradient of 1 in 2 is 6 m. Points C and D are drawn 6 m from the formation edge and joined to points E and F respectively to form the 70 m embankment contours.

All other contours at vertical intervals of 1 metre are parallel to these two lines, and are spaced at 2 metre horizontal intervals.

Example

4 Figure 5.22 shows ground contours at vertical intervals of 1 metre. AB is the centre line of a proposed roadway, which is to be constructed to the following specification:
  (i) Formation width, 5 metres
 (ii) Formation level A, 60.00 m AD
(iii) Gradient AB, 1 in 25 rising
 (iv) Side Slopes, 1 in 3

(a) Show clearly the outline of any earthworks formed between A and B.
(b) Draw a cross-section of the embankment on the line indicated.

Answer (Fig. 5.23)

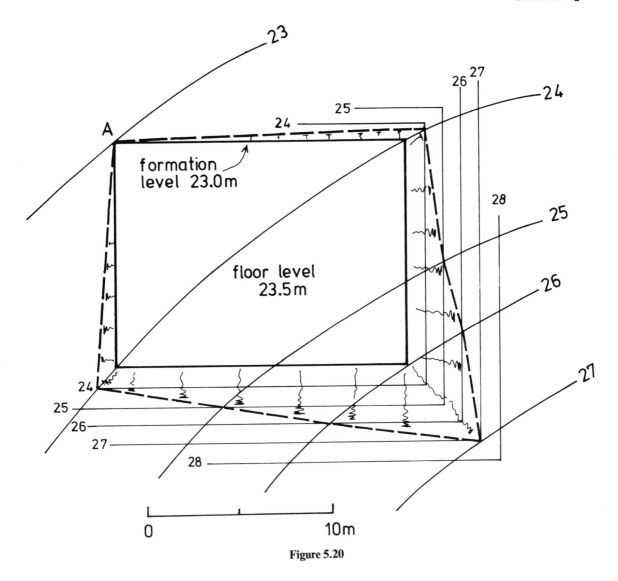

Figure 5.20

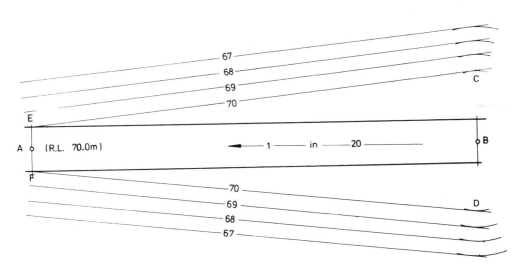

Figure 5.21

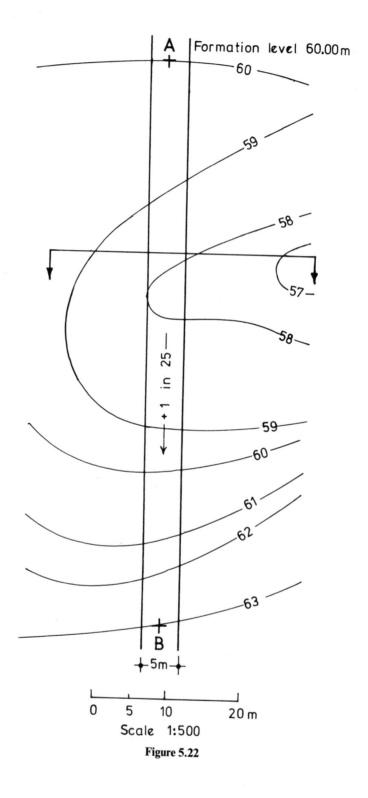

Figure 5.22

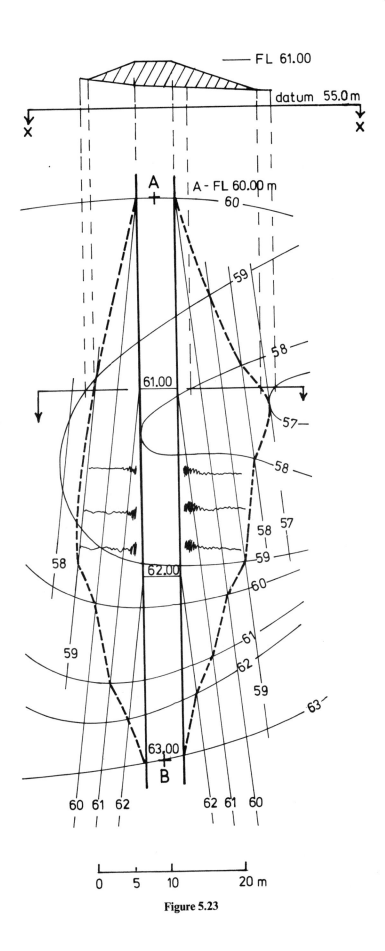

Figure 5.23

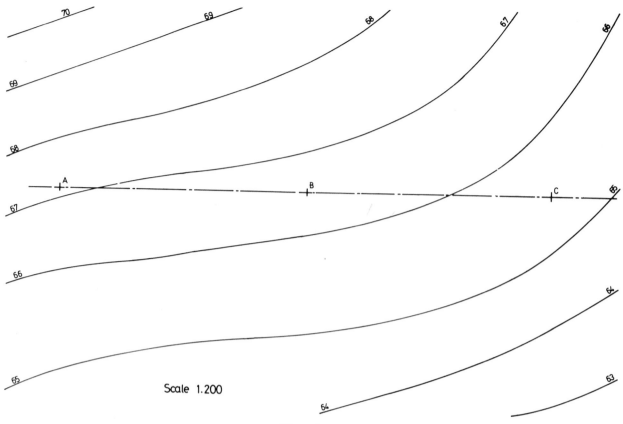

Figure 5.24

Exercise 5.4

1 Figure 5.24 shows ground contours at 1 metre vertical intervals. ABC is the line of a proposed level roadway having the following specification:
  (i) Formation level, 69.0 m
 (ii) Formation width, 5.0 m
(iii) Side slopes, 1 in 1 (45°)

(a) Show on the plan the outline of the earthworks required to form the roadway.
(b) Draw natural cross-sections at A, B and C to show the roadway embankment and the ground surface.
(c) From the plan or section, determine the ground levels at the base of the side slopes and on the centre line of the embankment across the sections at A, B and C.

Table for use with Exercise 5.4, question 2

BS	IS	FS	Rise or fall	Reduced level	Distance	Remarks
1.320				26.340		Bench mark
	0.290					A4
	0.910					B4
	2.170					C4
	3.620					D4
	1.920					A3
0.830		3.010				B3
	1.780					C3
	2.650					D3
	1.930					A2
	2.480					B2
0.610		3.240				C2
	1.570					D2
	1.840					A1
	1.840					B1
	1.850					C1
		3.320				D1

Table 5.5

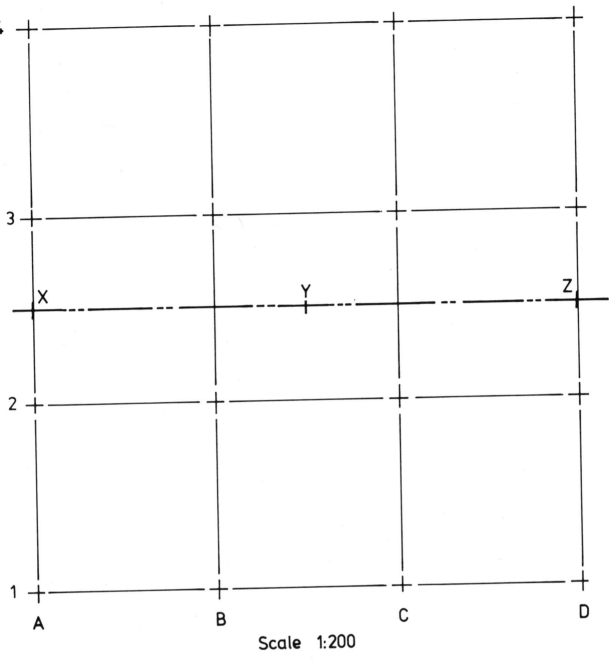

Scale 1:200

Figure 5.25

Note. This exercise will be used as an example in the areas and volumes section of this book, on page 346.

2 Figure 5.25 shows positions and Table 5.5 (opposite) shows the values of readings, taken on the levelling survey, of an area that is to be developed as part of a new road scheme. The new road is to be constructed on a centre line XYZ to the following specification:

(i) Formation level, 20.00 m
(ii) Road width, 5.00 m
(iii) Road gradient, level
(iv) Side slopes, 1 in 1 (45°)

(a) Reduce the levels and apply a check.
(b) Print the levels on the plan, in their appropriate locations.

(c) Interpolate the ground contours, at 1 metre vertical intervals.
(d) Show on the plan the outline of the cutting required to accommodate the roadway.
(e) Draw natural cross-sections, at points X, Y and Z, to show the roadway cutting and the ground surface.
(f) From either the plan or sections, determine the reduced levels at the tops of the cutting side slopes and on the centre line of the cutting, across each of the three sections at X, Y and Z.

Note. This exercise will be used as an example in the areas and volumes section of this book, on page 347.

4. Answers

Exercise 5.1

1 (a) Gradient AB = 1 in 25
Gradient CD = 1 in 20
Gradient EF = 1 in 21
(b) Height of wall = 3 m

Exercise 5.2

2 Table 5.6

BS	IS	FS	Rise or fall	Reduced level	Distance	Remarks
0.560				7.780		Peg B
	3.480		−2.920	4.860		0, 0
	2.110		1.370	6.230		0, 10
	1.190		0.920	7.150		0, 20
	0.340		0.550	8.000		0, 30
	3.530		−2.890	4.810		10, 0
	1.910		1.620	6.430		10, 10
	1.200		0.710	7.140		10, 20
	1.010		0.190	7.330		10, 28
	3.540		−2.530	4.800		20, 0
	2.230		1.310	6.110		20, 10
	1.530		0.700	6.810		20, 20
	1.360		0.170	6.980		20, 26
	3.570		−2.210	4.770		30, 0
	2.310		1.260	6.030		30, 10
	1.880		0.430	6.460		30, 20
	1.700		0.180	6.640		30, 24
	3.110		−1.410	5.230		40, 0
	2.450		0.660	5.890		40, 10
	1.870		0.580	6.470		40, 20
	1.730		0.140	6.610		40, 23
	3.750		−2.020	4.590		0, −2
	3.740		0.010	4.600		10, −2
	3.750		−0.010	4.590		20, −2
	3.940		−0.190	4.400		30, −2
	3.240		0.700	5.100		40, −2
	3.540		−0.300	4.800		−5, 0
	2.090		1.450	6.250		−5, 11
	1.160		0.930	7.180		−4, 21
	0.740		0.420	7.600		−5, 28
2.280		2.780	−2.040	5.560		Peg D
	2.440		−0.160	5.400		50, 0
	2.230		0.210	5.610		50, 10
	1.520		0.710	6.320		50, 20
	1.440		0.080	6.400		50, 22
	2.610		−1.170	5.230		60, 0
	1.920		0.690	5.920		60, 10
	1.320		0.600	6.520		60, 20
	2.710		−1.390	5.130		70, 0
	2.000		0.710	5.840		70, 10
	1.290		0.710	6.550		70, 18
	2.890		−1.600	4.950		60, −2
	2.940		−0.050	4.900		70, −2
	2.940		0.000	4.900		72, 0
	2.040		0.900	5.800		72, 10
	1.410		0.630	6.430		72, 20
		1.330	0.080	6.510		Peg F

Table 5.6

Exercise 5.3

1 Figure 5.26

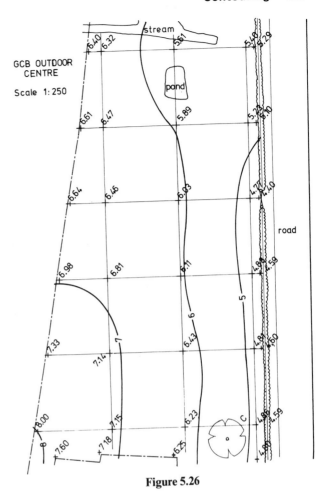

Figure 5.26

Exercise 5.4

1 (a) Figure 5.27

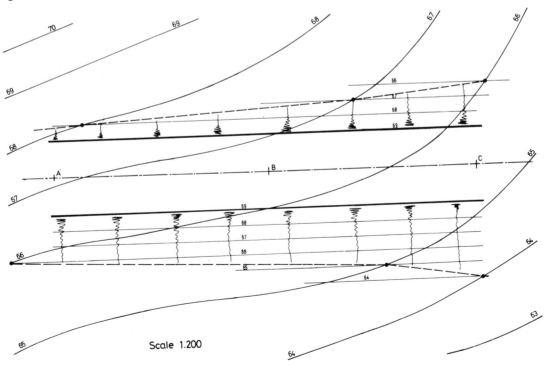

Scale 1:200

Figure 5.27

(b) (c) Figure 5.28

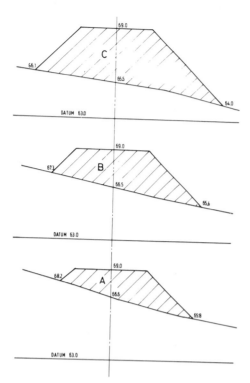

Figure 5.28

2 Table 5.7 and Figure 5.29

BS	IS	FS	Rise or fall	Reduced level	Distance	Remarks
1.320				26.340		Bench mark
	0.290		1.030	27.370		A4
	0.910		−0.620	26.750		B4
	2.170		−1.260	25.490		C4
	3.620		−1.450	24.040		D4
	1.920		1.700	25.740		A3
0.830		3.010	−1.090	24.650		B3
	1.780		−0.950	23.700		C3
	2.650		−0.870	22.830		D3
	1.930		0.720	23.550		A2
	2.480		−0.550	23.000		B2
0.610		3.240	−0.760	22.240		C2
	1.570		−0.960	21.280		D2
	1.840		−0.270	21.010		A1
	1.840		0.000	21.010		B1
	1.850		−0.010	21.000		C1
		3.320	−1.470	19.530		D1
2.76 −9.57 −6.81		9.57	−6.81	−6.81		

Table 5.7

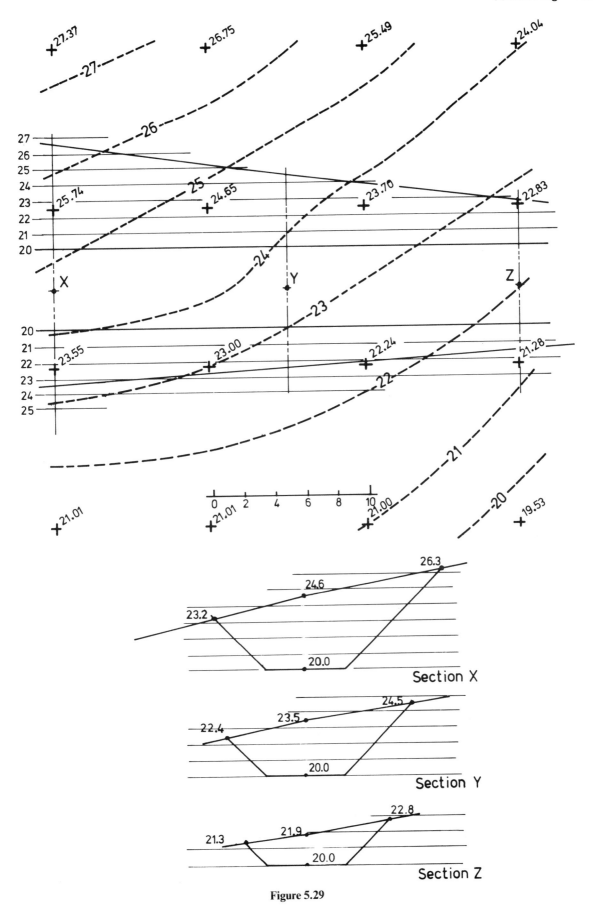

Figure 5.29

5. Project

Chapter 18 is a project covering all chapters of this textbook. It is intended that the project build into a complete portfolio of the surveying work required in the survey and the setting out of a building or engineering development.

If the reader wishes to continue work on the project or begin work at this stage, he or she should now turn to page 392 and attempt Section 4.

CHAPTER 6

Vertical sections

Objective

After studying this chapter, the reader should be able to make a levelling survey along a predetermined line set out on the ground, process the data and draw longitudinal sections and cross sections from the results.

One of the most important, and certainly one of the most common applications of levelling, is sectioning.

Whenever narrow works of long length, e.g. roads, sewers, drains, etc., are to be constructed, it is necessary to draw vertical sections showing clearly the profile of the ground. Two kinds are necessary:

(a) longitudinal sections, i.e. a vertical section along the centre line of the complete length of the works,

(b) cross sections, i.e. vertical sections drawn at right angles to the centre line of the works.

The information provided by the sections provides data for:

(a) determining suitable gradients for the construction works,

(b) calculating the volume of the earthworks,

(c) supplying details of depth of cutting or height of filling required.

1. Development plan

Figure 6.1 is the contoured plan of the GCB Outdoor Centre compiled from Fig. 3.34 (linear survey) and Fig. 5.26 (contouring). The architectural development has been added to the plan and shows the proposed positions of three houses and a private roadway serving them.

Roadway specification

Class:	Private
Width:	Road 4.0 m
	Footpath 1.2 m
Gradients:	(a) 1 in 30 rising from point R1 through R3 to R10
	(b) 1 in 100 rising from point R3 through R11 to R14
Side slopes:	1 horizontal to 1 vertical (45°)
Radii of curves:	To centre line of road 7.0 m
	To toe of kerb 5.0 m
	To heel of kerb 3.8 m

Building specification:

No. 12 Lochview:	2 storey, timber framed Floor level 6.00 m AOD
No. 11 Lochview:	2 storey, timber framed Floor level 6.00 m AOD
No. 10 Lochview:	single storey, timber framed Floor level 6.50 m AOD

Drainage specification:

Storm water drain:	150 mm pipe with manholes S1 to S4
	Invert level: manhole S1, 2.197 m AOD
Foul drain:	150 mm pipe with manholes F1 to F4
	Invert level: manhole F1, 2.950 m AOD

In dealing with drains and sewers, reference is made to the invert level rather than the formation level. The invert level is the inside of the bottom of the pipe, but for practical purposes it may be considered to be the bottom of the excavation.

Occasionally the pipes are laid on a bed of concrete, in which case the thickness of the concrete must be subtracted from the invert level to obtain the level of the bottom of the excavation.

2. Longitudinal sections

In construction work, the sewers and roadways are usually constructed first of all. They must be set out, on the ground, in their correct locations. Additional plans, in the form of vertical sections, are required.

In order to cost the development accurately, the volume of material required to construct the earthworks (cuttings or embankments) has to be calculated, generally by the quantity surveyor.

The first step in calculating the earthworks quantities and in preparing the setting-out information, is to

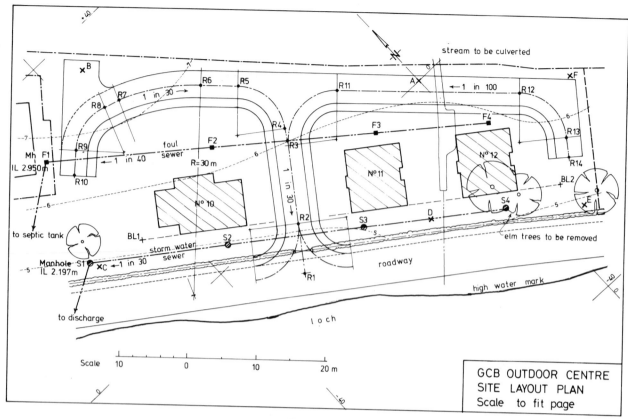

Figure 6.1

make longitudinal sections and cross sections along the lines of the proposed works.

(a) Fieldwork

1. The centre lines of the road and sewers are set out, on the ground, by a series of stakes. The setting out of the centre lines is usually done, using a theodolite. (Setting out is the subject of Chapter 13. It is sufficient for the moment to assume that the stakes have been set out in their correct positions.)
2. A levelling is made along the centre line with levels taken at all changes of gradient. A level is also taken at every tape length whether or not it signifies a change in gradient.
3. Horizontal measurements are made between all the points at which levels were taken. The measurements are accumulated from the first point such that

all points have a running chainage. Pegs are left at every tape length to enable cross sections to be taken later (Fig. 6.2).

4. *Procedure*

Generally a surveyor and three assistants are required if the section is long. The surveyor takes the readings and does the booking; one assistant acts as staffman, while the other two act as chainmen taking all measurements, lining-in ranging poles along the previously established centre line and leaving pegs at all tape lengths.

A flying levelling is conducted from some nearby bench mark to the peg denoting zero chainage. Thereafter the levelling is in series form with intermediate sights taken as necessary.

The tape is held at peg zero chainage and stretched out along the line of the section. The chainmen and staffman work together and while the latter holds the staff as a backsight at peg zero, the chainman marks

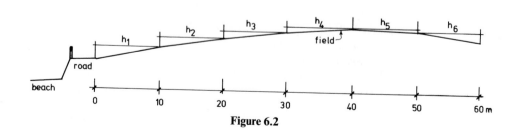

Figure 6.2

BS	IS	FS	Rise or fall	Reduced level	Distance	Remarks
1.350				5.200		OSBM (5.200 m)
	2.150				0.000	R1 centre line of road
	1.130				10.000	Centre line of road
1.650		0.285			20.000	Centre line of road
	1.020				30.000	Centre line of road
	0.665				40.000	Centre line of road
	0.595				50.000	Centre line of road
	1.330				60.000	R10 centre line of road
		0.125				Peg B (7.780)

Table 6.1

the changes in gradient and calls out the chainages of these points to the observer. The staffman follows up and holds the staff at all of the changes of gradient. When one tape length has been completed, a peg is left and the next length is observed. The procedure is repeated until the complete section has been levelled.

As in all surveying work a check must be provided. In sectioning this can be done by flying levelling from the last point of the section to the commencing bench mark, or to some other, closer, bench mark.

Example

1 Table 6.1 shows the readings obtained during the levelling along the line of the proposed road R1–R10. The levelling commenced on the nearby Ordnance Survey bench mark (OSBM), although it could have begun on any of the temporary bench marks, established on the site.

(a) Calculate the reduced levels of the ground at the various chainage points.

Answer (Table 6.2)

(b) Plotting

1. The field work is reduced and all checks applied. The result should be checked against Table 6.2 which shows the reduced levels.

It will be noticed that the levelling does not close exactly on to bench mark 8, the discrepancy being 0.010 m. This closure error is acceptable and the reduced levels are considered to be satisfactory.

2. The scales are chosen for the section drawing, such that the horizontal scale is the same as the scale of the plan view of the site. (In this book, however, the scale has been chosen to fit the width of the page.) Compared to the length of the section, the differences in elevation of the section points will always be comparatively small: consequently the vertical scale of the section is exaggerated to enable the differences in elevation to be readily seen. The horizontal scale is usually enlarged two to ten times, producing the following scales in Fig. 6.3:

Horizontal scale 1:500
Vertical scale 1:200

3. The reduced levels are scrutinized and the lowest point is found. A horizontal line is drawn to represent

BS	IS	FS	Rise or fall	Reduced level	Distance	Remarks
1.350				5.200		OSBM (5.200 m)
	2.150		−0.800	4.400	0.000	R1 centre line of road
	1.130		1.020	5.420	10.000	Centre line of road
1.650		0.285	0.845	6.265	20.000	Centre line of road
	1.020		0.630	6.895	30.000	Centre line of road
	0.665		0.355	7.250	40.000	Centre line of road
	0.595		0.070	7.320	50.000	Centre line of road
	1.330		−0.735	6.585	60.000	R10 centre line of road
		0.125	1.205	7.790		Peg B (7.780)
3.000		0.410	2.590	7.790		
−0.410				−5.200		
2.590				2.590		

Table 6.2

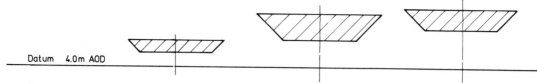

Cross sections; natural scale 1:200

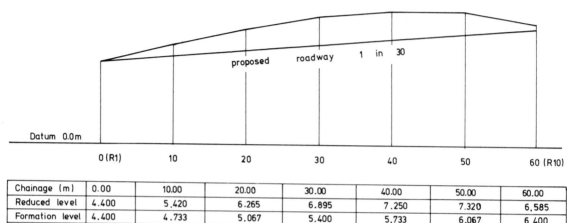

proposed   roadway   1 in 30

Datum 0.0 m

| 0 (R1) | 10 | 20 | 30 | 40 | 50 | 60 (R10) |

Chainage (m)	0.00	10.00	20.00	30.00	40.00	50.00	60.00
Reduced level	4.400	5.420	6.265	6.895	7.250	7.320	6.585
Formation level	4.400	4.733	5.067	5.400	5.733	6.067	6.400
Cut (+) fill (−)	0.000	0.687	1.198	1.495	1.517	1.253	0.185

Longitudinal section along road R1−R10   Scales:− horz...1:500   vert....1:200

Figure 6.3

a datum, some way below the lowest point of the section. In this case (Fig. 6.3), Ordnance Datum has been chosen, because the levels are all less than 10 m. In other circumstances, a multiple of 10 m would be selected as datum for the section.

4. The horizontal chainages are carefully measured along the datum line and perpendiculars are erected at each chainage point.

5. The reduced level of each station is carefully scaled off along the perpendicular. Each station so established is then joined to the next by a straight line to produce an exaggerated profile of the ground along the section. The lines joining the stations must *not* be drawn as curves since the levels have been taken at all changes of slope and the gradient is therefore constant between any two points.

6. The proposed works are added to the drawing. The formation level of any construction works is the level to which the earth is excavated or deposited to accommodate the works. In Fig. 6.3 the formation level of the new roadway at R1 is 4.400 m. The roadway is to rise at a gradient of 1 in 30 from R1 at chainage 0 m to R10 at chainage 60 m. Since the profile is exaggerated true gradients cannot be shown in the section.

Formation level R1 is plotted. The formation level of a second point, R10 (chainage 60 m), is calculated and also plotted:

Rise from R1 to R10 = 1/30th of 60 m = 2.000 m

Therefore

$$\text{formation level R10} = \text{FL.R1} + 2.000 \text{ m}$$
$$= 4.400 + 2.000 \text{ m}$$
$$= 6.400 \text{ m}$$

When joined, the line between the points represents the gradient. The depth of the excavation required to accommodate the roadway is scaled from the section, the depths being the distance between the surface and the line representing the roadway formation level. Since all scaling and calculations are performed on this section, it is called the working drawing.

7. The presentation drawing is then compiled by tracing the working drawing on to a plastic sheet or piece of tracing paper in black ink only. From this tracing any number of prints can be obtained.

(c) Calculation of cut and fill

In addition to being scaled from the section, the depths of cut or height to fill are also calculated. The method of calculation is similar for all vertical sections. Generally, it consists of:

1. Calculating the reduced level at each chainage point.
2. Calculating the proposed level of the new works at each chainage point.
3. Subtracting one from the other. Where the surface level is higher than the proposed level, there must be cutting and where the proposed level exceeds the surface level, filling will be required.

The complete calculation is as follows:

1. Surface levels, obtained from Table 6.2, i.e. the field book reduction.
2. Proposed levels.

Example (continued)

1 (b) Determine the formation levels of the roadway, given that:
  (i) the roadway gradient is 1 in 30 rising,
  (ii) formation level of the road is 4.400 m at chainage 0.00 m.

Answer
  Gradient of roadway R1–R10 = 1 in 30 rising

  Rise over any length = 1/30th of length
  Therefore rise over 10 m = 1/30 × 10.00
                            = 0.333 m

  Formation level at 0.00 m
              = 4.400   = 4.400
  Formation level at 10.00 m
              = 4.400 + 0.333 = 4.733
  Formation level at 20.00 m
              = 4.400 + 0.667 = 5.067
  Formation level at 30.00 m
              = 4.400 + 1.000 = 5.400
  Formation level at 40.00 m
              = 4.400 + 1.333 = 5.733
  Formation level at 50.00 m
              = 4.400 + 1.667 = 6.067
  Formation level at 60.00 m
              = 4.400 + 2.000 = 6.400

---

3. Cutting or filling. At each chainage point, the proposed level is less than, or equal to, the existing ground level. The road will therefore be in cutting for the whole of its length, e.g. at chainage point 20 m:

Depth of cutting = (ground level − formation level)
Therefore depth = 6.265 − 5.067
              = 1.198 m

All of the calculations are tabulated somewhere on the sectional drawing, either as a separate table (Table 6.3) or, more commonly, accompanying the longitudinal section (Fig. 6.3).

Example (continued)

1 (c) Determine the depths of cut or fill of the earthworks.

Answer
Table of cut and fill (Table 6.3)

Chainage	Reduced level	Formation level	Cut (+) and fill (−)
0.000	4.400	4.400	0.000
10.000	5.420	4.733	0.687
20.000	6.265	5.067	1.198
30.000	6.895	5.400	1.495
40.000	7.250	5.733	1.517
50.000	7.320	6.067	1.253
60.000	6.585	6.400	0.185

Table 6.3

Example

2 Table 6.4 shows the readings obtained during the levelling along the line of the proposed foul sewer F1–F4 (Fig. 6.1). The levelling commenced on TBM peg B (7.780 m AOD) and readings were taken at 8.0 m intervals along the line.
(a) Calculate the reduced levels of the ground at the various chainage points.
(b) Draw a longitudinal section along the line F1–F4, showing clearly:
  (i) the ground surface,
  (ii) the proposed sewer, rising from the existing invert level (2.950 m AOD) of manhole F1 at a gradient of 1 in 40.

Answer (Table 6.5 and Fig. 6.4)

3. Cross sections

(a) Plotting level cross sections

It may not be necessary actually to observe the levels in the field. The ground across the centre line at any chainage point may be level or nearly so, in which case the centre line level is assumed to apply across the line of the section.

The plotting is very similar to the plotting of longitudinal sections. One essential difference, however, is that the cross section is plotted to a natural scale, i.e. the horizontal and vertical scales are the same.

BS	IS	FS	Rise or fall	Reduced level	Distance	Remarks
0.120				7.780		Peg B (7.78 m AOD)
	1.240				0.000	Manhole F1 (existing)
	1.310				8.000	Centre line of proposed sewer
	1.150				16.000	Centre line of proposed sewer
1.250		1.490			24.000	Proposed manhole F2
	1.390				32.000	Centre line of proposed sewer
	1.460				40.000	Centre line of proposed sewer
	1.790				48.000	Proposed manhole F3
	1.860				56.000	Centre line of proposed sewer
0.850		1.640			64.000	Proposed manhole F4
		1.320				Peg D (5.555 m AOD)

Table 6.4

BS	IS	FS	Rise or fall	Reduced level	Distance	Remarks
0.120				7.780		Peg B (7.78 m AOD)
	1.240		−1.120	6.660	0.000	Manhole F1 (existing)
	1.310		−0.070	6.590	8.000	Centre line of proposed sewer
	1.150		0.160	6.750	16.000	Centre line of proposed sewer
1.250		1.490	−0.340	6.410	24.000	Proposed manhole F2
	1.390		−0.140	6.270	32.000	Centre line of proposed sewer
	1.460		−0.070	6.200	40.000	Centre line of proposed sewer
	1.790		−0.330	5.870	48.000	Proposed manhole F3
	1.860		−0.070	5.800	56.000	Centre line of proposed sewer
0.850		1.640	0.220	6.020	64.000	Proposed manhole F4
		1.320	−0.470	5.550		Peg D (5.555 m AOD)
2.220 −4.450 −2.230		4.450	−2.230	7.780 −5.550 2.230		

Table 6.5

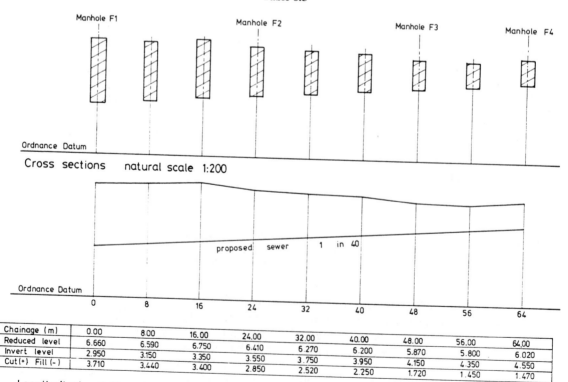

Cross sections   natural scale 1:200

Chainage (m)	0.00	8.00	16.00	24.00	32.00	40.00	48.00	56.00	64.00
Reduced level	6.660	6.590	6.750	6.410	6.270	6.200	5.870	5.800	6.020
Invert level	2.950	3.150	3.350	3.550	3.750	3.950	4.150	4.350	4.550
Cut (+) Fill (−)	3.710	3.440	3.400	2.850	2.520	2.250	1.720	1.450	1.470

Longitudinal section along sewer F1-F4   Scale   horz :-1:500   vert. :- 1:200

Figure 6.4

In Example 1 (Fig. 6.3), the proposed roadway is to be 4 metres wide and any earthworks are to be formed with sides sloping at 45° (i.e. 1 unit vertically to 1 unit horizontally). The cross sections are drawn as follows:

1. A line, representing the section datum (in this case, 4.00 m AOD), is drawn above the longitudinal section.
2. The construction lines of each chainage point of the longitudinal section are extended vertically to cut this datum and act as centre lines for the cross sections.
3. The ground level and the formation level of the roadway are plotted from the datum on the centre lines and horizontal lines drawn lightly through both points.
4. The width of the road is marked on the line representing the formation level and the sloping sides of the cutting drawn at 45° to cut the line representing ground level.
5. The trapezium representing the road cutting is drawn boldly (Fig. 6.3).

In Example 2 (Fig. 6.4), the sewer is to be 0.8 metre wide and the excavation is to have vertical sides. The sections are plotted as follows:

1. A line representing the section datum of 0.00 m is drawn lightly above the longitudinal section.
2. The construction lines of the chainage points of the longitudinal section are extended vertically to cut this datum and act as centre lines for the cross sections.
3. The ground level and the sewer invert level are plotted from the datum on the centre line and horizontal lines drawn lightly through both points.
4. The width of the sewer (0.80 m) is marked and the vertical sides of the drain drawn to cut the horizontals in (3) above.
5. The rectangle denoting the excavation is drawn boldly.

Exercise 6.1

1 Table 6.6 shows the table of levels obtained during the sectional levelling of the proposed roadway from point R3 to point R14 (Fig. 6.1).
(a) Reduce the levels.
(b) Determine the formation levels of the roadway at the various chainage points given that the roadway formation level at point R3, i.e. chainage 18.00 m, is 5.00 m AOD:

$$FL\ R3 = FL\ R2 + (1/30 \times 18)$$
$$= 4.400 + 0.600$$
$$= 5.000\ m$$

and determine the depth of cut or fill of the earthworks at the various chainage points given that the gradient of the road from R3 to R14 is to be 1 in 100 rising.
(c) Draw a longitudinal section of the roadway, from R3 to R14, at any suitable scale.
(d) Draw cross sections at chainages 20, 40 and 60 m, assuming that the ground is level across the section and given that the earthworks have side slopes of 45°.

2 Table 6.7 shows the reduced levels obtained during the levelling along the line of the proposed storm water sewer S1–S4 (Fig. 6.1). The levelling commenced on TBM C (5.125 m AOD) and readings were taken at 10.00 m intervals along the line.
(a) Reduce the levels.
(b) Draw a longitudinal section along the line S1–S4 showing clearly:
   (i) the ground surface and
   (ii) the proposed sewer, rising from the existing invert level (2.197 m AOD) of manhole S1, to manhole S3 at a gradient of 1 in 30 and then at such a gradient that the invert level of the pipe in manhole S4 is 1.200 m below the ground level at the manhole.
(c) Calculate the gradient of the pipe between manholes S3 and S4.

BS	IS	FS	Rise or fall	Reduced level	Distance	Remarks
1.630				5.555		Peg D (5.555 m AOD)
	1.055				18.000	Centre line of road R3
	0.920				20.000	Centre line of road
	0.715				30.000	Centre line of road
1.230		1.005			40.000	Centre line of road
	1.100				50.000	Centre line of road
	1.260				60.000	Centre line of road
	1.890				70.000	Centre line of road R14
		1.855				Peg D

Table 6.6

BS	IS	FS	Rise or fall	Reduced level	Distance	Remarks
1.325				5.125		TBM C (5.125 m AOD)
	1.550				0.000	Proposed manhole S1
	1.600				10.000	Centre line of sewer
	1.550				20.000	Proposed manhole S2
	1.675				30.000	Centre line of sewer
	1.150				40.000	Proposed manhole S3
	0.995				50.000	Centre line of sewer
	1.220				60.000	Proposed manhole S4
		0.895				TBM D (5.555 m AOD)

Table 6.7

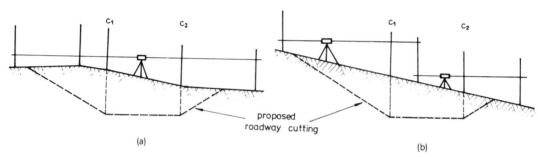

Figure 6.5

(d) Draw cross sections at every chainage point assuming that the ground is level across the section and given that:
  (i) the width of the sewer track it 0.8 m and
  (ii) the trench has vertical sides.

(b) Sloping cross sections

In cases where the ground across the centre line at any chainage point is obviously not flat, the following fieldwork is required to obtain levels for plotting the cross sections.

Fieldwork

Cross sections are taken at right angles to the longitudinal section at every point observed on the latter. Generally, this rule is not strictly observed and cross sections are taken usually at every tape length. The following fieldwork is necessary:

1. Right angles are set out, using a simple hand instrument, e.g. prism square or optical square. Where the ground is relatively flat, the right angle may be judged by eye. A ranging pole is inserted at either side of the centre line on the line of the cross section.
2. A levelling must be made from the peg previously established on the centre line of the longitudinal section, to every point where the gradient changes on the line of the cross section (Fig. 6.5). In addition, levels are always taken on either side of the centre line at points C_1 and C_2 denoting the formation width.

Where the ground is relatively flat, one instrument setting is usually sufficient (Fig. 6.5(a)). If the cross-gradient is steep, a short series levelling is required (Fig. 6.5(b)).

Each cross section is independent of every other. The peg on the centre line at each tape length is a temporary bench mark for its particular cross section.
3. Horizontal measurements must be made between all points at which levels are taken to cover the total width of the proposed works.
4. *Procedure*
The procedure is much the same as that for longitudinal sections. One surveyor and three assistants is the ideal number of personnel required. However, since the distances involved are short, the surveyor and one assistant frequently take the levels and measurements without further assistance.

Example

3 In Fig. 6.1, the ground appears to have a cross slope, from north to south, between chainages 40 and 60 m of the proposed roadway R3–R14. Cross-sectional readings were taken, in addition to the centre line readings, during the roadway levelling (Table 6.8), at chainages 40, 50 and 60 m. Reduce the levels for each cross section, in preparation for the plotting of these sections.

Answer (Table 6.9)

BS	IS	FS	HPC	Reduced level	Distance	Remarks
1.630			7.185	5.555		Peg D (5.555 m AOD)
1.230		1.005			40.000	Centre line of proposed road
	0.850					4 m left of centre line
	0.910					2 m left of centre line
	1.310					2 m right of centre line
	1.710					4 m right of centre line
	1.100				50.000	Centre line of proposed road
	0.890					4 m left of centre line
	1.110					2 m left of centre line
	1.390					2 m right of centre line
	1.600					4 m right of centre line
1.390		1.260			60.000	Centre line of proposed road
	1.025					4 m left of centre line
	0.960					2 m left of centre line
	1.490					2 m right of centre line
	1.590					4 m right of centre line
	2.020				70.000	Centre line of proposed road
		1.980				Peg D (5.555 m AOD)

Table 6.8

BS	IS	FS	HPC	Reduced level	Distance	Remarks
1.630			7.185	5.555		Peg D (5.555 m AOD)
1.230		1.005	7.410	6.180	40.000	Centre line of proposed road
	0.850			6.560		4 m left of centre line
	0.910			6.500		2 m left of centre line
	1.310			6.100		2 m right of centre line
	1.710			5.700		4 m right of centre line
	1.100			6.310	50.000	Centre line of proposed road
	0.890			6.520		4 m left of centre line
	1.110			6.300		2 m left of centre line
	1.390			6.020		2 m right of centre line
	1.600			5.810		4 m right of centre line
1.390		1.260	7.540	6.150	60.000	Centre line of proposed road
	1.025			6.515		4 m left of centre line
	0.960			6.580		2 m left of centre line
	1.490			6.050		2 m right of centre line
	1.590			5.950		4 m right of centre line
	2.020			5.520	70.000	Centre line of proposed road
		1.980		5.560		Peg D (5.555 m AOD)

Table 6.9

Plotting

The plotting is very similar to the plotting of longitudinal sections. One essential difference, however, is that the cross section is plotted to a natural scale, i.e. the horizontal and vertical scales are the same. The plotting is carried out in the following manner:

1. The levels are reduced (Table 6.9).
2. The longitudinal section is plotted (Fig. 6.6).
3. A line representing some chosen height above datum is drawn for each cross section and the measurements to left and right of the centre line are scaled off accurately (Fig. 6.6).
4. Perpendiculars are erected at each point so scaled and the reduced levels of each point plotted.

5. The points are joined to form a natural profile of the ground.
6. The formation level of the proposed works is obtained from the table accompanying the longitudinal section. In the data table of Fig. 6.6, the formation level at chainage 40 m is 5.22 m. This level is accurately plotted on the cross section and a horizontal line, representing the roadway, is drawn through the point.
7. The finished width of the road, called the formation width, is marked on the line and the side slopes of the cutting are added. In Fig. 6.6, the slope is 45°. The finished shape of the cutting is then drawn boldly.

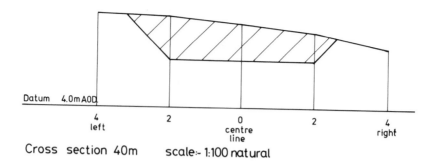

Datum 4.0m AOD

| 4 | 2 | 0 | 2 | 4 |
| left | | centre line | | right |

Cross section 40m   scale:- 1:100 natural

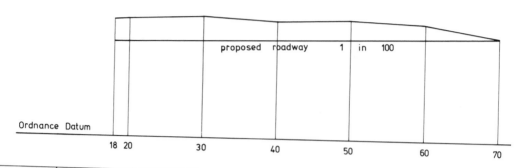

Ordnance Datum

| 18 20 | 30 | 40 | 50 | 60 | 70 |

Chainage (m)	18.00	20.00	30.00	40.00	50.00	60.00	70.00
Reduced level	6.130	6.265	6.470	6.180	6.310	6.150	5.520
Formation level	5.000	5.020	5.120	5.220	5.320	5.420	5.520
Cut (+) Fill (−)	1.130	1.245	1.350	0.860	0.990	0.630	0.000

Longitudinal section along roadway R3–R14   Scale   horz:- 1:500 vert. 1:200

Figure 6.6

(c) Embankments

The earlier parts of this section dealt with the plotting of cross sections. All of the examples and exercises have concentrated on the plotting of cuttings, in order to preserve continuity of theme. However, earthworks equally involve the plotting of embankments and are exemplified in Examples 4 and 5.

When earthworks involve embankments, the formation level of the proposed works is higher than the actual ground level, so the side slopes of the works must slope downwards from the formation level to the ground level. In all other aspects, the plotting procedure is as already described in the earlier parts of this section.

Exercise 6.2

1 Table 6.9 shows cross-sectional levels taken at chainages 40, 50 and 60 m of the proposed roadway, R3–R14. (Cross section 40 m is plotted in Fig. 6.6.)
(a) Plot the cross sections 50 and 60 m to a natural scale of 1:100.
(b) Add the cross sections of the proposed roadways to the drawing given that:
   (i) road width = 4.0 m,
   (ii) side slopes are at 45°,
   (iii) FL 50 m = 5.320 m AOD and FL 60 m = 5.420 m AOD.

(d) Further examples of levelling and sectioning

Examples

4 The following reduced levels were obtained along the centre line of a proposed road between two points A and B.

Chainage (m)	Reduced level (m)	Chainage (m)	Reduced level (m)
A 0	83.50	50	82.45
10	83.84	60	82.20
20	84.06	70	82.41
30	83.66	80	82.70
40	83.30	B 90	83.05

The roadway is to be constructed so that there is one regular gradient between points A and B.
(a) Draw a longitudinal section along the centre line at a horizontal scale of 1:1000 and a vertical scale of 1:100.
(b) Determine from the section the depth of cutting or height of fill required at each chainage point to form a new roadway.
(c) Check the answers by calculation.

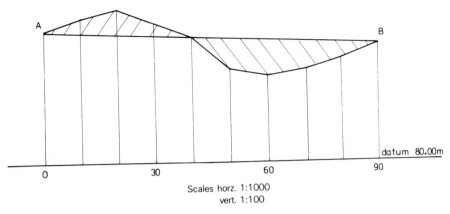

Figure 6.7

Answer
(a) The longitudinal section is shown in Fig. 6.7.

(b) *Gradient of AB*

Reduced level of A = 83.50 m
Reduced level of B = 83.05 m
Difference = 0.45 m
Distance A to B = 90 m
Therefore gradient A to B falling = 0.45 in 90
= 1 in 200

Proposed levels (formation levels):
Fall over 10 metres = 0.05 m
Therefore proposed level at chainage
10 m = 83.50 − 0.05 = 83.45 m
Proposed level at chainage
20 m = 83.45 − 0.05 = 83.40 m etc.

Chainage (m)	Reduced level (m)	Proposed level (m)	Cut (m)	Fill (m)
A 0	83.50	83.50		
10	83.84	83.45	0.39	
20	84.06	83.40	0.66	
30	83.66	83.35	0.31	
40	83.30	83.30	—	—
50	82.45	83.25		0.80
60	82.20	83.20		1.00
70	82.41	83.15		0.74
80	82.70	83.10		0.40
90 B	83.05	83.05		—

5 Cross sections are required at chainages 20, 40 and 60 m in Example 4 above. Levellings were made, and the results tabulated in the field book are shown in Table 6.10.

Using a natural scale of 1:100 draw cross sections at 20, 40 and 60 m chainages, given that the formation width of the roadway is 5 m and that the sides of any cuttings or embankments slope at 30° to the horizontal.

Answer
The cross sections are shown in Fig. 6.8. The method of plotting is briefly as follows:
(i) Draw arbitrary datum lines of 82.00 m.
(ii) Plot centre line, 5 m left and 5 m right along the datum lines
(iii) Plot the reduced levels of these points and join them to produce the surface profiles.
(iv) Plot the formation levels at 20, 40 and 60 m. These are 83.40, 83.30 and 83.20 respectively from the calculations of Example 4.
(v) Plot the formation widths, 2.5 m on both sides of the centre line.
(vi) Draw the side slopes at 30° to the horizontal. Where the reduced levels exceed the formation level, a cutting is produced as in Fig. 6.8(a). In Fig. 6.8(c) the formation level exceeds the reduced level producing

BS	IS	FS	HP collimation	Reduced level	Distance	Remarks
1.52			85.58	84.06	20	Peg on centre line
	0.88			84.70	20	5 m left of centre line
		2.02		83.56	20	5 m right of centre line
1.61			84.91	83.30	40	Peg on centre line
	0.56			84.35	40	5 m left of centre line
		2.34		82.57	40	5 m right of centre line
1.47			83.67	82.20	60	Peg on centre line
	0.67			83.00	60	5 m left of centre line
		1.66		82.01	60	5 m right of centre line

Table 6.10

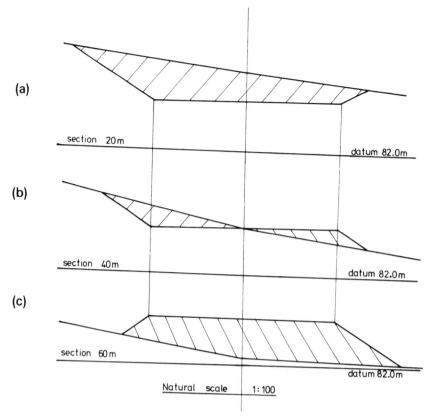

(a)

section 20m

datum 82.0m

(b)

section 40m

datum 82.0m

(c)

section 60m

datum 82.0m

Natural scale | 1:100

Figure 6.8

an embankment, while in Fig 6.8(b) the cross section is part cutting and part embankment.

Exercise 6.3

1 The following readings (Table 6.11) were taken on the ground surface, along the route of a proposed roadway and sewer XY.

The roadway is to be constructed on a regular gradient of 1 in 20 rising from the existing formation level of point X (chainage 0 m) towards point Y (chainage 80 m). A sewer is to be constructed along the centre line of the roadway and is to connect into

the manhole X (where the existing invert level is 91.20 m AD) at a gradient of 1 in 40.

The roadway is to be 4.0 m wide and earthworks have side slopes of 45°. The sewer is to be 0.8 m wide and the trench is to be vertical.

(a) Book and reduce the levels.
(b) Draw a longitudinal section along the roadway and sewer (scales: horizontal 1:500, vertical 1:100).
(c) Draw natural cross sections at the various chainage points (scale 1:100) to show separately:
   (i) the roadway earthworks and
   (ii) the sewer trench.

2 Table 6.12 levels taken along the centre line of a proposed sewer. The sewer at chainage 0 m has an

BS	1.370 m on BM 28 (RL 91.085 m AOD)
FS	0.035 m at X (manhole, start of section), 0.00 m chainage
BS	2.795 m to X
IS	3.040 m (chainage 20 m); IS 1.765 (chainage 40 m)
FS	0.565 m (chainage 60 m)
BS	3.865 m (chainage 60 m)
IS	2.095 m at Y (end of section) (chainage 80 m)
FS	1.075 m to change point
BS	2.240 m to change point
FS	0.385 m to BM 29 (RL 99.285 m AOD)

Table 6.11

BS	IS	FS	Rise	Fall	Reduced level	Distance	Remarks
1.670					92.550		BM
1.520		3.870				0	Ground level
	0.910					10	Ground level
	1.590					20	Ground level
	1.770					30	Ground level
	1.660					40	Ground level
	−4.200					50	Underside of bridge
		0.720				50	Ground level

Table 6.12

BS	IS	FS	Rise	Fall	Reduced level	Distance	Remarks
2.580						0	At A
	2.340					15	
	2.640					30	
	2.130					45	
0.830		2.580				60	
	0.930				57.750	62.5	At B
	0.340					75	
	1.790					90	
1.300		2.530				105	
	0.660					120	
	1.920					135	
		1.100				150	At C

Table 6.13

invert level of 88.900 and is to fall towards chainage 50 m at a gradient of 1 in 100.

(a) Reduce the levels and apply the appropriate checks.

(b) Draw a vertical section along the centre line of the sewer on a horizontal scale of 1:500 and vertical scale of 1:100.

(c) From the section determine the depth of cover at each chainage point.

(d) A mechanical excavator being used to form the sewer requires a minimum working height of 4.50 m. What clearance (if any) will it have when working below the bridge?

(e) Given that the sewer track is 0.60 m wide with vertical sides, draw cross sections to a scale of 1:100 to show the excavation at each chainage point.

3 (a) Table 6.13 shows notes made during a levelling for the construction of a new road. Complete the table and show all checks.

(b) The roadway is to be graded from A to C, both of whose present levels are to remain unaltered. Calculate the gradient AC and the depth of cut or fill required at B.

(c) Choose any suitable scales and draw a vertical section from A to C.

(d) Draw cross sections at the points of maximum cut and maximum fill to show the road construction given that the road is 5 m wide with 45° side slopes.

4. Answers

Exercise 6.1

1 (a) Table 6.14 and Fig. 6.9

BS	IS	FS	Rise or fall	Reduced level	Distance	Remarks
1.630				5.555		Peg D (5.555 m AOD)
	1.055		0.575	6.130	18.000	Centre line of road R3
	0.920		0.135	6.265	20.000	Centre line of road
	0.715		0.205	6.470	30.000	Centre line of road
1.230		1.005	−0.290	6.180	40.000	Centre line of road
	1.100		0.130	6.310	50.000	Centre line of road
	1.260		−0.160	6.150	60.000	Centre line of road
	1.890		−0.630	5.520	70.000	Centre line of road R14
		1.855	0.035	5.555		Peg D

Table 6.14

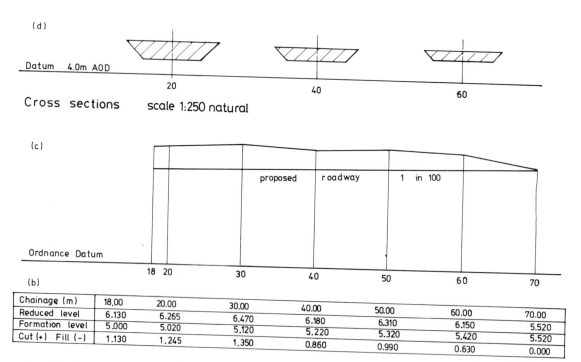

(d)

Datum 4.0m AOD

20   40   60

Cross sections   scale 1:250 natural

(c)

proposed   roadway   1 in 100

Ordnance Datum

18 20   30   40   50   60   70

(b)

Chainage (m)	18.00	20.00	30.00	40.00	50.00	60.00	70.00
Reduced level	6.130	6.265	6.470	6.180	6.310	6.150	5.520
Formation level	5.000	5.020	5.120	5.220	5.320	5.420	5.520
Cut (+) Fill (−)	1.130	1.245	1.350	0.860	0.990	0.630	0.000

Longitudinal section along roadway R3 - R14   Scale   horz. 1:500 vert. 1:200

Figure 6.9

2 (a) Table 6.15

BS	IS	FS	Rise or fall	Reduced level	Distance	Remarks
1.325				5.125		TBM C (5.125 m AOD)
	1.550		−0.225	4.900	0.000	Proposed manhole S1
	1.600		−0.050	4.850	10.000	Centre line of sewer
	1.550		0.050	4.900	20.000	Proposed manhole S2
	1.675		−0.125	4.775	30.000	Centre line of sewer
	1.150		0.525	5.300	40.000	Proposed manhole S3
	0.995		0.155	5.455	50.000	Centre line of sewer
	1.220		−0.225	5.230	60.000	Proposed manhole S4
		0.895	0.325	5.555		TBM D (5.555 m AOD)

Table 6.15

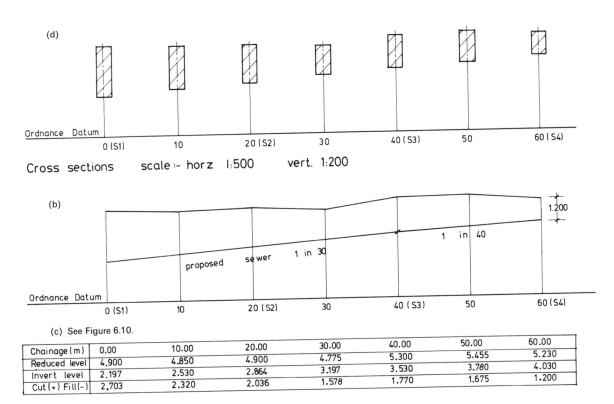

(d)

Ordnance Datum

0 (S1)   10   20 (S2)   30   40 (S3)   50   60 (S4)

Cross sections   scale :- horz 1:500   vert. 1:200

(b)

1.200

Ordnance Datum

proposed sewer 1 in 30   1 in 40

0 (S1)   10   20 (S2)   30   40 (S3)   50   60 (S4)

(c) See Figure 6.10.

Chainage (m)	0.00	10.00	20.00	30.00	40.00	50.00	60.00
Reduced level	4.900	4.850	4.900	4.775	5.300	5.455	5.230
Invert level	2.197	2.530	2.864	3.197	3.530	3.780	4.030
Cut (+) Fill (−)	2.703	2.320	2.036	1.578	1.770	1.675	1.200

Longitudinal section along sewer S1 to S4   Scale   horz 1:500   vert. 1:200

Note : Gradient of sewer from mh. S3 to mh. S4 = 0.5m rise in 20m = 1 in 40

Figure 6.10

Exercise 6.2

1 Figure 6.11

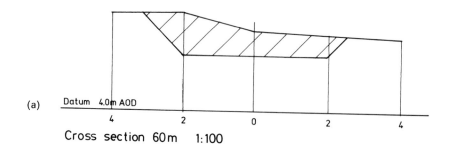

(a)   Datum   4.0m AOD

4   2   0   2   4

Cross section 60m   1:100

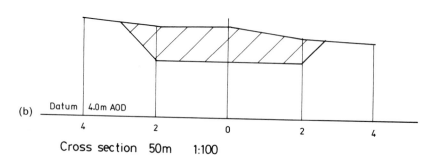

(b)   Datum   4.0m AOD

4   2   0   2   4

Cross section   50m   1:100

Figure 6.11

Exercise 6.3

1 (a) Table 6.16

BS	IS	FS	HPC	Reduced level	Remarks
1.370			92.455	91.085	BM 28
2.795		0.035	95.215	92.420	× manhole chainage 0.00 m
	3.040			92.175	Chainage 20 m
	1.765			93.450	Chainage 40 m
3.865		0.565	98.515	94.650	Chainage 60 m
	2.095			96.420	Chainage 80 m
2.240		1.075	99.680	97.440	Change point
		0.385		99.295	BM 29

Table 6.16

(c)(i)

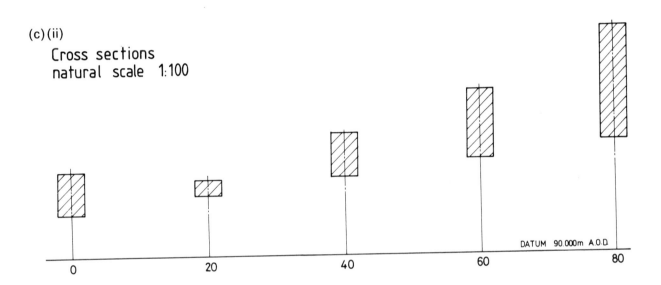

DATUM 90.000m A.O.D.

20   40   60

(c)(ii)

Cross sections
natural scale 1:100

DATUM 90.000m A.O.D.

0   20   40   60   80

(b)

Longitudinal section XY
scales: horz. 1:500
       vert. 1:100

proposed roadway

proposed sewer

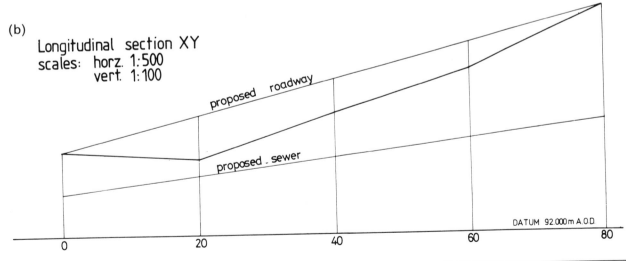

DATUM 92.000m A.O.D.

0   20   40   60   80

Chainage	0	20	40	60	80
Reduced level	92.420	92.175	93.450	94.650	96.420
Formation level	92.420	93.420	94.420	95.420	96.420
Fill	0	1.245	0.970	0.770	0
Invert level	91.200	91.700	92.200	92.700	93.200
Cut	1.220	0.475	1.250	1.950	3.220

Figure 6.12

2 (a)

Chainage	0	10	20	30	40	50 (Bridge)	50
Reduced level	90.35	90.96	90.28	90.10	90.21	96.07	91.15

(b) See Fig. 6.13

(c)

Chainage (m)	0	10	20	30	40	50
Cover (m)	1.45	2.16	1.58	1.50	1.71	2.75

(d) Clearance 0.42 m

(e) See Fig. 6.13.

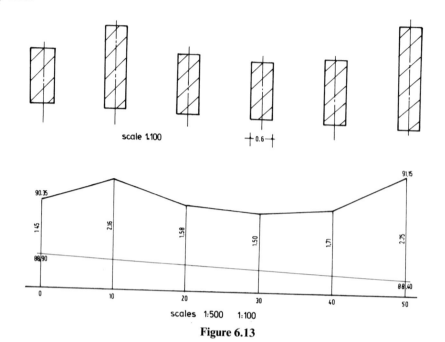

Figure 6.13

3 (a)

Chainage	0 (A)	15	30	45	60	62.5 (B)	75	90	105	120	135	150 (C)
Reduced level	57.85	58.09	57.79	58.30	57.85	57.75	58.34	56.89	56.15	56.79	55.53	56.35

(b) Gradient 1 in 100 falling; cut at B, 0.525 m

(c) See Fig. 6.14.

(d) See Fig. 6.14.

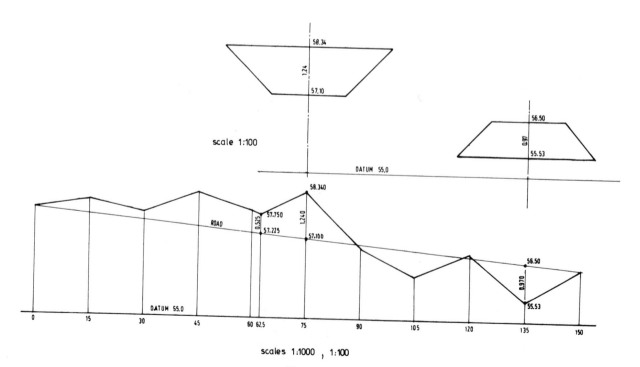

Figure 6.14

5. Project

Chapter 18 is a project covering all chapters of this textbook. It is intended that the project build into a complete portfolio of the surveying work required in the survey and the setting out of a building or engineering development.

If the reader wishes to continue work on the project or begin work at this stage, he or she should now turn to page 393 and attempt Section 5.

CHAPTER 7

Theodolites

Objective

After studying this chapter, the reader should be able to (a) select the most appropriate method of measuring horizontal and vertical angles, (b) measure and record these angles and determine their most probable values and (c) understand the errors that affect angle measurement and minimize their effects.

In succeeding chapters of this book, the topics of traversing, tacheometry, radial positioning and setting out structures will be considered in detail. In all of those topics, the subject of angular measurement will be a common factor. It is therefore essential that the surveyor has a sound knowledge of the instruments and methods used in the measurement of horizontal and vertical angles.

Until comparatively recently, angles were measured on surveys by compass as well as, though much less accurately than, by theodolite. The compass, however, is now obsolescent and is used only rarely, e.g. in heavily overgrown or in thickly wooded areas, where an approximate solution is sufficient.

The compass has one major advantage over the theodolite, namely that it can, of course, find the bearing, i.e. the direction, of a line relative to magnetic north. A conventional theodolite cannot be used to find north without the addition of a compass needle attachment which, of course, really turns it into a glorified compass.

The theodolite is an instrument that is used to measure angles, with an accuracy varying from 1 to 60 seconds of arc. It is initially a difficult instrument to use but a working knowledge can be gained, fairly quickly, if the instrument is, diagrammatically, broken into its component parts and each examined carefully.

1. Classification

A theodolite is generally classified according to the method used to read the circles. Broadly speaking, the methods are:

1. Vernier (now obsolete)
2. Direct reading
3. Optical scale
4. Optical micrometer
5. Opto-electronic

Despite this multiplicity of reading systems the basic principles of construction of the theodolite are similar.

2. Principles of construction

The main components of a theodolite are illustrated in Fig. 7.1. These are as follows.

(a) Tripod

The purpose of the tripod is to provide support for the instrument. Tripods may be telescopic, i.e. they have sliding legs, or may have legs of fixed length.

(b) Trivet stage

The trivet stage is the flat base of the instrument which screws on to the tripod and carries the feet of the levelling screws.

(c) Tribrach

The tribrach is the body of the instrument carrying all the other parts. The tribrach has a hollow, cylindrical, socket into which fits the remainder of the instrument (Fig. 7.1). All modern theodolites have hardened steel cylindrical axes. A cylindrical ball race takes the weight of the upper part of the instrument.

(d) Levelling arrangement

To enable the tribrach to be levelled, levelling screws are fitted between the tribrach and trivet stage. Movement of the footscrews centres the bubble of the plate spirit level, situated on the cover plate of the horizontal circle. The sensitivity of the spirit level is of the order of 2 mm = 40 seconds of arc.

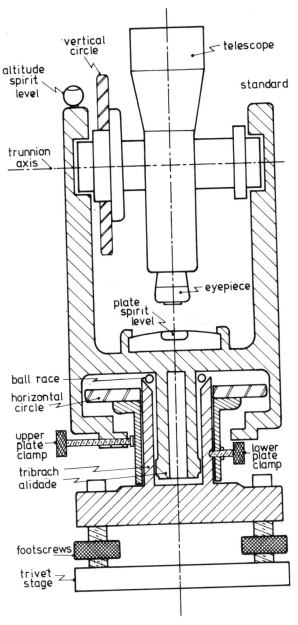

Figure 7.1

(e) Horizontal circle (lower plate)

The horizontal circle is in reality a protractor graduated in a clockwise direction and numbered from 0° to 360°. Modern instruments have glass circles. The horizontal circle is mounted on a cylindrical axis which fits around the outside of the tribrach (Fig. 7.1).

The circle is therefore free to rotate around the tribrach and can be stopped in any position by applying the lower plate clamp. When tightened, the clamp locks the horizontal circle (lower plate) to the tribrach. A very limited amount of horizontal movement is still possible via the slow motion screw

attached to the clamp. The slow motion screw only works when the lower plate clamp is locked.

(f) Alidade (upper plate)

The alidade is the remainder of the theodolite comprising the uprights which support the telescope and vertical circle and the spirit levels.

In Fig. 7.1, the alidade is carried on a central spindle, which fits within the hollow socket of the tribrach. It is therefore free to rotate with respect to the horizontal circle, which is itself free to rotate around the tribrach sleeve, as has already been explained.

(g) Controls for measuring horizontal angles

1. *Double centre system*

This uses upper and lower plate clamps. Figure 7.1 shows the double centre axis system and Fig. 7.2(b) shows the two clamps. It is essential that the function of the upper and lower plate clamps be understood since they control the entire operation of measuring a horizontal angle.

When both clamps are open the lower plate (horizontal circle) and upper plate (alidade) are free to move in any direction relative to the tribrach and to each other (Fig. 7.3).

When the lower plate clamp is closed, the horizontal circle is locked in position and the alidade is free to rotate inside the stationary circle. The reading on the horizontal circle will therefore change continuously.

When both clamps are closed, neither plate can rotate. If the lower plate clamp is now released the upper and lower plates will move together and there will be no change in reading on the horizontal circle. The instrument can be used in the repetition and reiteration measurement of angles.

2. *Circle-setting screw*

Some theodolites do not have a lower plate clamp. The alidade is clamped directly to the tribrach (e.g. Watts 1, Kern DKM 1). During measurement the horizontal circle remains stationary while the alidade moves over it in the usual way. However, the circle can be moved by a continuous-drive circle-setting screw and the instrument may therefore be easily set to, say, 00° 00′ 00″ for any pointing of the telescope (Fig. 7.4). The instrument cannot measure angles by the repetition method.

3. *Repetition clamp system*

Some theodolites (WILD T16) are fitted with a repetition clamp instead of a lower plate screw. When the clamp is in the closed position the horizontal circle rests against the tribrach. When the clamp is opened the circle is clamped to the alidade (Fig. 7.5). The upper plate clamp connects the alidade directly to the tribrach and the instrument may be used for either repetition or reiteration measurements.

(a) Watts ST 20 Vernier theodolite (now obsolete) included here for comparative purposes only

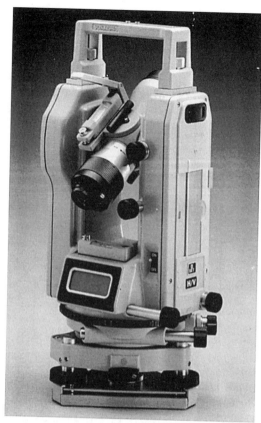

(b) Sokkia DT 20E electronic theodolite showing upper and lower clamps

(c) Leica KOS theodolite showing repetition clamp

(d) Watts ST 200 theodolite showing circle-setting screw

Figure 7.2

1 Alidade
2 Horizontal circle
3 Tribrach

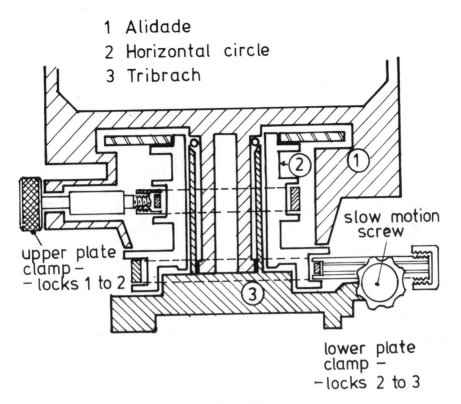

Figure 7.3

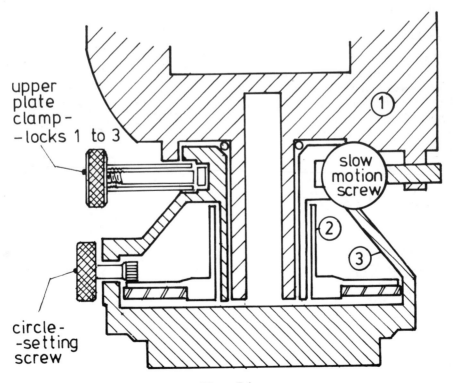

Figure 7.4

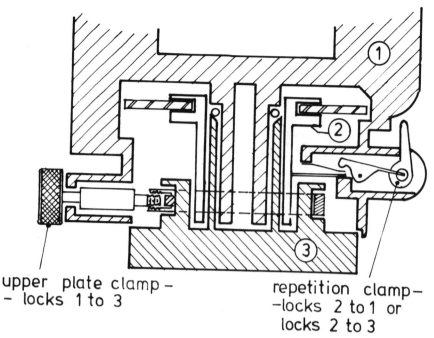

upper plate clamp –
– locks 1 to 3

repetition clamp–
–locks 2 to 1 or
locks 2 to 3

Figure 7.5

(h) Index marks

In order to read the circle for any pointings of the telescope it is convenient to imagine an index mark mounted on the alidade directly below the telescope. As the alidade is rotated, the index mark moves over the horizontal circle. When the alidade is locked, the index mark is read against the circle. Actually the index mark is a line etched on a glass plate, somewhere in the optical train of the theodolite.

When measuring a horizontal angle, e.g. angle PQR in Fig. 7.6, the theodolite is set over point Q and the lower plate is locked in any random position. The upper plate clamp is released and the telescope mounted on the alidade is turned to point in turn to stations P and R. A horizontal circle reading is taken for both telescope pointings and subtracted to give the horizontal angle.

(i) Transit axis or trunnion axis

The transit axis rests on the limbs of the standards and is securely held in position by a locknut. Attached to the transit axis are the telescope and vertical circle. All three are free to rotate in the vertical plane but can be clamped in any position in the plane by a clamp usually known as the telescope clamp (Fig. 7.7). Again, a certain amount of movement is permitted by a slow-motion device.

The telescope has been described in Chapter 4 and Fig. 4.4 shows the paths of the rays of light through

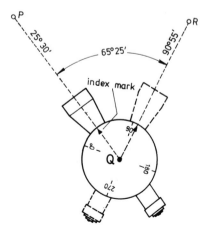

Figure 7.6

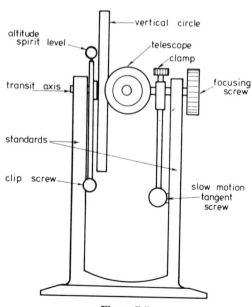

Figure 7.7

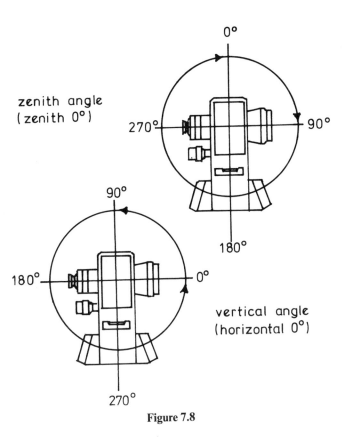

zenith angle
(zenith 0°)

vertical angle
(horizontal 0°)

Figure 7.8

On most modern theodolites automatic indexing is used. The spirit level is replaced with either a pendulum device which operates in similar fashion to an automatic level or the surface of a liquid is used. The reading of the vertical circle is reflected from the always true horizontal surface of the liquid and any deflection of the theodolite vertical axis is automatically compensated.

(k) Centring motion

Since the theodolite must be placed exactly over a survey station, it is fitted with a centring motion fitted usually above the tribrach, which allows the whole of the instrument above the tribrach to move relative to the latter. Since the total amount of movement is only 40 mm, the instrument must be placed very accurately over the survey mark before the centring motion is used.

(l) Optical plummet

On most theodolites an optical plummet is incorporated which greatly aids centring of the instrument, particularly in windy weather.

Figure 7.9 is a section through an optical plummet. When the theodolite is properly set up and levelled the observer is able to view the ground station through the eyepiece of the optical plummet, the line of sight being deflected vertically downwards by the 45° prism incorporated in the plummet. Movement of the centring motion allows the theodolite to be placed exactly over the survey station.

the telescope. A typical specification for a theodolite telescope is:

(a) internal focusing—damp and rust resistant;
(b) shortest focusing distance—2 metres;
(c) magnification of × 30;
(d) object glass diameter, 42 mm;
(e) field of view, 1° 12'.

The vertical circle, 80 mm in diameter, is attached to the telescope and is graduated in a variety of ways, two of which are shown in Fig. 7.8. The angle measured in the vertical plane can therefore be a vertical angle (where the zero degree reading indicates the horizontal position of the telescope) or a zenith angle (where the zero degree reading indicates the vertical position of the telescope).

(j) Altitude spirit level

Angles measured in a vertical plane must be measured relative to a truly horizontal line. The line is that which passes through the centre of the transit axis and is maintained in a horizontal position by the altitude spirit level (Fig. 7.7). The spirit level and index mark is attached to a 'T' frame which is made horizontal by activating the clip screw against the standards. The altitude spirit level is more sensitive than the plate spirit level.

Modern theodolites are fitted with a co-incidence bubble reader which is fully explained in Sec. 1(d) in Chapter 4. This device greatly increases the accuracy of the bubble setting.

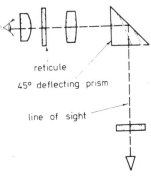

reticule
45° deflecting prism
line of sight

Figure 7.9

3. Reading the circles

The horizontal and vertical circles of modern theodolites are made of glass. The casing and standards are hollow, so it is possible to pass light through them using a suitable arrangement of prisms.

Four methods of reading the graduations on the circles are available. In Great Britain (and in most other parts of the world), the sexagesimal units are used in the measurement of angles; hence all subsequent examples use degrees, minutes and seconds.

(a) Direct reading

The horizontal circle is read through a reading eyepiece attached to one of the standards or alongside the main telescope eyepiece.

In the lower order theodolites, the horizontal circle is graduated at 5 or 10 minute intervals and read by estimation to the nearest minute. Figure 7.10(a) illustrates the reading of the horizontal and vertical circles of the Zeiss (Jena) Theo 080A minute reading theodolite. The circle has primary graduations at 1 degree intervals and secondary graduations at 5 minute intervals. The respective readings of vertical and horizontal circles are 96° 05′ and 359° 29′.

(b) Direct scale reading

Intermediate order theodolites employ an optical scale to read the horizontal and vertical circles.

Figure 7.10(b) illustrates the field of view of the Watts minute reading theodolite. The circles are graduated at 1 degree intervals. The image of any degree graduation is seen in the eyepiece super-imposed on a transparent scale graticule. This scale is graduated at 1 minute intervals and the circle reading is obtained by reading the degree mark against the scale. In the figure the readings of vertical and horizontal scales are 04° 47′ and 125° 19.5′ respectively.

Figure 7.10(c) shows the circle readings of the Kern KIS engineers theodolite. The scales are sub-divided into 0.5 minute intervals and reading is by estimation to 0.1 minute. The illustrated readings are vertical 78° 35.6′ and horizontal 68° 21.8′.

(c) Micrometer reading

Single reading micrometer

Figure 7.11 illustrates the optical train through a higher order theodolite, namely Watts 1. The optics are simple and the illumination is very clear. In the figure, the double open line follows the optical path through the horizontal circle while the dashed line follows that of the vertical circle.

The readings of both circles are seen through the circle eyepiece mounted on the outside of the standard upright. It can be rotated from one side to the other for comfortable viewing.

The eyepiece contains three apertures, the horizontal and vertical circle graduations appearing in those marked H and V respectively. As with the vernier theodolite, the circles are divided into 20-minute divisions. The horizontal circle, as read against the index arrow in Fig. 7.12(a) is therefore

$$35° 20′ + x$$

The fractional part x is read in the third aperture by means of a parallel plate optical micrometer inserted into the light path of the instrument.

A parallel plate micrometer is simply a glass block with parallel sides. In physics, the law of refraction states that a ray of light striking a parallel sided glass block at right angles will pass through the block without being refracted. If the block is tilted, however, the ray of light will be refracted but the emergent (i.e. exiting) ray will be parallel to the incident (i.e. entering) ray.

In Fig. 7.12(b) the ray of light from the horizontal circle is passing through the parallel plate when in the vertical position. The plate is directly geared to a drum mounted on the standard upright. Rotation of the drum causes the plate to tilt and the main scale reading of 35° 20′ is thereby made to coincide with the

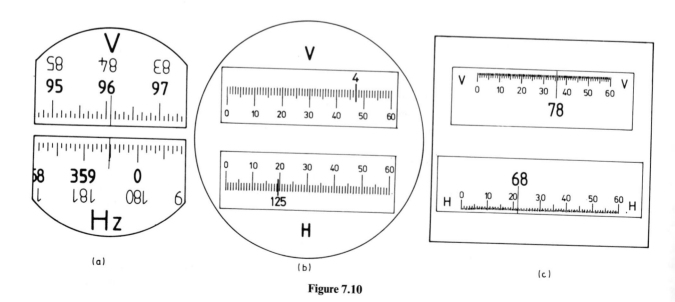

(a) (b) (c)

Figure 7.10

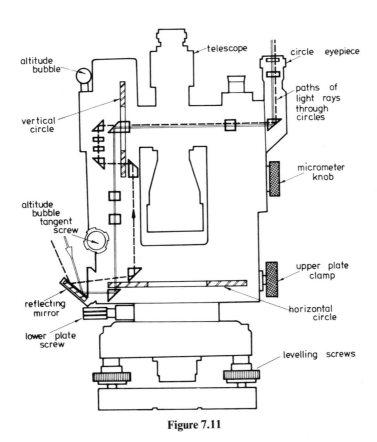

Figure 7.11

index mark. The resultant displacement, x, is read in minutes and seconds on the micrometer scale. The horizontal circle reading shown in Fig. 7.12(c) is therefore

$$
\begin{array}{r}
35^\circ\ 20' \\
+\ 06'\ 40'' \\
\hline
=\ 35^\circ\ 26'\ 40''
\end{array}
$$

Double reading micrometer

It is frequently necessary in engineering surveying to measure angles with a higher degree of accuracy

than can be obtained from a 20-second theodolite. In such cases a theodolite reading directly to 1 second is used.

It is possible to show that if the spindles of the upper and lower plates are eccentric, the measurement of the horizontal angles will be in error. The effects of eccentricity are entirely eliminated if two readings, 180° apart, are obtained and the mean of them taken.

On a 1-second theodolite, the reading eyepiece is again mounted on the upright. In most cases the horizontal and vertical circle readings are not shown

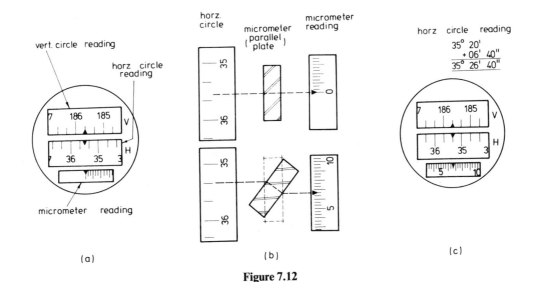

(a)   (b)   (c)

Figure 7.12

simultaneously. On the Hilger and Watts 2 microptic theodolite, a knob is situated below the reading eyepiece which, when turned to either H or V, brings into view the horizontal or vertical circle scale.

The images of divisions diametrically opposed to each other are automatically averaged when setting the micrometer (Fig. 7.13) and the reading direct to 1 second is free from circle eccentricity. The observer is actually viewing both sides of the circle simultaneously. In order to obtain a reading, the observer turns the micrometer drum until in the smallest aperture of the viewing eyepiece the graduations from one side of the circle are seen to be correctly superimposed on those from the other side, as in Fig. 7.13.

Both circles are read to 1 second, each circle having its own light path and separate micrometer.

The reading of the horizontal circle in Fig. 7.13 is

$$183° \, 20' + x \text{ (main scale)}$$
$$\underline{ 7' \, 26'' \text{ (micrometer scale)}}$$
$$183° \, 27' \, 26''$$

(d) Opto-electronic

Advances in the electronics industry have led to the development of the totally electronic theodolite. Most instrument makers produce a range of these theodolites, costing less than conventional instruments.

The horizontal and vertical circles are comprised of circular plates, incorporating angular encoders,

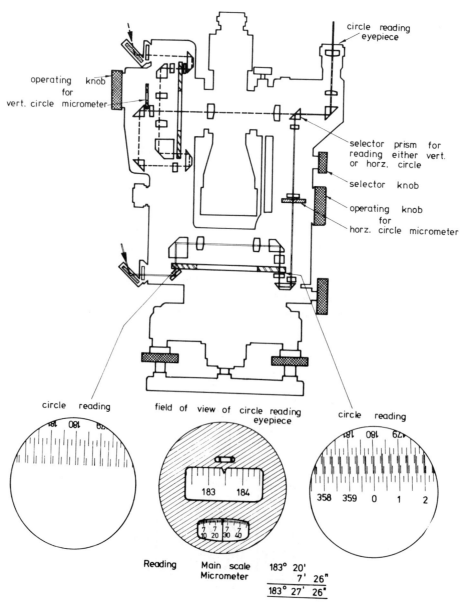

Figure 7.13

which, described in their simplest terms, are circular bar charts, which can be detected by a reading head in the theodolite.

The theodolite is controlled by a microprocessor prompted by the surveyor, using a control panel, built into the alidade. The code pattern is detected by the incremental encoder system, converted to a pulse signal, which is then changed to angular units by the microprocessor. The angular units are displayed in digital fashion on the screen of the control panel (Fig. 7.14).

The automatic display of the circle readings eliminates the need for scales and micrometers, and greatly reduces reading and booking errors. The time required for measuring an angle is also reduced because of the zero-set facility on most of these instruments (Topcon DT30, Sokkia DT5), whereby the circle can be set to read zero at the touch of a button.

Example

1 Figure 7.15 shows a view through the reading eyepiece of several theodolites in common use:
(a) By inspection determine the system employed to read the circles.
(b) Determine the readings of the various horizontal and vertical circles shown.

Answers
(a) Zeiss 080 direct	V, 96° 05′
	H, 359° 28′
(b) Wild T16 scale	V, 96° 06.5′
	H, 235° 56.4′
(c) Sokkia TM6 micrometer	H, 263° 15′ 24″
(d) Sokkia TS20A scale	H, 103° 2.5′, or
	H, 256° 57.5′
(e) Pentax TH06D micrometer	V, 58° 25′ 48″
(f) Sokkia DT5 electronic	V, 67° 05′ 10″
	H, 137° 08′ 00″

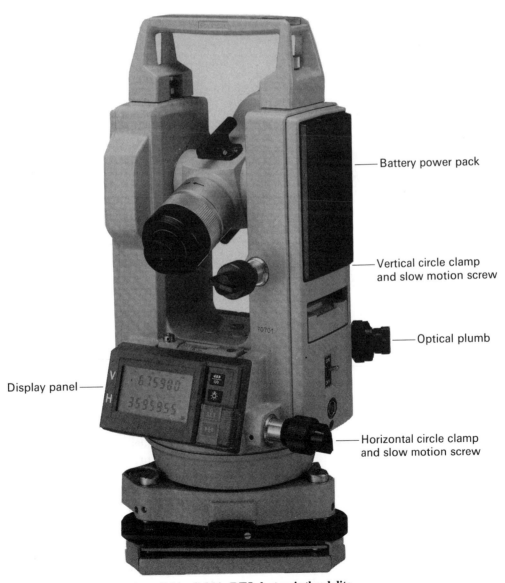

Battery power pack

Vertical circle clamp and slow motion screw

Optical plumb

Display panel

Horizontal circle clamp and slow motion screw

Figure 7.14   Sokkia DT5 electronic theodolite

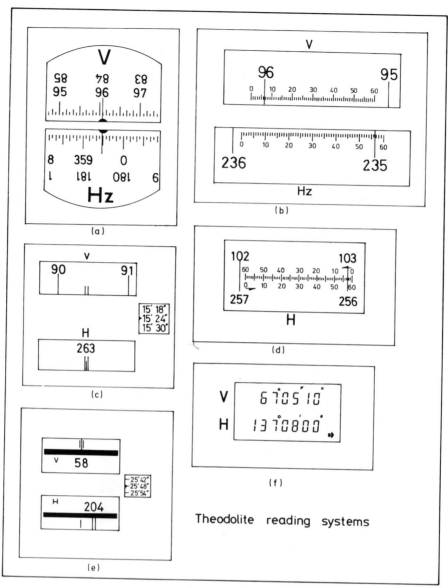

Figure 7.15

Exercise 7.1

1 Figure 7.16 shows the view through the reading eyepiece of several theodolites in common use.
(a) Determine the system used to read the circles.
(b) Determine the readings of the various horizontal and/or vertical circles shown in the figure.

4. Setting up the theodolite (temporary adjustments)

The sequence of operations required to prepare the instrument for measuring an angle is as follows.

(a) Plumb-bob method

1. *Setting the tripod*
This is probably the most important operation. If the tripod is not set properly, a great deal of time will have to be spent on subsequent operations 2 to 4.

The tripod legs are spread out and rested lightly on the ground around the mark. With the plumb-bob hanging from a hook fitted to the tripod head, the tripod is moved bodily until the plumb-bob is over the mark. The small circular spirit level mounted on the tripod is checked to ensure that the bubble is approximately central. If it is not, one leg of the tripod at a time is moved sideways until it is. Sideways movement of any leg will not greatly affect the position of the plumb-bob.

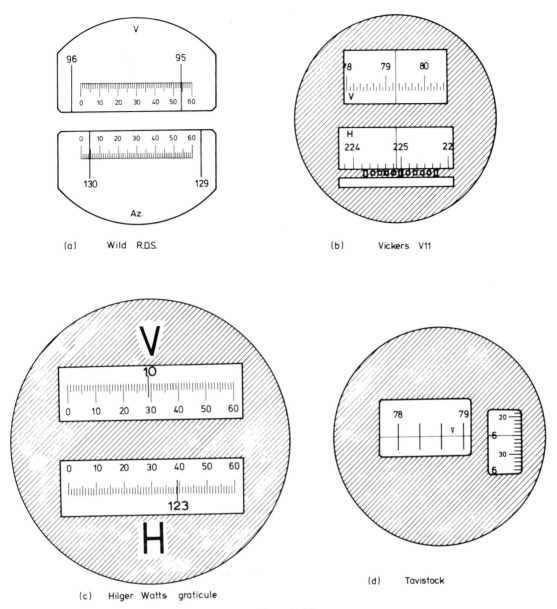

(a) Wild R.D.S.

(b) Vickers V11

(c) Hilger Watts graticule

(d) Tavistock

Figure 7.16

Only when the bubble is centred on the plumb-bob very close to the survey mark are the legs of the tripod pushed firmly home. This latter action will disturb the position of the plumb-bob and/or bubble and the telescopic sliding arrangements on the legs are used to reestablish the tripod's central position.

2. *Mounting the instrument*

The theodolite is then carefully removed from the box and firmly screwed to the tripod. It should be noted at this point that the theodolite should not be carried on the tripod. Not only is it wasteful of setting-up time but serious damage can be caused should the observer trip. Even discounting mishaps, damage is caused to the instrument's central axis because the weight of the instrument tends to bend it slightly.

3. *Levelling*

The levelling sequence is identical to that of the dumpy level, namely:

(a) Set the plate spirit level over two screws and centralize the bubble.

(b) Turn the instrument through 90° and recentralize the bubble.

(c) Repeat operations (a) and (b) until the bubble remains central for both positions.

4. *Centring*

The centring motion is released and the instrument moved until the plumb-bob is exactly over the survey mark. The centring motion is then tightened. This operation will have resulted in movement of the spirit level bubble from its central position; consequently, the operation of levelling and centring are repeated until both conditions are satisfied.

On a windy day it is very difficult to set the plumb-bob exactly over the survey point.

Most theodolites have an optical plummet which overcomes this disadvantage. Before use of the optical plummet, the theodolite is centred approximately using the plumb-bob and levelled using the plate spirit level. The centring motion is then released and the instrument head moved until the survey point is centred on the cross-wires of the optical plummet. Care should be taken to move the shifting head in the two directions used to level the spirit level.

5. *Parallax elimination*
A piece of paper is held in front of the telescope and the observer, sighting the paper through the telescope, turns the eyepiece carefully until the cross-wires of the reticule are sharply defined.

(b) Optical plumb method

1. *Setting the tripod*
The tripod legs are spread out and rested lightly on the ground around the survey point. Judging by eye, the legs are moved to bring the tripod head over the point, keeping the tripod head as level as possible.

2. *Mounting the instrument*
The theodolite is carefully removed from the box and screwed to the tripod, as before.

3. *Centring*
A sight is taken through the optical plummet to view the survey peg. The footscrews are turned to bring the instrument exactly over the peg. The theodolite is now centred but, of course, is not level.

4. *Levelling*
Levelling is accomplished by raising or lowering the legs of the tripod, using the sliding leg arrangement, until the circular spirit level is centred.

Refinements to the centring and levelling of the instrument must now be made, in the conventional manner, as described in Secs 4(a)(3) and (4).

5. *Parallax elimination*
A piece of paper is held in front of the telescope and the observer, sighting the paper through the telescope, turns the eyepiece carefully until the cross-wires of the reticule are sharply defined.

(c) Centring rod method

On some theodolites the centring is carried out using a telescope centring rod. The tip of the rod is placed over the survey point and the telescope tripod legs are used to centralize the bubble of a spherical spirit level at right angles to the rod. When the bubble is central, the rod is vertical and the theodolite itself approximately level.

5. Measuring horizontal angles

When exactly set over a survey mark and properly levelled, the theodolite can be used in two positions, namely:

(a) face left or circle left;
(b) face right or circle right.

The instrument is said to be facing left when the vertical circle is on the observer's left as an object is sighted. In order to sight the same object on face right, the observer must turn the instrument horizontally through 180° until the eyepiece is approximately pointing to the target. The telescope is then rotated about the transit axis, thus making the objective end of the telescope face the target. The vertical circle will now be found to be on the observer's right. This operation is known as transitting the telescope.

In Fig. 7.17, horizontal angle PQR is to be measured.

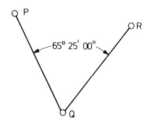

Figure 7.17

(a) Reiteration method

Without setting to zero
This method may be used with any type of theodolite and should be mastered before attempting the alternative method.

The possible horizontal control arrangements of a theodolite are: (a) upper and lower plate clamps, (b) repetition clamp and (c) circle setting screw (Figs 7.3, 7.4 and 7.5).

1. Using the field book (Table 7.1), enter theodolite station Q in column 1 line 1.

	1 Observation station	*2* Target station	*3* Face left reading	*4* Face right reading	*5* Accepted mean angle
1	Q	P	25° 30′	205° 30′	
2		R	90° 55′	270° 55′	
3			65° 25′	65° 25′	65° 25′

Table 7.1

2. Enter the left-hand target station P in column 2 line 1.

3. Enter the right-hand target station R in column 2 line 2.

4. Close the lower plate clamp or repetition clamp, if fitted, *and do not touch either of them again.*

5. Set the instrument on face left.

6. Open the upper plate clamp on the alidade and the telescope clamp.

7. Turn the instrument carefully towards the left-hand target P and sight the target using the auxiliary finder sights fitted to the telescope. Lock the upper plate clamp and telescope clamp.

8. Focus the telescope on the target. The cross-wires will not be on the target but should be close. Use the slow motion screws on the upper plate clamp and telescope clamp to bisect the target accurately.

9. Read the horizontal circle and note the reading (25° 30′) in column 3 line 1.

10. *Repeat operations 6, 7, 8 and 9* for the right-hand target R, booking the horizontal circle readings (90° 55′) in column 3 line 2.

11. Subtract reading P from reading R (90° 55′ − 25° 30′ = 65° 25′) and note in column 3 line 3.

In order to measure the angle above, sixteen manipulations of the theodolite controls, two circle readings and two bookings were required. Clearly an error could easily occur with an inexperienced operator. Besides, even if these operations were conducted perfectly, the theodolite might be in poor adjustment and the angle of 65° 25′ would be incorrect.

All of these possible sources of error are eliminated by remeasuring the angle on face right.

12. Transit the telescope to set the instrument to face right and make preparations to remeasure the angle.

13. *Repeat Operations 6, 7, 8 and 9*, noting the left-hand target reading P (205° 30′) in column 4 line 1. This reading should differ by 180° from that in column 3 line 1 if no errors are present.

14. *Repeat operations 6, 7, 8 and 9* for the right-hand target, noting the reading (270° 55′) in column 4 line 2.

15. Subtract reading P from reading R (270° 55′ − 205° 30′ = 65° 25′) and note in column 4 line 3.

16. Calculate the mean value of the angle and note in column 5 line 3.

Example

2 Table 7.2 shows the field measurements of four angles of a traverse. Using the table, calculate the values of the angles.

Answer (Table 7.3)

In Example 2 the angle DAB was calculated more easily than the other three angles simply because the initial reading was 00° 00′ 00″. Many surveyors prefer this method of measuring angles and in setting out work it is standard practice.

Setting to zero

The actual measurement procedure is the same as for method (a) except that the initial setting of the horizontal circle has to be 00° 00′ 00″. The mechanics of setting the circle varies with the type of theodolite.

Observation station	Target station	Face left reading	Face right reading	Accepted mean angle
B	A	89° 16′ 20″	269° 16′ 20″	
	C	185° 18′ 40″	05° 19′ 00″	
C	B	185° 39′ 40″	05° 39′ 20″	
	D	271° 38′ 20″	91° 38′ 40″	
D	C	275° 18′ 00″	95° 18′ 20″	
	A	01° 02′ 20″	181° 02′ 40″	
A	D	00° 00′ 00″	180° 00′ 00″	
	B	92° 15′ 30″	272° 15′ 30″	

Table 7.2

Observation station	Target station	Face left reading	Face right reading	Accepted mean angle
B	A	89° 16′ 20″	269° 16′ 20″	
	C	185° 18′ 40″	05° 19′ 00″	
		96° 02′ 20″	96° 02′ 40″	96° 02′ 30″
C	B	185° 39′ 40″	05° 39′ 20″	
	D	271° 38′ 20″	91° 38′ 40″	
		85° 58′ 40″	85° 59′ 20″	85° 59′ 00″
D	C	275° 18′ 00″	95° 18′ 20″	
	A	01° 02′ 20″	181° 02′ 40″	
		85° 44′ 20″	85° 44′ 20″	85° 44′ 20″
A	D	00° 00′ 00″	180° 00′ 00″	
	B	92° 15′ 30″	272° 15′ 30″	
		92° 15′ 30″	92° 15′ 30″	92° 15′ 30″

Table 7.3

(i) Theodolite with upper and lower plate clamps
Using a double centre theodolite, i.e. one fitted with upper and lower plate clamps, the procedure is:

1. Set the theodolite to face left position.
2. Set the micrometer (if fitted) to 00′ 00″.
3. Release the upper plate clamp only and set the index mark or vernier to zero degrees as closely as possible by eye. Close the clamp and using the slow motion screw, set the circle exactly to zero. The reading is now 00° 00′ 00″.
4. Release the *lower* plate clamp and telescope clamp. Sight the left-hand station P, lock the clamps and, using the *lower plate slow motion screw*, accurately bisect the target.

(ii) Theodolite with repetition clamp
Using a theodolite fitted with a repetition clamp:

1. Set the theodolite to face left position.
2. Set the micrometer (if fitted) to 00′ 00″.
3. Release the upper plate clamp and set the index mark to zero degrees as closely as possible by eye. Close the clamp and using the slow motion screw, set the circle exactly to zero. The reading is now 00° 00′ 00″.
4. Depress the repetition clamp. This action locks the horizontal circle to the alidade.
5. Release the upper plate clamp and telescope clamp. The alidade and circle will now move together and maintain a zero reading. Sight the left-hand station P, lock the clamps and, using the slow motion screws, accurately bisect the target.
6. Release the repetition clamp. The circle will now be free of the alidade.

(iii) Theodolite with circle-setting screw
1. Set the instrument to face left position.
2. Set the micrometer (if fitted) to 00′ 00″.
3. Release the upper plate clamp and telescope clamp. Sight the left-hand station P and, using the slow motion screws, accurately bisect the target.
4. Raise the hinged cover of the circle-setting screw and rotate the screw carefully until the horizontal circle reads exactly zero.

(iv) Electronic theodolite
1. Set the theodolite to face left position.
2. Release the upper plate clamp and telescope clamp. Sight the left-hand station P and, using the slow motion screws, accurately bisect the target.
3. Press the (zero set) key on the keypad. The horizontal circle reading will be reset to zero degrees.

In all four cases the situation has been reached where the theodolite circle reads zero and the left-hand target P is accurately bisected. The measurement of the angle is completed as follows:

1. In Table 7.4, enter the reading of 00° 00′ 00″ in column 3 line 1.
2. Open the upper plate clamp and telescope clamp.
3. Turn the instrument carefully towards the right and sight the right-hand target R.
4. Lock both clamps and, using the upper plate slow motion screw and telescope slow motion screw, accurately bisect the target.
5. Read the horizontal circle and enter the reading (65° 25′ 00″) in column 3 line 2.
6. Subtract reading P from reading R (65° 25′ 00″ − 00° 00′ 00″ = 65° 25′ 00″).

	Observation station	Target station	Horizontal FL	Angle FR	Accepted value
1	Q	P	00° 00′ 00″	180° 00′ 00″	
2		R	65° 25′ 00″	245° 25′ 00″	
3			65° 25′ 00″	65° 25′ 00″	65° 25′ 00″

Column numbers 1 2 3 4 5

Table 7.4

7. Transit the telescope to set the instrument on face right.

8. Resight the left-hand target and note the reading which should be 180° 00′ 00″ if no errors have been made and if the instrument is in adjustment. Note the reading in column 4 line 1.

9. Resight the right-hand station and note the reading 245° 25′ 00″ in column 4 line 2.

10. Subtract reading P from reading R (245° 25′ 00″ − 180° 00′ 00″ = 65° 25′ 00″).

11. Calculate the mean angle and enter the value in column 5 line 3.

Example

3 Table 7.5 shows the field measurements of four angles of a traverse. Using the table, calculate the values of the angles.

Answer (Table 7.6)

Exercise 7.2

1 Table 7.7 shows the field measurements of two angles of a traverse survey. Calculate the values of traverse angles.

(b) Repetition method

In order to reduce the number of times that the circle has to be read and thereby reduce a source of error, a method of measuring, known as repeated addition or repetition, is employed. The method is of particular value when small angles such as angle XYZ in Fig. 7.18 are to be measured.

Theodolites fitted with a circle-setting screw cannot be used to measure angles by repetition. The method of measuring is as follows:

1. Set the instrument to face left and close the lower plate clamp. Sight station X and note the reading as before.

2. Release the *upper* plate clamp, turn the instrument to sight station Z and close the clamp. The circle need

Observation station	Target station	Face left reading	Face right reading	Accepted mean angle
B	A	00° 00′ 00″	180° 00′ 00″	
	C	93° 14′ 20″	273° 14′ 20″	
C	B	00° 00′ 00″	180° 00′ 20″	
	D	161° 25′ 00″	341° 25′ 00″	
D	C	00° 00′ 00″	179° 59′ 40″	
	A	175° 31′ 40″	355° 31′ 40″	
A	D	00° 00′ 00″	179° 59′ 40″	
	B	195° 32′ 20″	15° 32′ 20″	

Table 7.5

Observation station	Target station	Face left reading	Face right reading	Accepted mean angle
B	A	00° 00′ 00″	180° 00′ 00″	
	C	93° 14′ 20″	273° 14′ 20″	
		93° 14′ 20″	93° 14′ 20″	93° 14′ 20″
C	B	00° 00′ 00″	180° 00′ 20″	
	D	161° 25′ 00″	341° 25′ 00″	
		161° 25′ 00″	161° 24′ 40″	161° 24′ 50″
D	C	00° 00′ 00″	179° 59′ 40″	
	A	175° 31′ 40″	355° 31′ 40″	
		175° 31′ 40″	175° 32′ 00″	175° 31′ 50″
A	D	00° 00′ 00″	179° 59′ 40″	
	B	195° 32′ 20″	15° 32′ 20″	
		195° 32′ 20″	195° 32′ 40″	195° 32′ 30″

Table 7.6

Observation station	Target station	Face left reading	Face right reading	Mean angle
D	C	00° 00′ 00″	179° 59′ 20″	
	E	189° 39′ 50″	9° 38′ 50″	
E	D	185° 14′ 30″	5° 13′ 50″	
	F	6° 15′ 10″	186° 14′ 10″	

Table 7.7

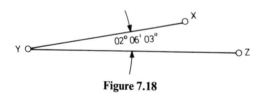

Figure 7.18

not be read but usually is to check the final value of the angle.

3. Release the lower plate clamp, resight station X and close the clamp.

4. Release the upper plate clamp, sight station Z for the second time and close the clamp.

The reading on the instrument is now double the value of the angle but again it is not noted.

5. Repeat operations 3 and 4 any number of times. The value of the angle is thereby added on the circle. If, after, say, six repetitions of the measurement, the circle is read and the value of 12° 36′ 18″ obtained, the mean value of angle XYZ is found by dividing by six:

$$\text{Angle XYZ} = 02° 06′ 03″$$

6. Repeat the angular measurements a further six times on face right and obtain a second mean value by subtracting the first reading on station X from the final reading on station Z and dividing by six:

$$\text{Say angle XYZ} = 02° 06′ 07″$$

7. Accepted $\text{XYZ} = \frac{1}{2}(02° 06′ 03″ + 02° 06′ 07″)$
$$= 02° 06′ 05″$$

6. Measuring angles in the vertical plane

In Section 2(i) and in Fig. 7.8, it was explained that, in the vertical plane, a theodolite can be used to measure either a vertical angle or a zenith angle. In a vertical angle, the zero reading of the vertical circle is in the horizontal position and in measuring a zenith angle, the zero reading is in the vertical position.

Figure 7.19 shows a survey line XY, measured on a slope. The angle of slope, measured on face left of a theodolite, set up at X, is either

15° 30′ 00″ vertical angle or
74° 30′ 00″ zenith angle.

(a) Measurement of vertical angles

It should be remembered that the construction of the theodolite is such that the vertical circle moves with the telescope and the index marker remains fixed. In a perfectly adjusted theodolite, the index reader of the vertical circle should read zero degrees (or ninety degrees, depending on the configuration of the vertical circle, Fig. 7.8) and the bubble of the altitude spirit

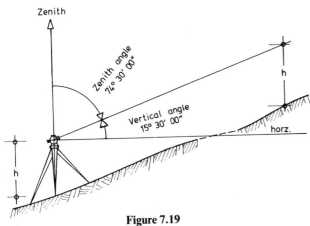

Zenith

Zenith angle
74° 30′ 00″

Vertical angle
15° 30′ 00″

horz.

h

h

Figure 7.19

level should be in the centre of its run when the telescope is in the horizontal position.

Very often this is not the case and all vertical angles *must* be measured on both face left and face right. The procedure is as follows when measuring a vertical angle:

1. Set the instrument to face left.
2. Release the telescope clamp and one of the horizontal plate clamps.
3. Sight the target using the finder sights. Lock the telescope clamp and plate clamp.
4. Focus the telescope on the target. The cross-wires will not be on the target but should be close. Use the slow motion screws to bisect the target accurately.
5. Set the altitude spirit level (if fitted) to the centre of its run and read the vertical circle.
6. Change the instrument to face right and repeat operations 2, 3, 4 and 5.

Figure 7.8 shows the method of graduating the vertical circle of several theodolites. Because of this system of graduation and since only one reading per face can be obtained, the face left reading of some vertical angles above might be 07° 49′ 56″.

When the face right reading is obtained, it should read 172° 10′ 04″ since the sum of face left and face right should total 180° 00′ 00″. Because of small maladjustments of the instrument, the face right reading will in all probability differ from the above value and the sum will differ from 180° 00′ 00″ by an amount known as the index error. If the actual reading on face right is 172° 09′ 56″, the index error and hence the correct vertical angle are obtained thus:

$$\begin{aligned}
\text{Face left reading} &= 07° \ 49′ \ 56″ \\
\text{Face right reading} &= 172° \ 09′ \ 56″ \\
\hline
\text{Sum} &= 179° \ 59′ \ 52″ \\
\text{Index error} &= \qquad -08″
\end{aligned}$$

The index error is halved and a correction of $+04″$ is applied to both angular measurements to bring their sum to 180° 00′ 00″:

$$\begin{aligned}
\text{Corrected face left reading} &= \quad 07° \ 50′ \ 00″ \\
\text{Corrected face right reading} &= 172° \ 50′ \ 00″ \\
\hline
\text{Sum} &= 180° \ 00′ \ 00″
\end{aligned}$$

The face left reading is taken as the vertical angle, i.e. 07° 50′ 00″ (elevation).

If an angle of depression is observed using this particular theodolite, the sum of face left and face right readings should total 540° 00′ 00″. For example:

$$\begin{aligned}
\text{Vertical angle FL reading} &= 330° \ 25′ \ 10″ \\
\text{FR reading} &= 209° \ 35′ \ 02″ \\
\hline
\text{Sum} &= 540° \ 00′ \ 12″ \\
\text{Index error} &= \qquad +12″
\end{aligned}$$

$$\begin{aligned}
\text{Correct face left reading} &= 330° \ 25′ \ 10″ - 06″ \\
&= 330° \ 25′ \ 04″ \\
\text{Vertical angle} &= 360° - 330° \ 25′ \ 04″ \\
&= 29° \ 34′ \ 56″ \text{ (depression)}
\end{aligned}$$

(b) Measurement of zenith angles

The actual measuring procedure is exactly the same as for the measurement of a vertical angle. The zenith angle readings corresponding to the vertical angle of elevation of part (a) above would therefore be:

$$\begin{aligned}
\text{Face left reading} &= \quad 82° \ 10′ \ 04″ \\
\text{Face right reading} &= 277° \ 50′ \ 04″ \\
\hline
\text{Sum} &= 360° \ 00′ \ 08″ \\
\text{Index error} &= \qquad +08″
\end{aligned}$$

The index error is halved and a correction of $-04″$ is applied to both angular measurements to bring their sum to 360° 00′ 00″.

Example

4 In Fig 7.19 the zenith angle measured, using a different theodolite, is as follows:

$$\begin{aligned}
\text{Face left reading} &= \quad 74° \ 29′ \ 30″ \\
\text{Face right reading} &= 285° \ 29′ \ 30″
\end{aligned}$$

Determine the correct zenith angle.

Answer

$$\begin{aligned}
\text{Face left reading} &= \quad 74° \ 29′ \ 30″ \\
\text{Face right reading} &= 285° \ 29′ \ 30″ \\
\hline
\text{Sum} &= 359° \ 59′ \ 00″ \\
\text{Index error} &= \qquad -01′ \ 00″ \\
\text{Correction to each reading} &= \qquad +00′ \ 30″ \\
\text{Corrected zenith angle} &= \quad 74° \ 30′ \ 00″
\end{aligned}$$

Exercise 7.3

1 Table 7.8 shows the field measurements of two angles of a traverse survey. Calculate the values of zenith angles.

Line	Face	Vertical circle reading
XY	L	79° 30′ 50″
	R	280° 30′ 10″
YZ	L	102° 13′ 50″
	R	257° 47′ 10″

Table 7.8

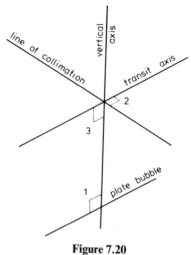

Figure 7.20

7. Errors affecting angular measurements

The errors that affect angular measurements can be considered under two headings:

(a) Instrumental maladjustments
(b) Human and other errors

(a) Instrumental maladjustments

Effect on horizontal angles If a theodolite is to be in perfect adjustment, the relationship between the various axes should be as shown in Fig. 7.20, namely:

1. The vertical axis should be truly vertical and at right angles to the plate bubble.
2. The line of collimation of the telescope should be at right angles to the transit axis.
3. The transit axis should be at right angles to the vertical axis of the instrument.

The relationships become disturbed through continuous use or misuse, and tests must be carried out before the start of any major contract and at frequent intervals thereafter to ensure that the instrument is in adjustment.

The various tests and adjustments are as follows:

1. The vertical axis must be truly vertical and at right angles to the plate bubble axis.

Test
(a) Erect the tripod firmly and screw on the theodolite. Set the plate spirit level over two screws and centralize the bubble. If the instrument is not in good adjustment the vertical axis will be inclined by the amount *e* as shown in Fig. 7.21(a).
(b) Turn the instrument through 90° and recentralize the bubble.
(c) Repeat these operations until the bubble remains central for positions (a) and (b).
(d) Turn the instrument until it is 180° from position (a). The vertical axis will still be inclined with an error *e* and the bubble of the spirit level will no longer be central. It will, in fact, be inclined to the horizontal at an angle of 2*e* (Fig. 7.21(b)). The number of divisions, *n*, by which the bubble is off-centre is noted.

Adjustment
(e) Turn the footscrews until the bubble moves back towards the centre by *n*/2 divisions, i.e. by half

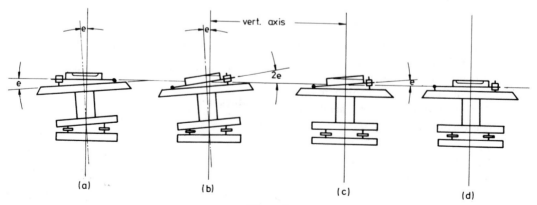

Figure 7.21

the error. The vertical axis is now truly vertical (Fig. 7.21(c)).

(f) Adjust the spirit level by releasing the capstan screws and raising or lowering one end of the spirit level until the bubble is exactly central. The other half $n/2$ of the error is thereby eliminated (Fig. 7.21(d)) and the spirit level is at right angles to the vertical axis.

Effect of maladjustment The effect of an inclined vertical axis is not serious and in fact the small maladjustment normally encountered has no effect on measurements made with a conventional theodolite.

There is no observational procedure that can be employed to eliminate this error. However, as already implied, the error can be ignored.

2. The line of collimation of the telescope must be at right angles to the transit axis. The line of collimation is defined as the line joining the optical centre of the object glass to the vertical cross-hair of the diaphragm. If the position of the diaphragm has been disturbed, the line of collimation will lie at an angle to the transit axis (Fig. 7.22), with e being the error in the line of collimation.

Test

(a) After properly setting up the instrument at a point which will be designated I, sight a well-defined mark A about 100 metres away and close both clamps. The line of sight will now be pointing to the target as in Fig. 7.23(a).

(b) Transit the telescope and sight a staff B laid horizontally about 100 metres away on the other side of the instrument from A.

Note the reading, 2.100 m in Fig. 7.23(b).

Since the transit axis and line of sight make an angle of $(90 - e)$ and since the transit axis maintains its position when the telescope is transmitted, the line of sight must diverge from the straight line by an amount $2e$.

(c) Change the instrument to face right and again sight mark A (Fig. 7.23(c)).

(d) Transit the telescope and sight staff B. The line of sight will again diverge from the straight line AB by a amount $2e$, but on the other side of the line. The staff reading is again noted and in Fig. 7.23(d) is 2.000 m.

(e) If the two staff readings are the same, the instrument is in adjustment and points A, I and B form a straight line.

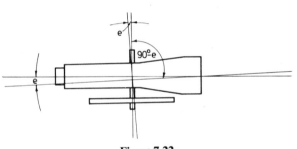

Figure 7.22

Adjustment

(f) Since the staff readings differ, the difference represents the error $4e$, that is $4e = (2.100 - 2.000) = 0.100$ m in this case. At the outset, it was shown that the error was e, therefore $e = (0.100/4) = 0.025$ m.

The error is eliminated by bringing the vertical cross-hair of the staff reading $(2.000 + e) = 2.025$ m.

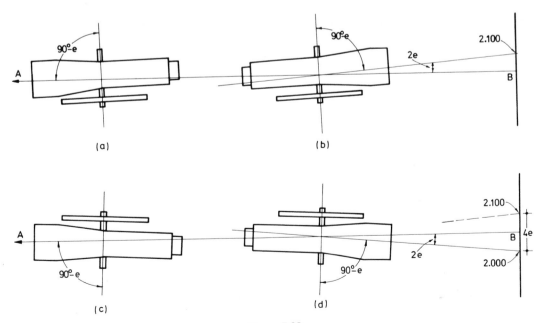

Figure 7.23

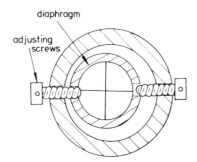

Figure 7.24

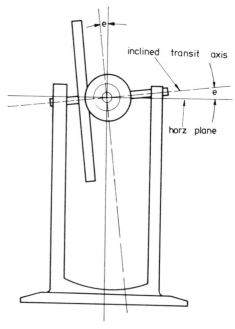

Figure 7.25

This is done by means of the antagonistic adjusting screws situated one on either side of the diaphragm.

In this case the diaphragm has to be shifted to the right (Fig. 7.24). The left screw is loosened slightly and the right screw tightened until the correct staff reading is obtained.

If the cross-hair is no longer vertical, it might be necessary to loosen the diaphragm and rotate it slightly until it is perfectly vertical. The verticality is usually checked against a plumb-line.

Effect of maladjustment If the instrument is used in its unadjusted state, every angle will be in error. However, as will be evident from Fig. 7.23, the mean of face left and face right readings is correct. For example, the mean staff reading $\frac{1}{2}(2.100 + 2.000) = 2.050$ m is the correct position of B since the line of sight diverges by $2e$ on either side of the mean position for face left and face right respectively.

3. The transit axis must be truly horizontal when the vertical axis is vertical. If the instrument is not in adjustment, the transit axis will not be horizontal when the instrument is correctly set up. In Fig. 7.25, the error is e. If the telescope is inclined, the cross-hair will travel along the plane shown by the dotted line, i.e. a plane at right angles to the transit axis but *not* a vertical plane.

Test

(a) Set up the instrument at I and level it properly. Sight a mark A at an elevation of about 60°. Close both upper and lower plate clamps (Fig. 7.26).

(b) Lower the telescope and sight a horizontal staff or scale laid on the ground at the base of the object A. Note the reading, 'b'.

(c) Repeat the operations on face right to obtain a second reading, 'c', on the scale. If the instrument is in adjustment both readings will be the same.

Adjustment

(d) If the readings differ, the mean is correct since the telescope will have traversed over planes each inclined at an angle e on either side of the vertical.

By means of either the upper or lower plate slow motion screw, set the instrument to the mean reading.

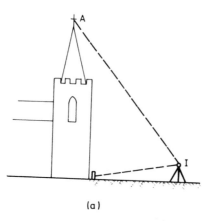

(a)

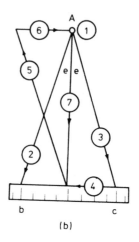

(b)

Figure 7.26

(e) Elevate the telescope until the horizontal cross-hair cuts mark A. The vertical cross-hair will lie to the side of the mark.

(f) Bring the vertical cross-hair on to mark A by means of the transit axis adjustment screw located on the standard immediately below the transit axis.

(g) Depress the telescope. If the adjustment has been carried out correctly the vertical cross-hair should read the mean staff reading.

Effect of maladjustment Angles measured between stations at considerably different elevations will be in error. However, as Fig. 7.26 clearly shows, the mean face left and face right is correct.

It must be noted at this juncture that most modern theodolites make no provision for this adjustment. The complex optical systems make the adjustment very difficult. However, since such instruments are assembled with a high degree of precision this adjustment is usually unnecessary. Besides, the mean of face left and face right observations cancels the error.

Effect on vertical angles

In a perfectly adjusted theodolite, the index reader of the vertical circle should read zero degrees (or ninety degrees, depending on the configuration of the vertical circle, Fig. 7.8) and the bubble of the altitude spirit level should be in the centre of its run when the telescope is in the horizontal position. This is the only maladjustment that materially affects the measurement of vertical angles. On modern theodolites, employing automatic vertical circle indexing to replace the altitude spirit level, no adjustment is usually required.

Test

(a) Set up the instrument at I on face left, centralize the altitude spirit level and read a vertical levelling staff at A with the vertical circle reading zero. Note the reading, say 1.500 (Fig. 7.27(a)).

(b) Transit the telescope and repeat operation (a) on face right. Note the staff reading, say 1.400 in Fig. 7.27(b).

Adjustment

(c) The mean staff reading is correct since the face left observation is elevated and the face right reading is depressed by equal errors, *e*. Using the vertical slow motion screw, the telescope is brought to the mean reading of 1.450 m (Fig. 7.27(c)).

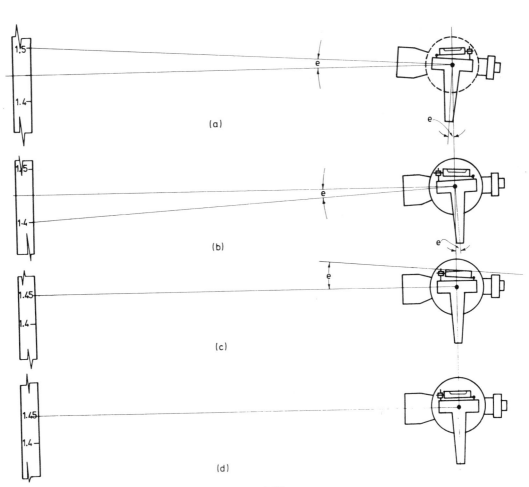

Figure 7.27

(d) The vertical circle will no longer read zero and must be made to do so by the clip screw which controls the altitude spirit level and vertical circle index.

(e) The movement of the clip screw, however, will move the bubble from its central position. The bubble is recentralized by adjusting the capstan screws situated at one end of the spirit level (Fig. 7.27(d)).

Effect of maladjustment If the instrument is in maladjustment all vertical angles will be in error. However, Fig. 7.27 clearly shows that the mean of face left and face right observations is correct.

(b) Human and other errors

Human errors can be considered under two headings:

1. Gross errors

These are mistakes on the part of the observer caused by ignorance, carelessness or fatigue. They include sighting the wrong target, measuring the anticlockwise angle, turning the wrong screw, opening the wrong clamp, reading the circles wrongly and booking incorrectly.

These errors can only be avoided by careful observation and by observing each angle at least twice. Only then will any error show up. It is absolutely useless to measure any angle on one face only as it is open to the wildest of errors, as already pointed out.

2. Random errors

Small errors cannot be avoided. They may be due to imperfections of human sight and touch which make it impossible to bisect the targets accurately or read the circles exactly. The errors, however, are small, and of little significance. They are minimized by taking several observations and accepting the mean.

Other errors arise from such sources as unequal expansion of the various parts of the instrument by the sun, instability of the instrument in windy weather, heat haze or mist affecting the sighting and lastly inaccurate centring of the instrument. They can only be avoided by shielding the instrument against the wind or sun and choosing times to observe which are favourable.

Lastly, if the instrument is not correctly centred, nothing can be done to eliminate or minimize the errors that must arise. Great care must be taken to position the instrument over the survey station with accuracy.

Summary

In general, gross errors cannot be eliminated or minimized by any observational system. A compensated measure will only show that there is an error. However, that should be sufficient as the observer should then take steps to trace the error. A mistake in subtraction will not necessitate any repetition of measuring but all other errors under this heading will, and in general another complete compensated measure should be made.

Systematic errors arise from instrumental maladjustments and defects. It has been shown that in almost every case a compensated measure of the angle cancels the error, the exception being that if the vertical standing axis is not truly vertical errors cannot be eliminated. However, they are second-order errors and do not affect measurements made by a conventional theodolite.

Small random errors are not eliminated by a compensated measurement but such action minimizes the error and the mean of a larger number of measurements would be very accurate.

It cannot be overemphasized that a compensated measure must be made of every angle regardless of its importance; otherwise the result is so uncertain as to be meaningless.

Exercise 7.4

1 Make a sketch of a theodolite to show clearly the principal parts.

2 Describe briefly the tests and adjustments that should be made in order to ensure that a theodolite is in good working order.

3 Using a diagram, describe briefly any form of optical plummet built into a theodolite.

4 Describe briefly any method of measuring a horizontal angle such that errors are minimized.

5 Outline the advantages and disadvantages of the following methods of reading the circles of a theodolite:

(a) Optical scale
(b) Optical micrometer

6 Make a list of errors that may arise in measuring angles, using the following headings:
(a) Gross errors
(b) Systematic errors
(c) Random errors

7 Table 7.9 shows the horizontal and vertical circle readings, recorded during the measurement of angles, at survey stations B, C and D. Determine the accepted values of the angles.

Instrument station	Target station	Face left reading	Face right reading	Mean horizontal angle	Survey line	Face	Zenith angle reading	Accepted zenith angle
B	A	00° 24′ 40″	180° 24′ 40″		AB	L	88° 20′ 20″	
	C	65° 36′ 20″	245° 36′ 20″			R	271° 39′ 00″	
C	B	00° 00′ 00″	179° 59′ 30″		BC	L	85° 50′ 50″	
	D	66° 34′ 20″	246° 33′ 50″			R	274° 08′ 30″	
D	C	332° 10′ 20″	152° 10′ 40″		CD	L	80° 33′ 00″	
	E	87° 08′ 00″	267° 08′ 20″			R	279° 26′ 20″	

Table 7.9

8. Answers

Exercise 7.1

1 (a) Optical scale   V, 95° 54′ 20″   H, 130° 04′ 40″
   (b) Optical scale   V, 79° 15′   H, 224° 54′
   (c) Optical scale   V, 10° 29′ 00″   H, 123° 39′ 20″
   (d) Micrometer   H, 78° 56′ 25″

Exercise 7.2

1 Table 7.10

Observation station	Target station	Face left reading	Face right reading	Mean angle
D	C	00° 00′ 00″	179° 59′ 20″	
	E	189° 39′ 50″	9° 38′ 50″	
		189° 39′ 50″	189° 39′ 30″	189° 39′ 40″
E	D	185° 14′ 30″	5° 13′ 50″	
	F	6° 15′ 10″	186° 14′ 10″	
		181° 00′ 40″	181° 00′ 20″	181° 00′ 30″

Table 7.10

Exercise 7.3

1 Table 7.11

Line	Face	Vertical circle reading	Index error	Zenith angle
XY	L	79° 30′ 50″	−30″	79° 30′ 20″
	R	280° 30′ 10″		
		360° 01′ 00″		
YZ	L	102° 13′ 50″	−30″	102° 13′ 20″
	R	257° 47′ 10″		
		360° 01′ 00″		

Table 7.11

Exercise 7.4

1 Answer page 145
2 Answer pages 162–166
3 Answer page 149
4 Answer page 156
5 (a) Optical scale—advantages
    (i) Most easily read
    (ii) Cheaper
    (iii) Can be illuminated for easier reading
  (b) Optical micrometer — advantages
    (i) easily read
    (ii) Can be illuminated
    (iii) Only one reading necessary
    (iv) double reading type cancels eccentricity errors
    (v) Reads to 1 second
   Optical micrometer — disadvantages
    (i) Difficult to read
    (ii) Cumbersome
6 (a) Gross errors: page 166
  (b) Systematic errors: pages 162–166
  (c) Random errors: page 166
7 See Table 7.12.

Instrument station	Target station	Face left reading	Face right reading	Mean horizontal angle	Survey line	Face	Zenith angle reading	Accepted zenith angle
B	A	00° 24′ 40″	180° 24′ 40″		AB	L	88° 20′ 20″	88° 20′ 40″
	C	65° 36′ 20″	245° 36′ 20″			R	271° 39′ 00″	
		65° 11′ 40″	65° 11′ 40″	65° 11′ 40″		Sum	359° 59′ 20″	
C	B	00° 00′ 00″	179° 59′ 30″		BC	L	85° 50′ 50″	85° 51′ 10″
	D	66° 34′ 20″	246° 33′ 50″			R	274° 08′ 30″	
		66° 34′ 20″	66° 34′ 20″	66° 34′ 20″		Sum	359° 59′ 20″	
D	C	332° 10′ 20″	152° 10′ 40″		CD	L	80° 33′ 00″	80° 33′ 20″
	E	87° 08′ 00″	267° 08′ 20″			R	279° 26′ 20″	
		114° 57′ 40″	114° 57′ 40″	114° 57′ 40″		Sum	359° 59′ 20″	

Table 7.12

CHAPTER 8

Traverse surveys

Objective
After studying this chapter, the reader should be able to make a traverse survey, reduce the field data and plot the results graphically.

In order to survey any parcel of ground, two distinct operations are required, namely

(a) a framework survey and
(b) a detail survey.

A framework survey consists of a series of straight lines, arranged in the form of triangles (linear surveys), polygons (closed traverse surveys) or vectors (open traverse surveys). A detail survey consists of a series of offsets, which are added to and supplement the framework survey.

In Sec. 1(a) (Fig. 3.4) in Chapter 3, the site at GCB Outdoor Centre was surveyed, using linear survey techniques. The finished plan is shown in Fig. 3.34. A linear survey, however, may not be accurate enough, in which case a traverse survey, utilizing a theodolite, would be carried out.

A traverse survey consists of a series of survey lines, connected to each other, each line having length and direction. They are, therefore, vectors. The vectors, may or may not close to form a polygon.

1. Types of traverse

Theodolite traverses are classified under the following three headings.

(a) Open traverse

In Fig. 8.1, points A, B, C and D are the survey stations of an open traverse, following a stream that is to be surveyed and plotted to scale. Lines AB, BC and CD are the measured legs of the traverse and angles ABC and BCD (shown hatched) are the clockwise angles, measured using a theodolite. Together they form the survey framework.

The detail survey is carried out from the legs of the traverse, by offsetting, in the same manner as for linear surveying. This type of traverse is not self-checking and errors in either angular or linear or both measurements may pass unchecked. The only check that can be provided is to repeat the complete traverse

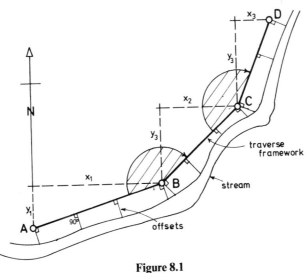

Figure 8.1

or resurvey the traverse in the opposite direction from D to A.

In order to plot the traverse stations, the horizontal length and direction relative to north of each line must be found. In other words, the vector data are required. These data are then transformed into rectangular coordinates x_1, y_1, etc., which are plotted to scale on plan.

(b) Closed traverse

Figure 8.2 shows a closed traverse round a building, where the vectors AB, BC, CD and DA form a closed polygon. The lengths of the lines and the values of the clockwise angles are again measured. The angles are the interior angles of the polygon and are measured as the theodolite is moved from station to station around the traverse.

The traverse is self-checking to some extent, since the sum of the interior angles should equal $[(2n - 4) \times 90°]$, where n is the number of angles. In this case, therefore, the sum should be 360°.

In a closed traverse, the exterior angles may be measured in preference to the interior angles, in which case the sum should be $(2n + 4) \times 90°$.

Example

1 Determine the sum of the *exterior* angles of the traverse of Fig. 8.2.

Answer
$$\text{Sum} = (2n + 4) \times 90°$$
$$= (8 + 4) \times 90°$$
$$= 1080°$$

Again the vector data, i.e. the horizontal length and direction of each line, relative to north, are required. These data are transformed into rectangular coordinates x_1, y_1, ..., x_4, y_4, which are plotted to scale.

(c) Traverse closed between previously fixed points

In this type of survey, the traverse begins on two points A and B of known bearing and ends on two different known points C and D (Fig. 8.3). The survey is once again self-checking in that the bearing of line CD deduced from the traverse angles should agree with the already known bearing of CD.

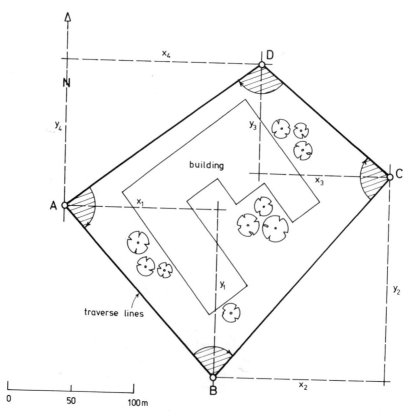

Figure 8.2

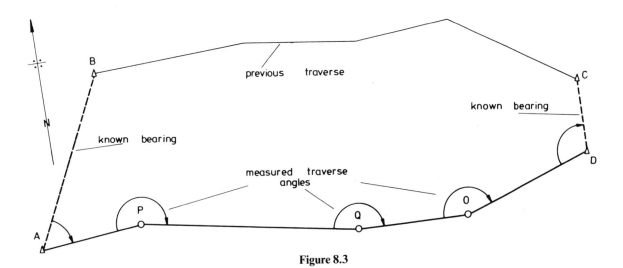

Figure 8.3

2. Basic principles of traversing

In all three traverses (Figs 8.1, 8.2 and 8.3) the survey fieldwork consists of:

(a) measuring the slope length of every line,
(b) measuring the angle of gradient of every line,
(c) measuring the clockwise horizontal angle between adjacent lines,
(d) measuring the bearing, i.e. the direction relative to *north* of *one* line of the traverse.

The office work consists of the following calculations:

(a) the horizontal length of every line derived from the slope length and angle of inclination,
(b) the bearing of every line derived from the observed bearing of one line and the horizontal angles,
(c) the partial rectangular coordinates (x, y) of each point relative to the previous point, derived from the horizontal lengths and bearings of the lines,
(d) the total rectangular coordinates of each point relative to the origin of the survey derived by accumulating algebraically the x and y partial coordinates.

3. Fieldwork

Figure 8.4 shows an open traverse ABCD following the route of a stream. The slope lengths, angles of inclination and horizontal angles required to complete the traverse are clearly shown on the figure.

It is generally accepted that a minimum of four surveyors is required to conduct a theodolite traverse. Their duties are:

(a) to select suitable stations,
(b) to measure the distances between the stations,
(c) to erect, attend and move the sighting targets from station to station,
(d) to measure and record the angles,
(e) to reference the stations for further use.

(a) Factors influencing choice of stations

On arrival at the site, the survey team's first task is to make a reconnaissance survey of the area with a view to selecting the most suitable stations. Generally the stations have to be fairly permanent and concrete blocks are generally formed *in situ*. A bolt or wooden peg is left in the concrete to act as a centre point. Where the survey mark is to be sited on a roadway a masonry nail is hammered into the surface and a circle painted around it.

The positions of the stations are governed by the following factors:

1. Easy measuring conditions. Since the angles can be measured very accurately with the theodolite, the linear measurements must be of comparable accuracy. Most measurements will be made along the surface using a steel tape; therefore the surface conditions must be conducive to good measuring. Roadways, paths and railways present good measuring conditions since they are smooth and have regular gradients. Conversely, measurements should not be made through long grass or undergrowth, over very undulating ground or heaps of rubble, etc.

2. Avoidance of short lines. Wherever possible the traverse should exclude short lines. The reason is simply that angular errors are introduced if targets at short range are not accurately bisected. For example, if the theodolite sights 2 millimetres off-target at a range of 10 metres, it is equivalent to sighting 40 millimetres off-target at a range of 200 metres.

3. Stations should be chosen so that the actual station mark can be sighted. If a pole has to be erected at a station it must be plumbed exactly above the mark using either a builder's spirit level or a plumb-bob. Failure to observe this simple precaution may result in a fairly substantial angular error. For example, in Fig. 8.5 a 2 metre pole has been erected on a station and is off-vertical by only 20 mm at the top.

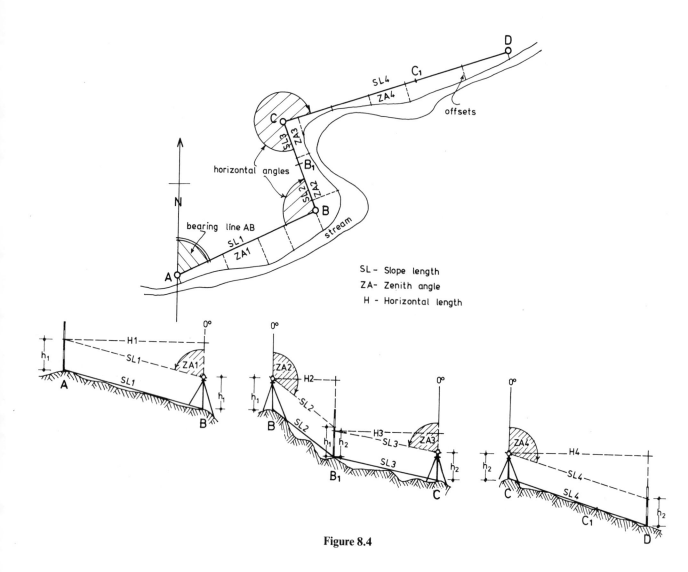

SL – Slope length
ZA – Zenith angle
H – Horizontal length

Figure 8.4

Figure 8.5

If a theodolite at a range of 50 metres sights the top of the pole instead of the ground station, the error in the measured angle will be 80 seconds.

4. If the other conditions can be readily satisfied the stations should be chosen near to some permanent objects, such as lamp standards, trees, etc., in order that they may be found readily at a later date by measuring from these objects.

(b) Linear measurement

In the calculation of rectangular coordinates, the true plan, i.e. horizontal length of every line, is required. Cognisance should therefore be taken of the various constant errors that affect linear measurements, and appropriate corrections should be applied to the measured lengths to produce true horizontal lengths. This suggests therefore that:

1. Measurements should be made on the slope and angles of inclination should be observed.
2. Each tape length should be aligned accurately using the theodolite.
3. Temperature should be recorded and the appropriate correction applied.
4. The steel tape should be standardized before use.
5. The correct tension should be applied to the tape and the tape should not be allowed to sag.

In engineering and construction surveying it is normal practice to align the tape by eye and dispense with temperature measurements. Frequently the tension is judged but this is not good practice and a spring balance or tension handle should be used.

Assuming that this practice is to be employed, the following equipment is necessary for measuring a traverse line:

1. A steel tape graduated throughout in metres and centimetres, or preferably, millimetres.
2. A spring balance calibrated in kilogrammes or a BS tension handle.
3. Measuring arrows and two ranging poles.

4. Marking plates or pegs to be used on soft ground. A marker plate can be made quite easily from a piece of metal, 100 mm square, if the corners are bent to form spikes. On hard surfaces, the surface is simply chalked and a mark made with a pencil.

Procedure

Before commencing the measurement of the lines of the traverse, the steel tape is checked against a known standard length (usually a new tape kept specifically for this purpose). The actual length of the tape is noted and all measured lengths are subsequently corrected using the method described in Sec. 2(c) in Chapter 3, where the correction c was shown to be

$$c = (L - l) \times \text{number of whole or}$$
$$\text{partial tape lengths}$$

in which formula

$$L = \text{actual length of tape}$$
$$l = \text{nominal length of tape}$$

Four persons are required to measure a line with high accuracy. Two are stationed at the rear and two at the forward end of the steel tape. Suppose the line AB in Fig. 8.4 is to be measured. The line is shorter than the length of the 30 metre tape. The procedure is as follows:

1. The tape is unwound and laid across the two pegs A and B.
2. One man anchors the forward end by putting a measuring pin through the handle into the ground. (Fig. 8.6).
3. A second man attaches the spring balance to the rear end, tightens the tape and anchors it by putting a second pin through the spring balance handle. The pin is levered back until the correct pull of 5 kg is registered on the spring balance. 'Read' is then called to the third and fourth men stationed at either end of the tape.
4. On receiving this command, they read the tape against the pencil crosses. Both readings are entered in the field book on line 1 (Table 8.1).

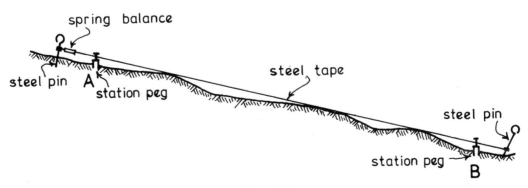

Figure 8.6

		Standard tension	5 kg
		Nominal tape length	30.000 m
		Actual tape length	30.003 m

	Line	Rear	Forward	Length	Mean (m)
1	AB	0.419	26.150	25.731	
2		0.365	26.098	25.733	
3		0.123	25.855	25.732	25.732
4	B–B1	0.511	11.611	11.100	
5		0.483	11.583	11.100	
6		0.276	11.376	11.100	11.100
7	B1–C	0.375	12.576		
8		0.675	12.876		
9		0.203	12.403		
10	C–C1	0.110	29.911		
11		0.122	29.923		
12		0.131	29.933		
13	C1–D	0.567	24.077		
14		0.523	24.034		
15		0.295	23.806		

Table 8.1

5. The difference between the readings is worked out.
6. The forward anchor man moves his pin slightly and the whole procedure is repeated to produce a second set of readings (line 2).
7. A third set of readings is obtained (line 3) and the mean of the three sets is calculated.
8. The gradient of the line is measured by theodolite. (A full description of the techniques involved is given in Sec. 3(c), following.)

Line BC (Fig. 8.4)
The line BC is again shorter than the length of the tape but the ground surface undulates considerably along its length. The line must be measured in sections or bays, the lengths of the bays being governed by the occurrence of the undulations.

Line BC is divided into two bays, B to B1 and B1 to C. A marker plate or peg is inserted on the line to denote point B1. Each bay is measured in exactly the same manner as line AB. The results of the measurement are noted in the field book (Table 8.1) on lines 4 to 9, where bay 1 of line BC has been reduced to produce a slope length of 11.100 metres.

The gradient of each bay is measured separately, using a theodolite (Sec. 3(c)).

Line CD (Fig. 8.4)
Line CD is longer than the length of the tape and the ground has a regular gradient along the whole length of the line. The line must again be divided into bays C to C1 and C1 to D. A marker plate or peg is inserted on the line at approximately 29 metres to denote point C1. Each bay is measured in exactly the same manner as line AB and the results noted in the field book (Table 8.1). The gradient of the whole line CD is

measured using a theodolite (Sec. 3(c)). It is not necessary to measure the gradient of each bay separately in this case, although this is often done.

An alternative method of measuring is commonly used. Complete 30 m tape lengths are marked on the plates and summed at the end of the line. This method is open to more errors than the former and the line should be remeasured in the opposite direction.

Example

2 Table 8.1 shows the results of the linear measurement of lines AB, BC and CD of the open traverse (Fig. 8.4). The table has been partially completed.

(a) Complete the reduction of the results, to produce the mean slope length of each bay of lines BC and CD.
(b) Correct the results for standardization.

Answer (Table 8.2)

		Standard tension	5 kg
		Nominal tape length	30.000 m
		Actual tape length	30.003 m

	Line	Rear	Forward	Length	Mean (m)
1	AB	0.419	26.150	25.731	
2		0.365	26.098	25.733	
3		0.123	25.855	25.732	25.732
4	B–B1	0.511	11.611	11.100	
5		0.483	11.583	11.100	
6		0.276	11.376	11.100	11.100
7	B1–C	0.375	12.576	12.201	
8		0.675	12.876	12.201	
9		0.203	12.403	12.200	12.201
10	C–C1	0.110	29.911	29.801	
11		0.122	29.923	29.801	
12		0.131	29.933	29.802	29.801
13	C1–D	0.567	24.077	23.510	
14		0.523	24.034	23.511	
15		0.295	23.806	23.511	23.511
	CD				53.312

Table 8.2

Standardization correction
Correction $= (L - l)$ per tape length
$\qquad = (30.003 - 30.000)$ per tape length

Line AB
Number of tape lengths $= 25.732/30 = 0.858$
$\qquad$ Correction $c = +0.003 \times 0.858$
$\qquad\qquad = +0.003$ m

Correct slope length $= 25.732 + 0.003$
$\qquad\qquad = 25.735$ m

Line B–B1

$$\text{Correction } c = + 0.003 \times 11.100/30$$
$$= + 0.003 \times 0.37$$
$$= + 0.001$$

$$\text{Correct slope length} = 11.100 + 0.001$$
$$= 11.101 \text{ m}$$

Similarly,

$$\text{correct slope length } C–B_1 = 12.201 + 0.001$$
$$= 12.202 \text{ m}$$

$$\text{and } CD = 53.312 + 0.005$$
$$= 53.317 \text{ m}$$

(c) Angular measurement

In Fig. 8.4, horizontal angles ABC and BCD are to be measured, together with the gradients of lines AB, B–B_1, B_1–C and CD. It is normal practice to measure the horizontal angle followed by the angles of inclination at one station, although experienced surveyors frequently measure them simultaneously.

Horizontal angles

In Fig. 8.4, angle ABC is first to be measured. The complete procedure for measuring a horizontal angle was described in Sec. 5 of Chapter 7. The section should be revised at this juncture since the method of measuring an angle depends upon the type of theodolite being used and whether or not the instrument is to be set to zero.

Briefly, a compensated measure of the angle ABC is made on face left and face right as follows:

1. The theodolite is plumbed over peg B and accurately levelled.
2. Peg A is sighted on face left with the theodolite set preferably to zero degrees, although any circle reading is acceptable. The reading is entered in the field book (Table 8.3) on line 1, column 3.
3. Peg C is sighted and the reading (93° 15′ 40″) noted on line 2, column 3.
4. The face left horizontal angle is computed and entered on line 3, column 3.
5. The instrument is set to face right (by transitting the telescope) and peg A is sighted again. The reading (179° 59′ 40″) is noted on line 1, column 4 of the field book (Table 8.3).
6. Peg C is sighted and the reading (273° 15′ 40″) noted on line 2, column 4.
7. The face right horizontal angle is computed and entered on line 3, column 4.
8. The mean of face left and face right horizontal angles is computed on line 3, column 5.
9. If the difference between the face left and face right values of the horizontal angle differ by more than 20 seconds, the angle should be remeasured.

Angles of inclination

The angle of gradient may be measured as a vertical angle or as a zenith angle, depending upon the configuration of the vertical circle of the theodolite. The complete procedure for the measurement of an angle

	1	2	3	4	5	6	7	8	9
	Instrument station	Station observed	Face left (FL)	Face right (FR)	Mean horizontal angle	Line	Face	Vertical circle reading	Zenith angle
1 2	B	A C	00° 00′ 00″ 93° 15′ 40″	179° 59′ 40″ 273° 15′ 40″		BA	L R	87° 20′ 40″ 272° 39′ 40′	87° 20′ 30″
3	AB̂C	=	93° 15′ 40″	93° 16′ 00″	93° 15′ 50″			360° 00′ 20″	
4 5	C	B D	00° 00′ 00″ 274° 31′ 00″	180° 00′ 00″ 94° 31′ 20″		B–B1	L R	105° 25′ 40″ 254° 34′ 20″	105° 25′ 30″
6	BĈD	=						360° 00′ 20″	
7 8						C–B1	L R	85° 15′ 20″ 275° 45′ 00″	
9									
10 11						CD	L R	93° 27′ 00″ 266° 33′ 20″	
12									

Table 8.3

of gradient was described in Sec. 6 in Chapter 7. The section should be revised at this juncture.

Briefly, the angle of gradient from B to A (Fig. 8.4) is measured as a zenith angle as follows:

1. The theodolite, already set over peg B, is accurately levelled. The height of the theodolite (h_1) above peg B is measured (using a tape or rule) from the top of the peg to the centre of the transit axis.
2. A levelling staff is held on peg A and the reading h_1 is sighted on face left. The vertical circle reading (87° 20' 40") is noted in line 1, column 8 (Table 8.3).
3. The instrument is set to face right (by transitting the telescope) and the staff again read at height h_1. The vertical circle reading (272° 39' 40") is noted on line 2, column 8.
4. The angles are added, the index error calculated and the corrected face left zenith angle is computed and entered in the field book on line 1, column 9.

$$
\begin{aligned}
\text{Face left reading} &= 87° \ 20' \ 40'' \\
\text{Face right reading} &= 272° \ 39' \ 40'' \\
\hline
\text{Sum} &= 360° \ 00' \ 20'' \\
\text{Index error} &= +20'' \\
\text{Correction} &= -10'' \\
\hline
\text{Corrected FL reading} &= 87° \ 20' \ 30''
\end{aligned}
$$

The zenith angle from B to B1 is then measured in exactly the same manner as for line AB. The results of the measurement are shown in the field book (Table 8.3) on lines 4, 5 and 6 where the zenith angle is computed as 105° 25' 30" (line 4, column 9).

The theodolite is returned to its box and carried to the next traverse station C, where horizontal angle BCD and zenith angles C to B1 and C to D are measured.

The complete set of fieldwork notes is shown in Table 8.3.

Example

3 Table 8.3 shows the results of the angular measurements of traverse ABCD. The table has been partially completed. Complete the reduction of the results to produce the horizontal angle BCD and zenith angles C–B1 and CD.

Answer (Table 8.4)

(d) Multiple tripod system

Figure 8.7 is an example of a modern traverse outfit, comprising a theodolite and several targets, each of which has its own tripod. The theodolite and targets detach from the tribrach and are interchangeable. The tribrach, of course, is fitted with an optical plummet, three levelling screws and plate bubble.

For the measurement of any angle, three plumbing operations are necessary regardless of the form of target being used, resulting in the expenditure of much time and effort and, of course, with such a system small plumbing errors must inevitably occur. When three tripods are available the errors

	1 Instrument station	2 Station observed	3 Face left (FL)	4 Face right (FR)	5 Mean horizontal angle	6 Line	7 Face	8 Vertical circle reading	9 Zenith angle
1	B	A	00° 00' 00"	179° 59' 40"		BA	L	87° 20' 40"	87° 20' 30"
2		C	93° 15' 40"	273° 15' 40"			R	272° 39' 40"	
3	AB̂C	=	93° 15' 40"	93° 16' 00"	93° 15' 50"			360° 00' 20"	
4	C	B	00° 00' 00"	180° 00' 00"		B–B1	L	105° 25' 40"	105° 25' 30"
5		D	274° 31' 00"	94° 31' 20"			R	254° 34' 40"	
6	BĈD	=	274° 31' 00"	274° 31' 20"	274° 31' 10"			360° 00' 20"	
7						C–B1	L	85° 15' 20"	85° 15' 10"
8							R	275° 45' 00"	
9								360° 00' 20"	
10						CD	L	93° 27' 00"	93° 26' 50"
11							R	266° 33' 20"	
12								360° 00' 20"	

Table 8.4

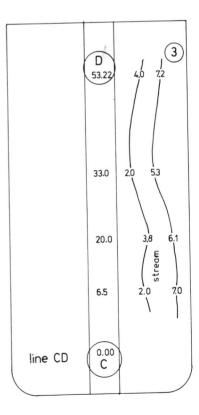

Figure 8.7

are eliminated. In the measurement of angle ABC in Fig. 8.7, the tripods are plumbed and levelled over the stations.

Targets are placed at stations A and C and the theodolite at station B. The plummet, theodolite and targets are completely interchangeable and are detachable from the levelling head.

The angle ABC is measured in the normal way. Angle BCD is measured by leap-frogging tripod A to station D; target A to B; theodolite B to C; and target C to D. There is no need to centre the theodolite at C since the tripod is already over the mark. Similarly, the target at back station B is correctly centred.

(e) Detail survey

The details that have to be added in the case of the traverse ABCD (Fig. 8.4) are simply the banks of the stream. These are surveyed by offsetting at frequent intervals along the various traverse lines. All of the offsets are shown on three pages of booking in Fig. 8.8. The method of offsetting was fully explained in Secs 1(b) and 3(b) in Chapter 3.

(f) Orientation of traverse surveys

The theodolite in its conventional form does not measure bearings directly. However, it is usually

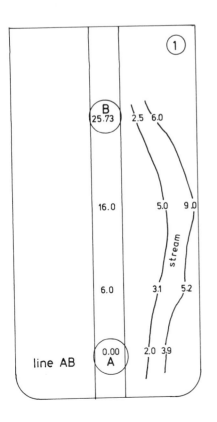
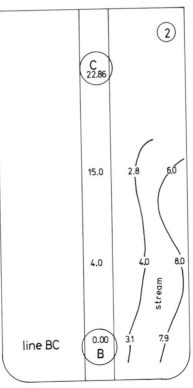
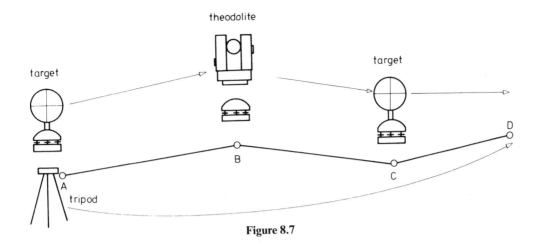

Figure 8.8

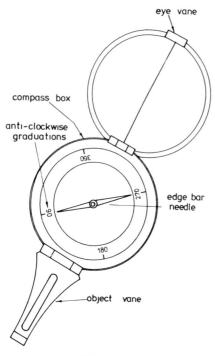

Figure 8.9

necessary to orientate any traverse survey to north. Two methods are available.

Use of compass
The most conventional way to find bearings is by the use of a compass, set up at any (but preferably the first) station of the traverse. Figure 8.9 shows a typical surveyor's compass. The compass needle is balanced on a pivot point and, when released, automaticlly points to magnetic north.

In use, the compass is set on a tripod (or may be hand-held) over the first point A of the traverse (Fig. 8.4). The needle is released, allowed to settle and a sight is taken through the sighting vanes of the compass to station B. The angle between the needle and the line AB is the magnetic bearing of the line.

Use of Ordnance Survey plan
The bearing of a survey line may be obtained from an Ordnance Survey plan. An OS plan is grided at 100 m intervals with the National Grid (Sec. 2(a) in Chapter 2), the lines of which point to Grid North.

On any traverse survey, one of the lines is plotted as accurately as possible on the OS plan and the angle between the grid line and the survey line is measured by protractor to the nearest half-degree.

Exercise 8.1

1 Table 8.5 and Fig. 8.10 show the layout and corresponding data of a closed traverse ABCDEF at the

GCB Outdoor Centre. (The station points are, in fact, those of the linear survey, detailed and plotted in Chapter 3, Sec. 3.) From the table, determine:

(a) the six clockwise measured angles, A to F, of the traverse,
(b) the angular error of the traverse,
(c) the slope lengths of the six lines AB, BC, CD, DE, EF and FA,
(d) the vertical angle of gradient of each line.

4. Plotting a traverse

(a) Introduction

On completion of the fieldwork of a traverse, the data, collected on the survey, must be processed into suitable units for plotting the survey on paper. Sections 3(b) and 3(c) of this chapter describe how the field results are obtained and Tables 8.2 and 8.4 show respectively the linear and angular data of the open traverse of Fig. 8.4. The reduced data, together with a sketch of the survey, is reproduced as Fig. 8.11.

Two methods are available for plotting the survey.

1. *Graphical method*
The survey is plotted by scale rule and protractor.
2. *Mathematical method*
The survey is plotted by rectangular coordinates.

The graphical method is not commonly used, since angles, measured in the field with an accuracy of a few seconds of arc, are plotted using a protractor with an accuracy of only half of a degree, at best. Nevertheless, it is a valuable step in the understanding of bearings and the survey will be plotted, using this method, in Sec. 4(d). The angles are plotted directly by protractor, without any further calculation. The slope lengths, however, must be converted into plan lengths before plotting the survey.

(b) Plan lengths from zenith angles

Figure 8.12 shows a theodolite, set up at a traverse station X. The height from the peg to the transit axis of the theodolite is *h* metres. A staff is held vertically on station Y and the height *h* is sighted. A zenith angle of θ degrees is recorded.

In the right-angled triangle formed by the vertical axis through the theodolite, the horizontal line through the target and the line joining the two points, length *S* equals the slope length XY and length *H* is the plan length of line XY. In the right-angled triangle.

$$H/S = \text{sine } \theta$$
$$\text{Therefore } H = S \times \sin \theta$$

i.e.

horizontal length = slope length × sin (zenith angle)

This formula holds good for any zenith angle.

Standard tape tension 5 kg
Nominal tape length 50.000 m Theodolite: Sokkia 10 s electronic
Actual tape length 50.000 m

Instrument station	Station observed	Face left	Face right	Mean observed angle	Line	Face	Vertical circle reading	Zenith angle	Measured length			
									Rear	Forward	Length	Mean
A	F	00° 00′ 00″	180° 00′ 20″		AB	L	91° 30′ 40″		0.150	48.796		
	B	183° 33′ 20″	3° 33′ 40″			R	268° 28′ 40″		0.325	48.969		
									0.200	48.842		
B	A	00° 00′ 00″	179° 59′ 40″		BC	L	95° 32′ 00″		0.350	27.874		
	C	83° 41′ 20″	263° 41′ 40″			R	264° 27′ 20″		0.750	28.272		
									0.523	28.049		
C	B	00° 00′ 00″	180° 00′ 20″		CD		89° 29′ 10″		0.360	48.790		
	D	86° 48′ 00″	266° 48′ 00″				270° 30′ 30″		0.820	49.250		
									0.600	49.030		
D	C	00° 00′ 00″	180° 00′ 00″		DE		90° 37′ 20″		0.200	22.350		
	E	182° 39′ 40″	2° 39′ 40″				269° 22′ 20″		0.090	22.240		
									0.325	22.475		
E	D	00° 00′ 00″	179° 59′ 40″		EF	L	86° 13′ 30″		1.115	19.236		
	F	89° 15′ 40″	269° 15′ 40″			R	273° 45′ 50″		1.300	19.416		
									0.800	18.911		
F	E	00° 00′ 00″	180° 00′ 00″		FA	L	90° 02′ 50″		0.326	22.121		
	A	94° 00′ 40″	274° 00′ 40″			R	269° 56′ 30″		0.250	22.045		
									0.200	21.995		

Table 8.5

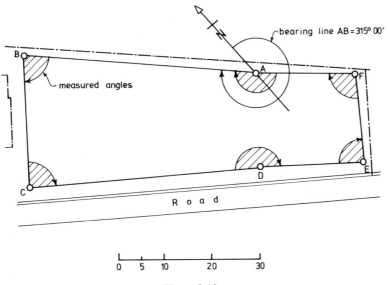

Figure 8.10

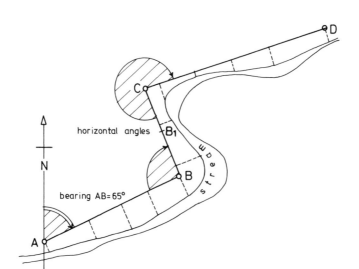

Horizontal Angle	Accepted Value	Line	Slope Length	Zenith Angle
ABC	93° 15′ 50″	AB	25.735	87° 20′ 30″
BCD	274° 31′ 10″	BB′	11.101	105° 25′ 30″
		B′C	12.202	85° 15′ 10″
		CD	53.317	93° 26′ 50″

Figure 8.11

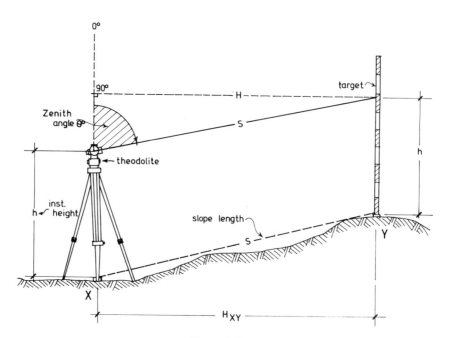

Figure 8.12

Example

4 In Fig. 8.11 the following data refer to lines AB, BC and CD:

Line	Slope length	Zenith angle
AB	25.735	87° 20′ 30″
BB′	11.101	105° 25′ 30″
B′C	12.202	85° 15′ 10″
CD	53.317	93° 26′ 50″

Calculate the plan lengths of lines AB, BC and CD.

Answer

Plan length = slope length × sin zenith angle

Plan AB = 25.735 × sin (87° 20′ 30″)
= 25.707 m

Plan BC = 11.101 × sin (105° 25′ 30″)
+ 12.202 × sin (85° 15′ 10″)
= 10.701 + 12.160
= 22.861 m

Plan CD = 53.317 × sin (93° 26′ 50″)
= 53.221 m

(c) Plan lengths from vertical angles

Some theodolites record the vertical angle rather than the zenith angle. The vertical angle is the angle of inclination measured on the vertical circle of the theodolite when the zero degree reading indicates the horizontal position of the telescope rather than the vertical position. (The configuration of the vertical circle depends upon the manufacturer's preference.)

In Fig. 8.13, the vertical angle is clearly shown. It is, of course, the complement of the zenith angle.

In the right-angled triangle, formed by the horizontal axis through the theodolite, the vertical axis through the staff and the line joining them, length S equals the slope length XY and length H is the plan length of the line XY. In the right-angled triangle,

$$H/S = \text{cosine } \alpha$$
$$\text{Therefore } H = S \times \cos \alpha$$

i.e.

horizontal length = slope length × cos (vertical angle)

Example

5 The following observations were obtained on a theodolite traverse (Fig. 8.14):

Line	Bay	Slope length (m)	Vertical angle
PQ	1	53.220	+03° 25′ 30″
	2	29.610	−00° 30′ 40″
	3	17.325	+04° 19′ 00″

When checked against a standard 30 metre length, the steel tape was found to measure 30.004 m. Calculate the true length of line PQ.

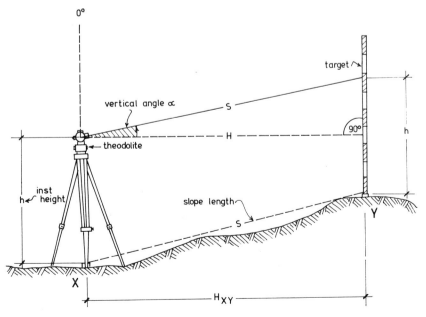

Figure 8.13

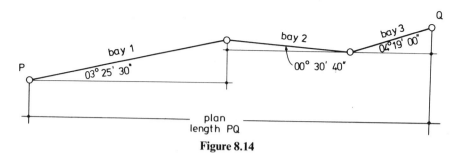

Figure 8.14

Solution

Correction for standardization

$$= (L - l) \text{ per tape length}$$
$$= (30.004 - 30.000) \text{ per tape}$$

Bay 1

Number of tapes $= 53.220/30 = 1.774$

Correction $= 0.004 \times 1.774$
$$= +0.007 \text{ m}$$

Correct slope length $= 53.220 + 0.007$
$$= 53.227 \text{ m}$$

Bay 2

Correction $= 0.004 \times 0.987$
$$= +0.004$$

Slope length $= 29.610 + 0.004$
$$= 29.614 \text{ m}$$

Bay 3

Correction $= 0.004 \times 0.5775$
$$= +0.002$$

Slope length $= 17.325 + 0.002$
$$= 17.327 \text{ m}$$

Plan length $=$ slope length $\times$ cos inclination

Bay 1

Plan length $= 53.227 \times \cos 03° 25' 30''$
$$= 53.227 \times 0.998\ 213\ 7$$
$$= 53.132 \text{ m}$$

Bay 2

Plan length $= 29.614 \times 0.999\ 960\ 2$
$$= 29.613 \text{ m}$$

Bay 3

Plan length $= 17.327 \times 0.997\ 148\ 7$
$$= 17.278 \text{ m}$$

So plan length PQ $= 53.132 + 29.613$
$$+ 17.278$$
$$= 100.023 \text{ m}$$

(d) Graphical method of plotting

Using this method, the survey is plotted by scale rule and protractor.

Equipment

The equipment required for plotting is as follows:

1. A large circular protractor graduated in half-degrees,
2. An appropriate scale rule,
3. An instrument for drawing parallel lines. A set of roller parallels is best, but for classroom and examination work, two set-squares suffice.
4. A 4H pencil and a needle.

Procedure

An open traverse along the north bank of a stream is shown in Fig. 8.4 (framework) and in Fig. 8.8 (offsets). The survey is to be plotted to scale 1:500. The corrected framework details are:

Bearing (direction) of line AB $= 65° 00'$
Angle ABC $= 93° 16'$
Angle BCD $= 274° 31'$
Lengths   AB $= 25.73$ m,
BC $= 22.86$ m,
CD $= 53.22$ m

1. A freehand drawing of the survey is made in order to locate the survey centrally on the drawing paper.
2. A line representing the magnetic meridian is drawn lightly through the first point A of the survey (Fig. 8.15).
3. The protractor is laid on this line with the 0° graduation facing to magnetic north and the bearing of line AB, namely 65°, is marked off.
4. A line is drawn through this latter plotting mark and the point A and the length 25.73 m is marked, thus establishing point B.
5. The protractor is centred on station B with the zero degree graduation pointing back to station A and angle ABC ($= 93° 16'$) is marked off (Fig. 8.15).
6. A line is drawn from point B through this point and the distance of 22.86 m is scaled to establish point C.
7. The procedures (5) and (6) are repeated, using angle BCD ($= 274° 31'$) and CD ($= 53.22$ m) to establish point D.
8. Circles of 2 mm diameter are drawn in ink around each survey station and the stations clearly designated.

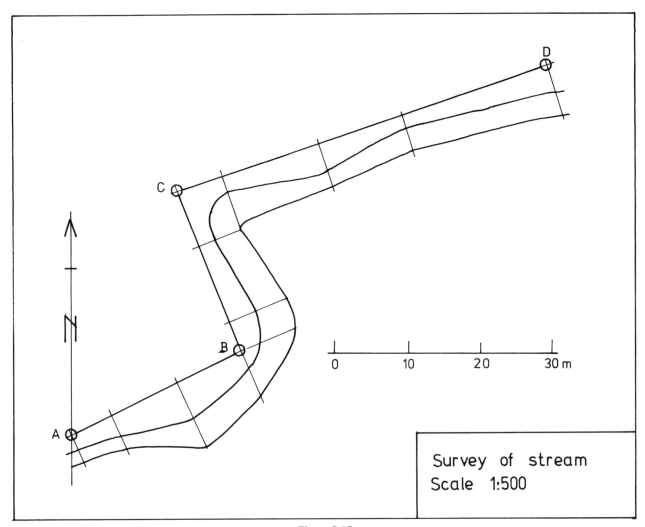

Figure 8.15

9. The offsets are plotted and the features drawn in ink.

Exercise 8.2

1 Figure 8.16 shows a closed traverse survey at the GCB Outdoor Centre. (The stations are actually those of the linear survey, detailed in Chapter 3, Fig. 3.34.)

(a) Calculate the plan lengths of the survey lines.
(b) Plot the survey, using scale rule and protractor, to scale 1:500.

2 Table 8.6 shows the plan lengths and bearings of five lines of a closed compass traverse. (The bearing is the clockwise angle measured from magnetic north to the survey line.)

(a) Plot the survey to a scale of 1:1000.
(b) Scale the closing error.

Line	AB	BC	CD	DE	EA
Bearing	29° 30′	100° 45′	146° 30′	242° 00′	278° 45′
Length (m)	83.50	61.80	62.00	51.20	90.40

Table 8.6

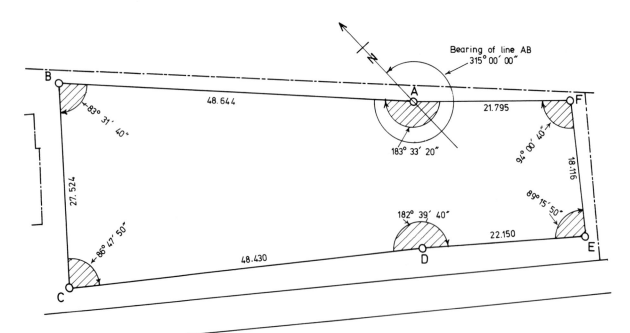

Horz. angle	Accepted value	Line	Slope length	Zenith angle
FAB	183° 33′ 20″	AB	48.644	91° 31′ 00″
ABC	83° 41′ 40″	BC	27.524	95° 32′ 20″
BCD	86° 47′ 50″	CD	48.430	89° 29′ 20″
CDE	182° 39′ 40″	DE	22.150	90° 37′ 30″
DEF	89° 15′ 50″	EF	18.116	86° 13′ 50″
EFA	94° 00′ 40″	FA	21.795	90° 03′ 10″

Figure 8.16

5. Answers

Exercise 8.1

1 Table 8.7

Standard tape tension   5 kg
Nominal tape length   50.000 m   Theodolite: Sokkia 10 s electronic
Actual tape length   50.000 m

Instrument station	Station observed	Face left	Face right	Mean observed angle	Line	Face	Vertical circle reading	Zenith angle	Rear	Forward	Length	Mean
A	F	00° 00′ 00″	180° 00′ 20″		AB	L	91° 30′ 40″	91° 31′ 00″	0.150	48.796	48.646	
	B	183° 33′ 20″	3° 33′ 40″			R	268° 28′ 40″		0.325	48.969	48.644	48.644
	FÂB	183° 33′ 20″	183° 33′ 20″	183° 33′ 20″			359° 59′ 20″		0.200	48.842	48.642	
B	A	00° 00′ 00″	179° 59′ 40″		BC	L	95° 32′ 00″	95° 32′ 20″	0.350	27.874	27.524	
	C	83° 41′ 20″	263° 41′ 40″			R	264° 27′ 20″		0.750	28.272	27.522	27.524
	AB̂C	83° 41′ 20″	83° 42′ 00″	83° 41′ 40″			359° 59′ 20″		0.523	28.049	27.526	
C	B	00° 00′ 00″	180° 00′ 20″		CD		89° 29′ 10″	89° 29′ 20″	0.360	48.790	48.430	
	D	86° 48′ 00″	266° 48′ 00″				270° 30′ 30″		0.820	49.250	48.430	48.430
	BĈD	86° 48′ 00″	86° 47′ 40″	86° 47′ 50″			359° 59′ 40″		0.600	49.030	48.430	
D	C	00° 00′ 00″	180° 00′ 00″		DE		90° 37′ 20″	90° 37′ 30″	0.200	22.350	22.150	
	E	182° 39′ 40″	2° 39′ 40″				269° 22′ 20″		0.090	22.240	22.150	22.150
	CD̂E	182° 39′ 40″	182° 39′ 40″	182° 39′ 40″			359° 59′ 40″		0.325	22.475	22.150	
E	D	00° 00′ 00″	179° 59′ 40″		EF	L	86° 13′ 30″	86° 13′ 50″	1.115	19.236	18.121	
	F	89° 15′ 40″	269° 15′ 40″			R	273° 45′ 50″		1.300	19.416	18.116	18.116
	DÊF	89° 15′ 40″	89° 16′ 00″	89° 15′ 50″			359° 59′ 20″		0.800	18.911	18.111	
F	E	00° 00′ 00″	180° 00′ 00″		FA	L	90° 02′ 50″	90° 03′ 10″	0.326	22.121	21.795	
	A	94° 00′ 40″	274° 00′ 40″			R	269° 56′ 30″		0.250	22.045	21.795	21.795
	EF̂A	94° 00′ 40″	94° 00′ 40″	94° 00′ 40″			359° 59′ 20″		0.200	21.995	21.795	

Sum = 719° 59′ 00″
Traverse error =   − 01′ 00″

Table 8.7

Exercise 8.2

1 (a) *Line*   *Plan length*
   AB   48.627
   BC   27.395
   CD   48.428
   DE   22.149
   EF   18.076
   FA   21.795

(b) Figure 8.17

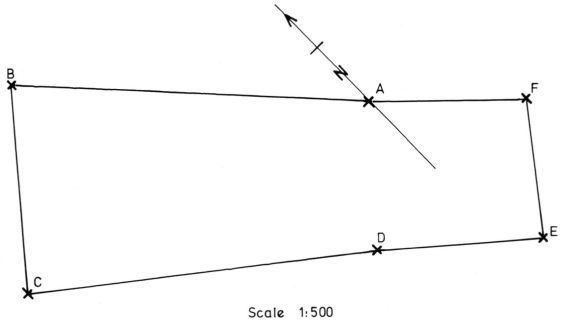

Scale   1:500

Figure 8.17

2 Figure 8.18

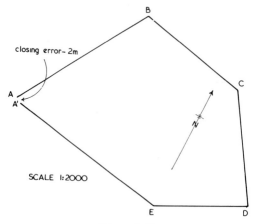

Figure 8.18

CHAPTER 9

Traverse survey computations

Objective

After studying this chapter, the reader should be able to (a) determine the bearings of the lines of a traverse survey, (b) compute the rectangular coordinates of the traverse stations and (c) plot a traverse survey to scale.

In every form of surveying, where angles are measured, the bearings of the survey lines usually have to be obtained. The bearing of a line is its direction relative to north.

1. Magnetic bearings

Magnetic bearings are measured using a compass. On all compasses, there is a magnetic needle, which, when suspended freely or when allowed to swing freely on a pivot will settle in the magnetic meridian, i.e. it will point to magnetic north. The scientific reasons for this do not really concern the building surveyor. Some theories suggest that it is due to the iron core of the earth's centre while others ascribe it to the electrical currents in the atmosphere caused by the earth's rotation. The fact is that the needle does settle in the magnetic meridian. Figure 9.1 shows the earth with the magnetic meridians meeting at the north and south magnetic poles.

If a compass were held at any point X the magnetic needle would settle with its north-seeking end pointing to the magnetic north pole. In Fig. 9.2, lines XA, XB and XC are survey lines radiating from point X.

The line XA makes an angle of 40° with the magnetic meridian while lines XB and XC make angles of 160° and 290° respectively. The maximum value of any angle is 360°; i.e. the angle can have any value on the whole circle of graduations provided it is measured in a clockwise direction. The angle of 160° formed by the meridian XN and the line XB is the whole circle magnetic bearing of line XB and is defined as the angle measured in a clockwise direction from the magnetic meridian to the line.

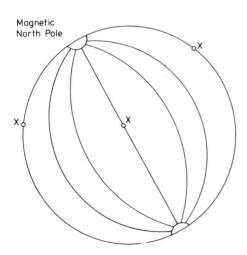

Figure 9.1

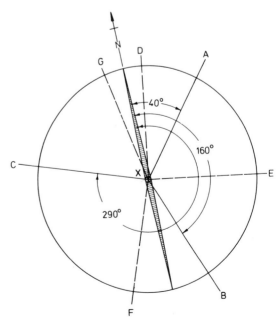

Figure 9.2

2. True bearings

Figure 9.3 shows the earth with the geographical meridians coinciding at the true north and true south poles. Any angle measured in a clockwise direction from the true meridian to a line is a true whole circle bearing (WCB). The true whole circle bearings of lines XA, XB and XC (Fig. 9.4) are 30°, 150° and 280° respectively.

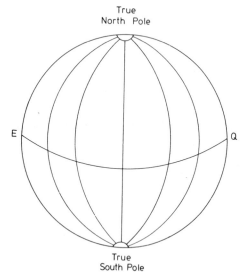

Figure 9.3

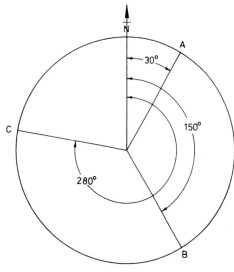

Figure 9.4

3. Grid bearings

In Sec. 2 in Chapter 2 it was pointed out that on Ordnance Survey maps only the central meridian of 2° west longitude points to true north. All other north grid lines are parallel to the central meridian and point to grid north. The grid whole circle bearing of any survey line is the clockwise angle between a grid north line and the survey line.

Figure 9.5 shows a part of a 1:2500 scale OS plan. The grid bearings of the centre lines of the roadways, measured from point X to points A, B and C, are 40°, 127° and 239° respectively.

4. Magnetic declination

When Figs 9.2 and 9.4 are combined, the result in Fig. 9.6 shows that the true and magnetic whole circle bearings of any line are different. The difference is called magnetic declination. Magnetic declination is said to be westerly when the magnetic meridian lies to the west of the true meridian.

Since the three quantities, magnetic whole circle bearing, true whole circle bearing and magnetic decli-

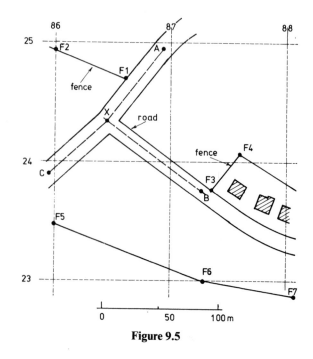

Figure 9.5

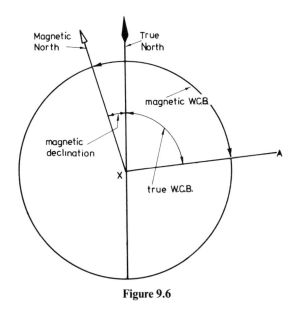

Figure 9.6

nation, are related, it is possible to calculate the third quantity if any two of them are known.

Example

3 Supply the missing quantities in the table below:

	Column 1	*Column 2*	*Column 3*
Magnetic WCB	60°	230°	—
True WCB	—	221°	166°
Magnetic declination	10° west	—	4° west

Answer (Fig. 9.7)

$$\text{True WCB} = 60° - 10°$$
$$= 50°$$
$$\text{Magnetic declination} = 230° - 221°$$
$$= 9° \text{ west}$$
$$\text{Magnetic WCB} = 166° + 4°$$
$$= 170°$$

This simple example has been used principally to show that the value of magnetic declination varies, the reasons for the variation being as follows.

(a) Geographical location

Figure 9.8 shows a plan view of the earth with the angle of declination drawn at three different locations. Clearly the value of the angle changes with geographical position. At point A the value is, say, 9° west; at point B it is 6° east, while at point C the value is almost zero.

(b) Annual variation

The magnetic poles of the earth are constantly changing their position relative to the true north and south poles, with the result that the value of declination at any point on the earth slowly changes its value throughout the year. The mean annual change in the value of declination, called the secular variation, is about one-tenth of a degree in Great Britain. At the present moment, i.e. 1995, the value of declination is about 6° west in the Midlands. Because of the secular variation, this value will decrease to zero by the year 2055.

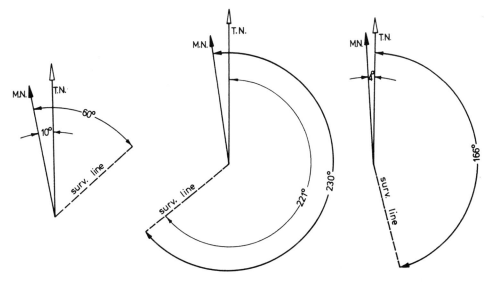

Figure 9.7

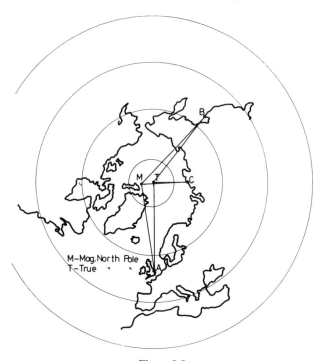

Figure 9.8

M—Mag.Nor th Pole
T—True " "

Example

4 Supply the missing quantities in the table below:

	Column 1	Column 2	Column 3
Magnetic WCB	40°	—	330°
True WCB	—	55°	315°
Magnetic declination	10° W	5° E	—

Answer (Fig. 9.9)

Column 1	Column 2	Column 3
30°	50°	15° W

---

5. Assumed bearings

On local surveys of small building sites and engineering works, it may not be necessary to relate the survey to either magnetic or true north. Some arbitrary point is chosen as reference object and treated as being the equivalent of the north pole. Common reference objects (ROs) are tall chimneys, church steeples and pegs hammered into the ground at some points where they can be easily found. Whole circle bearings are therefore the clockwise angles measured from the RO to any point. Alternatively, two points on the ground may be pinpointed on a large-scale OS map where the grid bearing of the line joining the points may be measured by protractor.

6. Forward and back bearings

It should be clear that any survey line can have four whole circle bearings, namely magnetic, true, grid or arbitrary. With the addition of certain other information these bearings can be interrelated as required.

Figure 9.10 shows a line joining two stations A and B where the whole circle bearing from A to B is 63°. If an observer were to stand at station A and look towards station B he or she would be looking in a direction 63° relative to the chosen north point. In surveying terms, the observer is looking *forward* towards station B from station A, and the forward whole circle bearing of the line AB is 63°. If the observer now stood at station B and looked towards station A he or she would be looking *back* along the line, i.e. in exactly the reverse direction.

This direction BA is known as the *back* whole circle bearing. The angular value is the clockwise angle from the north point to the line BA, namely (180° + 63°) = 243°.

From this simple example an equally simple rule can be deduced.

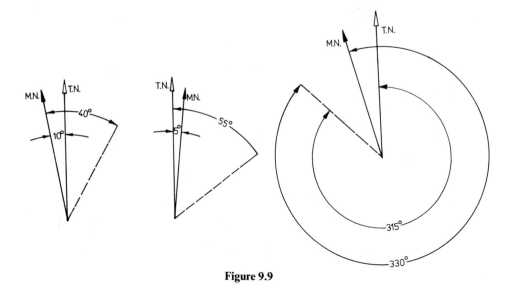

Figure 9.9

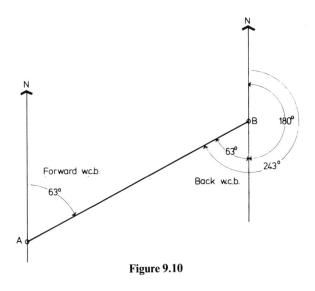

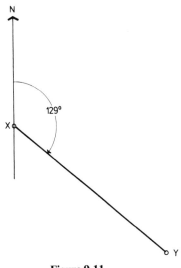

Figure 9.10

Figure 9.11

Rule
To obtain back bearings from forward bearings, or vice versa, add or subtract 180°.

In Fig. 9.11 the forward whole circle bearing of line XY is 129°. Using the rule above the back whole circle bearing YX is (180° + 129°) = 309°. The reader should now complete the diagram by drawing the north point through station Y and proving that 309° is the correct answer.

Example

5 The forward bearings of four lines are as shown below:

AB, 31°   BC, 157° 30'   CD, 200°   DE, 347° 15'

Calculate the back bearings.

Answer

Line	Forward		Back
AB	31° 00'	+180°	211° 00'
BC	157° 30'	+180°	337° 30'
CD	200° 00'	−180°	20° 00'
DE	347° 15'	−180°	167° 15'

---

7. Quadrant bearings

In subsequent calculations, whole circle bearings may require to be converted to quadrant bearings and vice versa.

If the cardinal points of the compass are drawn and labelled north, east, south and west respectively the whole 360° circle will have been divided into four quadrants of 90° (Fig. 9.12).

The quadrants are known as the north-east, south-east, south-west and north-west quadrants. The quadrant bearing of any line is the angle that it makes with the north–south axis. This angle is given the name of

the quadrant into which it falls. In Fig. 9.13, the whole circle bearings of four lines AB, AC, AD and AE are 40°, 121°, 242° and 303° respectively.

The angles that the lines make with the north–south axis are respectively:

AB	40°	quadrant bearing	N 40° E
AC	59°	quadrant bearing	S 59° E
AD	62°	quadrant bearing	S 62° W
AE	57°	quadrant bearing	N 57° W

The conversion of whole circle to quadrant bearings, sometimes called reduced bearings, can be readily carried out by the following rules:

1. When the whole circle bearing lies between 0° and 90°, its quadrant bearing has the same numerical value and lies in the NE quadrant.

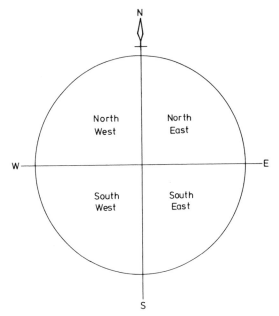

Figure 9.12

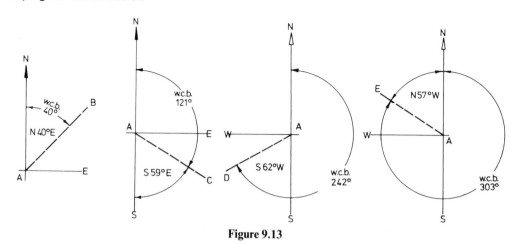

Figure 9.13

2. When the whole circle bearing lies between 90° and 180° the quadrant bearing is (180°—WCB) and lies in the SE quadrant.
3. When the whole circle bearing lies between 180° and 270° the quadrant bearing is (WCB—180°) and lies in the SW quadrant.
4. When the whole circle bearing lies between 270° and 360° the quadrant bearing is (360°—WCB) and lies in the NW quadrant.

Example

6 Convert the following whole circle magnetic bearings into quadrant bearings:

60° 30'   240° 10'   352° 10'   131° 00'

Answer
Whole circle

bearing	Quadrant bearing	
60° 30'		N60° 30' E
240° 10'	(240° 10' − 180°)	S 60° 10' W
352° 10'	(360° − 352° 10')	N07° 50' W
131° 00'	(180° − 131° 00')	S 49° 00' E

Exercise 9.1

1 The following magnetic whole circle bearings are known in a certain locality, where magnetic declination is 6° west:

Line	Magnetic WCB
AB	65° 00'
BC	132° 00'
XY	216° 30'
PQ	279° 30'

Calculate the true bearings of the lines.

2 The whole circle forward bearings of the lines of a traverse are as follows:

Line	Magnetic WCB
MN	35° 00'
NO	175° 36'
OP	214° 14'
PQ	267° 32'
QR	356° 20'

Calculate the back bearings of the lines.

3 Convert the following whole circle true bearings into quadrant bearings:

311° 10'   247° 30'   060° 10'   093° 00'   270° 00'
167° 50'   111° 10'   264° 50'   359° 10'   179° 00'

4 Convert the following quadrant bearings into whole circle bearings.

N 30° 10' W   S 60° 30' W   S 07° 45' E   N 10° 00' E
East   S 79° 10' W   S 89° 50' E   S 64° 30' E

5 The following true whole circle bearings are known in a certain area:

(a) 162° 10'   (b) 350° 00'   (c) 348° 30'   (d) 210° 10'

At a certain time the magnetic declination in the locality is found to be 10° 30' west.
Calculate the magnetic whole circle bearings of the lines and then convert them to magnetic quadrant bearings.

6 Complete the following table:

Magnetic WCB (forward)	26°	88°	233°	315°	8°
Magnetic declination	8°W	2°E	5°E	5°W	10°W
True WCB (forward)					
True WCB (back)					
True quadrant bearing (back)					

7 Convert the following whole circle bearings into quadrant bearings:

30° 45' 10"   163° 18' 20"   227° 36' 35"   343° 08' 09"

8 Convert the following quadrant bearings into whole circle bearings:

$$N\,40° \, 30' \, 20'' \, E \quad S\,75° \, 15' \, 45'' \, E$$
$$S\,39° \, 18' \, 17'' \, E \quad N\,64° \, 59' \, 58'' \, W$$

8. Conversion of angles to bearings

In order to plot the positions of traverse stations, using the method of rectangular coordinates, the forward bearing of every line of any traverse is required. In the field, the bearing of the first line only is measured, usually by compass. Alternatively, it may be obtained from a map, or it may simply be assumed. The bearing of every other line must be obtained by calculation.

(a) Open traverse

Figure 9.14 shows three lines of an open traverse. The forward bearing of line AB is 65° 00'. The traverse angles are

$$\text{Angle ABC} = 100° \, 00' \text{ and}$$
$$\text{Angle BCD} = 135° \, 00'$$

The forward bearing of line BC and line CD is required.

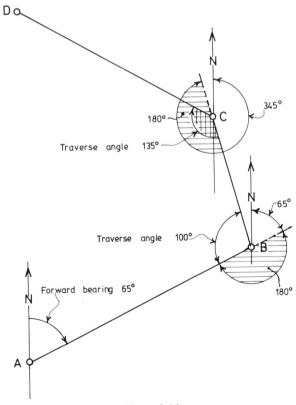

Figure 9.14

Line AB
The required forward bearing is angle NAB. It has been measured directly as 65° 00'.

Line BC
The required forward bearing is angle NBC. In Fig. 9.14, line AB has been projected forward, through station B.

$$
\begin{aligned}
\text{Angle NBC} &= 65° + 180° \text{ (hatched)} + 100° \\
&= (65° + 180°) + 100° \\
&= 245° + 100° \\
&= \text{back bearing BA} \\
&\qquad + \text{traverse angle ABC} \qquad (1) \\
&= 345° \, 00'
\end{aligned}
$$

Line BC
The required forward bearing is angle NCD. In Fig. 9.14, line BC has been projected forward, through station C.

$$
\begin{aligned}
\text{Angle NCD} &= 345° - 180° \text{ (hatched)} \\
&\qquad + 135° \text{ (cross-hatched)} \\
&= (345° - 180°) + 135° \\
&= 165° + 135° \\
&= \text{back bearing CB} \\
&\qquad + \text{traverse angle BCD} \qquad (2) \\
&= 300° \, 00'
\end{aligned}
$$

From these calculations, formulae (1) and (2) provide a general formula for use in bearings calculation.

> *Forward* bearing of *any* line
> = *back* bearing of *previous* line
> + *clockwise* angle between the lines

Frequently, it will be found that forward bearings, obtained from the above formula, will exceed 360°. In those cases, 360° is subtracted from the values to produce the correct bearings.

> **Example**
>
> **7** Figure 9.15 shows an open traverse along the pegs of a site boundary. The forward bearing of line AB of the traverse is 65° 34' 20''. Calculate the forward bearing of each line of the survey.
>
> *Answer*
>
> | | | Whole circle bearing |
> |---|---|---|
> | AB | | 65° 34' 20'' |
> | Back bearing BA | 245° 34' 20'' | |
> | + ABC | 110° 05' 20'' | |
> | BC | 355° 39' 40'' | 355° 39' 40'' |

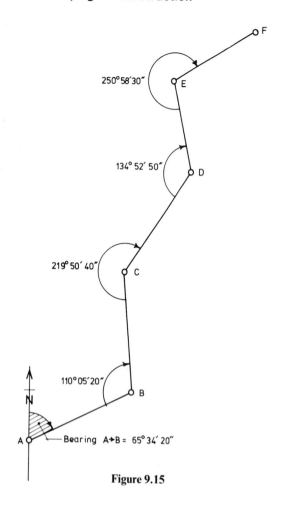

Figure 9.15

Back bearing CB	175° 39′ 40″	
+ BCD	219° 50′ 40″	
	395° 30′ 20″	
	− 360°	
CD	35° 30′ 20″	35° 30′ 20″
Back bearing DC	215° 30′ 20″	
+ CDE	134° 52′ 50″	
DE	350° 23′ 10″	350° 23′ 10″
Back bearing ED	170° 23′ 10″	
+ DEF	250° 58′ 30″	
	421° 21′ 40″	
	− 360°	
EF	61° 21′ 40″	61° 21′ 40″

Many surveyors prefer to use an amended version of the rule in which the calculation of the back bearing is omitted. Compensation is made at the end of the calculation for any line by adding or subtracting 180°.

The rule then becomes: 'To the *forward* bearing of the previous line add the next clockwise angle. If the sum is greater than 180°, subtract 180°. If the sum is less than 180°, add 180°. If the sum is greater than 540°, subtract 540°. The result is the forward bearing of the next line.'

In Fig. 9.15, the forward bearing of line BC is calculated as follows:

Forward bearing AB =	65° 34′ 20″
+ ABC	110° 05′ 20″
	175° 39′ 40″

Since the sum is less
than 180° add 180°   + 180° 00′ 00″

Forward bearing BC =	355° 39′ 40″

which agrees with the previous calculation.

Using this method the computerized solution of bearings is simplified.

(b) Computer solution

Bearings can be readily calculated using a computer program. Since some readers may not wish, or have the facilities, to consider this method of solution, the program (6) is included in Chapter 17.

Exercise 9.2

1 Figure 9.16 is a reproduction of Fig. 8.4, showing a short traverse along the north bank of a stream. The bearing of line AB, determined using a compass, is 65° 00′. Calculate the forward bearings of lines BC and CD.

(c) Closed traverse

The fieldwork of a closed traverse is self-checking to some extent. In any closed traverse the sum of the internal angles should equal $(2n - 4) \times 90°$, where n is the number of angles. When exterior angles are measured the sum should be $(2n + 4) \times 90°$.

In a closed traverse of six stations, therefore, the sum of the exterior angles should be

$$(2n + 4) \times 90° = (12 + 4) \times 90°$$
$$= 1440° \ 00′ \ 00″$$

It is very unlikely that the angles will sum to this amount because of the small errors inherent in angular measurements. However, the sum should be very close to this amount and should certainly be within the following limits:

$$\pm 40\sqrt{n} \text{ seconds, in the case of a single-second reading theodolite}$$

and

$$\pm \sqrt{n} \text{ minutes, in the case of a twenty-second reading theodolite}$$

In each case, n is the number of instrument stations. In all cases, where the sum of the observed angles is not $(2n \pm 4) \times 90°$ the angles must be adjusted so that they do sum to this figure.

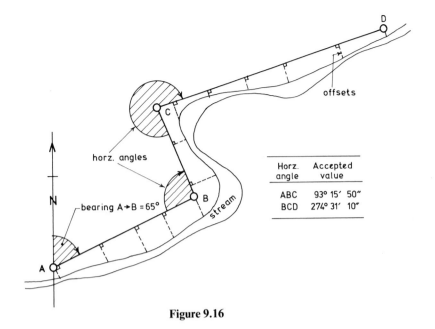

Figure 9.16

Example

8 In Fig. 9.17, the values of the exterior angles, measured by a one second theodolite, are:

Angle	Mean observed value
ABC	272° 03′ 10″
BCD	272° 05′ 51″
CDE	104° 50′ 31″
DEF	261° 11′ 06″
EFA	266° 10′ 15″
FAB	263° 38′ 25″

(a) Determine the sum of the measured angles.

(b) Determine the angular error of the traverse.
(c) Adjust the angles of the traverse to eliminate the error.
(d) Given that the whole circle bearing of line AB is 43° 40′ 45″, calculate the bearings of all other lines of the traverse.

Answer
(a) Sum of exterior angles $= (2n + 4) \times 90°$
$\phantom{\text{Sum of exterior angles }}= 16 \times 90°$
$\phantom{\text{Sum of exterior angles }}= 1440° \, 00′ \, 00″$
Sum of observed angles $= 1439° \, 59′ \, 18″$

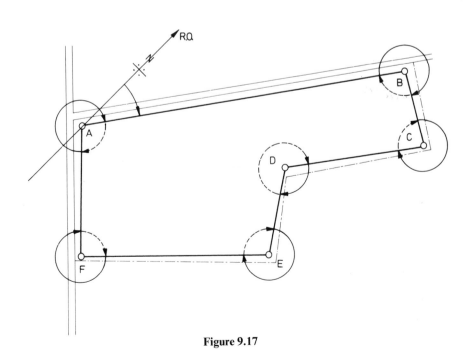

Figure 9.17

(b) Therefore angular error $= -42''$
Correction per angle $= +\frac{1}{6}$ of $42''$
$= +07''$

(c) Corrected angular values are:

Angle	Mean observed value	Correction	Corrected angle
ABC	272° 03′ 10″	+07″	272° 03′ 17″
BCD	272° 05′ 51″	+07″	272° 05′ 58″
CDE	104° 50′ 31″	+07″	104° 50′ 38″
DEF	261° 11′ 06″	+07″	261° 11′ 13″
EFA	266° 10′ 15″	+07″	266° 10′ 22″
FAB	263° 38′ 25″	+07″	263° 38′ 32″
	1439° 59′ 18″	+42″	1440° 00′ 00″

(d)

		Whole circle bearing
AB		43° 40′ 45″
Back bearing BA	223° 40′ 45″	
+ ABC	+272° 03′ 17″	
	495° 44′ 02″	
	−360°	
BC	135° 44′ 02″	135° 44′ 02″
Back bearing CB	315° 44′ 02″	
+ BCD	+272° 05′ 58″	
	587° 50′ 00″	
	−360°	
CD	227° 50′ 00″	227° 50′ 00″
Back bearing DC	47° 50′ 00″	
+ CDE	+104° 50′ 38″	
DE	152° 40′ 38″	152° 40′ 38″
Back bearing ED	332° 40′ 38″	
+ DEF	+261° 11′ 13″	
	593° 51′ 51″	
	−360°	
EF	233° 51′ 51″	233° 51′ 51″
Back bearing FE	53° 51′ 51″	
+ EFA	+266° 10′ 22″	
FA	320° 02′ 13″	320° 02′ 13″
Back bearing AF	140° 02′ 13″	
+ FAB	+263° 38′ 32″	
	403° 40′ 45″	
	−360°	
AB	43° 40′ 45″	

Agrees with initial bearing

Exercise 9.3

1 Figure 9.18 is a reproduction of Fig. 8.10, which shows the GCB Outdoor Centre closed traverse. In Exercise 8.2, the mean observed angles were calculated and are duplicated here, in Fig. 9.18.
(a) Determine the sum of the measured angles.
(b) Determine the angular error of the traverse.
(c) Adjust the angles of the traverse to eliminate the closing error.
(d) Given that the National Grid bearing of line AB is 315° 00′ 00″, calculate the forward bearings of all lines of the traverse.

2 Figure 9.19 shows the mean observed angles of traverse MNOP. Calculate the whole circle bearing and quadrant bearing of each line of the traverse.

(d) Traverse closed on previously fixed points

In Fig. 9.20 (see page 199), the bearings of lines BA and DC are known from a previous traverse to be 204° 11′ 05″ and 02° 10′ 47″ respectively. The observed angles of traverse BAPQODC are:

Angle	Observed value
BAP	72° 39′ 42″
APQ	187° 40′ 12″
PQO	169° 23′ 47″
QOD	161° 58′ 20″
ODC	106° 17′ 21″

Line BA is the opening line of the traverse from which the bearings of all other lines are calculated.

		Whole circle bearing
BA		204° 11′ 05″
Back bearing AB	24° 11′ 05″	
+ BAP	+72° 39′ 42″	
AP	96° 50′ 47″	96° 50′ 47″
Back bearing PA	276° 50′ 47″	
+ APQ	+187° 40′ 12″	
	464° 30′ 59″	
	−360°	
PQ	104° 30′ 59″	104° 30′ 59″
Back bearing QP	284° 30′ 59″	
+ PQO	+169° 23′ 47″	
	453° 54′ 46″	
	−360°	
QO	93° 54′ 46″	93° 54′ 46″
Back bearing OQ	273° 54′ 46″	
+ QOD	+161° 58′ 20″	
	435° 53′ 06″	
	−360°	
OD	75° 53′ 06″	75° 53′ 06″

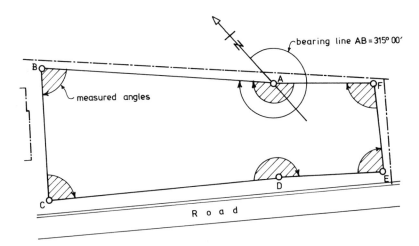

Figure 9.18

Horizontal angle	Accepted value
FAB	183° 33′ 20″
ABC	83° 41′ 40″
BCD	86° 47′ 50″
CDE	182° 39′ 40″
DEF	89° 15′ 50″
EFA	94° 00′ 40″

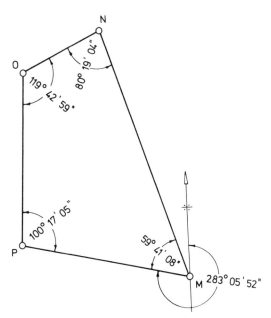

Figure 9.19

Known whole circle bearing line DC = 02° 10′ 47″

Traverse error = −20″

Correction per angle = +20/5″

= +04″

The observed angles could be recalculated by adding 04″ to each and the bearings could be found from the corrected angles as before. The corrected bearing of line AP will therefore be 04″ greater than the original while the bearing of the next line PQ will be 08″ greater. In fact the corrected bearings of each line will increase successively by 04″, until finally bearing DC will be (5 × 04)″ greater than the original uncorrected bearing.

The correction is therefore much more quickly done by simply finding the correction per angle and adding it successively to the uncorrected bearings.

Back bearing DO	255° 53′ 06″	
+ ODC	+106° 17′ 21″	
	362° 10′ 27″	
	−360°	
DC	02° 10′ 27″	02° 10′ 27″

Line	Uncorrected bearing	Correction	Corrected bearing
BA			204° 11′ 05″
AP	96° 50′ 47″	+04″	96° 50′ 51″
PQ	104° 30′ 59″	+08″	104° 31′ 07″
QO	93° 54′ 46″	+12″	93° 54′ 58″
OD	75° 53′ 06″	+16″	75° 53′ 22″
DC	02° 10′ 27″	+20″	02° 10′ 47″

Example

9 AB and FG are two survey lines with bearings fixed from a primary traverse as 39° 40′ 20″ and 36° 18′ 30″ respectively. The bearings are to be unaltered.

A secondary traverse run between the above stations produced the following results:

Angle	Mean observed value
ABC	179° 59′ 40″
BCD	210° 05′ 40″
CDE	149° 44′ 40″
DEF	177° 46′ 40″
EFG	179° 02′ 20″

Calculate the corrected bearings of each line.

Solution

		Whole circle bearing	Correction	Corrected bearing
AB		39° 40′ 20″	—	39° 40′ 20″
Back bearing BA	219° 40′ 20″			
+ ABC	+ 179° 59′ 40″			
	399° 40′ 00″			
	− 360°			
BC	39° 40′ 00″	39° 40′ 00″	− 10″	39° 39′ 50″
Back bearing CB	219° 40′ 00″			
+ BCD	+ 210° 05′ 40″			
	429° 45′ 40″			
	− 360°			
CD	69° 45′ 40″	69° 45′ 40″	− 20″	69° 45′ 20″
Back bearing DC	249° 45′ 40″			
+ CDE	+ 149° 44′ 40″			
	399° 30′ 20″			
	− 360°			
DE	39° 30′ 20″	39° 30′ 20″	− 30″	39° 29′ 50″
Back bearing ED	219° 30′ 20″			
+ DEF	+ 177° 46′ 40″			
	397° 17′ 00″			
	− 360°			
EF	37° 17′ 00″	37° 17′ 00″	− 40″	37° 16′ 20″
Back bearing FE	217° 17′ 00″			
+ EFG	+ 179° 02′ 20″			
	396° 19′ 20″			
	− 360°			
FG	36° 19′ 20″	36° 19′ 20″	− 50″	36° 18′ 30″

Correct bearing FG = 36° 18′ 30″
Traverse error   = + 50″
Angular correction = − (50/5)″
          = − 10″

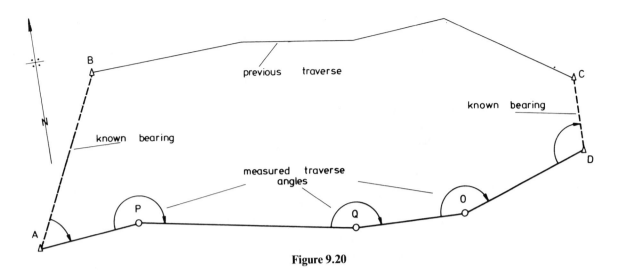

Figure 9.20

9. Obtaining bearings directly

It is possible to find the bearings of traverse lines directly in the field. In Fig. 9.21 the bearings of lines AB, BC and CD are required.

The theodolite is set at A on face left and, with the horizontal circle set to read zero, the reference object RO is sighted. The theodolite is in fact oriented along the north line of the survey.

When the upper plate clamp is released and the telescope is directed to station B, the reading of the horizontal circle, 70° 00′ 00″, is the bearing of line AB.

The theodolite is then taken to station B and the bearing of line BC is obtained by the following method. Using this method, the basic idea is to set up the theodolite so that it is correctly oriented with north.

Back bearing method

1. Calculate the back bearing of line AB:

Back bearing = 70° 00′ 00″ ± 180° = 250° 00′ 00″

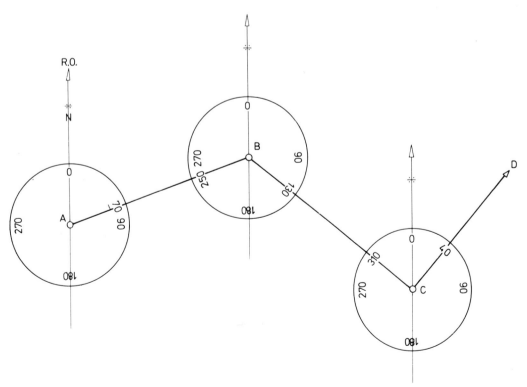

Figure 9.21

Set this bearing on the horizontal circle by releasing the upper plate screw and rotating the upper plate until the circle reads 250° 00′ 00″.

2. Release the lower plate screw and direct the telescope to station A, on face left; bisect the station and tighten the clamp. The 250° graduation is thereby pointing to station A, i.e. on its correct bearing. The zero graduation is therefore pointing along the north meridian and the instrument is correctly oriented.

3. Release the upper plate, sight station C, tighten the upper clamp and read the circle. The reading shown in the figure is 130° 00′ 00″ and this is the bearing of line BC.

4. Transport the instrument to station C, set up and bisect station B, with the circle reading the back bearing of line BC, namely 130° 00′ 00″ ± 180° = 310° 00′ 00″.

5. Operation (3) above is repeated by sighting D and the circle reading of 40° 00′ 00″ is the bearing of line CD.

10. Rectangular coordinates— terminology

If it can be imagined, for the moment, that a large sheet of graph paper could be unrolled over a site, with the y axis coinciding with the north direction, then it would be possible to define any point on the site by conventional X and Y coordinates.

In Fig. 9.22, station A is the origin of the survey and has the coordinates (0.00, 0.00). The coordinates of station B are (XB, YB), while those of station E are (XE, YE). It should be remembered that X coordinates are always given before Y coordinates.

In civil engineering and construction surveying, the X coordinates are called *eastings*, while Y coordinates are called *northings*. In other branches of surveying, the X and Y coordinates are called *departures* and *latitudes* respectively. All of the terms are equally acceptable.

Following logically from these are the terms *difference in eastings* and *difference in northings*, denoted in

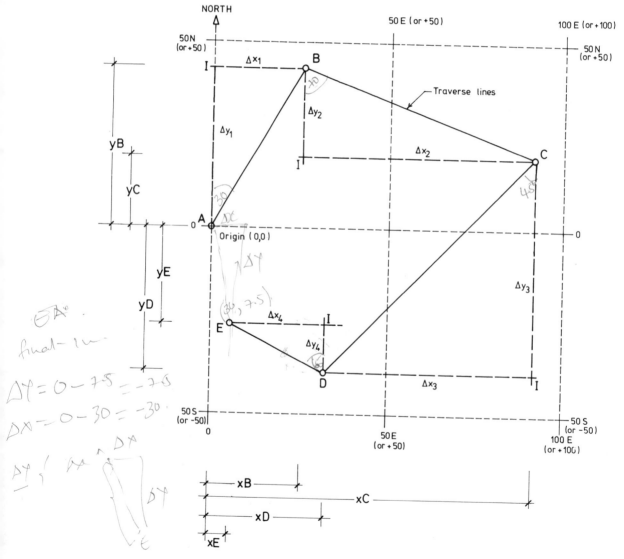

Figure 9.22

Fig. 9.22 as Δx and Δy respectively. Alternative titles for these terms are *departure differences* and *latitude differences*, or *partial departures* and *partial latitudes*. Again all of these terms are equally acceptable.

The terms partial departure, partial latitude, total departure and total latitude are used in Sec. 11(a), which involves the use of quadrant bearings.

11. Coordinate calculation—open traverse

(a) Method 1: using quadrant bearings

The following data refer to the open traverse ABCDE of Fig. 9.22. The rectangular coordinates of each station are required:

Line	AB	BC	CD	DE
WC bearing	30°	110°	225°	295°
Length	50.0	70.0	82.0	31.2

Basic calculation

1. The first step in the calculation is to obtain the quadrant bearing of every line. These are as follows:

AB:N 30°E   BC:S 70°E   CD:S 45°W   DE:N 65°W

2. The x and y axes are drawn through two stations at a time such that the quadrant bearing of each line is enclosed in a right-angled triangle (Fig. 9.22); e.g. on line BC the y axis is drawn through B and the x axis through C to enclose quadrant bearing S 70°E.

3. In IAB:

$$\Delta x_1 = IB \qquad\qquad \Delta y_1 = IA$$
$$= AB \sin 30° \qquad = AB \cos 30°$$
$$= 50.0 \times \sin 30.0 \qquad = 50.00 \times \cos 30.0$$
$$= +25.0 \text{ m} \qquad\quad = +43.3 \text{ m}$$

Note that both signs are positive since the line travels eastwards (to the right) along the x axis and northwards (upwards) along the y axis.

4. In IBC:

$$\Delta x_2 = IC \qquad\qquad \Delta y_2 = IB$$
$$= BC \sin 70° \qquad = BC \cos 70°$$
$$= 70.0 \sin 70° \qquad = 70° \cos 70°$$
$$= +65.8 \text{ m} \qquad\quad = -23.9 \text{ m}$$

Note that x_2 has a positive sign since the line travels eastwards along the x axis and y_2 has a negative sign since the line is travelling southwards (down) along the y axis.

5. In ICD:

$$\Delta x_3 = ID \qquad\qquad \Delta y_3 = IC$$
$$= CD \sin 45° \qquad = CD \cos 45°$$
$$= 82.0 \sin 45° \qquad = 82° \cos 45°$$
$$= -58.0 \text{ m} \qquad\quad = -58.0 \text{ m}$$

The signs are both negative since the line travels westwards (to the left) along the x axis and southwards (down) along the y axis.

6. In IDE:

$$\Delta x_4 = IE \qquad\qquad \Delta y_3 = ID$$
$$= DE \sin 65° \qquad = DE \cos 65°$$
$$= 31.2 \sin 65° \qquad = 31.2 \cos 65°$$
$$= -28.3 \qquad\qquad = +13.2$$

Note that x_4 has a negative sign since the line travels westwards along the r axis and y_4 has a positive sign since the line travels northwards along the y axis.

Partial coordinates

The differences in x coordinates (Δx_1, Δx_2, Δx_3, Δx_4) are known as 'departure' differences or 'partial departures' and are *always* found from the formula:

> Partial departure = length of line
> × sin quadrant bearing

'Easterly' departures are always positive and 'westerly' departures are always negative.

The differences in y coordinates (Δy_1, Δy_2, Δy_3, Δy_4) are known as latitude differences or 'partial latitudes' and are *always* found from the formula:

> Partial latitude = length of line
> × cos quadrant bearing

'Northerly' latitudes are always positive and 'southerly' latitudes are always negative.

Collectively the partial departures and partial latitudes are known as partial coordinates. These coordinates could be plotted on a rectanglar grid, each station being plotted from the previous station. This system has the distinct disadvantage that if one station is plotted wrongly the succeeding stations will automatically be wrong.

Total coordinates

To overcome this disadvantage, all the stations are plotted from the first station or origin of the survey by converting the partial coordinates to 'total coordinates'. The total coordinates of any station are obtained by algebraically adding the previous partial coordinates.

Treating station A as the origin of the survey, i.e. its coordinates are (0.0, 0.0), the total departures of successive stations, B, C, D and E, are found as in Table 9.1, with the numerical values as given in Table 9.2. The total latitude of each station is found in an identical manner (Table 9.3), with the numerical values as given in Table 9.4. The total coordinates of each station are therefore as shown in Table 9.5.

Traverse table

All of the calculations are set out in a traverse table as in Table 9.6. It is completed as follows:

Line	Partial departure	Total departure	Station
		0.0	A
AB	Δx_1	$0.0 + \Delta x_1$	B
BC	Δx_2	$0.0 + \Delta x_1 + \Delta x_2$	C
CD	Δx_3	$0.0 + \Delta x_1 + \Delta x_2 + \Delta x_3$	D
DE	Δx_4	$0.0 + \Delta x_1 + \Delta x_2 + \Delta x_3 + \Delta x_4$	E

Table 9.1

Line	Partial departure	Total departure		Station
		0.0	$=$ 0.0	A
AB	$+25.0$	$0.0 + 25.0$	$= +25.0$	B
BC	$+65.8$	$0.0 + 25.0 + 65.8$	$= +90.8$	C
CD	-58.0	$0.0 + 25.0 + 65.8 + (-58.0)$	$= +32.8$	D
DE	-28.3	$0.0 + 25.0 + 65.8 + (-58.0) + (-28.3)$	$= + 4.5$	E

Table 9.2

Line	Partial latitude	Total latitude	Station
		0.0	A
AB	Δy_1	$0.0 + \Delta y_1$	B
BC	Δy_2	$0.0 + \Delta y_1 + \Delta y_2$	C
CD	Δy_3	$0.0 + \Delta y_1 + \Delta y_2 + \Delta y_3$	D
DE	Δy_4	$0.0 + \Delta y_1 + \Delta y_2 + \Delta y_3 + \Delta y_4$	E

Table 9.3

Line	Partial latitude	Total latitude		Station
		0.0	$=$ 0.0	A
AB	$+43.3$	$0.0 + 43.3$	$= +43.3$	B
BC	-23.9	$0.0 + 43.3 + (-23.9)$	$= +19.4$	C
CD	-58.0	$0.0 + 43.3 + (-23.9) + (-58.0)$	$= -38.6$	D
DE	$+13.2$	$0.0 + 43.3 + (-23.9) + (-58.0) + 13.2$	$= +25.4$	E

Table 9.4

Station	Total departure	Total latitude
A	0.0	0.0
B	$+25.0$	$+43.3$
C	$+90.8$	$+19.4$
D	$+32.8$	-38.6
E	$+4.5$	-25.4

Table 9.5

	1	2	3	4	5	6	7	8	9	10
				Partial coordinates				Total coordinates		
	Line	Quadrant bearing	Distance (m)	E +	W −	N +	S −	Departure	Latitude	Station
1								00.0	00.0	A
2	AB	N 30° E	50.0	25.0	—	43.3	—	+ 25.0	+ 43.3	B
3	BC	S 70° E	70.0	65.8	—	—	23.9	+ 90.8	+ 19.4	C
4	CD	S 45° W	82.0	—	58.0	—	58.0	+ 32.8	− 38.6	D
5	DE	N 65° W	31.2	—	− 28.3	13.2	—	+ 4.5	− 25.4	E
				+ 90.8	− 86.3	+ 56.5	− 81.9	+ 4.5	− 25.4	
				− 86.3		− 81.9				
				+ 4.5		− 25.4				

Table 9.6

1. Enter the station designations in column 10 (i.e. enter letters A to E in successive lines). Otherwise leave line 1 blank.
2. Enter the survey lines AB, BC, CD and DE in column 1 beginning on line 2.
3. Enter the quadrant bearing of each line in column 2.
4. Enter the plan length of each line in column 3.
5. Calculate the partial coordinates as described. Enter the partial departures in columns 4 and 5 and the partial latitudes in columns 6 and 7.
6. Calculate the total coordinates from the partial coordinates. Enter the departures in column 8 and latitudes in column 9.
7. A check is provided on the arithmetic as follows. The algebraic sum of the partial departure columns 4 and 5 should equal the difference between the total departures of the first and last stations in column 8.

The algebraic sum of the partial latitude columns 6 and 7 should equal the difference between the total latitudes of the first and last stations in column 9.

Note. Throughout the remainder of this book, civil engineering and construction terminology will be used in coordinate calculations. The terms are, once again, difference in eastings (ΔE), difference in northings (ΔN), eastings (E) and northings (N).

(b) Method 2: using whole circle bearings

Although a few surveyors still calculate coordinates using quadrant bearings, the most common practice is to compute the coordinates directly from the whole circle bearings, thus eliminating a potential source of error in calculating the quadrant bearing.

In Fig. 9.22 the partial coordinates of line BC are

$$\Delta E = \text{distance} \times \sin \text{quadrant bearing}$$
$$= 70.0 \times \sin 70°$$
$$= 70.0 \times 0.939\,69$$
$$= 65.8 \text{ m east}$$
$$= + 65.8 \text{ m}$$

$$\Delta N = \text{distance} \times \cos \text{quadrant bearing}$$
$$= 70.0 \times \cos 70°$$
$$= 70.0 \times 0.342\,02$$
$$= 23.9 \text{ m south}$$
$$= - 23.9 \text{ m}$$

Since the whole circle bearing 110° is a second quadrant angle in mathematical terms, then $\sin 110° = + \sin 70°$ and $\cos 110° = - \cos 70°$.

The partial coordinates of line BC, using the whole circle bearing, are therefore

$$\Delta E = \text{length BC} \times \sin \text{WCB}$$
$$= 70.0 \times \sin 110°$$
$$= 70.0 \times 0.939\,69$$
$$= + 65.8 \text{ m}$$

$$\Delta N = \text{length} \times \cos \text{WCB}$$
$$= 70.0 \times \cos 110°$$
$$= 70.0 \times - 0.342\,02$$
$$= - 23.9 \text{ m}$$

Similarly the partial coordinates of line CD are

$$\Delta E = 82.0 \times \sin 225°$$
$$= 82.0 \times - 0.707\,11$$
$$= - 58.0 \text{ m}$$

$$\Delta N = 82.0 \times \cos 225°$$
$$= 82.0 \times - 0.707\,11$$
$$= - 58.0 \text{ m}$$

and the partial coordinates of line DE are

$$\Delta E = 31.2 \times \sin 295°$$
$$= 31.2 \times - 0.906\,31$$
$$= - 28.3 \text{ m}$$

$$\Delta N = 31.2 \times \cos 295°$$
$$= 31.2 \times + 0.422\,62$$
$$= + 13.2 \text{ m}$$

Using the whole circle bearing method, the traverse Table 9.6 is shortened from ten to eight columns in Table 9.7. In this table the partial coordinates occupy

		Whole circle bearing	Distance (m)	Partial coordinates		Total coordinates		Station
	Line			ΔE	ΔN	Eastings	Northings	
1						00.0	00.0	A
2	AB	30°	50.0	+ 25.0	+ 43.3	+ 25.5	+ 43.3	B
3	BC	110°	70.0	+ 65.8	− 23.9	+ 90.8	+ 19.4	C
4	CD	225	82.0	− 58.0	− 58.0	+ 32.8	− 38.6	D
5	DE	295°	31.2	− 28.3	+ 13.2	+ 4.5	− 25.4	E
				+ 90.8	+ 56.5	+ 4.5	− 25.4	
				− 86.3	− 81.9			
				+ 4.5	− 25.4			

Table 9.7

only two instead of four columns. Using the data of Sec. 11(a), the table is completed as follows:

1. Enter the station designations in column 8.
2. Enter the survey lines in column 1 beginning on line 2.
3. Enter the whole circle bearing of each line in column 2.
4. Enter the plan length of each line in column 3.
5. Enter the difference in eastings in column 4 and the difference in northings in column 5.
6. Calculate the total coordinates by successively adding the partial coordinates. Enter the eastings in column 6 and the northings in column 7.

(c) Method 3: use of $\boxed{P \rightarrow R}$ function of a calculator

Figure 9.23(a) shows the mathematical concept of polar (r, θ) and rectangular (x, y) coordinates of a line AB, referred to the X and Y axes.

On all scientific calculators, e.g. CASIO fx, TEXAS TI models, the rectangular coordinates can be found directly from the polar coordinates and vice versa by using the $\boxed{P \rightarrow R}$ and $\boxed{R \rightarrow P}$ function buttons. These are secondary functions on a calculator and have to be used in conjunction with a button usually marked $\boxed{INV}$ or $\boxed{2nd}$, shown in Fig. 9.23(c).

When the X, Y axes are rotated through 90°, then reflected through 180°, the positive X axis points north and the positive Y axis points east. Thus they form the coordinate axis system for surveying in Fig. 9.23(b).

The polar coordinates (r, θ) become the length and whole circle bearing of the line AB while the x value is the surveying ΔN and the y value is the surveying ΔE. A slight disadvantage in a $\boxed{P \rightarrow R}$ conversion calculation is that the x value (ΔN) is produced before the y value (ΔE) by the calculator.

In Sec. 11(a) the polar coordinates of line AB are length 50.0 m and whole circle bearing 30°. The rec-

tangular coordinates, i.e. ΔN and ΔE are computed on the CASIO fx calculator as follows:

Instruction	Calculator operation
1. Enter length	Input 50.0
2. Operate buttons	$\boxed{INV\ P \rightarrow R}$
3. Enter WCB (decimalized)	Input 30.0
4. Operate button	$\boxed{=}$
5. Read answer (ΔN)	43.3
6. Operate button	$\boxed{X \rightarrow Y}$
7. Read answer (ΔE)	25.0

8. Repeat these steps for line BC (length: 70.0 m; WCB: 110°)

Input 70.0, $\boxed{INV}$ $\boxed{P \rightarrow R}$
Input 110°, $\boxed{=}$, read − 23.9,
$\boxed{X \rightarrow Y}$, read + 65.8

9. Repeat for remaining lines CD and DE

(d) Method 4: use of computer program

All of the calculations of the previous sections may be carried out on a microcomputer using a computer program.

Using the $\boxed{P \rightarrow R}$ conversion of the scientific calculator, the length and bearing of a line were input by hand. The calculator then computed the partial coordinates and displayed the answer. The simple computer program works on the same principle.

It is realized that some readers will not wish to or have the facilities to consider further methods of solution. Therefore the computer solutions have been placed in Chapter 17. The programs, explanations and running instructions are provided in programs 1 to 5.

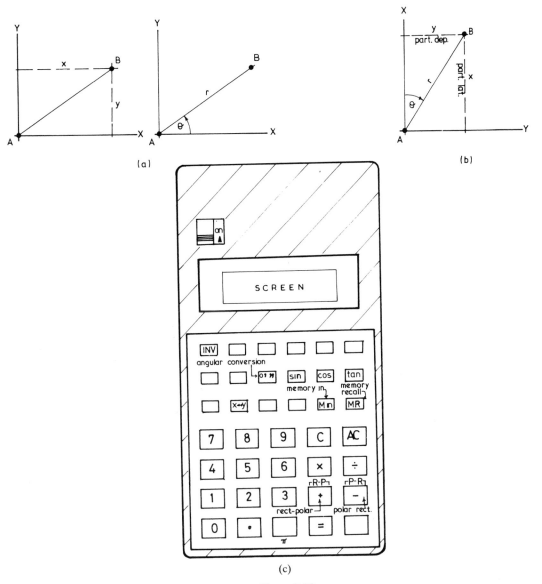

(a)

(b)

(c)

Figure 9.23

(e) Method 5: use of spreadsheets

Spreadsheets are commercially available software packages on which tabular solutions for repetitive calculations may be devised. The spreadsheets, instructions and examples are included with the basic computer programs in Chapter 17.

12. Plotting rectangular coordinates

Figure 9.24 is a portion of a rectangular grid drawn to a scale of 1:1000. The lengths of the sides of the squares are 50 m long.

Point A is the origin of the survey and is therefore the intersection of the zero departure (easting) and zero latitude (northing) lines.

Point C (90.8, 19.4) is plotted thus:

1. The square in which the point will fall is determined by inspection.
2. The total departure (easting) 90.8 m, is scaled along the zero latitude (northing) line and the point marked c_1.
3. Similarly the departure (easting) is scaled and marked c_2 along the 50 m latitude line.
4. The points c_1 and c_2 are joined and the line is checked to ensure that it is parallel to the north–south grid lines.
5. The total latitude (northing), 19.4 m, is similarly marked along the 50 m and 100 m departure (easting) lines at c_3 and c_4, and the line joining them is checked for parallelism against the east–west grid lines.
6. The intersection of the lines c_1c_2 and c_3c_4 forms the point C.

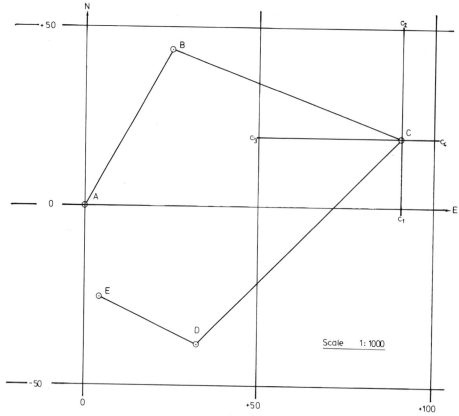

Figure 9.24

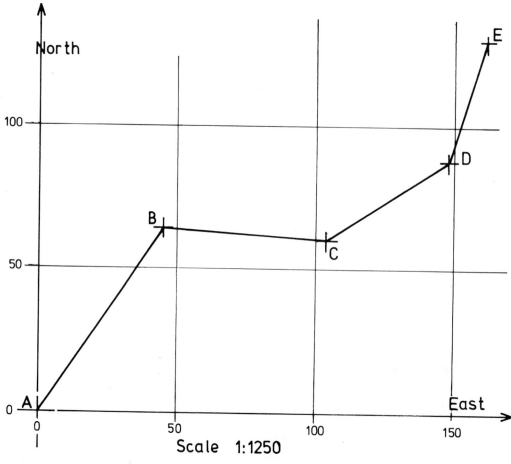

Scale 1:1250

Figure 9.25

All other survey points are similarly plotted and clearly marked by a small circle or triangle. The exact position of the survey station is marked by a needle.

Example

10 Calculate and plot the coordinates of the following traverse, using (a) quadrant bearings and (b) whole circle bearings. The relevant traverse data are as follows:

Line	AB	BC	CD	DE
WC bearing	35°	94°	58°	17°
Length (m)	79.0	59.0	52.0	44.0

Answer (Fig. 9.25, opposite)
(a) Using quadrant bearings (Table 9.8)
(b) Using whole circle bearings (Table 9.9)

Exercise 9.4

1 Figure 9.26 shows the traverse of Fig. 8.15, which was plotted by protractor and scale rule. The traverse details are as follows:

Line	Whole circle bearing	Plan length (m)
AB	65° 00′ 00″	25.707
BC	338° 15′ 50″	22.861
CD	72° 47′ 00″	53.221

Given that the coordinates of station A are 100.00 m east and 100.00 m north, calculate the coordinates of stations B, C and D and then plot the traverse to scale 1:500.

13. Calculation of a closed traverse

Basically there is no difference between the calculation of a closed traverse and an open traverse. In a closed traverse, the initial and final coordinates of the first station should be identical. This will very seldom occur and the resultant closing error must be eliminated.

Example

11 Calculate the total coordinates of all stations in the following traverse given that station A is the origin:

Line	AB	BC	CD	DE	EA
WC bearing	29° 30′	100° 45′	146° 30′	242° 00′	278° 45′
Length (m)	83.50	59.40	62.00	50.30	90.40

This is an example of a closed traverse which has a closing error. In this example the sexagesimal measure of bearings occurs for the first time in trigonometrical calculations. All of the bearings must be converted to decimal form. Using the scientific calculator a bearing of, say,

---

Line	Quadrant bearing	Distance (m)	Partial coordinates				Total coordinates		Station
			E +	W −	N +	S −	Departure	Latitude	
							00.0	00.0	A
AB	N 35° E	79.0	45.3	—	64.7	—	+ 45.3	+ 64.7	B
BC	S 86° E	59.0	58.9	—	—	4.1	+ 104.2	+ 60.6	C
CD	N 58° E	52.0	44.1	—	27.6	—	+ 148.3	+ 88.2	D
DE	N 17° E	44.0	12.9	—	42.1	—	+ 161.2	+ 130.3	E
			161.2		134.4 −4.1 + 130.3	− 4.1	+ 161.2	+ 130.3	

Table 9.8

Line	Whole circle bearing	Distance (m)	ΔE	ΔN	Eastings	Northings	Station
					0.0	0.0	A
AB	35°	79.0	+ 45.3	+ 64.7	+ 45.3	+ 64.7	B
BC	94°	59.0	+ 58.9	− 4.1	+ 104.2	+ 60.6	C
CD	58°	52.0	+ 44.1	+ 27.6	+ 148.3	+ 88.2	D
DE	17°	44.0	+ 12.9	+ 42.1	+ 161.2	+ 130.3	E
			+ 161.2	+ 130.3	+ 161.2	+ 130.3	

Table 9.9

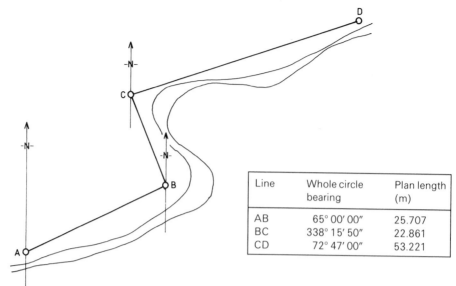

Figure 9.26

Line	Whole circle bearing	Plan length (m)
AB	65° 00′ 00″	25.707
BC	338° 15′ 50″	22.861
CD	72° 47′ 00″	53.221

30° 53′ 15″ is decimalized using the converter button in Fig. 9.23(c) as follows:

Press: 30 $\boxed{° ′ ″}$ 53 $\boxed{° ′ ″}$ 15 $\boxed{° ′ ″}$

The answer appears on the screen as 30.8875°.

Note. The method of converting sexagesimal measure to centesimal measure varies with the type of calculator.

Answer (Table 9.10)

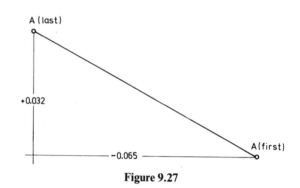

Figure 9.27

(a) Closed traverse—closing error

The difference between the initial and final coordinates of station A determines the errors in eastings and northings. Figure 9.27 shows the initial and final coordinates plotted on a very much exaggerated scale.

The error of closure AA, is found by the theorem of Pythagoras to be

$$AA = \sqrt{0.065^2 + 0.032^2}$$
$$= 0.072 \text{ m}$$

The total length of the traverse is 345.60 metres; therefore the fractional accuracy of the survey is

0.072 m in 345.6 m

which equals

1 in 4800 approximately

(b) Adjustment by Bowditch's rule

The closing error of the survey is due to the accumulation of errors along each draft of the

Line	Whole circle bearing	Length (m)	Difference in eastings	Difference in northings	Eastings	Northings	Station
					0.000	0.000	A
AB	29° 30′ 0″	83.500	41.117	72.675	41.117	72.675	B
BC	100° 45′ 0″	59.400	58.358	−11.080	99.475	61.595	C
CD	146° 30′ 0″	62.000	34.220	−51.701	133.695	9.894	D
DE	242° 00′ 0″	50.300	−44.412	−23.614	89.283	−13.720	E
EA	278° 45′ 0″	90.400	−89.348	13.752	−0.065	0.032	A

Table 9.10

Line	Whole circle bearing	Length	ΔE	ΔN	ΔE correction	ΔN correction	Corrected ΔE	Corrected ΔN	Eastings	Northings	Station
									0.000	0.000	A
AB	29° 30′ 0″	83.500	41.117	72.675	0.016	−0.008	41.133	72.667	41.133	72.667	B
BC	100° 45′ 0″	59.400	58.358	−11.080	0.011	−0.006	58.369	−11.086	99.502	61.581	C
CD	146° 30′ 0″	62.000	34.220	−51.701	0.012	−0.006	34.232	−51.707	133.734	9.874	D
DE	242° 00′ 0″	50.300	−44.412	−23.614	0.009	−0.004	−44.403	−23.618	89.331	−13.744	E
EA	278° 45′ 0″	90.400	−89.348	13.752	0.017	−0.008	−89.331	13.744	0.000	0.000	A
		345.600									
	Errors		−0.065	0.032	0.065	−0.032	0.000	0.000			
	Corrections		0.065	−0.032							

Correction $(k_1) = +1.885 \times 10^{-4}$
Correction $(k_2) = -9.197 \times 10^{-5}$
Accuracy 1 in 4800

Table 9.11

traverse. It should be remembered that the angular and linear errors are assumed to be equally inaccurate. As a result the closing error is distributed throughout the traverse in proportion to the lengths of the various drafts.

In the rectangular coordinate method, it is the separate ΔE and ΔN errors that have to be distributed and not the closing error of distance. Each error is treated separately and is distributed throughout the traverse in proportion to the lengths of the drafts, as given in Table 9.11.

Using Example 11, the error in eastings $= -0.65$ m and therefore the correction to eastings $= +0.65$ m.

Easting correction at station B

$$= \frac{\text{length of line AB}}{\text{total length of survey}} \times \text{total correction}$$

and in fact at any station the easting correction (c_E) is

$$c_E = \left(\frac{\text{length of draft}}{\text{total length of survey}} \right.$$

$$\left. \times \text{total correction in departure} \right)$$

$$= \frac{l}{L} \times C_E$$

$$= \frac{C_E}{L} \times l$$

Therefore $c_E = k_1 \times l$ (since C_E and L are constants for every line).

The northing correction (c_N) for every station is derived exactly as above:

$$c_N = \frac{l}{L} \times C_N$$

$$= \frac{C_N}{L} \times l$$

$$= k_2 \times l$$

In Example 11, the procedure for correction is as follows:

1. Total easting correction $(C_E) = +0.065$ m
   Total length of survey $(L) = 345.60$ m

   Therefore $k_1 = \dfrac{+0.065}{345.60}$

   $= 1.885 \times 10^{-4}$

2. Easting correction at

   $\text{B} = 1.885 \times 10^{-4} \times 83.50 = 0.016$ m
   $\text{C} = 1.885 \times 10^{-4} \times 59.40 = 0.011$ m
   $\text{D} = 1.885 \times 10^{-4} \times 62.00 = 0.012$ m
   $\text{E} = 1.885 \times 10^{-4} \times 50.30 = 0.009$ m
   $\text{A} = 1.885 \times 10^{-4} \times 90.40 = \underline{0.017}$ m
   Sum $= \overline{\underline{0.065}}$ m

3. Total northing correction $(C_N) = +0.032$ m
   Total length of survey $(L) = 345.60$ m

   Therefore $k_2 = \dfrac{-0.032}{345.60}$

   $= -9.197 \times 10^{-5}$

4. Northing correction at

   $\text{B} = -9.197 \times 10^{-5} \times 83.50 = -0.008$ m
   $\text{C} = -9.197 \times 10^{-5} \times 59.40 = -0.006$ m
   $\text{D} = -9.197 \times 10^{-5} \times 62.00 = -0.006$ m
   $\text{E} = -9.197 \times 10^{-5} \times 50.30 = -0.004$ m
   $\text{A} = -9.197 \times 10^{-5} \times 90.40 = \underline{-0.008}$ m
   Sum $= \overline{\underline{-0.032}}$ m

5. The corrected partial coordinates are obtained from the algebraic addition of the calculated partial coordinates (Table 9.10) and the corrections. For example, corrected partial coordinates of B are

$$\Delta E$$
$$+41.117 + 0.016 = 41.133$$

$$\Delta N$$
$$+72.675 - 0.008 = 72.667$$

The calculation is more conveniently performed in tabular fashion, as in Table 9.11. It will be immedi-

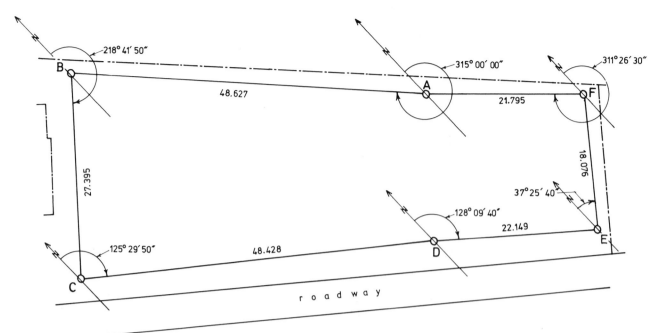

Line	Length	W.C. Bearing
AB	48.627	315° 00′ 00″
BC	27.395	218° 41′ 50″
CD	48.428	125° 29′ 50″
DE	22.149	128° 09′ 40″
EF	18.076	37° 25′ 40″
FG	21.795	311° 26′ 30″

Figure 9.28

ately apparent that the table is an extension of the original traverse table.

Exercise 9.5

1 Figure 9.28 shows the reduced data of the GCB Outdoor Centre closed traverse. Calculate:
(a) the partial coordinates of the traverse,
(b) the closing error of the traverse.
Adjust the traverse, using Bowditch's rule.

14. Miscellaneous coordinate problems

(a) Closing bearing and distance

In work involving coordinates it is frequently necessary to calculate the bearing and distance between two points whose coordinates are known. If, for example, the coordinates of the traverse C to H (Fig. 9.29) are calculated, the bearing and distance between the first and last stations would be found as follows.

Coordinates of Station C:

$$300.20 \text{ E}$$
$$320.00 \text{ N}$$

Coordinates of station H:

$$463.30 \text{ E}$$
$$595.30 \text{ N}$$

In triangle CHX, the difference in easting between the points C and H is ΔE

$$\Delta E = 463.30 - 300.20$$
$$= +163.10$$

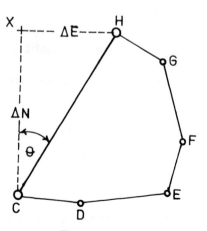

Figure 9.29

The difference in northing between the points C and H is ΔN.

$$\Delta N = 595.30 - 320.00$$
$$= 275.30$$

The quadrant bearing of line CH is angle θ:

$$\tan \theta = \frac{\Delta E}{\Delta N} = \frac{\Delta x}{\Delta y}$$

$$= \frac{163.10}{275.30}$$

Therefore $\theta = 30° \ 38' \ 40''$

Therefore quadrant bearing CH is N 30° 38′ 40″ E. The distance between points C and H is, by Pythagoras,

$$= \sqrt{163.10^2 + 275.30^2}$$
$$= \sqrt{102\,391.70}$$
$$= 319.99 \ m$$

Alternatively, once the bearing has been calculated, the distance between the points C and H is found in either of the following ways:

$$\frac{\Delta N}{CH} = \cos \theta \qquad\qquad \frac{\Delta E}{CH} = \sin \theta$$

$$\text{Therefore } CH = \frac{\Delta N}{\cos \theta} \qquad \text{and } CH = \frac{\Delta E}{\sin \theta}$$

$$= \frac{275.30}{0.860\,347} \qquad\qquad = \frac{163.10}{0.509\,709}$$

$$= 319.99 \ m \qquad\qquad = 319.99 \ m$$

Closing bearing and distance using $\boxed{R \to P}$ ***function on calculator***

The calculation may be carried out on a scientific calculator using the rectangular–polar conversion facility (Fig. 9.23(c)) as follows. (*Note.* The calculation varies with the type of calculator but fundamentally the method is correct.)

To find the bearing and distance from C to H, subtract the coordinates of C from those of H:

$$N = 595.30 - 320.00 = +275.30 \ (\text{as before})$$
$$E = 463.30 - 300.20 = +163.10 \ (\text{as before})$$

On the calculator.

Enter 275.30, press $\boxed{INV}$ $\boxed{R \to P}$, enter 163.10, press $\boxed{=}$

Display shows: 319.99 m = length CH

Press $\boxed{X \to Y}$

Display shows: 30.644 392° = bearing C to H (dec.)

Press $\boxed{INV}$ $\boxed{°\ '\ ''}$

Display shows: 30° 38′ 40″ = bearing C to H

Using rectangular–polar angular conversion and memory facilities, the complete calculation is as follows:

On the calculator:

Enter 463.30 − 300.20

Display shows: + 163.10

Press $\boxed{Min}$, i.e. enter into memory

Enter 595.30 − 320.00

Display shows: + 275.30 Press $\boxed{INV}$ $\boxed{R \to P}$

Press $\boxed{MR}$ $\boxed{=}$

Display shows: 319.99 = length CH

Press $\boxed{X \to Y}$

Display shows: 30.644 392° = bearing from C to H (dec.)

Press $\boxed{INV}$ $\boxed{°\ '\ ''}$

Display shows: 30° 38′ 40″ = bearing from C to H

(b) Solution of triangles

In Fig. 9.30, ABCD is an open traverse along the route of a sewer. The total coordinates of the stations have been found to be:

	A	B	C	D
	00.00 E	9.04 E	91.78 E	141.52 E
	00.00 N	59.31 S	146.51 S	154.09 S

It is proposed to connect a point X (25.32E, 110.70S) to the sewer as economically as possible. Calculate the bearing and length of the shortest route.

The shortest route is the line XE at right angles to the line BC. The length XE can only be found by solving either the triangle BXE or triangle CXE.

Solution of triangle BXE:

1. Bearing BC = S 43° 30′ 00″ E (given)
$$= 136° \ 30' \ 00''$$

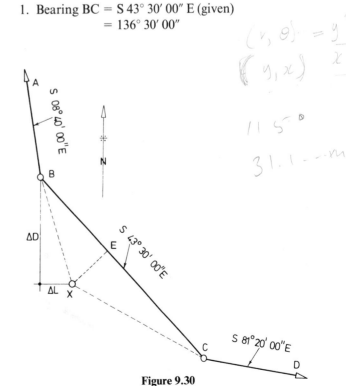

Figure 9.30

2. Therefore bearing EX = 136° 30′ 00″ + 90°
$$= 226° 30′ 00″$$

3.
Departure B	9.04 E	Latitude B	59.31 S
Departure X	25.32 E	Latitude X	110.70 S
$\Delta D =$	16.28	$\Delta L =$	51.39

Tan bearing BX $= \dfrac{\Delta D}{\Delta L}$

$$= \frac{16.28}{51.39}$$

$$= S \ 17° 34′ 40″ \ E$$

$$= 162° 25′ 20″$$

4. Distance BX $= \dfrac{\Delta L}{\cos 17° 34′ 40″}$

$$= \frac{51.39}{0.953\,307\,9}$$

$$= 53.90 \text{ m}$$

5. Angle EBX = bearing BX − bearing BE
$$= \quad 162° 25′ 20″$$
$$= -136° 30′ 00″$$
$$= \quad \underline{25° 55′ 20″}$$

6. Distance EX

$$\frac{EX}{BX} = \sin 25° 55′ 20″$$

Therefore EX = 53.90 × 0.437 150 7
$$= 23.56 \text{ m}$$

It should be noted that the majority of problems connected with coordinates finish with an unsolved triangle in which the coordinates of two points and the bearing of one or more sides is known. Generally the closing bearing and distance between the coordinated points must be found before the triangle can be solved.

Example

12 In Fig. 9.31, two tunnels AB and EF are being driven forward until they meet in order to accommodate telephone cables.

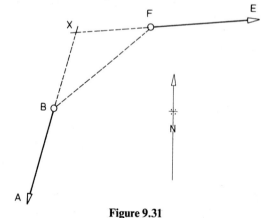

Figure 9.31

Calculate the distance still to be driven in each tunnel given the following information:

Bearing AB = 15° 00′
    Coordinates B   624.30 E
                    1300.50 N

Bearing EF = 265° 00′
    Coordinates F   845.90 E
                    1482.30 N

Solution

(a) Produce EF and AB until they meet at X. Join BF to form triangle FBX.

(b) Bearing XB = bearing BA
$$= 195° 00′$$

Bearing XF = bearing FE
$$= 85° 00′$$

Therefore angle FXB = 195° 00′ − 85° 00′
$$= 110° 00′$$

(c)
Easting F = 845.90	Northing F = 1482.30
B = 624.30	B = 1300.50
$\Delta E =$ 221.60	$\Delta N =$ 181.80

tan bearing BF $= \dfrac{221.60}{181.80}$

$$= 50° 38′ 05″$$

Distance BF $= \dfrac{221.60}{\sin 50° 38′ 05″}$

$$= 286.63 \text{ m}$$

(d) Angle XBF = bearing BF − bearing BX
$$\quad\; 50° 38′ 05″$$
$$-15° 00′ 00″$$
$$= \quad \underline{35° 38′ 05″}$$

(e) Angle BFX = bearing FX − bearing FB
$$= \quad 265° 00′ 00″$$
$$-230° 38′ 05″$$
$$= \quad \underline{34° 21′ 55″}$$

Check angles of triangles FXB

$$\text{Sum} = 110° 00′ 00″$$
$$35° 38′ 05″$$
$$34° 21′ 55″$$
$$\overline{180° 00′ 00″}$$

(f) In triangle FXB, by sine rule,

$$\frac{XB}{\sin F} = \frac{BF}{\sin X}$$

Therefore $XB = \dfrac{BF \sin F}{\sin X}$

$= \dfrac{286.63 \times \sin 34° 21' 55''}{\sin 110° 00' 00''}$

$= \dfrac{286.63 \times 0.564\,466\,9}{0.939\,692\,6}$

$= 172.17 \text{ m}$

$\dfrac{FX}{\sin B} = \dfrac{BF}{\sin X}$

Therefore $FX = \dfrac{BF \sin B}{\sin X}$

$= \dfrac{286.63 \times \sin 35° 38' 05''}{\sin 110° 00' 00''}$

$= \dfrac{286.63 \times 0.582\,615\,6}{0.939\,692\,6}$

$= 177.71 \text{ m}$

Exercise 9.6

1 The following angles were measured on a closed theodolite traverse:

Instrument station	Target station	Face left
A	D	00° 10′ 20″
	B	89° 26′ 40″
B	A	89° 26′ 40″
	C	189° 05′ 20″
C	B	189° 05′ 20″
	D	269° 29′ 40″
D	C	269° 29′ 40″
	A	00° 09′ 00″

Calculate:
(a) the measured angles,
(b) the angular correction,
(c) the corrected angles,
(d) the quadrant bearing of each line given that bearing AD is 45° 36′ 00″.

2 Calculate the closing error and accuracy of the following traverse:

Line	Length	Bearing
AB	110.20	156° 40′ 00″
BC	145.31	75° 18′ 00″
CD	98.75	351° 08′ 00″
DE	163.20	276° 29′ 00″
EA	52.34	187° 27′ 00″

3 The partial coordinates of a closed traverse are:

Line	Length	Difference in Easting	Difference in Northing
PQ	252.41	0.00	252.41
QR	158.75	−110.76	−113.82
RS	153.50	−25.24	−151.41
SP	136.74	+136.15	12.67

Distribute the closing error by Bowditch's rule and calculate the corrected partial coordinates.

4 The following data refer to an open theodolite traverse round an old farmhouse which is to be demolished at a future date (Fig. 9.32):

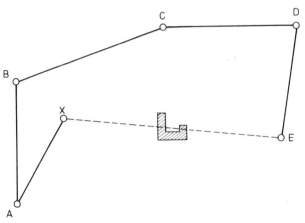

Figure 9.32

Line	Plan length (m)	Angle	Observed value
XA	160.00		
AB	186.40	XAB	330° 00′
BC	234.00	ABC	251° 30′
CD	170.60	BCD	198° 30′
DE	138.00	CDE	280° 45′

Given that the coordinates of station X are 100 m east, 100 m north and that the forward bearing of line XA is 210° 00′ 00″, calculate:
(a) the total coordinates of each station,
(b) the length of and bearing of a proposed sewer that is to be laid between the points X and E,
(c) the values of the angles to be set out at X and E by the theodolite to establish the line of the sewer.

5 The following lengths and angles were obtained on a closed theodolite traverse using a 20 m steel band and theodolite:

Line	Plan length	Quadrant bearing
AB	167.25	North
BC	228.34	N 30° 24′ 00″ E
CD	367.50	S 18° 16′ 40″ E
DA	220.70	N 89° 28′ 40″ W

It is suspected that there is a gross error in one of the linear measurements. Calculate the total coordinates of each station; thence determine the erroneous line and the probable reason for the wrong measurement.

6 Points on the boundary of a proposed building site have the following coordinates:

Station	Easting (m)	Northing (m)
A	314	762
B	263	801
C	293	849
D	354	871
E	394	793

The points are to be set out by theodolite and tape from the line A to D. Calculate the length and theodolite reading required to set out B, C and E.

(SCOTVEC, Ordinary National Diploma in Building)

7 A closed traverse ABCDEA produced results shown below:

Back station	Instrument station	Fore station	Clockwise horizontal angle
A	B	C	283° 31′ 40″
B	C	D	329° 06′ 50″
C	D	E	90° 47′ 20″
D	E	A	299° 43′ 00″
E	A	B	256° 50′ 20″

(a) Calculate the error of closure of the angles.
(b) Adjust the angles for complete closure.
(c) Calculate the corrected bearings of lines BC, CD, DE and EA, given that the bearing of AB is S 152° 24′ 40″ E.

15. Answers

Exercise 9.1

1 *Line*   *True bearing*

AB	59 00
BC	126 00
XY	210 30
PQ	273 30

2 *Line*   *Back bearings*

MN	215 00
NO	355 36
OP	34 14
PQ	87 32
QR	176 20

3 N 48° 50′ W;   S 67° 30′ W;   N 60° 10′ E;
S 87° 00′ E;   West;
S 12° 10′ E;   S 68° 50′ E;   S 84° 50′ W;
N 00° 50′ W;   S 01° 00′ E

4 329° 50′;   240° 30′;   172° 15′;   10° 00′;
90° 00′;   259° 10′;   270° 10′;   115° 30′

5 (a) 172° 40′;   (b) 00° 30′;
S 07° 20′ E;   N 00° 30′ E;
(c) 359° 00′;   (d) 220° 40′;
N 01° 00′ W;   S 40° 40′ W

6
26°	88°	233°	315°	8°
8° W	2° E	5° E	5° W	10° W
18°	90°	238°	310°	358°
198°	270°	58°	130°	178°
S 18° W	West	N 58° E	S 50° E	S 2° E

7 N 30° 45′ 10″ E
S 16° 41′ 40″ E
S 47° 36′ 35″ W
N 16° 51′ 51″ W

8   40° 30′ 20″;   104° 44′ 15″
140° 41′ 43″;   295° 00′ 02″

Exercise 9.2

Forward AB =	65° 00′ 00″
Back BA =	245° 00′ 00″
+ Angle B	93° 15′ 50″
Forward BC =	338° 15′ 50″
Back CB =	158° 15′ 50″
+ Angle C	274° 31′ 10″
	432° 47′ 00″
Forward CD =	72° 47′ 00″

Exercise 9.3

1

Angle	Measured angle	Adjustment	Adjusted angle
FAB	183° 33′ 20″	+10″	183° 33′ 30″
ABC	83° 41′ 40″	+10″	83° 41′ 50″
BCD	86° 47′ 50″	+10″	86° 48′ 00″
CDE	182° 39′ 40″	+10″	182° 39′ 50″
DEF	89° 15′ 50″	+10″	89° 16′ 00″
EFA	94° 00′ 40″	+10″	94° 00′ 50″
Sum =	719° 59′ 00″	+01′ 00″	720° 00′ 00″

Forward AB =	315° 00′ 00″
Back BA =	135° 00′ 00″
+ ABC	83° 41′ 50″
Forward BC =	218° 41′ 50″
Back CB =	38° 41′ 50″
+ BCD	86° 48′ 00″
Forward CD =	125° 29′ 50″

Back DC = 305° 29′ 50″
+ CDE   182° 39′ 50″
        = 488° 09′ 40″
Forward DE = 128° 09′ 40″

Back ED = 308° 09′ 40″
+ DEF = 89° 16′ 00″
         397° 25′ 40″
Forward EF = 37° 25′ 40″

Back FE = 217° 25′ 40″
+ EFA = 94° 00′ 50″
Forward FA = 311° 26′ 30″

Back AF = 131° 26′ 30″
+ FAB   183° 33′ 30″
Forward AB = 315° 00′ 00″

2   Sum of interior angles = $(2n - 4) \times 90°$
                    = 360° 00′ 00″
Sum of observed angles = 360° 00′ 16″
Angular error = +16″
Correction per angle = 16/4
                     = −04″

Angle	Observed value	Correction	Correction value
PMN	59° 41′ 08″	−04″	59° 41′ 04″
MNO	80° 19′ 04″	−04″	80° 19′ 00″
NOP	119° 42′ 59″	−04″	119° 42′ 55″
OPM	100° 17′ 05″	−04″	100° 17′ 01″
Sum =	360° 00′ 16″	−16″	360° 00′ 00″

Line	Whole circle bearing	Quadrant bearing
MN	342° 46′ 56″	N 17° 13′ 04″ W
NO	243° 05′ 56″	S 63° 05′ 56″ W
OP	182° 48′ 51″	S 02° 48′ 51″ W
PM	103° 05′ 52″	S 76° 54′ 08″ E

Exercise 9.4

1 Table 9.12 and Fig. 9.33

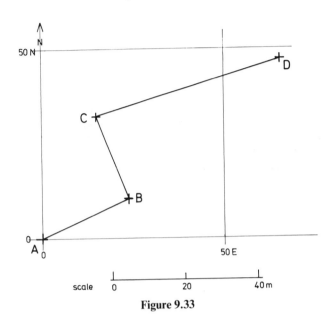

Figure 9.33

Exercise 9.5

1 Table 9.13

Line	Whole circle bearing	Length	Difference in eastings	Difference in northings	Eastings	Northings	Station
					0.000	0.000	A
AB	65° 0′ 0″	25.707	23.298	10.864	23.298	10.864	B
BC	338° 15′ 50″	22.861	−8.466	21.236	14.832	32.100	C
CD	72° 47′ 0″	53.221	50.836	15.753	65.668	47.853	D

Table 9.12

Line	Length	Whole circle bearing	Difference in eastings	Difference in northings	ΔE correction	ΔN correction	Corrected ΔE	Corrected ΔN	Eastings	Northings	Station
									0.000	0.000	A
AB	48.627	315° 0′ 0″	−34.384	34.384	0.006	0.006	−34.379	34.390	−34.379	34.390	B
BC	27.395	218° 41′ 50″	−17.127	−21.381	0.003	0.003	−17.124	−21.377	−51.503	13.013	C
CD	48.428	125° 29′ 50″	39.427	−28.120	0.006	0.006	39.433	−28.115	−12.070	−15.102	D
DE	22.149	128° 9′ 40″	17.415	−13.685	0.003	0.003	17.418	−13.683	5.348	−28.784	E
EF	18.076	37° 25′ 40″	10.986	14.355	0.002	0.002	10.988	14.357	16.336	−14.428	F
FA	21.795	311° 26′ 30″	−16.338	14.425	0.003	0.003	−16.336	14.428	0.000	0.000	A
	186.470	Errors	−0.022	−0.022	0.022	0.022	0.000	0.000			
		Corrections	0.022	0.022							

Correction $(k_1) = 0.000\,116\,26$
Correction $(k_2) = 0.000\,119\,21$

Accuracy of survey is 1 in 6005.301 62

Table 9.13

Exercise 9.6

1

	(a) Observed angle	(b) Correction	(c) Corrected angle
DAB	89° 16′ 20″	+20″	89° 16′ 40″
ABC	99° 38′ 40″	+20″	99° 39′ 00″
BCD	80° 24′ 20″	+20″	80° 24′ 40″
CDA	90° 39′ 20″	+20″	90° 39′ 40″
	359° 58′ 40″	+1′ 20″	360° 00′ 00″

(d) AD N 45° 36′ 00″ E
    AB S 45° 07′ 20″ E
    BC N 54° 31′ 40″ E
    CD S 45° 03′ 40″ W

2 Closing error
$$\Delta E = +0.036 \text{ m}$$
$$\Delta N = -0.212 \text{ m}$$
$$\text{Distance} = \sqrt{0.036^2 + 0.212^2}$$
$$= 0.215 \text{ m}$$
$$\text{Accuracy} = \overline{1 \text{ in } 2650}$$

3

ΣE	ΣW	ΣN	ΣS
136.15	136.00	265.08	265.23
$\Delta E = +0.15$		$\Delta N = -0.15$	

Corrected partial coordinates

Line	Easting	Northing
PQ	−0.054	252.464
QR	−110.794	−113.686
RS	−25.273	−151.477
SP	136.121	12.699

4 (a) Total coordinates:

Eastings	Northings	Station
100.000	100.000	X
19.955	−38.642	A
19.955	147.758	B
241.967	222.042	C
412.417	222.042	D
386.654	86.346	E

(b) Length XE = 287.20 m
    Bearing XE = 92° 44′
(c) Angle set out at X = 242° 44′
    Angle set out at E = 261° 59′

5 Total coordinates:

Departure	Latitude
0.00	0.00 A
0.00	167.25 B
115.55	364.20 C
230.81	15.24 D
10.12	17.25 A

Closing error $= \sqrt{10.12^2 + 17.25^2}$
$$= 19.99 \text{ m}$$

Tan bearing closing error $= \dfrac{10.12}{17.25}$
$$= \text{N } 30° 23′ 55″ \text{ E}$$

Conclusion: line BC is 20 m too long, caused by miscounting the tape lengths.

6

Line	Setting out angle	Length
AE	48° 40′ 10″	85.80
AB	287° 15′ 15″	64.20
AC	326° 16′ 40″	89.50

7

Angle	Measured value	Adjustment	Adjusted angle
ABC	283° 31′ 40″	+10″	283° 31′ 50″
BCD	329° 06′ 50″	+10″	329° 07′ 00″
CDE	90° 47′ 20″	+10″	90° 47′ 30″
DEA	299° 43′ 00″	+10″	299° 43′ 10″
EAB	256° 50′ 20″	+10″	256° 50′ 30″
	1259° 59′ 10″	+50″	1260° 00′ 00″

Line	Whole circle bearing	Quadrant bearing
AB	152° 24′ 40″	S 27° 35′ 20″ E
BC	255° 56′ 30″	S 75° 56′ 30″ W
CD	45° 03′ 30″	N 45° 03′ 30″ E
DE	315° 51′ 00″	N 44° 09′ 00″ W
EA	75° 34′ 10″	N 75° 34′ 10″ E

16. Project

Chapter 18 is a project covering all chapters of this textbook. It is intended that the project build into a complete portfolio of the surveying work required in the survey and the setting out of a building or engineering development.

If the reader wishes to continue work on the project or begin work at this stage, he or she should now turn to page 395 and attempt Section 6.

CHAPTER 10

Tacheometry

Objective

After studying this chapter, the reader should be able to (a) explain the principles of distance measurement by stadia tacheometric and electromagnetic methods, (b) use a theodolite and an EDM instrument to obtain survey data in the field and (c) process and plot the results to a specified scale.

On any survey, the fieldwork consists of measuring horizontal angles, vertical angles, distances and heights. Most surveyors agree that the most difficult operation in survey work is the measurement of distance. Consequently, much time, effort and money has been spent on developing methods of measuring distances instrumentally and several ingenious methods, incorporating the optical properties of a telescope, have been devised. Recently, the advent of electromagnetic methods of measuring distance has revolutionized survey methods. All of the methods are classified as tacheometry. The term stems from the Greek *tacheos* meaning fast and *metros* meaning measurement. Tacheometry is literally a fast method of measuring distances without the use of a tape. The accuracy attainable with tacheometric methods varies from about 1 part in 500, in the case of stadia tacheometry, to about 1 part in 20 000 in the case of electromagnetic distance measurement (EDM).

In each system of tacheometry, except EDM, a small angle called the parallactic angle is measured by a theodolite or purpose-built tachymeter, to a short base line, defined on a staff, which is held either vertically or horizontally. The distance between the theodolite and staff is then some function of the parallactic angle. The angle may be fixed or variable, resulting in the following systems of tacheometry:

Parallactic angle	Staff position	Tacheometric system
Variable	Vertical	(1) Tangential
Fixed	Vertical	(2) Stadia
Variable	Horizontal	(3) Subtense bar
Fixed	Horizontal	(4) Optical wedge

Almost all of these methods have been rendered obsolete by the widespread use of EDM and only the stadia method is used to any extent as an alternative, in circumstances where a high degree of accuracy is not important.

1. Stadia tacheometry

(a) Basic concept

Stadia tacheometry makes use of the optical properties of the telescope and may be carried out using either a theodolite or level. Figure 10.1 is a diagram of an externally focusing telescope. Such telescopes are no longer used in surveying instruments but since the classic tacheometric formula cannot be derived easily for a modern internally focusing telescope, it is derived here for the externally focusing telescope and suitably amended for use with modern instruments.

Figure 10.1 shows the rays of light from a staff A_1B_1, passing through the object glass of a theodolite telescope to the diaphragm at AB. The distance FO is the focal length, f, of the lens while the distances u and v are known as conjugate focal lengths. The focal length, f, can be calculated from the lengths of the conjugate focal lengths since

$$\frac{1}{f} = \frac{1}{u} + \frac{1}{v}$$

Triangles ABO and A_1B_1O are similar. Therefore

$$\frac{\text{Staff intercept } (S)}{\text{Image } (i)} = \frac{OC_1}{OC}$$

$$\text{i.e.} \quad \frac{S}{i} = \frac{u}{v}$$

$$\text{Also} \quad \frac{1}{f} = \frac{1}{u} + \frac{1}{v}$$

Multiply both sides by fu:

$$u = f + f\left(\frac{u}{v}\right)$$

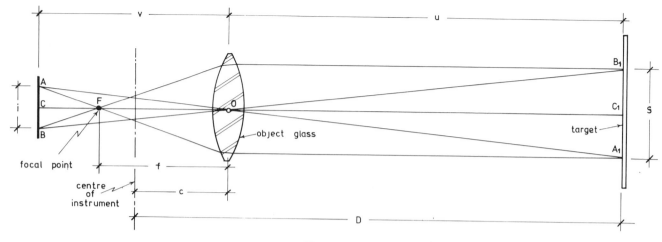

Figure 10.1

Substitute S/i for u/v:

$$u = f + \frac{fS}{i}$$

The distance $OC_1 = u$ is the horizontal distance from the object glass to staff, but the distance D from the centre of the instrument to the staff is required; therefore a constant c has to be added:

$$u + c = f + c + \frac{fS}{i}$$

Therefore $D = \left(\frac{f}{i}\right)S + (f+c)$

The ratio f/i is a constant by which the staff intercept S is multiplied while the quantity $(f + c)$ is a quantity that must be added for each observation. Calling f/i the multiplying constant, m and $(f + c)$, the additive constant, k, the horizontal distance D becomes

$$D = mS + k$$

The values of m and k are supplied by the manufacturer of the instrument. Generally f/i is 100 to 1 (occasionally 50 to 1) while $(f + c)$ varies for each instrument depending on the focal length of the telescope. For a telescope 250 mm long, $(f + c)$ is 375 mm.

The simple theory outlined above is only possible for externally focusing telescopes since the focal length f is constant for that type only. In an internally focusing telescope the focal length f varies, owing to the action of the internal double concave lens. As a result the multiplying constant m varies slightly, but is compensated by the variable k. The discrepancy caused by treating both of these variables as constant is, however, so small that it may be neglected.

The multiplying factor m is taken as being 100 while the additive constant is zero, thus reducing the basic formula to

$$D = mS \qquad (1)$$

It is of academic interest to note that in 1823 an Italian engineer named Porro succeeded in producing a tacheometric theodolite that did not have any additive constant $(f + c)$. An extra lens called an anallatic lens was introduced into the theodolite between the object glass and the vertical axis of the instrument. It had the effect of making the focal point f coincide with the vertical axis of the instrument and so obviated the necessity for adding the constant $(f + c)$.

Modern internally focusing telescopes are not anallatic but, as has already been explained, the error in treating them as being so is negligible.

(b) Stadia tacheometry (horizontal sights)

Stadia tacheometry may be carried out using a level or theodolite. In the simple case illustrated in Fig. 10.2, a levelling instrument (or a theodolite with the telescope set in the horizontal position) is being used. The horizontal distance and the difference in level between A and B are required.

The procedure is as follows:

1. Set up the level over station A, level it accurately and record the instrument height i (1.300 m) and the reduced level of the instrument station (14.110 m).
2. Sight a vertical staff held on B and read the centre cross-hair C as for normal levelling. Book the reading (2.340 m).
3. Read the other cross-hairs, called the stadia lines, at D and E and note the readings (2.660 and 2.020 m).
4. The difference between readings D and E is the staff intercept S:

$$S = 2.660 - 2.020$$
$$= 0.640 \text{ m}$$

5. For a modern internal focusing telescope:

$$H = 100 \times S \qquad (2)$$
$$= 64.0 \text{ m}$$

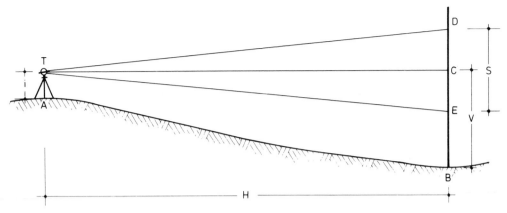

Figure 10.2

6. Calculate the collimation height of the instrument, i.e. HPC:

$$\text{HPC} = 14.110 + 1.300$$
$$= 15.410 \text{ m}$$

7. Calculate the reduced level (RL) of the staff station:

$$\text{RL} = 15.410 - 2.340$$
$$= 13.070 \text{ m}$$

Example

1 Figure 10.3 and Table 10.1 show field data, observed during the tacheometric survey of a copse of trees, which are the subject of a preservation order. An automatic level was used with a multiplying constant of $m = 100$.

(a) Calculate the horizontal distance from station A to each tree numbered 1, 3 and 5.

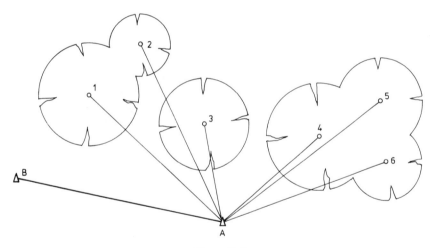

Figure 10.3

| Instrument station | A (RL 25.27 m AOD) |
| Instrument height | 1.35 m |

| *Target* | *Horizontal circle reading* | *Stadia readings* | | | *Remarks* |
		Upper	*Mid*	*Lower*	
B	00° 00′				Reference point
1	31° 10′	0.598	0.549	0.500	Oak tree 3 m radius
2	53° 00′	0.864	0.812	0.760	Oak tree 1.8 m radius
3	66° 30′	1.180	1.154	1.128	Beech 2.8 m radius
4	128° 10′	1.192	1.157	1.122	Sycamore 3.0 m radius
5	133° 10′	0.304	0.250	0.196	Sycamore 2.8 m radius
6	149° 20′	2.106	2.059	2.012	Sycamore 2.0 m radius

Table 10.1

| Instrument station | A (RL 25.27 m AOD) | | | | | | | | |
| Instrument height | 1.35 m | | | | | | | | |

Target	Horizontal circle reading	Stadia readings			S (upper–lower)	Horizontal distance (= 100 S)	HPC	Reduced level of target	Remarks
		Upper	Mid	Lower					
B	00° 00′								Reference point
1	31° 10′	0.598	0.549	0.500	0.098	9.80	26.62	26.07	Oak tree 3.0 m radius
2	53° 00′	0.864	0.812	0.760					Oak tree 1.8 m radius
3	66° 30′	1.180	1.154	1.128	0.052	5.20	26.62	25.47	Beech 2.8 m radius
4	128° 10′	1.192	1.157	1.122					Sycamore 3.0 m radius
5	133° 10′	0.304	0.250	0.196	0.108	10.80	26.62	26.37	Sycamore 2.8 m radius
6	149° 20′	2.106	2.059	2.012					Sycamore 2.0 m radius

Table 10.2

(b) Calculate the reduced level of the ground at trees numbered 1, 3 and 5.

(c) Plot the tree positions to scale 1:100, using a scale rule and protractor.

Answer (Table 10.2)

(a) the horizontal distance from station A to each of the trees numbered 2, 4 and 6,

(b) the reduced level of the ground at those trees.

Plot the complete survey to scale 1:100.

(c) Inclined sights

On most surveys, it is not possible to observe all target points using a horizontal line of sight. Consequently, vertical angles (or zenith angles, depending upon the configuration of the vertical circle) must be observed to every target point. In Fig. 10.4, the

Exercise 10.1

1 Table 10.2 of Example 1 shows the partially completed field data of a tacheometric survey. Complete the table by calculating:

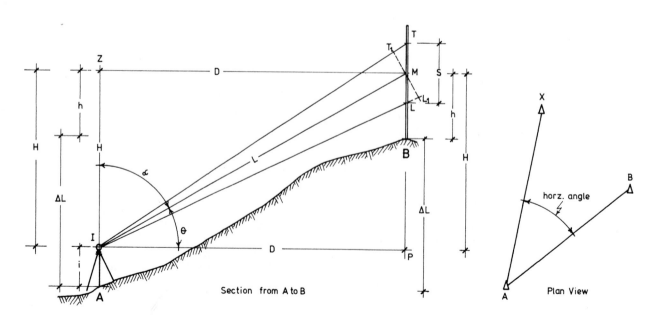

Figure 10.4

	Target	Horizontal circle reading	Vertical (or zenith) angle	Stadia readings			Remarks
				Upper	Mid	Lower	
1	X	00° 00′ 00″	—	—	—	—	Reference point
2	B	40° 05′ 00″	21° 00′ 00″	1.410	1.270	1.130	Peg

Instrument station — A
Instrument height — 1.27 m
Reduced level of station — 123.45 m AOD

Table 10.3

horizontal distance from A to B and the reduced level of station B are required. Point X is assumed to be the north direction and the reduced level of station A is known (or assumed).

1. *Field procedure*

(a) Set the theodolite over station A; measure instrument height i and note the height in the field book (Table 10.3).

(b) Sight the reference station X; set the horizontal circle to read zero degrees and note the reading in the field book on line 1.

(c) Hold a staff vertically on point B, utilizing the spirit level attached to the back of the staff (if fitted).

(d) Sight the staff and read all three stadia lines (T, M, L), plus the vertical angle θ or the zenith angle α. Note all of these readings in the field book on line 2.

2. *Basic formulae*

Supposing, for the moment, that the staff was held at right angles to the line of sight through the theodolite and the stadia readings were T_1L_1M; then T_1L_1 would be the staff intercept.

$$\text{Therefore } L = (m \times T_1L_1)$$

However, the staff is vertical and the intercept is in fact $TL = S$. Consider the triangle TT_1M. Angle $T_1MT = \theta$ and angle T_1 is a right angle (almost).

$$\text{Therefore } \frac{T_1M}{TM} = \cos \theta$$
$$\text{and } T_1M = TM \cos \theta$$

Also in the triangle LL_1M,

$$L_1M = LM \cos \theta$$
$$\text{Therefore } T_1L_1 = TL \cos \theta$$

Now, since

$$L = m \times T_1L_1$$
$$L = m \times TL \cos \theta$$
$$= mS \cos \theta$$

In triangle IMP,

$$\frac{IP}{IM} = \cos \theta$$

Therefore IP = IM $\cos \theta$
i.e. $D = L \cos \theta$
$$= (mS \cos \theta) \cos \theta$$
$$= mS \cos^2 \theta \qquad (3)$$

Also

$$\frac{MP}{IM} = \sin \theta$$

Therefore MP = IM $\sin \theta$
i.e. $H = L \sin \theta$
$$= (mS \cos \theta) \sin \theta$$
$$= mS \cos \theta \sin \theta \qquad (4)$$

Alternatively, in triangle IMP,

$$\frac{MP}{IP} = \tan \theta$$

Therefore MP = IP $\tan \theta$
i.e. $H = D \tan \theta$

Difference in level A to B = ΔL:

$$\Delta L = i + H - h \qquad (5)$$

For angles of depression,

$$\Delta L = i - H - h \qquad (6)$$

Where the zenith angle α has been measured, rather than the vertical angle θ, formulae (3) and (4) will change. The angles α and θ are complementary,

i.e. $(\alpha + \theta) = 90°$
Therefore $\theta = (90° - \alpha)$

Formula (3) shows that

Horizontal distance $D = mS \cos^2 \theta$
$$= mS \cos^2 (90 - \alpha)$$
$$= mS \sin^2 \alpha \qquad (7)$$

Formula (4) shows that

Difference in height $= mS \cos \theta \sin \theta$
$$= mS \cos (90 - \alpha) \sin (90° - \alpha)$$
$$= mS \sin \alpha \cos \alpha \qquad (8)$$

Examples

2 Stations M, N and O form a right-angled triangle at station M (Fig. 10.5). A theodolite was used to determine the tacheometric data in Table 10.4.

Instrument station	M		
Height of instrument	1.410 m		
Reduced level of station	129.600 m		

Target	Vertical	Stadia reading		
station	angle	Upper	Mid	Lower
N	−5° 40′	1.830	1.500	1.170
O	+2° 30′	2.810	2.610	2.410

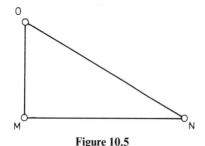

Figure 10.5

Calculate:

(a) horizontal lengths MN and MO,
(b) horizontal length NO,
(c) reduced levels of N and O.

Answer:
Line MN

Staff intercept at N = 1.830 − 1.170
$$= 0.660 \text{ m}$$

Horizontal length MN
$$= mS \cos^2 \theta$$
$$= 66.0 \cos^2 5° 40'$$
$$= 65.36 \text{ m}$$

$$
\begin{aligned}
H &= \text{(instrument height M} \\
&\quad - \text{mid staff reading N} \\
&= -mS \cos \theta \sin \theta \\
&= -66.0 \cos 5° 40' \sin 5° 40' \\
&= -6.49 \text{ m}
\end{aligned}
$$

$$
\begin{aligned}
\text{Reduced level of N} &= 129.600 + i - H - h \\
&= 129.600 + 1.410 \\
&\quad - 6.490 - 1.500 \\
&= 123.020 \text{ m}
\end{aligned}
$$

Line MO

Staff intercept at O = 2.810 − 2.410
$$= 0.400 \text{ m}$$

Horizontal length MO
$$= mS \cos^2 \theta$$
$$= 40.0 \cos^2 2° 30'$$
$$= 39.92 \text{ m}$$

$$
\begin{aligned}
H &= \text{(mid staff reading O} \\
&\quad - \text{instrument height M)} \\
&= \text{MO} \tan \theta \\
&= 39.92 \tan 2° 30' \\
&= +1.74 \text{ m}
\end{aligned}
$$

$$
\begin{aligned}
\text{Reduced level of O} &= 129.600 + i + H - h \\
&= 129.600 + 1.410 + 1.74 \\
&\quad - 2.610 \\
&= 130.140 \text{ m}
\end{aligned}
$$

Horizontal distance NO
$$= \sqrt{\text{MN}^2 + \text{MO}^2}$$
$$= \sqrt{65.36^2 + 39.92^2}$$
$$= 76.59 \text{ m}$$

3 Table 10.5 and Fig. 10.6 show a tacheometric survey of the pegs of a site boundary.
(a) Calculate the horizontal distance from station A to pegs numbered 1 and 3.
(b) Calculate the reduced levels of the ground at those pegs.

Answer (Table 10.6)

Instrument station	A (RL 25.27 m AOD)				
Instrument height	1.35 m				

Target	Horizontal circle reading	Vertical angle	Stadia readings			Remarks
			Upper	Mid	Lower	
B	00° 00′ 00″					Reference point
1	61° 11′ 00″	4° 40′ 00″	1.399	1.350	1.301	Peg 1
2	83° 00′ 00″	2° 57′ 40″	1.402	1.350	1.298	Peg 2
3	96° 00′ 00″	1° 02′ 20″	1.404	1.350	1.296	Peg 3
4	158° 09′ 00″	1° 34′ 40″	1.385	1.350	1.315	Peg 4
5	198° 05′ 00″	7° 07′ 30″	1.550	1.500	1.450	Peg 5
6	179° 18′ 00″	−6° 07′ 40″	1.047	1.000	0.953	Peg 6

Table 10.5

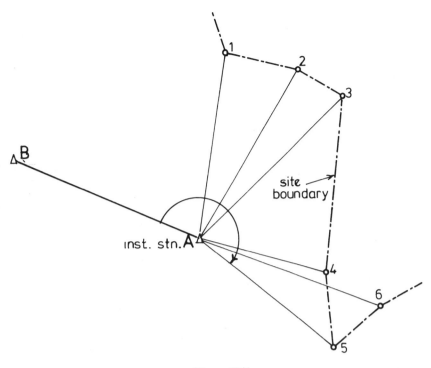

Figure 10.6

			Stadia readings								
Instrument station	A (RL 25.27 m AOD)										
Instrument height	1.35 m										
Target	Horizontal circle reading	Vertical angle	Upper	Mid	Lower	S (upper–lower)	D ($mS \cos^2 \theta$)	H ($mS \cos \theta \sin \theta$)	ΔH (i+H−h)	Reduced level	Remarks
B	00° 00′ 00″									25.27	Reference point
1	61° 11′ 00″	4° 40′ 00″	1.399	1.350	1.301	0.098	9.74	+0.79	+0.79	26.06	Peg 1
2	83° 00′ 00″	2° 57′ 40″	1.402	1.350	1.298						Peg 2
3	96° 00′ 00″	1° 02′ 20″	1.404	1.350	1.296	0.108	10.80	+0.20	+0.20	25.47	Peg 3
4	158° 09′ 00″	1° 34′ 40″	1.385	1.350	1.315						Peg 4
5	198° 05′ 00″	7° 07′ 30″	1.550	1.500	1.450						Peg 5
6	179° 18′ 00″	−6° 07′ 40″	1.047	1.000	0.953						Peg 6

Table 10.6

Exercise 10.2

1 Table 10.6 of Example 3 shows the partially completed field data of a tacheometric survey. Complete the table by calculating

(a) the horizontal distances to pegs numbered 2, 4, 5 and 6,

(b) the reduced level of the ground at those pegs.

Using a scale rule and protractor, plot the survey to scale 1:100.

2. Errors in stadia tacheometry

The sources of gross error are those that result from

(a) wrong reading of the staff,
(b) wrong reading of vertical angle,
(c) wrong booking.

In tacheometry the observer should not act as booker since there are so many observations to be recorded. With an experienced booker, booking errors are rare. The vertical angle is very seldom checked on both faces in tacheometry; consequently great care must be taken in observing and reading the vertical angles and staff readings.

Systematic errors are those that result from:

1. Non-perpendicularity of the staff

In Fig. 10.7(a) the staff is leaning backwards from the normal position. The intercept AB used in the calculation is therefore greater than it would be with the staff in its correct vertical position with the result that H is greater. Similarly, H is smaller in Fig. 10.7(b), though it could be greater if the error in verticality were sufficiently great.

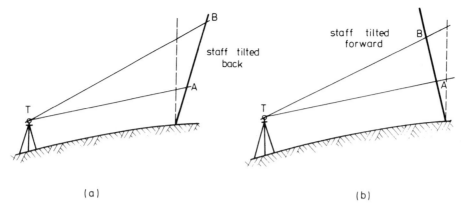

Figure 10.7

Tacheometric staffs should always be fitted with handles and a good spirit level to ensure that the staff is vertical for every sight.

2. *Differential refraction*

Refraction has an important bearing on the accuracy of tacheometric observations. During observations, rays of light are bent when passing from one substance to another of different density. The density of the atmosphere changes fairly rapidly close to the earth, with the result that the lower line of sight is refracted more than the upper in stadia tacheometry.

In order to minimize this effect the lower line of sight should be kept as high as possible. In practice this means that the lower staff reading should not be less than 1 metre.

3. Methods of simplifying calculations

The basic formulae for vertical staff stadia tacheometry as shown in Sec. 1(c) are

Horizontal distance $D = mS \cos^2 \theta$
Difference in level $\Delta L = i + D \tan \theta - h$
or $\Delta L = i + (mS \cos \theta \sin \theta) - h$

All of these formulae produce fairly lengthy calculations and the following methods are designed to simplify the calculations:

1. *Use of tacheometric tables*

Tacheometric tables, the work of D. T. F. Munsey MA, FRICS, have been produced entirely in metric units (Table 10.7).

Values of $m = 100$ and $k = 0$ are used throughout the tables. The range of vertical angles is from 0° 10′ to 10° 00′ at 10′ intervals and from 10° 20′ to 20° 00′ at 20′ intervals. For each vertical angle the horizontal component $mS \cos^2 \theta$ and vertical component $mS \cos \theta \sin \theta$ are given for values of mS ranging from 10 to 210 m at intervals of 1 metre.

In the tables, the distance mS is called the generating number G while the values $G \cos^2 \theta$ and

$G \cos \theta \sin \theta$ are respectively called D for distance and H for height.

In Example 2, the calculations showed the horizontal distance MN to be 65.36 while the corresponding height was -6.49. In Table 10.7, D and H for the vertical angle 5° 40′ are underlined where $G = 66$, the values agreeing with those obtained above. Where G is not an integral number the values of H and D have to be interpolated.

Example

4 Find the values of H and D corresponding to a staff intercept of 0.634 and vertical angle 5° 40′.

Answer

	G	D	H
From the tables	63	62.39	6.19
	64	63.38	6.29
By interpolation	63.4	62.79	6.23

In the field, the vertical angle must be set to some multiple of 10 minutes if under 10° 00′ or set to some multiple of 20 minutes if the angle exceeds this value.

---

2. The calculation of differences in level is reduced if, in the field, the centre cross-hair is set to read on the staff the value of the instrument height i. The formula for difference in level then becomes simply $\Delta L = D \tan \theta$ since h and i cancel each other.

Using this method, the tacheometric tables cannot be used since the vertical angle might not be a multiple of 10 minutes.

3. The calculation of the staff intercept S is simplified if the lower cross-wire is set to a round value of 0.1 metre, on the staff. The other two wires are observed in the usual manner.

G	D	H	G	D	H	G	D	H	G	D	H
10	9.90	0.98	60	59.41	5.90	110	108.9	10.81	160	158.4	15.72
11	10.89	1.08	61	60.41	5.99	111	109.9	10.91	161	159.4	15.82
12	11.88	1.18	62	61.40	6.09	112	110.9	11.00	162	160.4	15.92
13	12.87	1.28	63	62.39	6.19	113	111.9	11.10	163	161.4	16.02
14	13.86	1.38	64	63.38	6.29	114	112.9	11.20	164	162.4	16.11
15	14.85	1.47	65	64.37	6.39	115	113.9	11.30	165	163.4	16.21
16	15.84	1.57	66	65.36	6.49	116	114.9	11.40	166	164.4	16.31
17	16.83	1.67	67	66.35	6.58	117	115.9	11.50	167	165.4	16.41
18	17.82	1.77	68	67.34	6.68	118	116.8	11.59	168	166.4	16.51
19	18.81	1.87	69	68.33	6.78	119	117.8	11.69	169	167.4	16.61
20	19.80	1.97	70	69.32	6.88	120	118.8	11.79	170	168.3	16.70
21	20.80	2.06	71	70.31	6.98	121	119.8	11.89	171	169.3	16.80
22	21.79	2.16	72	71.30	7.07	122	120.8	11.99	172	170.3	16.90
23	22.78	2.26	73	72.29	7.17	123	121.8	12.09	173	171.3	17.00
24	23.77	2.36	74	73.28	7.27	124	122.8	12.18	174	172.3	17.10
25	24.76	2.46	75	74.27	7.37	125	123.8	12.28	175	173.3	17.20
26	25.75	2.55	76	75.26	7.47	126	124.8	12.38	176	174.3	17.29
27	26.74	2.65	77	76.25	7.57	127	125.8	12.48	177	175.3	17.39
28	27.73	2.75	78	77.24	7.66	128	126.8	12.58	178	176.3	17.49
29	28.72	2.85	79	78.23	7.76	129	127.7	12.68	179	177.3	17.59
30	29.71	2.95	80	79.22	7.86	130	128.7	12.77	180	178.2	17.69
31	30.70	3.05	81	80.21	7.96	131	129.7	12.87	181	179.2	17.78
32	31.69	3.14	82	81.20	8.06	132	130.7	12.97	182	180.2	17.88
33	32.68	3.24	83	82.19	8.16	133	131.7	13.07	183	181.2	17.98
34	33.67	3.34	84	83.18	8.25	134	132.7	13.17	184	182.2	18.08
35	34.66	3.44	85	84.17	8.35	135	133.7	13.26	185	183.2	18.18
36	35.65	3.54	86	85.16	8.45	136	134.7	13.36	186	184.2	18.28
37	36.64	3.64	87	86.15	8.55	137	135.7	13.46	187	185.2	18.37
38	37.63	3.73	88	87.14	8.65	138	136.7	13.56	188	186.2	18.47
39	38.62	3.83	89	88.13	8.74	139	137.6	13.66	189	187.2	18.57
40	39.61	3.93	90	89.12	8.84	140	138.6	13.76	190	188.1	18.67
41	40.60	4.03	91	90.11	8.94	141	139.6	13.85	191	189.1	18.77
42	41.59	4.13	92	91.10	9.04	142	140.6	13.95	192	190.1	18.87
43	42.58	4.23	93	92.09	9.14	143	141.6	14.05	193	191.1	18.96
44	43.57	4.32	94	93.08	9.24	144	142.6	14.15	194	192.1	19.06
45	44.56	4.42	95	94.07	9.33	145	143.6	14.25	195	193.1	19.16
46	45.55	4.52	96	95.06	9.43	146	144.6	14.35	196	194.1	19.26
47	46.54	4.62	97	96.05	9.53	147	145.6	14.44	197	195.1	19.36
48	47.53	4.72	98	97.04	9.63	148	146.6	14.54	198	196.1	19.46
49	48.52	4.81	99	98.03	9.73	149	147.5	14.64	199	197.1	19.55
50	49.51	4.91	100	99.02	9.83	150	148.5	14.74	200	198.0	19.65
51	50.50	5.01	101	100.0	9.92	151	149.5	14.84	201	199.0	19.75
52	51.49	5.11	102	101.0	10.02	152	150.5	14.94	202	200.0	19.85
53	52.48	5.21	103	102.0	10.12	153	151.5	15.03	203	201.0	19.95
54	53.47	5.31	104	103.0	10.22	154	152.5	15.13	204	202.0	20.04
55	54.46	5.40	105	104.0	10.32	155	153.5	15.23	205	203.0	20.14
56	55.45	5.50	106	105.0	10.42	156	154.5	15.33	206	204.0	20.24
57	56.44	5.60	107	106.0	10.51	157	155.5	15.43	207	205.0	20.34
58	57.43	5.70	108	106.9	10.61	158	156.5	15.52	208	206.0	20.44
59	58.42	5.80	109	107.9	10.71	159	157.4	15.62	209	207.0	20.54
60	59.41	5.90	110	108.9	10.81	160	158.4	15.72	210	208.0	20.63

Table 10.7

This is probably the least useful simplification. Since tables cannot be used the difference in level, ΔL, has to be worked out in full since h will probably not equal i.

4. A compromise field method is sometimes used which does not introduce any appreciable error.

The vertical angle is set to some multiple of 10 minutes and the centre hair reading is noted. The vertical slow motion screw is touched to bring the lower cross-hair on to some round 0.1 m value and the upper and lower cross-hairs read. S is then easily found and the tacheometric tables can be used.

Instrument station	A
Height of instrument	1.390 m
Reduced level	116.210

1	2	3	4	5	6	7	8	9	10
				Stadia					
Target station	Horizontal circle	Vertical angle	Top/ bottom	S	Mid	D	H	ΔL	Reduced level target
RO	00° 00′								
B	12° 30′	+ 5° 40′	2.040 1.600	0.440	1.820	43.57	4.32	+ 3.89	120.100
C	34° 15′	+ 2° 26′	1.670 1.110	0.560	1.390	55.90	+ 2.38	+ 2.38	118.590
D	63° 26′	− 8° 10′	1.380 1.000	0.380	1.190	37.23	− 5.34	− 5.14	111.070

Table 10.8

The staff intercept S is unchecked by this method since the centre cross-hair is not the mean of the upper and lower readings.

Table 10.8 illustrates the first three simplification methods outlined above. Columns 1 to 6 are completed in the field and 7 to 9 are filled in later.

Calculations
Line AB

From tables $D = 43.57$ m
(when G is 44) $H = 4.32$ m
$$\Delta L = 1.390 + 4.32 - 1.820 = + 3.89 \text{ m}$$

Line AC

$$D = 56 \cos^2 2° 26' = 55.90 \text{ m}$$
$$H = 56 \cos 2° 26' \sin 2° 26' = + 2.38 \text{ m}$$
$$\Delta L = 1.390 + H - 1.390 = + 2.38 \text{ m}$$

Line AD

$$D = 38 \cos^2 8° 10' = 37.23$$
$$H = - 37.23 \tan 8° 10' = - 5.34$$
$$\Delta L = 1.390 - 5.340 - 1.190 = - 5.14$$

4. Use of stadia tacheometry

The maximum accuracy obtainable by this method is 1/1000. Consequently the uses are limited to obtaining minor details and spot heights for contouring.

Figure 10.8 shows the route of a proposed roadway along which contours are required. Since several instrument stations are necessary the survey is laid out in the form of an open traverse, wherein the horizontal clockwise angles between stations RO and D are measured. The distances are measured tacheometrically in both directions and the traverse calculated, plotted and adjusted as described in Chapter 9.

At each station several rays are observed and oriented to the RO with a view to determining the position and spot level of a number of points. The order of observing is shown by the numbers in the figure. The spot levels are plotted on to the traverse plan generally by protractor and scale, and contours are interpolated from them.

The stadia method is generally adopted when the terrain is unsuitable for linear measuring and when the result is required to a fairly low degree of

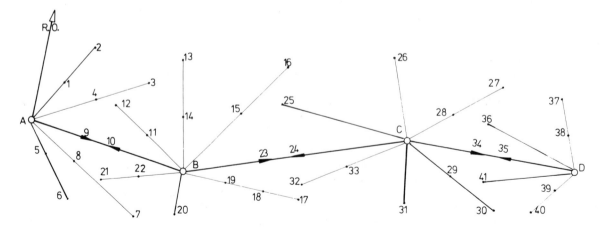

Figure 10.8

Instrument station	A
Instrument height	1.400 m
Reduced level of station	52.75 m AOD

Observation station	Horizontal reading	Zenith angle	Stadia readings Top	Stadia readings Mid	Stadia readings Bottom	S	Plan length D	Vertical height H	Difference height ΔH	Reduced level	Remarks
RO	0° 0′ 0″										Reference
1	30° 5′ 0″	82° 23′ 0″	1.466	1.400	1.334						Spot height
2	30° 5′ 0″	82° 23′ 0″	1.530	1.400	1.270						Spot height
3	61° 30′ 0″	84° 20′ 0″	1.222	1.050	0.878						Spot height
4	61° 30′ 0″	84° 20′ 0″	1.270	1.200	1.130						Spot height
5	143° 32′ 0″	97° 10′ 0″	1.450	1.400	1.350						Spot height
6	143° 32′ 0″	97° 10′ 0″	1.515	1.400	1.285						Spot height
7	20° 28′ 0″	94° 23′ 0″	1.592	1.400	1.208						Spot height
8	120° 28′ 0″	94° 23′ 0″	1.478	1.400	1.322						Spot height
(B)9	96° 25′ 0″	89° 5′ 0″	1.622	1.400	1.178						Traverse station

Instrument station	B
Instrument height	1.350 m
Reduced level of station	53.46 m AOD

			Top	Mid	Bottom	S	D	H	ΔH		Remarks
A	0° 0′ 0″	90° 55′ 0″	1.572	1.350	1.128						Traverse station
C	154° 33′ 0″	88° 35′ 0″	1.655	1.350	1.045						Traverse station
13	72° 2′ 0″	80° 50′ 0″	1.500	1.350	1.200						Spot level
14	72° 2′ 0″	80° 50′ 0″	1.420	1.350	1.280						Spot level
20	265° 1′ 0″	106° 5′ 0″	1.410	1.350	1.290						Spot level

Table 10.9

accuracy. The method also has the advantage of being speedy and requires the minimum amount of equipment and least number of personnel.

5. Use of computer program

All the calculations shown in the previous sections of this chapter may be carried out using a simple BASIC program.

It is realized that some readers will not wish to or have the facilities to consider this further method of calculation. The computer programs, instructions and examples are therefore included in Chapter 17, program 7.

Exercise 10.3

1 Table 10.9 above shows some of the observations, taken during a contour survey, along the route of a proposed roadway (Fig. 10.8 opposite). The complete survey was carried out using stadia tacheometry.

(a) Calculate the coordinates of traverse stations B and C, given that station A is the origin of the survey (100.00 m east, 100.00 m north).
(b) Calculate the reduced levels of points 1 to 9 from the reduced level of station A (52.75 m AOD) and the reduced levels of points 13 to 20 from the reduced level of station B.
(c) Plot the points to scale 1:500.
(d) Interpolate the contour line positions at 1 metre vertical intervals.

2. The tacheometric observations in Table 10.10 were taken along the line of a proposed roadway at approximately 20 metre intervals.

The new roadway is to begin at point A and is to rise at a gradient of 1 in 50 towards B.

Instrument station	D (approx. chainage 60 m)
Instrument height	1.45 m
Reduced level of station	234.21 m AOD

Target station	Vertical angle	Stadia Top	Stadia Mid	Stadia Bottom	Remarks
A	−4° 20′	1.750	1.450	1.140	Chainage 0 m
B	−4° 20′	1.400	1.200	1.000	Approx. 20 m
C	−4° 20′	1.210	1.105	1.000	Approx. 40 m
E	−3° 00′	1.200	1.100	1.000	Approx. 80 m
F	−3° 00′	1.650	1.450	1.250	Approx. 100 m

Instrumental constant $m = 100$.

Table 10.10

Instrument station	Instrument height	Staff station	Whole circle bearing	Zenith angle	Staff readings		
					Top	Mid	Bottom
A	1.35 m	B	350° 10′	87° 20′	4.200	3.000	1.800
		C	40° 10′	86° 00′	3.760	2.500	1.240
		D	78° 10′	84° 40′	3.050	2.000	0.950

Table 10.11

Calculate, for each station:

(a) the reduced level,
(b) the true chainage,
(c) the proposed roadway level,
(d) the depth of cutting required.

3 The tacheometric observations in Table 10.11 were made using a theodolite, where the staff was held vertically. The instrument constant is $m = 100$.
Calculate:

(a) the horizontal distances AB, AC and AD,
(b) the coordinates of stations B, C and D, relative to A as origin,
(c) the reduced levels of B, C and D, relative to station A (25.31 m AOD).

6. Electromagnetic distance measurement (EDM)

(a) Introduction

Modern developments in electronics have now made possible the measurement of distance using an electromagnetic signal. The measurement is accomplished in seconds with a very high degree of accuracy.

The instruments were first introduced during the fifties and every manufacturer of surveying equipment produces a variety of EDM equipment.

(b) Basic concept of measurement

The basic concept is simple. An EDM instrument capable of transmitting an electromagnetic signal is set up over a survey station at one end of a survey line (Fig. 10.9). The signal is directed to a reflector or second transmitter at the other end of the line, where it is reflected or instantly retransmitted back to the transmitter. The transit time of the double journey is measured by the transmitter and, since the speed of light is accurately known, the distance is calculated from the formula:

Distance between stations (m)
  = velocity of signal (m/s) × transit time (s)

$$\text{i.e. } D = V \times t \tag{9}$$

The electromagnetic signal that is transmitted is in the form of radio waves, infra-red light, visible light or laser beam—all of which have different properties, although they travel at the same speed.

In order to understand the complexities of EDM, even in this simplified version, it is necessary to have a basic understanding of the properties of electromagnetic radiation and the methods employed in measuring the time interval.

(c) Properties of the signal

Light, infra-red rays and radio waves are all forms of electromagnetic radiation and, like heat and sound, are forms of energy.

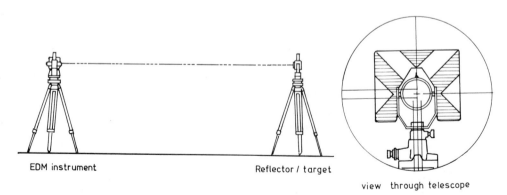

EDM instrument   Reflector / target

view through telescope

Figure 10.9

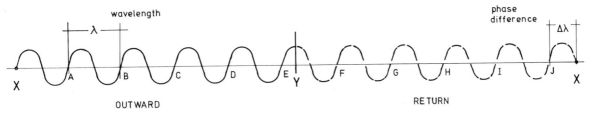

Figure 10.10

Wavelength and phase difference

Figure 10.10 shows an electromagnetic wave being transmitted from a survey point X to a second point Y, where it is reflected back along a parallel path and is received at station X. There are five whole waves and a fraction of a wave in the double journey.

The distance between any two similar points on the wave is the wavelength (λ), so XA = AB, etc. = 1 wavelength. Waves are said to be in-phase when a complete number occurs between the starting and finishing points. In this case there is an incomplete portion of a wave finishing on point X and this particular wave is said to exhibit a phase difference ($\Delta\lambda$).

Frequency

The wave takes a finite but very short time to travel from X to Y. Its frequency is the number of complete wavelengths, or cycles, that would occur in one second of time. The SI unit of frequency meaning one cycle per second is the hertz. The various multiples of one hertz are derived in the usual manner by prefixing kilo, mega and giga to the word hertz:

$$1 \text{ hertz} = 1 \text{ Hz}$$
$$10^3 \text{ hertz} = 1 \text{ kilohertz} = 1 \text{ kHz}$$
$$10^6 \text{ hertz} = 1 \text{ megahertz} = 1 \text{ MHz}$$
$$10^9 \text{ hertz} = 1 \text{ gigahertz} = 1 \text{ GHz}$$

The time taken by the wave to travel from X to Y is the number of wavelengths (n) divided by the frequency (f) of occurrence of the wave. In the basic formula (9),

$$D = V \times T$$

Therefore, by replacing T by (n/f),

$$D = V \times (n/f) \tag{10}$$

The fieldwork in EDM therefore consists of counting the number of wavelengths and measuring the phase difference. From formula (10), the transit time of the wave is computed, which, when multiplied by the velocity of the signal, produces the slope distance between X and Y.

Figure 10.11 shows part of the electromagnetic spectrum. The wavelengths of the various bands vary from 10 000 m long waves to 0.001 mm visible light waves, the corresponding frequencies being 30 kHz and 30×10^{10} kHz respectively. Only a narrow band

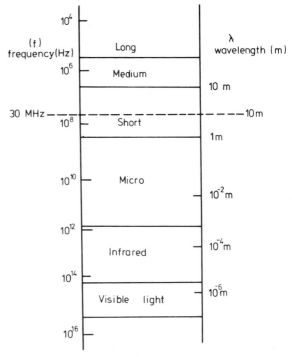

Figure 10.11

of these waves can be used to measure distances to the standard of accuracy required in surveying.

In the sections following, it will be pointed out that a phase difference of one-thousandth part of a wavelength can be resolved by EDM instruments. For most construction surveys, an accuracy of ± 1 cm is acceptable; therefore the derived wavelength is 1000×1 cm = 10 m, which from Fig. 10.10 corresponds to a frequency of 30 MHz.

Table 10.12 lists a few of the EDM instruments in common use and shows the frequency and approximate wavelength of the measuring signal.

As previously stated, the range of suitable measuring frequencies shown in Table 10.9 represents only a small section of the complete electromagnetic spectrum. Unfortunately this range of frequencies is not suitable for direct transmission through the atmosphere by EDM instruments, because the waves tend to fade and scatter, and suffer from interference.

Very high frequency waves are not so prone to these effects, and it is possible to impress a low-frequency measuring wave on to a wave of higher

Instrument	Frequency	Approximate wavelength
Wild MD 60		
Fine measurement	149.848 3 MHz	2 m
Coarse measurement	14.977 337 MHz	20 m
Kern DM 500		
Fine measurement	14.985 4 MHz	20 m
Coarse measurement	149.854 kHz	2000 m
Wild DI 3		
Fine measurement	7.492 7 MHz	40 m
Coarse measurement	74.927 kHz	4000 m
Geodimeter 6	30 MHz	10 m
Tellurometer CA 1000	(19–25) MHz	(16–12) m

Table 10.12

frequency and transmit them together. The high-frequency wave acts as a carrier for the low-frequency wave and is said to be modulated by this process. Among others, infra-red and visible light waves are suitable carriers.

In simpler language, the visible light wave is analogous to the thin strip of steel from which a tape is manufactured. The steel is 'modulated' by the metric graduations stamped on it and carries them when the tape is stretched during linear measurement.

Velocity

All electromagnetic waves travel through outer space with the same velocity (c) of 299 792.5 km/s. When travelling through earth's atmosphere, the velocity (v) is retarded. Variations in temperature, pressure and humidity affect the velocity, with the result that the value of v is not quite constant.

This is analogous to measuring lines with a steel tape which continually changes its length, so some standard must be set for the EDM instrument in the same way as standards of 20°C temperature and 44.5 N tension are set for steel tapes.

The normal standardizing values are 760 mmHg pressure and 12°C temperature, and under those conditions it can be shown that electromagnetic signals are travelling at 99.97 per cent of their velocity (c) in vacuum. The velocity (v) through earth's atmosphere is therefore (299 792.5 × 99.97 per cent) = 299 708.0 km/s approximately.

If at the time of measurement the values of temperature, pressure and humidity differ from the standard values, corrections must be made to the measurement of the line.

7. Principle of distance measurement

The basic concept of measurement already outlined now becomes a definite principle in which a modulated electromagnetic wave of known frequency (f) is transmitted to a distant reflector and returned to the transmitter. The transmitting instrument is capable of counting the number (n) of wavelengths with an accuracy of one-thousandth part of a wavelength.

The value of (n/f) is computed automatically by the instrument, and multiplied by the 'standardized' speed of the signal through the atmosphere. The result is the slope length of the line measured.

8. EDM systems

The systems developed from the transmission of electromagnetic waves can be conveniently divided into two classes, namely:

(a) microwave system (long range).
(b) electro-optical system (medium and short range).

(a) Microwave system

As the name suggests, this group of EDM instruments uses microwaves to measure distances from 20 m to a maximum of 150 km, with an accuracy of about 3–4 mm per km. An instrument typical of this class is the Wild DI 60 (Fig. 10.12) which operates on frequencies of about 15 MHz.

These instruments are used primarily for geodetic purposes. They are seldom used in construction survey work, except perhaps in major road construction, stretching for many kilometres, where geodetic techniques would be used in any case.

(b) Electro-optical system

The instruments employed in this system of measurement can be conveniently divided into two classes, depending upon which part of the spectrum they use

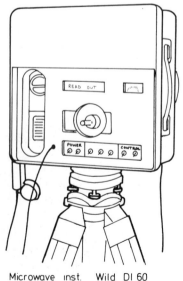

Microwave inst.   Wild DI 60

Figure 10.12

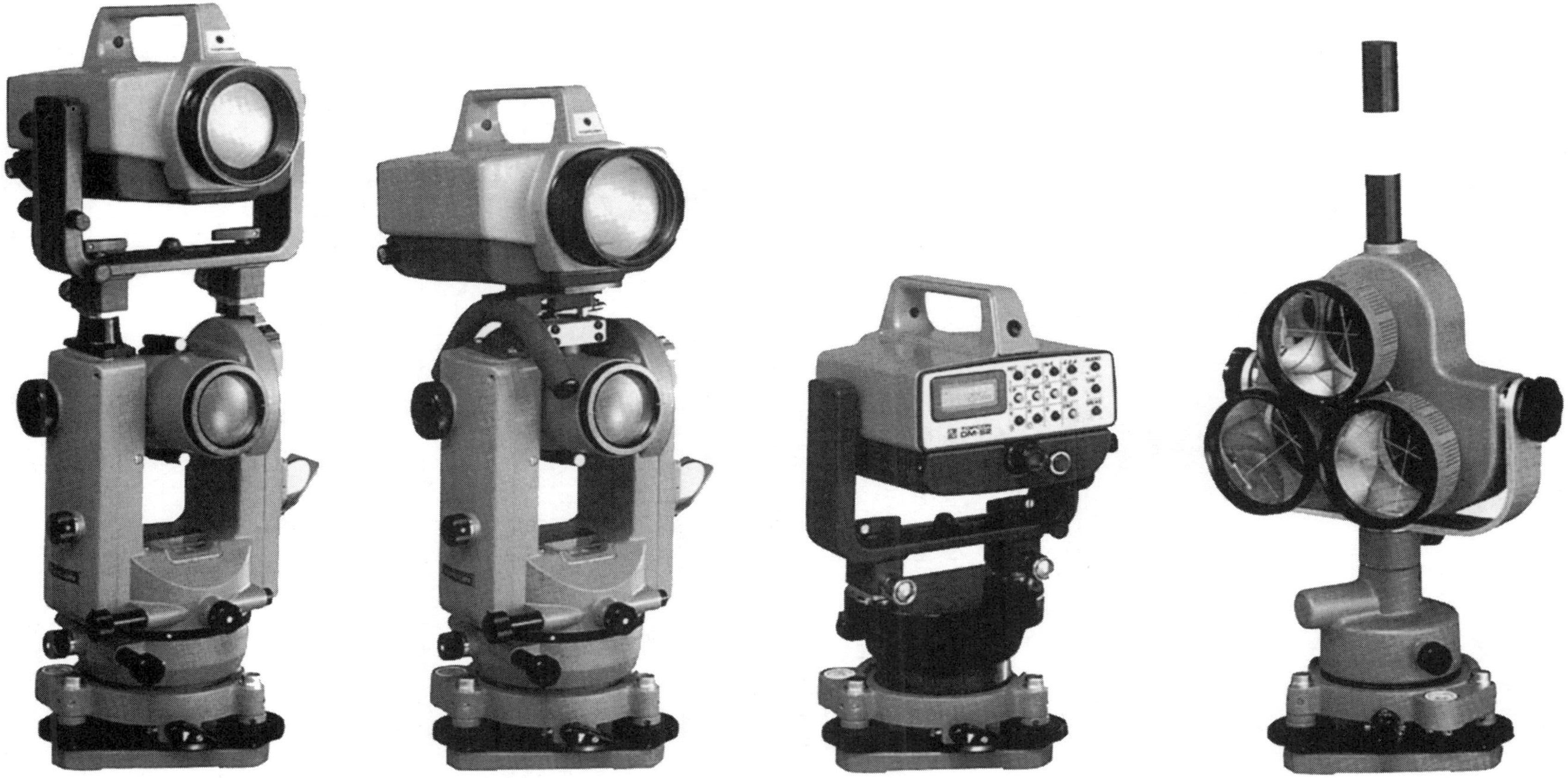

(a) DM-S2 yoke setting

(b) DM-S3 telescope setting

(c) DM-S2 independent setting

(d) Triple prism with tilting holder

Figure 10.13 Topcon DM S2/S3

for signal transmission. The instruments which use visible light form the medium-range class, while those using infra-red light form the short-range class. The classes have many common elements.

The measuring signals are carried on a narrow, highly focused beam of light which has to be directed optically to the distant target by means of an in-built telescope. Alternatively, the short-range EDM units may be mounted on the telescope of a theodolite by means of a bracket specially designed to point the EDM unit exactly along the line of the theodolite telescope, wherever it is pointed (Fig. 10.13, page 231).

Fieldwork
The fieldwork is substantially the same for all kinds of electro-optical distance measurement:

1. Measurements of the prevailing atmospheric conditions are made before EDM measurement starts, in order to determine the correction to be applied to the measured slope lengths. The correction is necessitated by the fact that the velocity of the signal is affected by changes in pressure and temperature. Pressure is measured by a barometer, with an accuracy of ±4 millibars. Temperature is measured by a thermometer, with an accuracy of ±1°C. These values are read against a nomograph (usually supplied with the instrument) and a correction in millimetres per kilo-

metre (parts per million) is entered into the instrument, using a settable switch. Changes in humidity have little effect at infra-red frequencies.

Example

5 An EDM survey was carried out on a day when the prevailing temperature was 25°C and atmospheric pressure was 1000 mb. Using the nomograph (Fig. 10.14), determine the correction to be applied to measurements in parts per million, i.e. mm per km.

Answer
Draw a horizontal line at 25°C and a vertical line at 1000 mb. The intersection of the lines shows the correction to be + 13 ppm.

---

2. The transmitter is placed at one end of the line being measured and is accurately aligned by means of the telescope on to a corner-cube reflector at the other end (Fig. 10.9). The reflector is reciprocally aligned back to the transmitter. The alignment is not critical. The important property of the reflector is that it reflects light back along any line on an exactly parallel course. In general one corner-cube reflector is effective

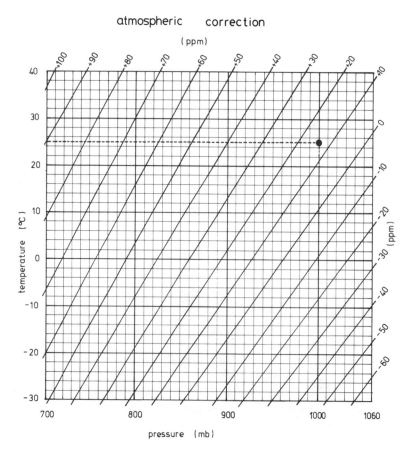

atmospheric   correction
(ppm)

Figure 10.14

for ranges up to about 600 m. For longer lengths a bank of three, six or nine prisms is required.

3. The signal is transmitted on a known frequency to the reflector, from which it is returned to the instrument and the phase difference is measured. The frequency is changed automatically by the instrument, and the measuring procedure repeated, enabling the number of wavelengths to be counted and the slope distance to be calculated.

4. The vertical angle between the instrument and reflector is measured in the usual manner to enable the plan length of the line to be calculated.

Instruments

1. *Classification*

In construction work, the vast majority of EDM instruments are classified under the heading of short-range infra-red instruments. Their maximum range is of the order of two to three kilometres, which is, of course, long enough for construction work.

Modern infra-red EDM instruments are compact and lightweight and can be readily telescope mounted or yoke mounted on a theodolite or can be independently mounted on a detachable theodolite tribrach (Fig. 10.13(a) to (c)).

The reflector is mounted on a telescopic rod which can be extended and set to instrument height or may be tripod mounted. One, three or nine prism assemblies are used for different distances (Fig. 10.13(d)).

2. *Signal generation*

All modern short-range instruments emit an infra-red carrier wave generated by a gallium arsenide (GaAs) diode. The wavelength is less than 1 micrometre. The power is supplied by nickel–cadmium dry-cell batteries or by a 12 volt car battery. The beam is invisible and harmless, and will produce the correct distance even when it is broken by traffic.

3. *Wavelength count*

All EDM instruments measure part of one cycle only, i.e. the phase difference, but the distance D, measured electromagnetically, is given by the formula

$$D = n\lambda + \Delta\lambda$$

where n is an unknown number of wavelengths. Hence some means of determining the value of the integer n must be built into the instrument.

Construction surveyors need not concern themselves unduly about how the number of wavelengths is counted, since the whole operation is carried out automatically by the instrument. An in-depth knowledge of the concepts of electrical theory is required to understand fully the actual mechanics of the operation.

One method of counting the number is to measure the line on three slightly different, but closely related, frequencies. Two frequencies are sufficient if the maximum range of the instrument is less than two kilometres. If the chosen frequencies were around

30 MHz, the wavelength would be 10 metres for the double journey.

Using half the wavelength gives an equivalent value of 5 m for a 'one way' journey. The three wavelengths used for the measurement are

$$w_1 = 5.000\,000 \text{ m}$$
$$w_2 = 4.987\,532 \text{ m}$$
$$w_3 = 4.761\,904 \text{ m}$$

These lengths are chosen such that

$$400w_1 = 401w_2 = 2000 \text{ m}$$
and
$$20w_1 = 21w_3 = 100 \text{ m}$$

Assuming that the length of the survey line AB is 835.300 m, the phase differences resulting from measurements on the wavelengths w_1, w_2 and w_3 are $\Delta w_1 = 0.300$, $\Delta w_2 = 2.382$ and $\Delta w_3 = 1.967$ respectively.

$$\text{Distance AB} = nw_1 + \Delta w_1 \quad (11)$$
$$= nw_2 + \Delta w_2 \quad (12)$$
$$= nw_3 + \Delta w_3 \quad (13)$$

From (11) and (12):

$$n(w_1 - w_2) = \Delta w_2 - \Delta w_1$$
and since
$$400w_1 = 401w_2$$
$$w_2 = (400/401)w_1$$
then
$$n\left(w_1 - \frac{400}{401}w_1\right) = 2.382 - 0.300$$

Therefore
$$\frac{nw_1}{401} = 2.082$$
$$nw_1 = 834.9$$
$$= 835$$
and so
$$n = 167$$

This value will be repeated every 2000 m.

From (11) and (13),

$$n(w_1 - w_3) = \Delta w_3 - \Delta w_1$$
and since
$$20w_1 = 21w_3$$
$$w_3 = (20/21)w_1$$
then
$$n\left(w_1 - \frac{20}{21}w_1\right) = 1.967 - 0.300$$

Therefore
$$\frac{nw_1}{21} = 1.667$$
$$nw_1 = 35$$
and so
$$n = 7$$

This value will be repeated every 100 m.

The total number of wavelengths is therefore $167 + \Delta w_1 = (167 \times 5 + 0.30) \text{ m} = 835.30 \text{ m}$.

4. *Phase difference measurement*

Until comparatively recently, the phase difference was measured in short-range instruments by an electromechanical resolver, but nowadays it is measured using digital methods.

The transmitted signal triggers a counting mechanism within the instrument, which is stopped by the return of the reflected ray. The number of pulses allowed through the counting gate, while it is open, is counted and displayed. Each pulse represents a known short length, usually 1 millimetre. A large number of counts is made automatically by the instrument and averaged to produce the correct result.

It is now possible to resolve phase difference with an accuracy of 1/10 000th part of a cycle and a 1 millimetre readout is now common.

5. *Range and accuracy*

Reference has been made throughout this chapter to the range of various classes of instrument. These are summarized in Table 10.13.

The accuracy of EDM equipment comprises two elements, namely:

(a) the instrumental limitation,
(b) the influence of atmospheric irregularities.

Most instruments have an instrumental error of about ± 5 mm, although a development of the Kern Mekometer has reduced this error to ± 0.2 mm. Atmospheric vagaries of temperature, pressure and humidity produce errors varying from 1 to 10 mm per km. The errors are summarized in Table 10.13.

9. Types of EDM instrument

Three different types of instrument are currently available. The variety within the groups is great and currently there are about 60 different EDM instruments marketed by various manufacturers.

(a) Theodolite and EDM working independently

Virtually every manufacturer of surveying instruments produces a simple EDM instrument, which can be either yoke mounted or telescope mounted on a conventional optical theodolite (Fig. 10.13(a) and (b)), or may be used independently (Fig. 10.13(c)). These instruments measure only the slope distance. The vertical angle of gradient and the horizontal bearing of the survey line are measured, using the theodolite. Typical of this class are the Sokkia Red

mini and Red 1A (Fig. 10.15(b)), the Leica Wild DI1001 and the Leica Kern DM 104 (Fig. 10.15(a)).

This is the cheapest form of angle/distance instrument. The measuring range is of the order of 800 to 1100 m using a single prism, with an accuracy of 5 mm + 5 ppm. The measuring time per distance varies between 2 seconds for the Wild DI1001 and 5 seconds for the Red mini.

The disadvantage in using this combination is the time required to measure the vertical and horizontal angles and the fact that the readings cannot be entered automatically on a data recorder (Sec. 9(d)). The use of a data recorder is not precluded, however, since data may be entered manually.

(b) Theodolite and EDM in combination

When an EDM unit is mounted on an electronic theodolite, the operation of the unit can be controlled from the theodolite keyboard, thus forming an effective angle/distance measuring combination. Typical examples of this class are the Sokkia Red 2L and Red mini 2, Pentax MD14/MD20 (Fig. 10.16), Leica Wild DI1600 and Topcon DM-S2/S3 (Fig. 10.13). These instruments have a higher specification and can measure distances up to 9.8 km in favourable conditions (Table 10.13).

When the appropriate key (each instrument differs) is pressed on the theodolite keyboard, the distance is fed directly into the theodolite. The distance, horizontal angle and vertical angle can be read either simultaneously or in sequence. They may be recorded manually or, more likely, automatically, on a data recorder.

The measuring time varies between two seconds in the case of the Leica Wild DI1600 and six seconds in the case of the Sokkia Red 2L. All of these instruments can calculate the horizontal distance and height difference from the theodolite station to the target station and display the results at the press of the appropriate button. The coordinates of the target station may also be displayed. It would be pointless to give specific instructions for any one instrument, since

Instrument	Maximum range	Number of prisms	Measuring conditions	Accuracy
Sokkia Red 2L	3800–	1	Average	5 mm + 3 ppm
	9800 m	9	Good	
Pentax MD20	1600–	1	Average	5 mm + 5 ppm
	3500 m	9	Good	
Leica Wild DI1600	2500–	1	Average	3 mm + 2 ppm
	7000 m	11	Good	
Topcon DM-S3	2100–	1	Average	5 mm + 3 ppm
	5500 m	9	Good	

Condition average: ordinary haze with visibility about 7 km, sunny with slight heat shimmer.
Condition good: no haze with visibility over 30 km, overcast with no heat shimmer.

Table 10.13

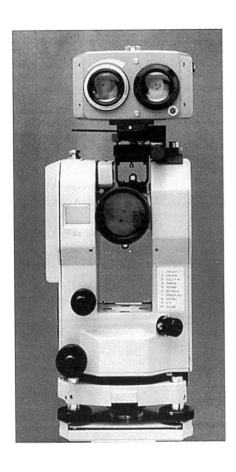

(a) Leica Kern DM 104

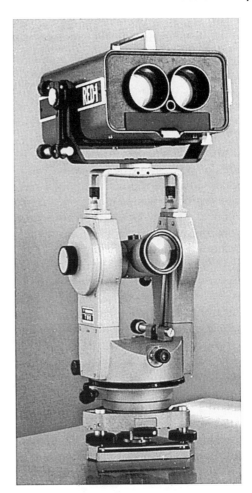

(b) Sokkia 'RED' series

Figure 10.15

they all vary slightly in their mode of operation. Each manufacturer produces comprehensive instructions, which should be consulted if more information on any particular instrument is required.

(c) Theodolite and incorporated EDM

These instruments are commonly called total stations. They are technically electronic tacheometers and are really a combination of a theodolite and an EDM unit, controlled by the same microprocessor. Typical examples of these instruments are the Sokkia SET range: Topcon ET-2 and GTS range (Fig. 10.17), Leica Wild TC range, Pentax PTS range and Geotronics Geodimeter range (Fig. 10.18).

All of these instruments measure angles and distances electronically, both of which are continuously updated and displayed in real time from the sighting point. Angular accuracy varies between one and seven seconds. The maximum distance measurable is about four kilometres with an accuracy of 5 mm + 5 ppm.

In general these instruments have simple keyboard commands allowing the surveyor access to a range of onboard software including selection of horizontal and vertical angles, slope and horizontal distances, difference in height, target coordinates and setting-out information. The telescope and distance meter optics are coaxial on most instruments, thus providing a single line of sight. In order to facilitate sighting, the strength of the signal is indicated by an acoustic buzzer, the pitch of which is variable.

All recognized instrumental errors are tested during the set-up period and compensatory adjustments are made automatically by the central microprocessor.

Electronic field books (data recorders) can be linked to all total stations, controlling all functions of the instrument and providing a complete range of data collection, calculation and transfer functions.

(d) Electronic field book (data recorder)

Most electronic tacheometers can be linked to some form of data recorder or electronic field book. The angles and distance are stored automatically in the field book, resulting in a saving of time and reduction of errors. Typical examples of current data recorders

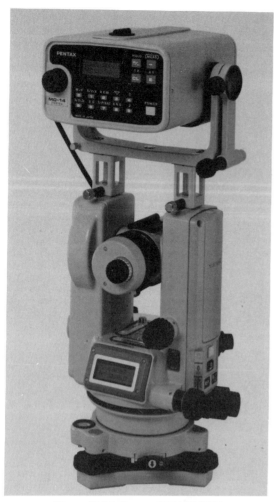

Figure 10.16 Pentax MD14/MD20

are the Sokkia SDR range (Fig. 10.19), the Leica GRE4, the Pentax DC 1Z, the Topcon FC-2/2C and the Geotronics Geodat.

The instruments are programmed to record the instrument station and instrument height; the horizontal angle, vertical angle and slope distance to a target; the target station and target height. The surveyor is prompted by the data recorder and enters information at the touch of a button. The instruments can record the information of hundreds of points.

The data may be processed in the field utilizing the survey programs of the data collector or may be down-loaded into a computer system via an interface.

10. Applications of EDM

Electro-optical instruments are used in all surveying fields except that of long-range geodetic surveying. Their versatility, accuracy and saving in time and manpower have made them invaluable for all short-range survey measurements.

They are used for purposes of traversing and trigonometrical levelling in and around construction sites;

for baseline measurement in minor triangulation schemes; for measuring shaft depths and tunnels in mining; and for measuring previously inaccessible distances across rivers, etc. They have virtually replaced tacheometry as a means of detail surveying.

Most instruments have a tracking device which allows them to be used for setting out all manner of construction and engineering works.

11. Operation of an EDM instrument

The variety of instruments available is very wide indeed and the task of describing all of them is impossible in any one textbook, unless the book is specifically written for that purpose. However, the following description is a general guide to the operation of most of these instruments.

(a) Theodolite/EDM combinations

The EDM unit is either yoke or telescope mounted on the theodolite. Horizontal and vertical angles are measured by the theodolite in the prescribed manner (Secs 5 and 6 in Chapter 7).

The EDM unit is switched on and a test signal shows whether the instrument is functioning correctly. The telescope of the EDM unit is horizontally collimated with the theodolite, such that, when the target is sighted through the telescope of the theodolite, the EDM need only be tilted vertically in order to sight the reflector. This is accomplished using a clamp and slow motion screw. In order to assist fast sighting, most EDM units transmit an audible signal, which may be switched off, if required.

The EDM measurement key is pressed. The infrared signal passes to the reflector, is returned to the unit and, after a period of a few seconds, the distance is displayed on the readout panel.

Figure 10.20 shows the complete operation, using the Sokkia Red mini instrument.

(b) Electronic tacheometer (total station)

The instruments are fitted with some form of keyboard, which controls all instrument functions. Figure 10.21 shows the keyboard of the Sokkia Set 4 model, while Figs 10.17 and 10.18 show the control panels of a Topcon and Geotronics instrument respectively.

Instruments work on different modes for the measurement of angles and distances. The mode usually has to be selected from the keyboard, though some instruments measure angles and distances simultaneously, without keyboard selection. When first switched on, the instrument defaults to angle measurement mode.

Assuming that the instrument has been set up and levelled correctly over a survey station B, the

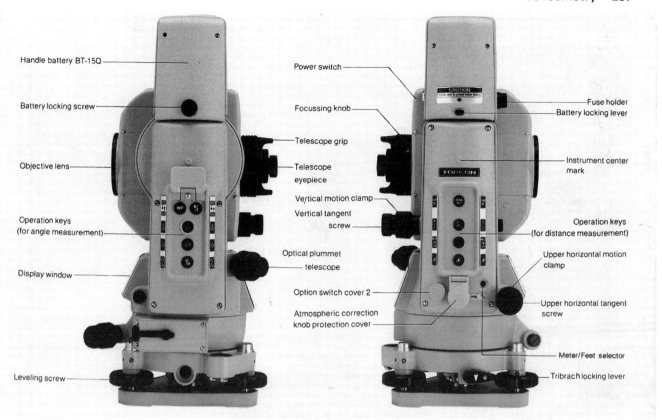

Handle battery BT-15Q

Battery locking screw

Objective lens

Operation keys
(for angle measurement)

Display window

Leveling screw

Power switch

Focussing knob

Telescope grip

Telescope
eyepiece

Vertical motion clamp

Vertical tangent
screw

Optical plummet
telescope

Option switch cover 2

Atmospheric correction
knob protection cover

Fuse holder

Battery locking lever

Instrument center
mark

Operation keys
(for distance measurement)

Upper horizontal motion
clamp

Upper horizontal tangent
screw

Meter/Feet selector

Tribrach locking lever

Figure 10.17 Topcon GTS 3B

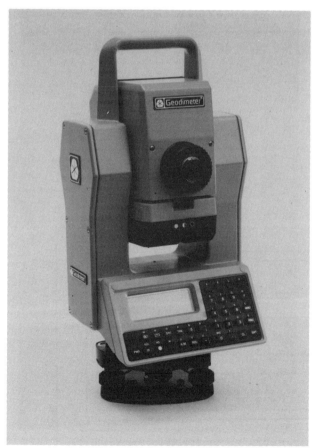

Figure 10.18 Geodimeter 600

Figure 10.19 Sokkia SDR 33

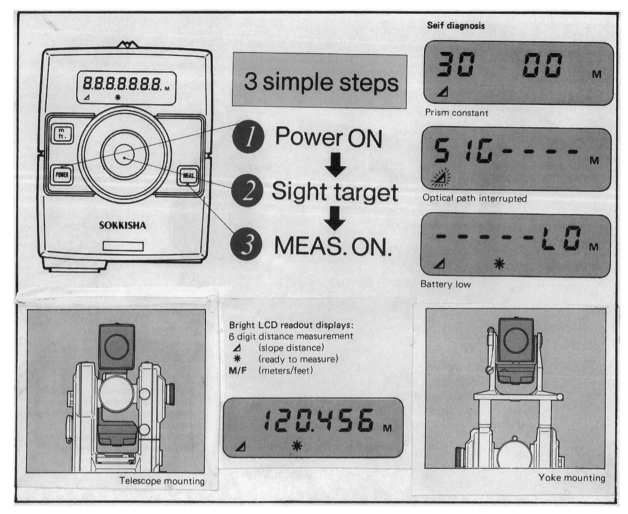

Figure 10.20 Sokkia Red mini

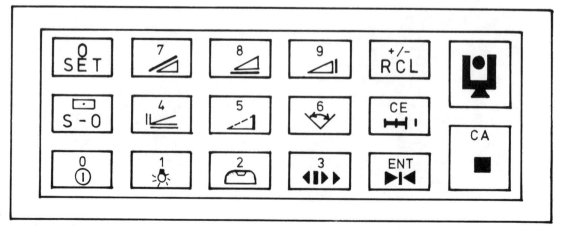

Figure 10.21

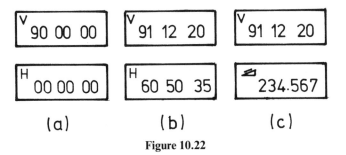

Figure 10.22

procedure for measuring an angle ABC and a distance BC is along the following lines:

1. Switch on the instrument and observe the panel display.
2. Release the horizontal and vertical circle clamps and set the instrument to observe the first station A. Bisect the target and tighten the clamp(s).
3. Using the zero-set key (Fig. 10.21), set the horizontal circle to read 00° 00′ 00″ and enter the value, i.e. press keys [SET 0] and [ENT]. The display panel will show the horizontal circle reading of zero degrees (Fig. 10.22(a)).
4. Release the upper plate clamp and sight target station B. The display panel shows the horizontal circle and zenith angle readings (Fig. 10.22(b)).
5. Change the mode from angle to basic by selecting the cancel key [CA].
6. Switch on the EDM power [⌀] and sight the reflector. The instrument is now in theodolite/EDM mode.
7. Select the distance key (slope, plan or height) to measure the required distance. The slope distance key is key 7. After a few seconds the display panel shows the slope length and the zenith (or vertical) angle (Fig. 10.22(c)).
8. Press the cancel key to return to basic mode.
9. Select the theodolite key (Fig. 10.21) to observe the next point and measure a further horizontal angle.

As already stated, many instruments measure angles and distances simultaneously. Hence they have the advantage that the whole operation is faster, though no more accurate.

When a data recorder is employed, all measurements are recorded automatically by pressing one key of either the data recorder or of the theodolite keyboard.

Exercise 10.4

1 Distinguish between the following electromagnetic systems of measurement: (a) electro-optical system, (b) microwave system.

2 Explain briefly the basic principle of electro-optical distance measurement as used with infra-red instruments.

3 Explain how atmospheric changes affect the measurement of distances in electromagnetic distance measurement.

4 List three applications of electromagnetic distance measurement in surveying. Give an example of an instrument that would be used in each case, explaining the reasons for your choice. Indicate the range and accuracy of the instrument.

12. Answers

Exercise 10.1

1 Table 10.14 and Fig. 10.23

Exercise 10.2

1 Table 10.15 and Fig. 10.24

Exercise 10.3

1 (a) Table 10.16
  (b) Table 10.17
  (c) Figure 10.25
  (d) Figure 10.25

Instrument station	A (RL 25.27 m AOD)								
Instrument height	1.35 m								

Target	Horizontal circle reading	Stadia readings Upper	Mid	Lower	S (upper–lower)	Horizontal distance (= 100 S)	HPC	Reduced level of target	Remarks
B	00° 00′								Reference point
1	31° 10′	0.598	0.549	0.500	0.098	9.80	26.62	26.07	Oak tree 3.0 m radius
2	53° 00′	0.864	0.812	0.760	0.104	10.40	26.62	25.81	Oak tree 1.8 m radius
3	66° 30′	1.180	1.154	1.128	0.052	5.20	26.62	25.47	Beech 2.8 m radius
4	128° 10′	1.192	1.157	1.122	0.070	7.00	26.62	25.46	Sycamore 3.0 m radius
5	133° 10′	0.304	0.250	0.196	0.108	10.8	26.62	26.37	Sycamore 2.8 m radius
6	149° 20′	2.106	2.059	2.012	0.094	9.40	26.62	24.56	Sycamore 2.0 m radius

Table 10.14

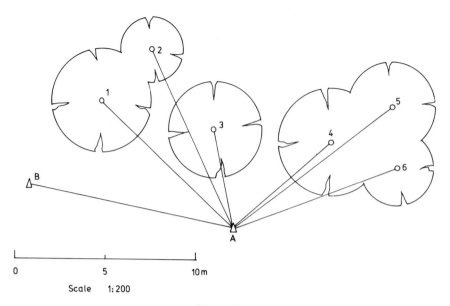

Figure 10.23

Target	Horizontal circle reading	Vertical angle	Stadia readings			S (upper − lower)	D ($mS\cos^2\theta$)	H ($mS\cos\theta\sin\theta$)	ΔH ($i+H-h$)	Reduced level	Remarks
			Upper	Mid	Lower						
B	00° 00′ 00″									25.27	Reference point
1	61° 11′ 00″	4° 40′ 00″	1.399	1.350	1.301	0.098	9.74	+ 0.79	+ 0.79	26.06	Peg 1
2	83° 00′ 00″	2° 57′ 40″	1.402	1.350	1.298	0.104	10.37	+ 0.54	+ 0.54	25.81	Peg 2
3	96° 00′ 00″	1° 02′ 20″	1.404	1.350	1.296	0.108	10.80	+ 0.20	+ 0.20	25.47	Peg 3
4	158° 09′ 00″	1° 34′ 40″	1.385	1.350	1.315	0.070	6.99	0.19	+ 0.19	25.46	Peg 4
5	198° 05′ 00″	7° 07′ 30″	1.550	1.500	1.450	0.100	9.85	+ 1.23	+ 1.08	26.35	Peg 5
6	179° 18′ 00″	−6° 07′ 40″	1.047	1.000	0.953	0.094	9.29	− 1.00	− 0.65	24.62	Peg 6

Instrument station   A (RL 25.27 m AOD)
Instrument height   1.35 m

Table 10.15

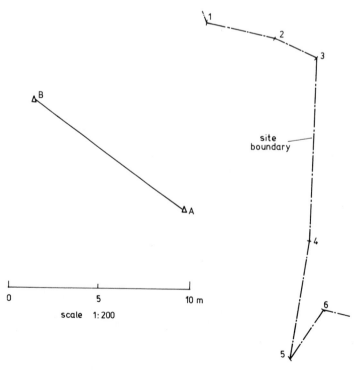

Figure 10.24

Line	Plan length	Whole circle bearing	ΔE	ΔN	Eastings	Northings	Station
					100.00	100.00	A
AB	44.39	96° 25′ 00″	+ 44.11	− 4.96	144.11	95.04	B
BC	60.96	70° 58′ 00″	+ 57.63	+ 19.88	201.74	114.92	C

Table 10.16

Instrument station   A
Instrument height   1.400 m
Reduced level of station   52.75 m AOD

Observation station	Horizontal reading	Zenith angle	Vertical angle (dec.)	Stadia readings			S	Plan length D	Vertical height H	Difference height ΔH	Reduced level	Remarks
				Top	Mid	Bottom						
RO	0° 0′ 0″	82° 23′ 0″	7.616 666 7									Reference
1	30° 5′ 0″	82° 23′ 0″	7.616 666 7	1.466	1.400	1.334	0.132	12.97	1.73	1.73	54.48	Spot height
2	30° 5′ 0″	82° 23′ 0″	7.616 666 7	1.530	1.400	1.270	0.260	25.54	3.42	3.42	56.17	Spot height
3	61° 30′ 0″	84° 20′ 0″	5.666 666 7	1.222	1.050	0.878	0.344	34.06	3.38	3.73	56.48	Spot height
4	61° 30′ 0″	84° 20′ 0″	5.666 666 7	1.270	1.200	1.130	0.140	13.86	1.38	1.58	54.33	Spot height
5	143° 32′ 0″	97° 10′ 0″	− 7.166 667	1.450	1.400	1.350	0.100	9.84	− 1.24	− 1.24	51.51	Spot height
6	143° 32′ 0″	97° 10′ 0″	− 7.166 667	1.515	1.400	1.285	0.230	22.64	− 2.85	− 2.85	49.90	Spot height
7	120° 28′ 0″	94° 23′ 0″	− 4.383 333	1.592	1.400	1.208	0.384	38.18	− 2.93	− 2.93	49.82	Spot height
8	120° 28′ 0″	94° 23′ 0″	− 4.383 333	1.478	1.400	1.322	0.156	15.51	− 1.19	− 1.19	51.56	Spot height
(B)9	96° 25′ 0″	89° 5′ 0″	0.916 666 7	1.622	1.400	1.178	0.444	44.39	0.71	0.71	53.46	Traverse station

Instrument station   B
Instrument height   1.350 m
Reduced level of station   53.46 m AOD

Observation station	Horizontal reading	Zenith angle	Vertical angle (dec.)	Top	Mid	Bottom	S	Plan length D	Vertical height H	Difference height ΔH	Reduced level	Remarks
A	0° 0′ 0″	90° 55′ 0″	− 0.916 667	1.572	1.350	1.128	0.444	44.39	− 0.71	− 0.71	52.75	Traverse station
C	154° 33′ 0″	88° 35′ 0″	1.416 666 7	1.655	1.350	1.045	0.610	60.96	1.51	1.51	54.97	Traverse station
13	72° 2′ 0″	80° 50′ 0″	9.166 666 7	1.500	1.350	1.200	0.300	29.24	4.72	4.72	58.18	Spot level
14	72° 2′ 0″	80° 50′ 0″	9.166 666 7	1.420	1.350	1.280	0.140	13.64	2.20	2.20	55.66	Spot level
20	265° 1′ 0″	106° 5′ 0″	− 16.083 33	1.410	1.350	1.290	0.120	11.08	− 3.19	− 3.19	50.27	Spot level

Table 10.17

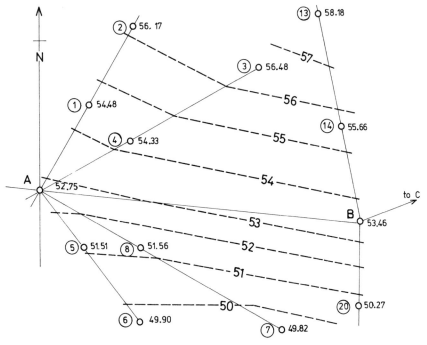

Figure 10.25

From station A (52.78) to point 1 (54.48),
height = 1.70 m

From station A (52.78) to 53 m contour,
height = 0.22 m

From station A to point 1,
horizontal distance = 12.97 m

From station A to contour position,
distance = x m

$$\text{Proportionally } \frac{1.70}{0.22} = \frac{12.97}{x}$$

$$\text{Therefore } x = \frac{12.97 \times 0.22}{1.70} = 1.68 \text{ m}$$

2

	A	B	C	D	E	F
Reduced levels	229.61	231.45	232.98	234.21	233.51	232.12
Chainages	0.00	20.88	39.77	60.65	80.60	100.54
Proposed levels	229.61	230.03	230.41	230.82	231.22	231.62
Cutting	0.00	1.42	2.57	3.39	2.29	0.50

All other contour positions are calculated in the same way.

3

Line	Horizontal length
AB	239.48
AC	250.77
AD	208.19

Station	Easting	Northing	Height
A	0.00	0.00	25.31
B	−40.90	235.96	34.81
C	161.75	191.63	41.70
D	203.77	42.69	44.10

Exercise 10.4

1 See pages 230–232.
2 See page 230.
3 See page 232.
4 See pages 234, 236.

13. Project

Chapter 18 is a project covering all chapters of this textbook. It is intended that the project build into a complete portfolio of the surveying work required in the survey and the setting out of a building or engineering development.

If the reader wishes to continue work on the project or begin work at this stage, he or she should now turn to page 396 and attempt Section 7.

CHAPTER 11

Radial positioning

Objective
After studying this chapter, the reader should be able to (a) obtain the three-dimensional coordinates of a series of points, process the results and manually produce a contoured plan, (b) number and code the points for inputting to a microcomputer-based mapping system and (c) process the results to produce a contoured plan automatically, using the mapping system.

The objective of most surveys is to determine the three-dimensional coordinates x, y and z (east, north and height) of a series of points of detail. These coordinates are plotted and a contoured plan is compiled from the results. The field methods used so far in this book, namely linear survey, traversing and offsetting plus levelling, are all viable, traditional and well-proved methods. However, they tend to be slow, often cumbersome and are manpower intensive.

Another method of three-dimensional positioning, which is speedy, very accurate and frugal in its use of manpower, is that of radial positioning using EDM. By this method, the three-dimensional coordinates of any point may be determined, with an accuracy of a few millimetres, in a few seconds of time.

1. Principles of radial positioning

(a) Orientation of survey

Figure 11.1 illustrates the principles. Point A is a survey station with known (or assumed) coordinates. The reference object (RO) is the point chosen as the starting point, from which the bearings of all lines of the survey will be measured. The RO may be a second survey station or some prominent object such as the spire of a church, a pylon or a road sign. The direction (bearing) from station A to the RO must be known or assumed. The direction may be relative to magnetic, true or grid north, or may simply be assumed to point to north, in which case the bearing from station A to the RO is zero degrees.

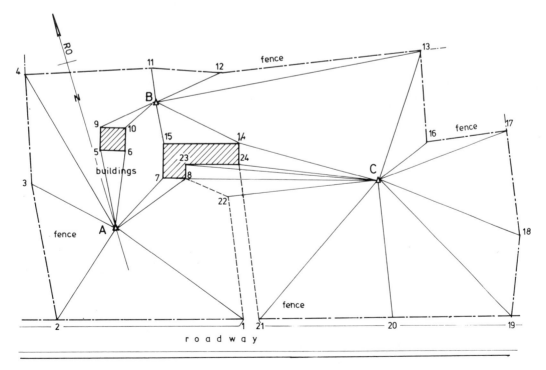

Figure 11.1

Points 1 to 24 are the points of detail of the survey, e.g. spot heights, manholes, etc. The three-dimensional coordinates of the points are required.

(b) Objectives

A minimum of two surveyors is required to conduct a radial survey. Their duties are:

1. To select suitable survey stations. The only criterion for the selection is that as many detail points as possible should be visible from any one survey station. If all of the points of detail are not visible from a single station, a traverse is made to establish more survey stations, e.g. in Fig. 11.1 the required traverse stations are A, B, C.

2. To measure the horizontal angle, the vertical or zenith angle and slope distance, using EDM means, to every point (numbered 1 to 24), of the survey.

3. To record the survey data, either manually or automatically, on an electronic field book and to process these data. The data may be calculated manually or post-processed on computer.

(c) Field work

The RO is assumed to be the north point, bearing zero degrees (00° 0′ 00″).

1. The theodolite/EDM is set over the point A and accurately centred and levelled.

2. The height of the instrument, *i*, from pet A to the transit axis is measured, using a tape or rule and noted in the field book (Table 11.1).

3. The coloured target, or the actual reflector if the EDM telescope is coaxial, is set to this height and taken to point B (Fig. 11.2) by an assistant, where it is held vertically by utilizing the attached spirit level. The target height is noted in the field book on the same line as station B (line 2).

4. The RO is sighted by the observer and the instrument is set to zero by means of the zero-set key (electronic theodolite or total station) or by means of the

upper and lower plate clamps (optical theodolite). This reading of 00° 00′ 00″ is recorded in the field book on line 1. The horizontal circle reading to any other point will therefore be the whole circle bearing from station A to the point.

5. Point B is then sighted and the horizontal circle reading and vertical circle reading are noted in the field book on line 2. The slope distance is then measured to the reflector, using the EDM and also noted on line 2. This completes the field work for point B.

6. The assistant is directed to the next point and the procedures 3 and 4 are repeated for that point. The field data for points 1 and 2 are shown in Table 11.1 on lines 3 and 4.

2. Calculation and plotting of survey (manually)

The field data may be calculated and plotted manually or may be post-processed on computer.

(a) Derivation of formulae

Calculations are carried out in the following order since each part-calculation depends upon the result from the previous part.

1. *Plan length*
In the right-angled triangle (Fig. 11.2(b)) formed by the slope length (S), the plan length (P) and the vertical (V),

$$P/S = \sin \theta$$
$$\text{Therefore } P = S \sin \theta \tag{1}$$

i.e. plan length = slope length × sin zenith angle

2. *Vertical height V*
In the triangle, the vertical dimension (V) is the height from the transit axis of the theodolite/EDM to the target:

$$V/S = \cos \theta$$
$$\text{Therefore } V = S \cos \theta \tag{2}$$

i.e. vertical height = slope length × cos zenith angle

Instrument station	A
Easting	100.000 m
Northing	200.000 m
Reduced level	35.210 m AD
Instrument height	1.350 m

	Target point	Target height	Horizontal circle reading	Zenith angle	Slope distance	Remarks
1	RO		00° 00′ 00″	—	—	Reference object
2	B	1.350	36° 30′ 00″	71° 30′ 10″	18.325	Survey station
3	1	1.350	147° 29′ 10″	92° 04′ 00″	21.110	Fence
4	2	0.850	231° 15′ 30″	87° 13′ 40″	14.676	Fence

Table 11.1

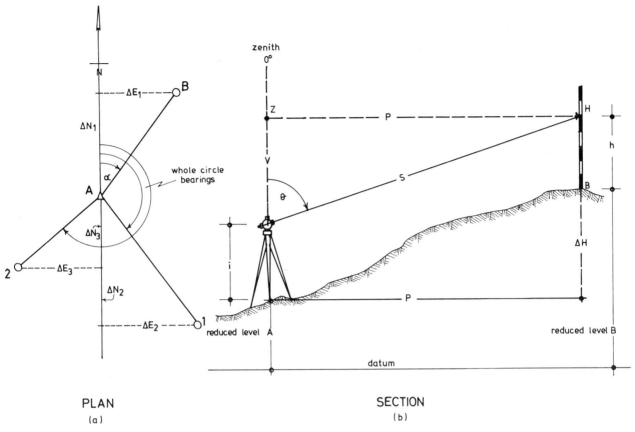

Figure 11.2

3. *Difference in height* (ΔH)

The difference in height between the ground at instrument station A and the ground at target station B is ΔH. In Fig. 11.2(b),

$$\Delta H = i + V - h \qquad (3)$$

i.e. difference in height between stations
$$= \text{instrument height} + \text{vertical height}$$
$$- \text{target height}$$

For angles of depression (i.e. downhill sights) the vertical dimension (V) is negative. Since the zenith angle (θ) in such a case would be greater than $90°$, the calculation $V = P \cos \theta$ would naturally result in a negative answer.

In order to reduce the calculation of $\Delta H = i + V - h$, the target height (h) is made equal to the instrument height (i), as in this case.

$$\text{Therefore } \Delta H = V \qquad (3a)$$

4. *Reduced level of target station*

In Fig. 11.2(b) station B lies at a higher level than station A.

$$\text{Therefore, RL(B)} = \text{RL(A)} + \Delta H \qquad (4)$$

i.e. reduced level target station = reduced level of
instrument station
$$+ \text{rise} (\Delta H)$$

Station B could, of course, be at a lower level than station A, in which case

$$\text{Reduced level (B)} = \text{reduced level (A)} - \text{fall} (\Delta H)$$

5. *Coordinates of target station*

In Fig. 11.2(a) the whole circle bearing of the line from station A to station B is α. From previous theory on traversing (Sec. 11 in Chapter 9) the difference in eastings between the stations is ΔE and the difference is northings is ΔN:

$$\Delta E_\text{I} = \text{AB} \times \sin \alpha \qquad (5)$$
$$= plan \text{ length AB} \times sin \text{ bearing AB}$$

and

$$\Delta N_\text{I} = \text{AB} \times \cos \alpha \qquad (6)$$
$$= plan \text{ length} \times cos \text{ bearing AB}$$

These formulae hold good for any line between the instrument station and target station.

Again from previous theory on traversing (Sec. 11 in Chapter 9), the total coordinates $(E$ and $N)$ of any target station are found by adding the difference in easting (ΔE) between the instrument station and target station to the easting of the instrument station, and likewise for northings, i.e.

$$\text{Easting (target station)} = \text{easting (instrument station)}$$
$$+ \text{difference in eastings}$$

$$E_\text{T} = E_\text{I} + \Delta E \qquad (7)$$

Formula number	Formula	Calculation	Result
1	$P = S \sin \theta$	$P = 18.325 \sin 71° \ 30' \ 10''$	17.378
2	$V = S \cos \theta$	$V = 18.325 \cos 71° \ 30' \ 10''$	5.814
3	$\Delta H = i + V - h$	$\Delta H = 1.350 + 5.814 - 1.350$	5.814
4	$RL(B) = RL(A) + \Delta H$	$RL(B) = 35.210 + 5.814$	41.024
5	$\Delta E_{AB} = P \sin \alpha$	$\Delta E_{AB} = 17.378 \sin 36° \ 30' \ 00''$	10.337
6	$\Delta N_{AB} = P \cos \alpha$	$\Delta N_{AB} = 17.378 \cos 36° \ 30' \ 00''$	13.969
7	$E_B = E_A + \Delta E_{AB}$	$E_B = 100.000 + 13.969$	110.337
8	$N_B = N_A + \Delta N_{AB}$	$N_B = 200.000 + 10.337$	213.969

Table 11.2

and

Northing (target station)

$$= \text{northing (instrument station)} + \text{difference in northings}$$

$$N_T = N_I + \Delta N \tag{8}$$

(b) Summary of formulae

Plan length = slope length × sin zenith angle
$$P = S \sin \theta \tag{1}$$

Vertical V = slope length × cos zenith angle
$$V = S \cos \theta \tag{2}$$

Difference in height ΔH
$$= \text{instrument height} + \text{vertical } V - \text{target height}$$
$$\Delta H = i + V - h \tag{3}$$

RL target station = RL instrument station + difference in height
$$RL(T) = RL(I) + \Delta H \tag{4}$$

Difference in eastings
$$= \text{plan length} \times \sin \text{bearing}$$
$$\Delta E = P \sin \alpha \tag{5}$$

Difference in northings
$$= \text{plan length} \times \cos \text{bearing}$$
$$\Delta N = P \cos \alpha \tag{6}$$

Eastings of target = easting instrument + difference in eastings
$$E_T = E_I + \Delta E \tag{7}$$

Northings of target = northing instrument + difference in northings
$$N_T = N_I + \Delta N \tag{8}$$

(c) Calculations

The calculations of point B are shown in Table 11.2.

(d) Field data/calculation table

Usually, the calculations are processed on the field data table as in Table 11.3, which should be self-explanatory.

Example

1 Table 11.1 shows the tacheometric data observed to points B, 1 and 2. (The coordinates of point B are calculated in Tables 11.2 and 11.3.) Calculate the coordinates (east, north, reduced level) of points 1 and 2.

Instrument station	A
Easting	100.000 m
Northing	200.000 m
Reduced level	35.210 m
Instrument height	1.350 m

Target point	Target height (m)	Horizontal circle (α) reading	Zenith angle (θ)	Slope distance S (m)	P (S sin θ)	V (S cos θ)	ΔH (i+V−h)	RL target (RL_I+ΔH)	ΔE (P sin α)	ΔN (P cos α)	E (E_I+ΔE)	N (N_I+ΔN)	Remarks
RO	—	00° 00' 00"	—	—	—	—	—	—	—	—	—	—	Reference object
B	1.350	36° 30' 00"	71° 30' 10"	18.325	17.378	5.814	5.814	41.024	10.337	13.969	110.337	213.969	Survey Station
1	1.350	147° 29' 10"	92° 04' 00"	21.110	21.096	−0.761	−0.761	34.449	11.339	−17.789	111.339	182.211	Fence
2	0.850	231° 15' 30"	87° 13' 40"	14.676	14.659	0.710	1.210	36.420	−11.434	−9.174	88.563	190.826	Fence

Note. The calculations, required in a radial positioning survey are lengthy and fairly complex. In order to minimize errors, the calculations are usually made on a computer. The calculations of Example 1 are made using BASIC programming techniques, in Sec. 8 of Chapter 17.

Table 11.3

	Point 1	*Point 2*
$P = S \sin \theta$	$21.110 \sin 92° \, 04' \, 00'' = 21.096$	$14.676 \sin 87° \, 13' \, 40'' = 14.659$
$V = S \cos \theta$	$21.110 \cos 92° \, 04' \, 00'' = -0.761$	$14.676 \cos 87° \, 13' \, 40'' = 0.710$
$\Delta H = i + V - h$	$1.350 + (-0.761) - 1.350 = -0.761$	$1.350 + 0.710 - 0.850 = 1.210$
$RL_T = RL_A + \Delta H$	$35.210 - 0.761 = 34.449$	$35.210 + 1.210 = 36.420$
$\Delta E_{AT} = P \sin \alpha$	$21.096 \sin 147° \, 29' \, 10'' = 11.339$	$14.659 \times \sin 231° \, 15' \, 30'' = -11.437$
$\Delta N_{AT} = P \cos \alpha$	$21.096 \cos 147° \, 29' \, 10'' = -17.789$	$14.659 \times \cos 231° \, 15' \, 30'' = -9.174$
$E_T = E_A + \Delta E_{AT}$	$100.00 + 11.339 = 111.339$	$100.000 - 11.437 = 88.563$
$N_T = N_A + \Delta N_{AT}$	$200.00 - 17.789 = 182.211$	$200.000 - 9.174 = 190.826$

Table 11.4

Answer (Table 11.4)
These answers are summarized on the field data/calculation table (Table 11.3).

Exercise 11.1

1 Figure 11.3 shows the layout of a radial positioning survey of part of the GCB Outdoor Centre site and Table 11.5 shows the partially completed field data of the survey.

(a) Calculate the three-dimensional coordinates (east, north, height) of each point, given that station A

is the origin of the survey and the bearing from A to the RO is zero degrees.

(b) Plot the survey to scale 1:250.

3. Calculation and plotting using microcomputer-based mapping systems

(a) Introduction

The previous calculations are lengthy, laborious and repetitious—exactly the disadvantages that computers were designed to overcome. Every manufacturer

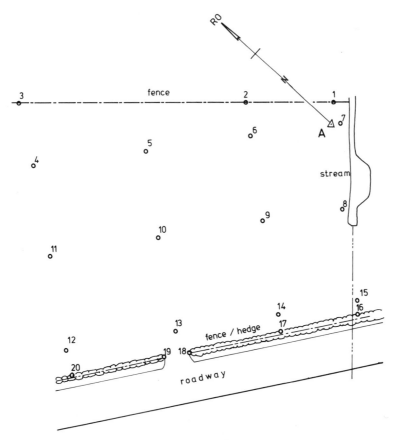

Figure 11.3

Instrument station	A	Backsight station RO	Whole circle bearing A–RO = 00° 00′ 00″
Instrument height	1.35 m		
Easting	0.00 m		
Northing	0.00 m		
Reduced level	6.48 m		

	Field data											
Target station	Target height (m)	Slope length (m)	Whole circle bearing	Zenith angle	P (m)	ΔH (m)	ΔE (m)	ΔN (m)	Easting (m)	Northing (m)	Reduced level (m)	Target station
1	1.35	3.02	72° 7′ 30″	92° 6′ 35″	3.02	−0.11	2.87	0.93	2.87	0.93	6.37	1
2	1.35	11.10	324° 9′ 45″	89° 21′ 0″	11.10	0.13	−6.50	9.00	−6.50	9.00	6.61	2
3	2.35	30.02	318° 55′ 20″	87° 8′ 35″	29.98	1.50	−19.70	22.60	−19.70	22.60	6.98	3
4	1.35	29.36	306° 49′ 50″	89° 21′ 50″	29.36	0.33	−23.50	17.60	−23.50	17.60	6.81	4
5	1.35	19.34	307° 13′ 30″	90° 4′ 16″	19.34	−0.02	−15.40	11.70	−15.40	11.70	6.46	5
6	0.35	9.35	310° 12′ 0″	96° 13′ 30″	9.29	−1.01	−7.10	6.00	−7.10	6.00	6.47	6
7	1.35	0.85	69° 26′ 40″	90° 16′ 5″								
8	1.35	9.59	210° 51′ 40″	95° 13′ 40″								
9	1.35	13.41	260° 7′ 9″	92° 32′ 20″								
10	2.00	21.63	279° 35′ 35″	89° 28′ 50″								
11	1.00	30.89	287° 8′ 40″	91° 20′ 20″								
12	1.35	35.37	272° 26′ 0″	92° 43′ 45″								
13	1.35	27.59	261° 1′ 0″	93° 33′ 40″								
14	1.35	21.66	240° 56′ 45″	93° 19′ 10″								
15	1.35	19.67	213° 22′ 0″	93° 9′ 35″								
16	1.35	21.25	209° 8′ 37″	93° 15′ 0″								
17	1.35	23.38	238° 29′ 20″	93° 23′ 35″								
18	1.35	28.16	256° 24′ 35″	94° 14′ 35″								
19	1.35	30.11	260° 13′ 25″	93° 58′ 10″								
20	1.35	36.55	269° 31′ 45″	92° 58′ 10″								

Table 11.5

of surveying instruments provides a microcomputer-based mapping system which solves these problems speedily, accurately and efficiently. Typical of these computer packages are Sokkia SDR mapping, Geotronics System 4000, Leica Liscad, Hall and Watts LSS, as well as those of the Eclipse, Moss and Optimal software houses.

Each of the packages requires suitable hardware in the form of a PC running at 386 or 486 MHz, 640k RAM minimum, Maths Co-processor, 20 Megabyte hard disc, floppy disc drive, graphics board, serial port, parallel port, mouse port, printer and HPGL plotter.

All of the packages are very similar, though manufacturers would claim differently. Each package has its own particular advantages and disadvantages but they all perform the same function, namely the production of a contoured plan.

It would not be possible nor desirable to describe the operation of all or even one of the packages completely. Some package manuals are hundreds of pages long. The following dissertation therefore describes the fundamental features of all packages and their effects on surveying procedure.

(b) Coding

On any radial positioning survey the data relating to a great number of points have to be recorded either manually or automatically using a data recorder. In manual recording, a sketch should always be made showing the point numbers and locations. The sketch enables the surveyor to join the various points correctly on the final plotted drawing.

When plotting by computer, however, the computer packages naturally cannot recognize a sketch, so the surveyor has to instruct the computer regarding the description of a point and how it is to be joined to other points. The points must be coded in some way such that the computer will recognize them. Thus any point of the survey will have a unique point number and a feature code, which may be a one-, two- or three-point code.

Point number

Every point on the survey is given a reference to differentiate it from any other point on the survey. The reference may be alphabetic or numeric or alphanumeric. Where a particular package accepts only numeric characters, it is usual to separate the survey stations from the points of detail by allocating numbers (say 1 to 100) to the traverse stations and numbers from 101 onwards to the detail points.

Feature codes

In a feature coding system there are three types of code, namely, point codes, control codes and control codes with parameters:

1. Point code

A point code is related specifically to the point being observed and gives a description of the point. Codes are kept simple and restricted as far as possible to

Feature	Code
Boundary	BDY
Bank	BANK
Building	BLD
Fence	FCE
Footpath	FP
Hedge	HEDGE
Instrument station	IS
Kerb	KERB
Lamp post	LP
Manhole	MH
Pond	POND
Wall	WALL
Spot height	SH
Tree	TREE

Table 11.6

about four or five letters. Typical codes of a few common points of detail are shown in Table 11.6.

Each package has its own feature code library which must be used in the field when recording point data, otherwise the package will not compute the data.

2. *Control code*

(a) Control codes control the way in which point codes are implemented. In general, in any computer package, points with the same code are joined in sequence by straight lines unless a control code instructs otherwise. Thus three points coded as 101 FCE, 102 FCE and 103 FCE will be computed and plotted by the software and joined by straight lines of a certain line type, as defined in the package library. If another point on the fence is observed later in the

survey, e.g. point 109, the latter point will be joined to point 103 by a straight line unless prevented from doing so by the insertion of a control code.

(b) It is therefore usual on a radial point survey to observe all points that have the same code before observing points of a different code. This practice is called 'stringing' and is particularly advantageous when using a data recorder. Recorders automatically increment the point numbers by adding one but retain the previous feature code until the surveyor changes it. Thus, when observing points along a fence or a kerb, the observer simply sights the reflector held at the first point of the string, enters the point number and code and then triggers the recorder. The slope distance, zenith angle and horizontal angle are recorded automatically. At the second and subsequent points of a string, the observer sights the reflector and triggers the recorder. The point number is increased by one, the code is retained and the three measurement parameters are recorded automatically.

Example

2 Figure 11.4 shows a new town parking area and public toilet block with boundaries comprising various hedges and fences. The area is to be radially surveyed from station 1 using line 1–2 as a reference line. Spot heights are to be observed at every point. Make a list of the coding required to enter points 101–106 into a computer package.

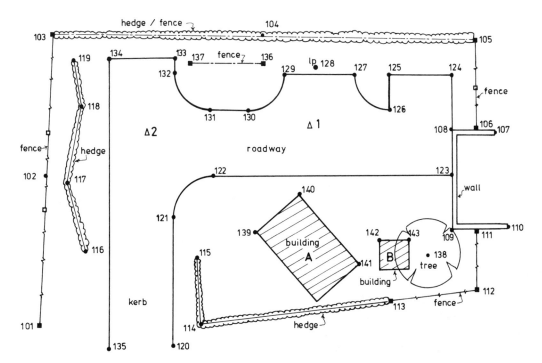

Figure 11.4

Answer

Point number	Code
101	FCE SH
102	FCE SH
103	FCE SH HEDGE
104	FCE SH HEDGE
105	FCE SH HEDGE
106	FCE SH

(c) Whenever a new string of points of the same feature commences, a control code must be added to prevent the string from joining to the previous string. The control code is usually the word START or ST. Thus in Fig. 11.4 point 111 is the start of a new fence and must be coded as 111 FCE ST. If the code START had been omitted, the package would have drawn a fence between point 106 and point 111.

Example

3 Using Fig. 11.4 make a list of the coding required to enter points 107–119 into a computer package. Spot heights are required at every point.

Answer

Point number	Code
107	WALL SH
108	WALL SH
109	WALL SH
110	WALL SH
111	FCE START SH
112	FCE SH
113	FCE HEDGE SH
114	HEDGE SH
115	HEDGE SH
116	HEDGE START SH
117	HEDGE SH
118	HEDGE SH
119	HEDGE SH

(d) It has already been pointed out that, in a computer package, a feature code, e.g. 'kerb', will be joined to the previous occurrence of the point having the code 'kerb' by a straight line. Frequently, however, kerbs are not straight; hence, a control code must be introduced to indicate the start and finish of a curve. As always, the feature code libraries of different packages vary and so the following control codes are only examples, but the principles are the same.

In Fig. 11.4 point 120 is the first occurrence of the feature 'kerb' and is coded as 120 KERB. If a spot height is required the code becomes 120 KERB SH. Point 121 is the start of a curve of unknown radius.

Line 120–121 is tangential to the curve. A typical control code is STCV, meaning 'start a smooth curve tangentially'; likewise point 122 is the end of the curve and line 122–123 is again tangential to the curve.

A typical control code to finish the curve is ENDCV or FINCV, meaning 'end the curve tangentially'. Some packages require a point about halfway around the curve, in order to be able to draw the curve smoothly.

Example

4 Using Fig. 11.4 make a list of the coding required to enter points 120–123 into a computer package. Spot heights are required at every point.

Answer

Point number	Code
120	KERB SH
121	KERB STCV SH
122	KERB ENDCV SH
123	KERB SH

(e) In Fig. 11.4 the kerb between points 126 and 127 is again a smooth curve but is not tangential to any line. It therefore requires a different code from the previous curve code. A typical coding for point 126 is NEWCV and for point 127 is ENDONCV. These codes start and finish a curve, without being tangential to a specified tangent line.

Example

5 Using Fig. 11.4 make a list of the coding required to enter points 123–128 into a computer package. Spot heights are required at all points.

Answer

Point number	Code
123	KERB SH
124	KERB SH
125	KERB SH
126	KERB NEWCV SH
127	KERB ENDONCV SH
128	LP SH

Exercise 11.2

1. Using Fig. 11.4 make a list of the coding required to enter points 128–137 into a computer package. Spot heights are required at all points.

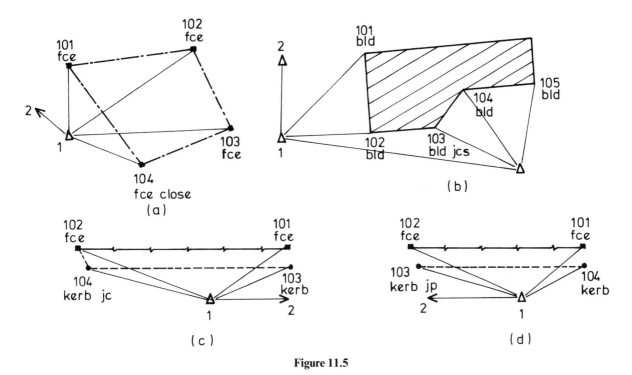

Figure 11.5

(f) In the introduction to coding it was pointed out that surveyors should use a diagram when recording points manually so that they know which points are to be joined when they plot the survey. These relationships can be completely defined through the coding system, although it is advisable to keep diagrams until complete proficiency in coding is gained.

There are numerous 'join' commands in coding. Figure 11.5 illustrates typical situations where these commands are used. In Fig. 11.5(a) a four-sided enclosure is being surveyed. If the code FCE is used only, point 104 will not close back to point 101. The control code CLOSE is used with point 104, which is coded as 104 FCE CLOSE in order to effect a closure on point 101. Other 'join' commands include JCS, meaning 'join to closest point of same code', which is useful when the theodolite/EDM has to be moved to another set-up point and the continuation of a string is carried out from the second point (Fig. 11.5(b))— JC meaning 'join to the closest point regardless of code' (Fig. 11.5(c)) and JP meaning 'join to the previous point regardless of code' (Fig. 11.5(d)).

3. *Control code with parameter*

(a) A control code with a parameter allows the surveyor to pass additional information to the plot. It is really a control code with a value attached to it.

(b) The branch spread of a tree can be entered with the control code, SIZE, followed by the diameter, or radius in some packages, of the branches. Thus the tree in Fig. 11.4 is coded as 138 TREE SIZE 10.

(c) From any one instrument position it is impossible to observe the four corners of a building. The problem can be overcome by observing three corners of the building and adding a control code, usually in the form of a note, which will close the rectangle.

Example

6 In Fig. 11.6(a) only three corner points, 101–103 can be observed from station 1. List the coding required to draw the building on the plot.

Answer

Point number	Code
101	BLD
102	BLD
103	BLD CLSRECT

The code CLSRECT will form a rectangle, depicting the building. A new point is coordinated in the databases of the computer package, so that the new sides are parallel and equal in length to the sides computed from the three observation points. Unfortunately, any errors in the survey of the three points will be transferred to the new fourth point. This particular type of coding produces a parallelogram rather than a true rectangle. This situation can be avoided by observing only two corners and using a tape to measure the other side.

In Fig. 11.6(b) points 101 and 102 form the baseline of the first building. The sides perpendicular to the baseline measure 4.0 metres. The control code is

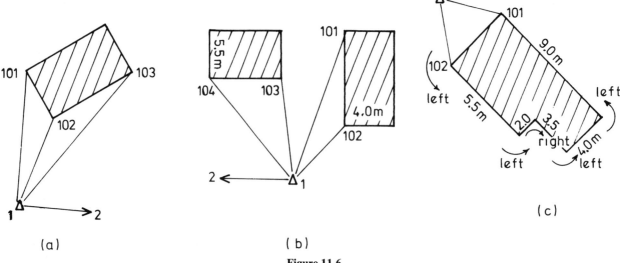

Figure 11.6

RECT followed by −4.0. The dimension is entered as a negative because the measured side makes a left turn from the base line.

Points 103 and 104 form the base line of the second building. The other side of the building measures 5.5 m and from the base line the measured side turns right. The 5.5 m parameter is added to the control code RECT as a positive dimension.

Example

7 In Fig. 11.6(b) two buildings have been observed from a survey point, by observing only two corners of each building. List the coding required to draw the buildings on the plot.

Answer

Point number	Code
101	BLD
102	BLD RECT −4.0
103	BLD START
104	BLD RECT 5.5

Note. Point 103 begins a new string and requires the control code START.

Finally, in Fig. 11.6(c) a multi-sided building is to be surveyed. Some form of coordinate option is used as a control code. One solution, typical of most packages, is to enter a control code in the form of a 'distance' note followed by all the measured sizes of the building. Positive sizes are used for right turns, negative for left turns.

Example

8 List the coding required to draw the building shown in Fig. 11.6(c).

Answer

Point number	Code
101	BLD
102	BLD DIST −5.5 −2.0 3.5
	−4.0 −9.0

Exercise 11.3

1 Using Fig. 11.4, list the coding required to enter points 138–143 into a computer package, given that building A is rectangular, building B is 3.0 metres square and the tree has a 7.0 m branch spread.

(c) Using a mapping system package

Section 1 dealt with the fieldwork and orientation of radial positioning surveys while in Sec. 2 the various formulae required to calculate the three-dimensional coordinates were derived and demonstrated in Example 1. Section 3(b) explained the need for, and use of, coding in the fieldwork when surveys are to be computed, plotted and drawn by an integrated micro-computer-based mapping system, commonly called the computer package.

This section explains how the fieldwork is translated into a plan using the computer package.

It has already been emphasized and must now be repeated that it is not possible nor desirable to describe the operation of any one particular computer package since every package works differently. The net result, however, is the same and this dissertation therefore describes the fundamental features of any package. Naturally, there is no substitute for 'hands-on' experience, in which case the manual for any particular package must be consulted and followed throughout.

With a few exceptions, most of the mapping systems are user friendly and in conjunction with the relevant manual may be followed fairly easily. The basis for the following description is of the Sokkia SDR mapping system which is one of the best currently available.

1. Figure 11.7 shows the layout of a radial point survey at the Glasgow College of Building and Table 11.7 shows the relevant field data. The survey was carried out using a Geodimeter 420 Total Station instrument. The data are to be input manually to a mapping system package and a plan is to be produced by the interconnected plotter to a scale of 1:200.

2. The computer, VD unit, printer and plotter are all switched on and at the C > prompt the appropriate mapping system identification is entered to call up the program, e.g. 'SDRMAP'.

3. All mapping systems require a Job Identification, usually in the form of a three-letter identifier. The mapping systems will not work without proper initialization. All field administration data (job description, date, time, operator, etc.) are stored in this job file.

Table 11.8 is a typical example in which the surveyor simply enters the data relevant to the current job.

4. The survey field notes must now be entered into a data file, which is created as the next step in the program. The user is guided through the creation of the file in a series of easy-to-follow steps.

The information required in the data file is as follows:

(a) Instrument station 1 details (east, north, reduced level, height of instrument, code).

(b) Back bearing to reference object 2 and actual reading of the horizontal circle of the theodolite. The theodolite may be set to the actual back bearing or may be zero set, in which case the package will compute the bearings to all points at the processing stage.

Alternatively, the coordinates of reference object 2 may be entered instead of the back bearing to this station.

(c) The height of the reflector at all observed detail points.

(d) The slope distance, vertical circle reading, horizontal circle reading to every detail point and the coding of the point. Once the details have been entered, it is good practice to obtain a hard copy printout of the data for checking. Table 11.9 shows the data file of the GCB patio area survey.

5. The field data, now contained in the data file, is transferred to the database and all of the records

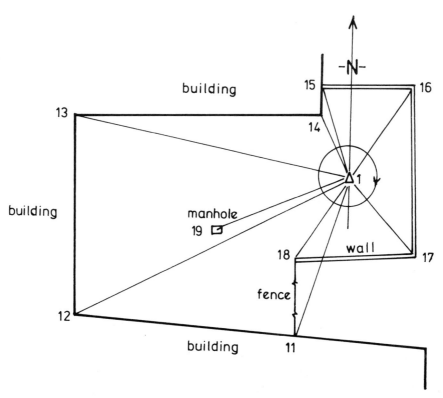

Figure 11.7

Tri-dimensional coordinates			Instrument station 1 Instrument height 1.52 Departure 100.00 Latitude 100.00 Reduced level 20.00		Backsight station 2
Target station	Target height (m)	Slope length (m)	Whole circle bearing	Zenith angle	Remarks
2			00° 00′ 00″		Reference station
11	1.52	11.94	197° 34′ 56″	89° 31′ 22″	BLD FCE SH
12	1.52	23.08	245° 25′ 8″	89° 54′ 12″	BLD SH
13	1.52	21.57	282° 36′ 4″	89° 45′ 20″	BLD SH
14	1.52	5.26	333° 42′ 48″	90° 11′ 2″	BLD SH
15	1.52	7.27	341° 3′ 16″	89° 48′ 32″	BLD WALL SH
16	1.52	8.74	37° 14′ 2″	89° 28′ 18″	WALL SH
17	1.52	7.44	134° 31′ 4″	89° 20′ 26″	WALL SH
18	1.52	6.41	212° 53′ 54″	89° 7′ 40″	WALL FCE SH
19	1.52	12.12	245° 27′ 20″	90° 19′ 16″	MH SH

Table 11.7

Job Identifier	GCB
Directory	C:\SDRDATA\
Job Name	GCB detail survey
Job Description	Patio area at College
Job Reference	18 April 1995
Surveyor	HND1 Building

Table 11.8

	Angle (deg) Temperature (°C)	Distance (m) Coordinates E–N–Elv	Pressure (mmHg) H.obs Right	
JOB		Job ID GCB		
STN	0001	East 100.000	North 100.000	Elv 20.000
		Theo ht 1.520	Code IS	
TRGET		Target ht 1.520		
BKB	0001–0002	Azmth 0° 00′ 00″	H.obs 0° 00′ 00″	
OBS	0001–0011	Dist 11.935	V.obs 89° 31′ 22″	H.obs 197° 34′ 56″
		Code BLD FCE SH		
OBS	0001–0012	Dist 23.075	V.obs 89° 54′ 12″	H.obs 245° 25′ 08″
		Code BLD SH		
OBS	0001–0013	Dist 21.572	V.obs 89° 45′ 20″	H.obs 282° 36′ 04″
		Code BLD SH		
OBS	0001–0014	Dist 5.261	V.obs 90° 11′ 02″	H.obs 333° 42′ 48″
		Code BLD SH		
OBS	0001–0015	Dist 7.270	V.obs 89° 48′ 32″	H.obs 341° 03′ 16″
		Code BLD WALL SH		
OBS	0001–0016	Dist 8.742	V.obs 89° 28′ 18″	H.obs 37° 14′ 02″
		Code WALL SH		
OBS	0001–0017	Dist 7.442	V.obs 89° 20′ 26″	H.obs 134° 31′ 04″
		Code WALL SH		
OBS	0001–0018	Dist 6.414	V.obs 89° 07′ 40″	H.obs 212° 53′ 54″
		Code WALL FCE SH		
OBS	0001–0019	Dist 12.124	V.obs 90° 19′ 16″	H.obs 245° 27′ 20″
		Code MH SH		

Table 11.9

Point number	Easting	Northing	Height	Code
1	100.000	100.000	20.000	IS
11	96.395	88.623	20.099	SH BLD FCE
12	79.016	90.401	20.039	SH BLD
13	78.948	104.706	20.092	SH BLD
14	97.670	104.717	19.983	SH BLD
15	97.640	106.876	20.024	SH BLD WALL
16	105.289	106.960	20.081	SH WALL
17	105.306	94.783	20.086	SH WALL
18	96.517	94.615	20.098	SH WALL FCE
19	88.972	94.964	19.932	SH MH

Table 11.10

processed into the three-dimensional coordinates of every point. The formulae and methods used are those detailed in Sec. 2(a). Table 11.10 shows the hard copy of the coordinates of the survey detail points.

6. The feature codes entered into the database via the data file are processed and checked against the program library. The user is warned if any codes are out with the library and is given an opportunity to return to the data file to edit the information.

7. When all coding errors have been corrected the option to begin plotting is given (Table 11.11).

At this stage, it is wise to select to plot on screen only. Once the selection is made the plot appears on the screen, after a short time, at a scale that fits the screen. Errors, if any, in the survey are usually evident at this stage. If the errors are in the field work, the survey may have to be repeated.

Usually errors are caused by wrongful implementation of the codes. The codes themselves were proved correct in operation 6 above by checking against the library, but they may well be applied wrongly. An opportunity is given to return to the data file to make amendments.

8. When all errors have been eliminated, the plot will appear correctly on the screen but, being scaled to fit the screen, will have to be edited.

The plot editor option allows the plot parameters (scale, rotation, origin, grid interval) to be changed to suit the user's requirements. The size of paper and layout of the drawing may also be selected at this stage.

9. When satisfied that all of the parameters have been correctly chosen, the user is returned to the plotting options (Table 11.11) and the options to plot on screen and on plotter are chosen.

Plotting options	
Plot on plotter	No
Plot on screen	Yes
Create Autocad DXF file	No

Table 11.11

A sheet of drawing paper of the correct size is loaded into the plotter and the plotter is brought on line. The instruction to plot is given and the plot should appear on the screen and, shortly afterwards, on the drawing paper.

10. In the plotting options (Table 11.11) the user may also choose to create an Autocad DXF file, in which case the complete job is saved and may be enhanced and refined using a CAD package.

11. Figure 11.8 shows the GCB patio area survey plotted to scale 1:200 as it was drawn on the Sokkia SDR mapping system. Spot heights are shown at all detail points and the grid network is drawn at 20 m intervals.

12. When a survey is recorded using an electronic field book, the whole process of the data collection is of course accelerated. The transfer of the field data to the computer takes only a few seconds. The data recorder is connected to the computer at the correct communications port, normally COM1; the baud rate and the parity of the computer are set to match those of the data recorder and the mapping system, set to 'Receive data file' or similar. The transfer is then effected very quickly.

Once the transfer of data is complete the remainder of the operations 5 to 11 above are carried out as described.

Exercise 11.4

1 Table 11.12 and Fig. 11.9 show the field data of the *western* half of the GCB Outdoor Centre survey. (The *eastern* half was the subject of Exercise 11.1.)
(a) Enter the data into any mapping system to which you have access and devise field codes to suit the library of that system.
(b) Using the printer produce hard copies of the data file and coordinate file.
(c) Produce a plotted survey to any suitable scale.

Note. The exercise may equally well be computed and plotted manually.

4. Digital ground modelling

(a) Introduction

Digital ground modelling (DGM) is a term used by surveyors to describe a method of representing the surface of the earth numerically, using the three-dimensional coordinates of a series of closely spaced points. In effect, these coordinates are stored in the database of a mapping system and are manipulated by the system to produce a contoured plan of the ground. Other terms used to describe this process are digital terrain modelling (DTM) and digital height modelling (DHM).

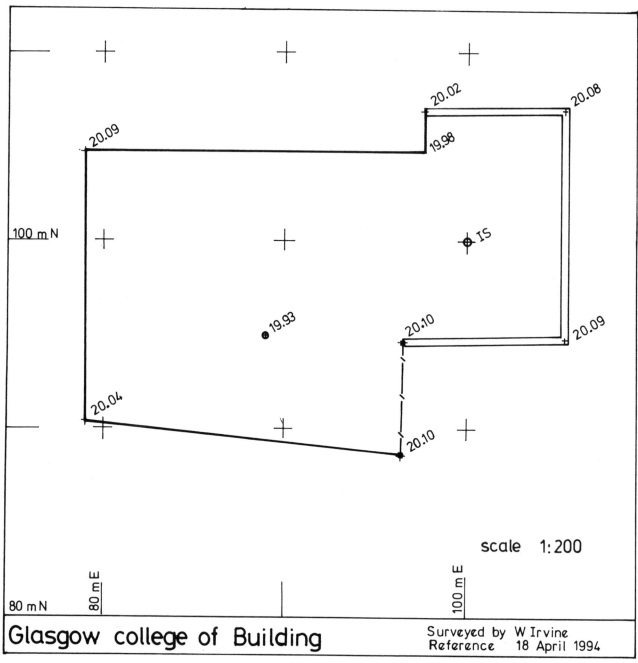

Figure 11.8

The data for the information of the model are acquired by a variety of methods, one of which is the radial positioning method, fully explained in the preceding sections 1 to 3. Figure 11.10 shows a database of eight coordinated points, acquired using a total station and plotted isometrically. A contoured plan of the area is to be constructed.

(b) Modelling techniques

It was stated in Sec. 3 that a large number of mapping systems are available today and that fundamentally they produce a plan view of the ground in similar fashion to each other. The program packages are also capable of contouring the plans. Each package uses its own particular method, but once again the characteristics of each are fundamentally very similar. Two methods are commonly used in mapping system packages.

Regular grid method

On a point location survey the points are usually surveyed in random fashion (Fig. 11.11). The software within the program package selects a regular grid of squares or rectangles to fit over the randomly coordinated points and, by a process of interpolation, calculates the height of the ground at each intersection. These grid intersections are called nodes. The

Instrument station	3		Backsight station 1	
Instrument height	1.35		Bearing 3–1 = 104° 10′ 39″	
Easting	−51.50		Easting	0.00
Northing	13.01		Northing	0.00
Reduced level	5.12			

Target station	Target height (m)	Slope length (m)	Horizontal circle reading	Zenith angle
1			0° 0′ 0″	
201	1.35	33.27	329° 2′ 25″	86° 47′ 0″
202	1.35	29.98	311° 11′ 30″	85° 46′ 40″
203	1.35	30.08	291° 35′ 35″	84° 29′ 30″
204	1.35	31.33	283° 55′ 25″	84° 21′ 45″
205	1.35	29.92	281° 50′ 20″	84° 54′ 35″
206	1.35	21.93	278° 10′ 10″	84° 37′ 0″
207	1.35	21.48	280° 43′ 15″	84° 30′ 15″
208	1.35	11.76	266° 54′ 5″	84° 30′ 5″
209	1.35	20.22	291° 44′ 25″	84° 14′ 45″
210	1.35	9.96	290° 59′ 25″	83° 37′ 0″
211	1.35	4.07	265° 46′ 20″	83° 59′ 0″
212	1.35	5.00	199° 44′ 5″	89° 7′ 5″
213	1.35	5.05	179° 28′ 20″	92° 31′ 56″
214	1.35	2.29	110° 52′ 55″	114° 4′ 0″
215	1.35	10.21	32° 27′ 10″	92° 56′ 15″
216	1.35	20.06	27° 24′ 20″	91° 31′ 20″
217	1.35	10.01	22° 44′ 20″	91° 47′ 30″
218	1.35	19.88	21° 12′ 55″	90° 55′ 50″
219	1.35	22.37	355° 54′ 0″	87° 28′ 15″
220	1.35	14.14	336° 28′ 0″	85° 3′ 45″
221	1.35	20.80	315° 41′ 20″	84° 26′ 0″
222	1.35	28.42	336° 30′ 45″	86° 35′ 50″

Table 11.12

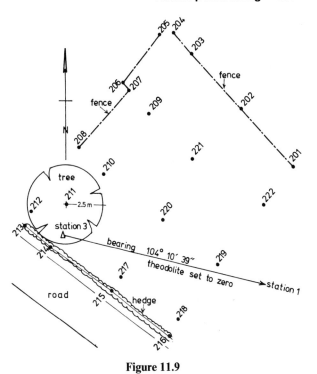

Figure 11.9

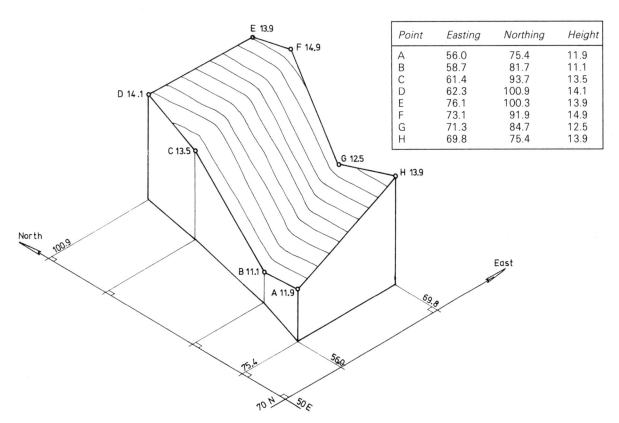

Point	Easting	Northing	Height
A	56.0	75.4	11.9
B	58.7	81.7	11.1
C	61.4	93.7	13.5
D	62.3	100.9	14.1
E	76.1	100.3	13.9
F	73.1	91.9	14.9
G	71.3	84.7	12.5
H	69.8	75.4	13.9

Figure 11.10

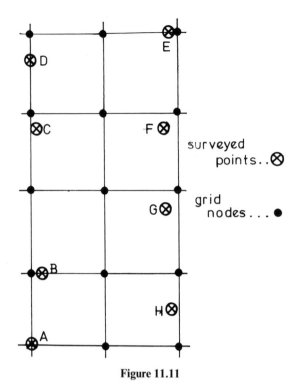

Figure 11.11

interpolation process is very complex but essentially the package searches for the nearest five or six points to each node and uses their heights to compute a height for the node. When the height of every node of a square has been calculated a simple linear interpolation is carried out along all four sides of the square and contour positions are thereby computed. The principle is therefore identical to the indirect method of contouring, fully explained and exemplified in Sec. 2(d) in Chapter 5.

Ambiguities may occur in the indirect method of contouring, however, as evidenced in Fig. 11.12, where the 7 metre contour is to be interpolated within a grid square or cell. In this case there are four possible solutions to the problem, none of which can be

guaranteed as correct. Packages solve the problem quite arbitrarily or resort to mathematical solutions using polynomial equations to represent the model surface.

Triangular irregular modelling (TIN)

This method has been adopted almost universally in ground modelling systems. The coordinated points themselves are used as the basis of a triangular grid, to form the surface of the ground model.

The program package software uses the coordinated points to create a set of triangles which are as near to equilateral as possible in the given circumstances. The method, known as the Delaunay triangulation method, is shown in operation in Fig. 11.13(a), where the coordinated points are joined to form a series of triangles.

The contour positions are determined automatically by the computer software using linear interpolation techniques in the same manner as for indirect contouring (Sec. 2(d) in Chapter 5). For example, in Fig. 11.13(b) the 12 and 13 m contour positions are to be plotted along line BC. In Sec. 2(d) in Chapter 5, Table 5.4 showed how the interpolation method could be laid out in simple tabular form and in Fig. 11.13 such a table is used to interpolate the positions of the required 12 and 13 m contour positions.

The program package first calculates the positions of the contour lines along the boundaries. These are of course the entry points of the contour lines. The exit points are then calculated along the internal sides of the triangles. These points are then the entry points for subsequent triangles. Finally the interpolated points are joined by smooth curves to form the final contour positions.

(c) Plotting the plan

In Sec. 3(c) a full explanation was given of how a mapping system package is used to produce a topo-

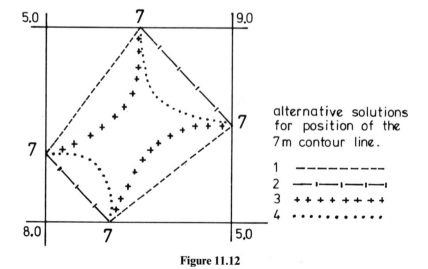

Figure 11.12

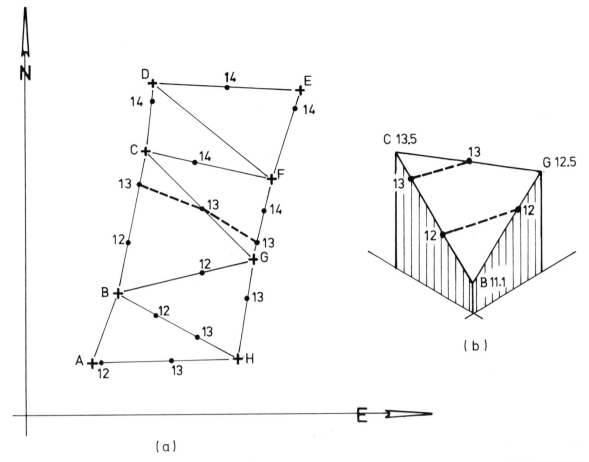

Figure 11.13

Lower level point 1	Higher level point 2	Contour value	Horizontal distance (point 1 to point 2)	Difference in level (high point 2 to low point 1)	Difference in level (contour to low point 1)	Horizontal distance from low point to contour
B(11.1)	C(13.5)	12.0	12.3	2.4	0.9	0.9/2.4 × 12.3 = 4.5
B(11.1)	C(13.5)	13.0	12.3	2.4	1.9	1.9/2.4 × 12.3 = 9.7

graphic plan from field data. In Exercise 11.4 (page 255) the field data of the survey of part of the GCB Outdoor Centre was given and the exercise required the manipulation of a mapping package to produce a database of coordinated points and a plan of these points plotted to some scale.

Figure 11.14 is the answer to the exercise and shows the database and plan plotted to a scale to fit the page format.

At this stage in the manipulation of the package, the plan has been plotted on the video display unit and the package will be awaiting further instructions from the user. If contour lines are not required, the user simply quits the program at this stage.

Assuming that contour lines are required, the further steps, which are basically the same in any package, are as follows:

1. Formation of boundaries and breaklines. The boundary is a set of points forming the perimeter of the database area. The user is required to define these points, otherwise the program may create contour lines outwith the confines of the surveyed area.

Likewise, the user must define any break lines. A break line is a line having a constant grade, e.g. the top or bottom of a bank. These have to be defined, in order to stop the package from making incorrect interpolations across them, otherwise the resultant contour plan will not accurately reflect these abrupt changes in the ground surface.

2. Next, the program selection is made to form the digital ground model. The software creates the Delaunay triangulation (Fig. 11.15(a)) and interpolates the positions of the contour lines along the triangle sides. The user has no control over this operation.

3. The surface plotting parameters are chosen by the user from a menu such as Table 11.13. This table shows the range of heights in the database and the lowest and highest possible values of contour lines using any selected contour interval. The user may

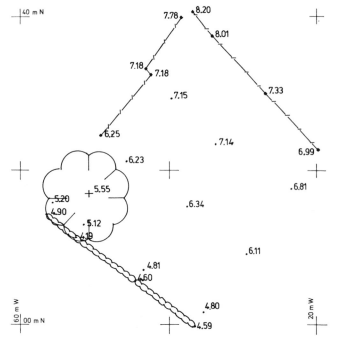

Pt No.	Easting	Northing	Height	Code
1	−0.000	−0.000	6.484	IS
3	−51.500	13.010	5.123	IS
201	−19.701	22.600	6.990	SH FCE
202	−26.900	30.000	7.330	SH FCE
203	−34.001	37.300	8.010	SH FCE
204	−36.816	40.509	8.200	SH FCE
205	−38.426	39.790	7.778	SH FCE ST
206	−43.200	33.200	7.180	FCE
207	−42.500	32.400	7.180	SH FCE
208	−49.250	24.500	6.250	SH FCE
209	−39.700	29.300	7.150	SH
210	−45.800	21.100	6.230	SH
211	−50.800	17.000	5.550	SH TREE
212	−55.650	15.800	5.200	SH
213	−56.400	14.200	4.900	SH HEDGE
214	−52.700	11.300	4.190	SH HEDGE
215	−44.500	5.600	4.600	SH HEDGE
216	−36.500	−0.300	4.590	SH HEDGE
217	−43.500	7.000	4.810	SH
218	−35.300	1.500	4.800	SH
219	−29.500	9.100	6.110	SH
220	−37.600	15.300	6.340	SH
221	−33.600	23.400	7.140	SH
222	−23.500	17.600	6.810	SH

Figure 11.14

choose to plot any or all of the following features: (a) contours, (b) triangles, (c) boundaries, (d) break lines. The user may also choose the minor and major contour intervals and choose to have straight or curved contour lines on the plans.

In Table 11.13, the selection shows that curved contour lines are to be drawn at 0.5 m vertical intervals and contours and triangles are both to be shown on the plan, though it would be unusual to show the triangles.

4. The plot editor option is chosen to set the plot parameters, namely scale of plan, origin of grid, grid interval, size of paper and layout of drawing to the user's requirements.

5. When satisfied that all of the plotting parameters have been correctly chosen, the user generates the plot and the package causes the plan to be drawn to these specifications either on the screen and/or plotter. Figure 11.15(b) shows the resultant contoured plan.

Exercise 11.5

1 Continue Exercise 11.4 to produce a contoured plan of the site.

Surface plotting parameters				
Plot contours	Yes	Plot triangles		Yes
		Plot boundaries		No
		Plot breaklines		No
Point heights range between		4.190	and	8.200 (m)
Plotting contours range between		4.200	and	8.200 (m)
	Minor	Major		
Contour interval (m)	0.500	1.000	Contour line	Curved
[F1] Generate plot		[F3] Edit plot parameters		
[F2] Begin plotting		[F4] Plot editor		

Table 11.13

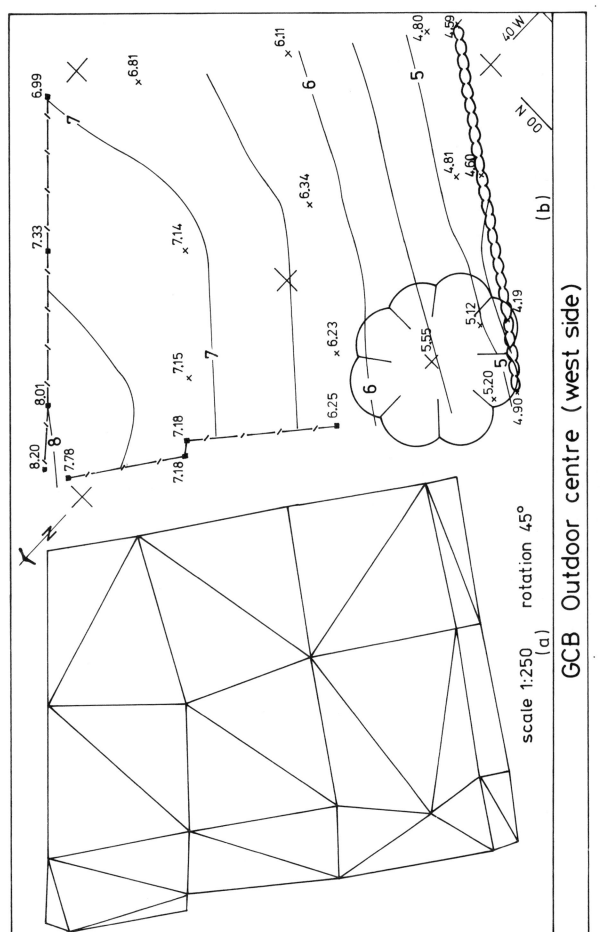

GCB Outdoor centre (west side)

scale 1:250   rotation 45°
(a)

(b)

Figure 11.15

Answers

Exercise 11.1

1 Table 11.14 and Fig. 11.16

Target station	Target height (m)	Slope length (m)	Whole circle bearing	Zenith angle	P (m)	ΔH (m)	ΔE (m)	ΔN (m)	Easting (m)	Northing (m)	Reduced level (m)	Target station
1	1.35	3.02	72° 7′ 30″	92° 6′ 35″	3.02	−0.11	2.87	0.93	2.87	0.93	6.37	1
2	1.35	11.10	324° 9′ 45″	89° 21′ 0″	11.10	0.13	−6.50	9.00	−6.50	9.00	6.61	2
3	2.35	30.02	318° 55′ 20″	87° 8′ 35″	29.98	1.50	−19.70	22.60	−19.70	22.60	6.98	3
4	1.35	29.36	306° 49′ 50″	89° 21′ 50″	29.36	0.33	−23.50	17.60	−23.50	17.60	6.81	4
5	1.35	19.34	307° 13′ 30″	90° 4′ 16″	19.24	−0.02	−15.40	11.70	−15.40	11.70	6.46	5
6	0.35	9.35	310° 12′ 0″	96° 13′ 30″	9.29	−1.01	−7.10	6.00	−7.10	6.00	6.47	6
7	1.35	0.85	69° 26′ 40″	90° 16′ 5″	0.85	0.00	0.80	0.30	0.80	0.30	6.48	7
8	1.35	9.59	210° 51′ 40″	95° 13′ 40″	9.55	−0.87	−4.90	−8.20	−4.90	−8.20	5.61	8
9	1.35	13.41	260° 7′ 9″	92° 32′ 20″	13.40	−0.59	−13.20	−2.30	−13.20	−2.30	5.89	9
10	2.00	21.63	279° 35′ 35″	89° 28′ 50″	21.63	0.20	−21.33	3.60	−21.33	3.60	6.03	10
11	1.00	30.89	287° 8′ 40″	91° 20′ 20″	30.89	−0.72	−29.51	9.10	−29.51	9.10	6.11	11
12	1.35	35.37	272° 26′ 0″	92° 43′ 45″	35.33	−1.68	−35.30	1.50	−35.30	1.50	4.80	12
13	1.35	27.59	261° 1′ 0″	93° 33′ 40″	27.54	−1.71	−27.20	−4.30	−27.20	−4.30	4.77	13
14	1.35	21.66	240° 56′ 45″	93° 19′ 10″	21.62	−1.25	−18.90	−10.50	−18.90	−10.50	5.23	14
15	1.35	19.67	213° 22′ 0″	93° 9′ 35″	19.64	−1.08	−10.80	−16.40	−10.80	−16.40	5.40	15
16	1.35	21.25	209° 8′ 37″	93° 15′ 0″	21.22	−1.20	−10.33	−18.53	−10.33	−18.53	5.28	16
17	1.35	23.38	238° 29′ 20″	93° 23′ 35″	23.34	−1.38	−19.90	−12.20	−19.90	−12.20	5.10	17
18	1.35	28.16	256° 24′ 35″	94° 14′ 35″	28.08	−2.08	−27.30	−6.60	−27.30	−6.60	4.40	18
19	1.35	30.11	260° 13′ 25″	93° 58′ 10″	30.04	−2.08	−29.60	−5.10	−29.60	−5.10	4.40	19
20	1.35	36.55	269° 31′ 45″	92° 58′ 10″	36.50	−1.89	−36.50	−0.30	−36.50	−0.30	4.59	20

Table 11.14

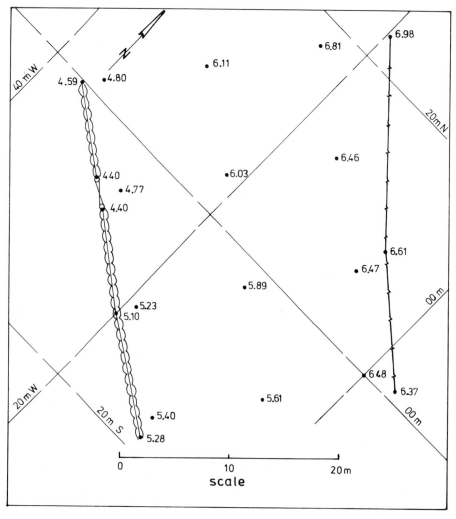

Figure 11.16

Exercise 11.2

1 *Point number*	*Code*
128	LP SH
129	KERB NEWCV SH
130	KERB ENDCV SH
131	KERB STCV SH
132	KERB ENDCV SH
133	KERB SH
134	KERB SH
135	KERB SH
136	FCE START SH
137	FCE SH

Exercise 11.3

1 *Point number*	*Code*
138	TREE Size 7
139	BLD
140	BLD
141	BLD
	Note CLS RECT
142	BLD START
143	BLD RECT 3.0

Exercise 11.4

1 Figure 11.17 and Tables 11.15 and 11.16

Figure 11.17

	Angle (deg) Temperature (°C)	Distance (m) Coordinates E–N–Elv	Pressure (mmHg) H.obs Right	
JOB		Job ID BK3		
STN	0003	East −51.500	Nrth 13.010	Elv 5.123
		Theo ht 1.350	Code IS	
TRGET		Target ht 1.350		
BKB	0003–0001	Azmth 104° 10′ 39″	H.obs 0° 00′ 00″	
POS	0001	East 0.000	Nrth 0.000	Elv 6.484
		Code IS		
OBS	0003–0201	Dist 33.266	V.obs 86° 47′ 00″	H.obs 329° 02′ 25″
		Code FCE SH		
OBS	0003–0202	Dist 29.978	V.obs 85° 46′ 40″	H.obs 311° 11′ 30″
		Code FCE SH		
OBS	0003–0203	Dist 30.076	V.obs 84° 29′ 30″	H.obs 291° 35′ 35″
		Code FCE SH		
OBS	0003–0204	Dist 31.325	V.obs 84° 21′ 45″	H.obs 283° 55′ 25″
		Code FCE SH		
OBS	0003–0205	Dist 29.919	V.obs 84° 54′ 35″	H.obs 281° 50′ 40″
		Code FCE START SH		
OBS	0003–0206	Dist 21.926	V.obs 84° 37′ 00″	H.obs 278° 10′ 10″
		Code FCE		
OBS	0003–0207	Dist 21.476	V.obs 84° 30′ 15″	H.obs 280° 43′ 15″
		Code FCE SH		
OBS	0003–0208	Dist 11.762	V.obs 84° 30′ 05″	H.obs 266° 54′ 05″
		Code FCE SH		
OBS	0003–0209	Dist 20.217	V.obs 84° 14′ 45″	H.obs 291° 44′ 25″
		Code SH		
OBS	0003–0210	Dist 9.958	V.obs 83° 37′ 00″	H.obs 290° 59′ 25″
		Code SH		
OBS	0003–0211	Dist 4.073	V.obs 83° 59′ 00″	H.obs 265° 46′ 20″
		Code TREE SIZE 50 SH		
OBS	0003–0212	Dist 5.001	V.obs 89° 07′ 05″	H.obs 199° 44′ 05″
		Code SH		
OBS	0003–0213	Dist 5.047	V.obs 92° 31′ 56″	H.obs 179° 28′ 20″
		Code SH HEDGE		
OBS	0003–0214	Dist 2.288	V.obs 114° 04′ 00″	H.obs 110° 52′ 55″
		Code SH HEDGE		
OBS	0003–0215	Dist 10.207	V.obs 92° 56′ 15″	H.obs 32° 27′ 10″
		Code SH HEDGE		
OBS	0003–0216	Dist 20.061	V.obs 91° 31′ 20″	H.obs 27° 24′ 20″
		Code SH HEDGE		
OBS	0003–0217	Dist 10.011	V.obs 91° 47′ 30″	H.obs 22° 44′ 20″
		Code SH		
OBS	0003–0218	Dist 19.875	V.obs 90° 55′ 50″	H.obs 21° 12′ 55″
		Code SH		
OBS	0003–0219	Dist 22.367	V.obs 87° 28′ 15″	H.obs 355° 54′ 00″
		Code SH		
OBS	0003–0220	Dist 14.140	V.obs 85° 03′ 45″	H.obs 336° 28′ 00″
		Code SH		
OBS	0003–0221	Dist 20.795	V.obs 84° 26′ 00″	H.obs 315° 41′ 20″
		Code SH		
OBS	0003–0222	Dist 28.424	V.obs 86° 35′ 50″	H.obs 336° 30′ 45″
		Code SH		
END OF REPORT				

Table 11.15

Point number	Easting	Northing	Height	Code
1	0.000	−0.000	6.484	IS
3	51.500	13.010	5.123	IS
201	−19.701	22.600	6.990	SH FCE
202	−26.900	30.000	7.330	SH FCE
203	−34.001	37.300	8.010	SH FCE
204	−36.816	40.509	8.200	SH FCE
205	−38.426	39.790	7.778	SH FCE ST
206	−43.200	33.200	7.180	FCE
207	−42.500	32.400	7.180	SH FCE
208	−49.250	24.500	6.250	SH FCE
209	−39.700	29.300	7.150	SH
210	−45.800	21.100	6.230	SH
211	−50.800	17.000	5.550	SH TREE
212	−55.650	15.800	5.200	SH
213	−56.400	14.200	4.900	SH HEDGE
214	−52.700	11.300	4.190	SH HEDGE
215	−44.500	5.600	4.600	SH HEDGE
216	−36.500	−0.300	4.590	SH HEDGE
217	−43.500	7.000	4.810	SH
218	−35.300	1.500	4.800	SH
219	−29.500	9.100	6.110	SH
220	−37.600	15.300	6.340	SH
221	−33.600	23.400	7.140	SH
222	−23.500	17.600	6.810	SH

Table 11.16

Exercise 11.5

1 See Fig. 11.15(b).

6. Project

Chapter 18 is a project covering all chapters of this textbook. It is intended that the project build into a complete portfolio of the surveying work required in the survey and setting out of a building or engineering development.

If the reader wishes to continue work on the project or begin work at this stage, he or she should now turn to page 397 and attempt Section 8.

CHAPTER 12

Curve ranging

Objective
After studying this chapter, the reader should be able to make the necessary calculations to fix the positions of points forming a horizontal or vertical curve.

In construction surveying, curves have to be set out on the ground for a variety of purposes. A curve may form the major part of a roadway, it may form a kerb line at a junction or may be the shape of an ornamental rose bed in a town centre. Obviously different techniques would be required in the setting out of the curves mentioned above, but in all of them a few geometrical theorems are fundamental and it is wise to begin the study of curves by recalling those theorems.

1. Curve geometry

In Fig. 12.1, A, B and C are three points on the circumference of a circle.

1. AB and AC are chords of the circle subtending angles θ and α respectively at the centre O. ADB and AEC are arcs of the circle. Their lengths are

$$2\pi R\left(\frac{\theta}{360}\right)^{\circ} \quad \text{and} \quad 2\pi R\left(\frac{\alpha}{360}\right)^{\circ} \text{ respectively}$$

More conveniently their lengths are $R\theta$ and $R\alpha$ respectively, where θ and α are expressed in radians.

2. In Fig. 12.2, lines ABC and ADE are tangents to the circle at B and D respectively. AB = AD and angles ABO and ADO are right angles.

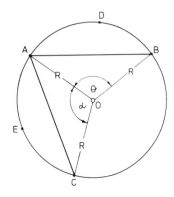

Figure 12.1

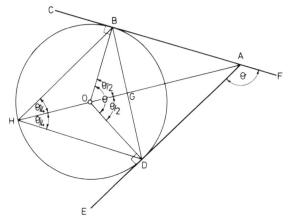

Figure 12.2

3. Since the opposite angles of a cyclic quadrilateral are supplementary, the figure ABOD must be cyclic as angles ABO and ODA together make 180°.

4. The exterior angle of a cyclic quadrilateral equals the interior opposite angle; therefore angle FAD = angle BOD = θ.

5. Join OA, the perpendicular bisector of chord BD. Angle OGB is therefore a right angle and angle BOG = $\theta/2$.

Angle ABG + angle GBO = 90°
and angle BOG + angle GBO = 90°
Therefore angle ABG = angle BOG = $\theta/2$

i.e. the angle ABG between the tangent AB and chord BD equals half the angle BOD at the centre.

6. Produce AO to the circumference at H and join HB. Angle BOG is the exterior angle of triangle BOH.

Therefore angle BOG = angle OHB + angle OBH

However, angles OHB and OBH are equal since triangle BOH is isosceles.

Therefore angle OHB $= \frac{1}{2}$ angle BOG
$$= \theta/4$$

Similarly angle OHD $= \theta/4$
Therefore angle BHD $= \theta/2$

i.e. the angle BHD at the circumference subtended by the chord BD equals half the angle BOD at the centre subtended by the same chord

Also the angle ABD between the tangent and chord equals the angle BHD at the circumference.

2. Curve elements

In Fig. 12.3, the centre lines AI and BI of two straight roadways, called simply the straights, meet at a point I called the intersection point. The roadways may actually exist on the ground or may simply be proposals on a roadway development plan. In either case, the two straights deviate by the angle θ, which is called the deviation angle. Alternatively, the angle may be called the deflection angle or intersection angle.

Clearly it is desirable to avoid having a junction at I, so the straights are joined by a circular curve of radius R.

The straights are tangential to the curve at the tangent points T_1 and T_2 and lengths IT_1 and IT_2, known as the tangent lengths, are equal. Before setting the curve on the ground, the exact location of the tangent points must be known.

If the two straights are existing roadways, then, in order to locate the tangent points, a theodolite is set at point I and the deviation angle θ is measured together with the lengths of the lines AI and IB.

If the roadway scheme exists only on a development plan, the angle θ and the distances AI and IB must be measured by protractor and scale rule or by calculation from the coordinates of A, I and B.

In either case, station A is treated as the point of zero chainage and the chainage of point I is the distance AI.

The radius R is usually a multiple of 50 metres and is supplied by the architect or designer. Knowing only the deviation angle and radius, the tangent lengths and curve length are derived thus:

$$\text{Angle } IT_1O = \text{angle } OT_2I = 90°$$

Therefore IT_1OT_2 is a cyclic quadrilateral and

$$\text{angle } T_1OT_2 = \theta$$

Join I to O.

$$\text{Angle } T_1OI = \text{angle } IOT_2 = \theta/2$$

(a) Tangent lengths IT_1 and IT_2

In triangle IT_1O,

$$\frac{IT_1}{R} = \tan \theta/2$$
$$\text{Therefore } IT_1 = R \tan \theta/2$$
$$\text{Chainage of } T_1 = \text{chainage of } I - IT_1$$

(b) Length of curve T_1T_2

$$\text{Curve } T_1T_2 = R \times \theta \text{ radians}$$
$$\text{Chainage of } T_2 = \text{chainage } T_1 + \text{curve length}$$

(The chainage of the second tangent point is *always* derived via the curve.) This information enables the beginning and end of the curve to be located.

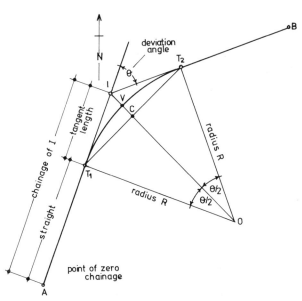

Figure 12.3

Example

1 In Fig. 12.3, the bearings and lengths of AI and IB are

AI	N20°E	450.30 m
IB	N70°E	275.00 m

The radius of the curve joining the straights is 300 m. Calculate the chainages of the tangent points.

Solution
(a) Deviation angle $\theta = 70° - 20°$
$$= 50°$$

(b) Chainage I $= 450.30$ m

(c) Tangent length $IT_1 = R \tan \theta/2$
$$= 300 \tan 25°$$
$$= 139.89 \text{ m}$$

(d) Chainage $T_1 = 450.30 - 139.89$
$$= 310.41 \text{ m}$$

(e)   Curve length $= R \times \theta$ radians
$$= 300 \times 0.872\,66$$
$$= 261.80 \text{ m}$$

(f)   Chainage $T_2 = 310.41 + 261.80$
$$= 572.21 \text{ m}$$

Exercise 12.1

1 Two straight roadways AB and BC meet at junction B. The junction is to be replaced by a circular curve of 300 metres radius, which is to be tangential to straights AB and BC. The coordinates of points A, B and C are as follows:

Point	Easting	Northing
A	0.000	0.000
B	+859.230	+151.505
C	+1423.046	−53.707

Calculate:

(a) the lengths of the straight roadways AB and BC,
(b) the deviation angle between straights AB and BC,
(c) the lengths of the tangents to the straights,
(d) the length of the curve joining the straights,
(e) the chainages of the tangent points, assuming station A is the origin of the survey.

Other curve elements are frequently required and are calculated from the values of R and θ.

(c) Long chord T_1T_2 (Fig. 12.3)

The long chord is the straight line joining T_1 and T_2. The line IO is the perpendicular bisector of T_1T_2 at C. In triangle T_1CO,
$$\frac{T_1C}{R} = \sin \theta/2$$
Therefore $T_1C = R \sin \theta/2$
and $T_1T_2 = 2R \sin \theta/2$

(d) Major offset CV (Fig. 12.3)

Frequently called the mid-ordinate or versine, the length CV is the greatest offset from the long chord to the curve:
$$CV = R - OC$$
In triangle T_1CO,
$$\frac{CO}{R} = \cos \theta/2$$
Therefore $CO = R \cos \theta/2$
and $CV = R - R \cos \theta/2$
$$= R(1 - \cos \theta/2)$$

(e) External distance VI (Fig. 12.3)

The length of VI is the shortest distance from the intersection point to the curve:
$$VI = IO - R$$

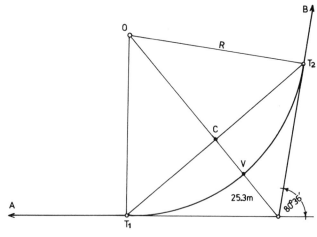

Figure 12.4

In triangle IT_1O,
$$\frac{IO}{R} = \sec \theta/2$$
Therefore $IO = R \sec \theta/2$
and $VI = R \sec \theta/2 - R$
$$= R(\sec \theta/2 - 1)$$

Example

2 Two straights AI and IB deviate to the left by 80° 36′. They are to be joined by a circular curve such that the shortest distance between the curve and intersection point is 25.3 m (Fig. 12.4). Calculate:

(a) the radius of the curve,
(b) the lengths of the long chord and major offset.

Solution
(a) $VI = R(\sec \theta/2 - 1)$
i.e. $25.3 = R(\sec 40° 18′ - 1)$
$$= R(1.311\,186 - 1)$$
Therefore $R = \dfrac{25.3}{0.311\,186}$ m
$$= 81.30 \text{ m}$$

(b) Long chord $T_1T_2 = 2R \sin \theta/2$
$$= 162.60 \sin 40° 18′$$
$$= 162.60 \times 0.646\,790$$
$$= 105.17 \text{ m}$$

Major offset $VC = R(1 - \cos \theta/2)$
$$= 81.30(1 - 0.762\,668)$$
$$= 81.30 \times 0.237\,332$$
$$= 19.30 \text{ m}$$

Exercise 12.2

Two straights XY and YZ deviate to the right by 47° 09′ 20″. They are to be joined by a circular curve of 50 metres radius. Calculate:

(a) the length of the long chord joining the tangent points,
(b) the major offset CV,
(c) the shortest distance YV from the curve to the intersection point,
(d) the tangent lengths IT_1 and IT_2,
(e) the length of the curve.

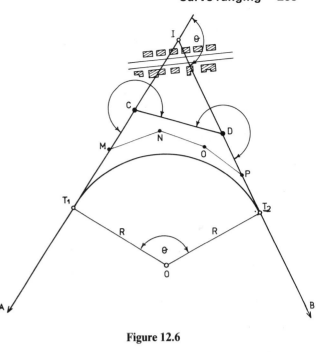

Figure 12.6

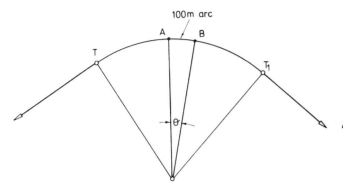

Figure 12.5

3. Designation of curves

In the United Kingdom, curves are designated by the length of the radius. Since there is generally some scope in the choice, the radius is usually a multiple of 50 metres.

The curve can also be designated by the degree of curvature which is defined as the number of degrees subtended at the centre by an arc 100 m long. The degree of curvature is given as a number of whole degrees. In Fig. 12.5, the angle $\theta = 5°$; i.e. the degree of curvature is 5°. The relationship between radius and degree of curvature is as follows:

$$\text{Arc length AB} = R \times \theta \text{ radians}$$

$$= R \times \theta \times \frac{\pi}{180}(\theta \text{ in degrees})$$

$$\text{Therefore } R = \frac{100 \times 180}{\theta \times \pi} \text{ m}$$

$$= \frac{5729.8}{\theta} \text{ m}$$

$$= 1145.96 \text{ m}$$

4. Problems in curve location

(a) Inaccessible intersection point

It happens frequently on site that the intersection point cannot be occupied because of some obstacle. In Fig. 12.6 the intersection point I is in a built-up area.

The relative position of the straights AI and IB must be obtained by traversing between them. The

best conditions for surface taping determine the route of the traverse. In Fig. 12.6 the route AMNOP might provide the best conditions in which the traverse coordinates would be calculated and the position of station I deduced from a solution of triangle IMP.

The simplest traverse between the straights is one straight line between two stations C and D. Angles ACD and CDB are measured together with distances AC and CD. The chainage of C is therefore known.

Example

3 In Fig. 12.6 the following data were derived from traverse ACDB.

Angle ACD 252° 15′ 00″	AC = 559.28 m
Angle CDB 227° 25′ 00″	CD = 256.50 m

Calculate the chainage of the tangent point T_1 if the straights are to be joined by a 300 m radius curve.

Solution
In triangle ICD,

$$\text{Angle C} = 72° 15′ 00″$$
$$\text{Angle D} = 47° 25′ 00″$$
$$\text{Therefore angle I} = 60° 20′ 00″$$
$$\text{By sine rule} \frac{IC}{\sin 47° 25′} = \frac{CD}{\sin 60° 20′}$$
$$\text{Therefore IC} = \frac{256.50 \times 0.738\,259}{0.868\,920}$$
$$= 217.35 \text{ m}$$
$$\text{Chainage of I} = AC + 217.35$$
$$= 776.63 \text{ m}$$

Deviation angle $\theta = 180°\ 00'\ 00''$
$$- 60°\ 20'\ 00''$$
$$= 119°\ 40'\ 00''$$

Tangent length $IT_1 = R \tan \theta/2$
$$= 300 \tan 59°\ 50'$$
$$= 300 \times 1.720\ 474$$
$$= 516.142\ \text{m}$$

Therefore chainage of T_1
$$= \text{chainage I} - 516.142$$
$$= 260.49\ \text{m}$$

Exercise 12.3

1 A horizontal curve of 100 metres radius is to be set out between two straights AX and XB, but the intersection point X is inaccessible. In order to overcome the problem, two points P and Q were selected on lines AX and BX respectively and the following information was recorded:

$$\text{Length PQ} = 55.00\ \text{m}$$
$$\text{Angle QPA} = 151°\ 20'$$
$$\text{Angle BQP} = 143°\ 10'$$

Determine:

(a) the values of the three angles of triangle PXQ and hence the lengths PX and QX, using the sine formula,
(b) the tangent lengths of the curve,
(c) the distances of A and B from P and Q respectively,
(d) the length of the curve AB.

(b) Curve tangential to three straights

In Fig. 12.7 three straights are to be joined by a circular curve, the radius of which is unknown and has to be calculated. One condition must be fulfilled, namely that each straight be a tangent to the curve of radius R.

First consider straights AB and BC only. BT_1 and BT_2 are equal tangent lengths deviating by angle θ. Therefore $BT_1 = BT_2 = R \tan \theta/2$.

Considering straights BC and CD only, CT_2 and CT_3 are equal tangent lengths deviating by angle α. Therefore $CT_2 = CT_3 = R \tan \alpha/2$.

The length BC is known and is also equal to $(BT_2 + CT_2)$.

$$\text{Therefore BC} = (BT_2 + CT_2)$$
$$= R \tan \theta/2 + R \tan \alpha/2$$
$$\text{Hence } R = BC \div (\tan \theta/2 + \tan \alpha/2)$$

Example

4 The following data refer to Fig. 12.7:

Straight	Bearing	Distance (m)
AB	34°	735.70
BC	74°	210.50
CD	124°	640.40

Calculate:

(a) the radius of the curve joining the straights,
(b) the length of curve.

Solution

(a) $\quad$ Angle $\theta = 74° - 34° = 40°$
$\quad\quad$ Angle $\alpha = 124° - 74° = 50°$
$\quad\quad R = BC \div (\tan \theta/2 + \tan \alpha/2)$
$\quad\quad\quad$ as before
$\quad\quad\quad = 210.50$
$\quad\quad\quad\quad \div (\tan 20° + \tan 25°)$
$\quad\quad\quad = 210.50 \div 0.830\ 277\ 9$
$\quad\quad\quad = 253.53\ \text{m}$

(b) Angle $T_1OT_3 = (40° + 50°) = 90°$
$\quad$ Therefore curve
$\quad\quad\quad = \pi/2 \times R$ metres
$\quad\quad\quad = 398.245\ \text{m}$

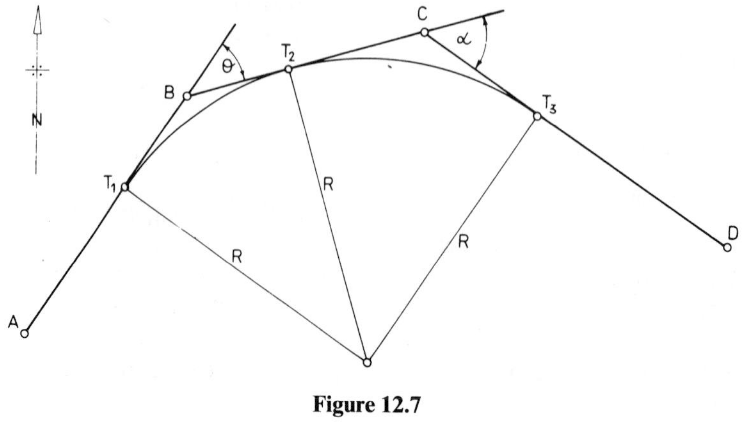

Figure 12.7

Exercise 12.4

1 A circular curve is to be set out so that it is tangential to three straight lines AB, BC and CD. The following data are given:

Line	Whole circle	Length
AB	37° 12′	—
BC	91° 02′	312.70 m
CD	147° 14′	—

(a) Calculate the radius of the curve.
(b) Making the necessary calculations, describe how the initial tangent point on line AB should be located on site.

(c) Curve passing through three known points

In Fig. 12.8, P, Q and R are three points whose co-ordinates are known. They are to be joined by a curve of unknown radius, the length of which is required. The circle that passes through the points is the circumscribing circle of triangle PQR. Therefore,

Angle QPR (at the circumference)
  $= \frac{1}{2}$ angle QOR (at the centre)
  $=$ angle SOR (SO bisects QR)

Angle QPR and SR can be calculated from the co-ordinates.

Therefore OR $=$ SR cosec $\hat{SOR}$

A simpler solution is provided by the sine rule, which states that

$$\frac{p}{\sin P} = \frac{q}{\sin Q} = \frac{r}{\sin R}$$
$$= 2 \times \text{radius of circumscribing circle}$$

Side QR $= p$ and $\hat{QPR}$ can be found from the co-ordinates.

Example

5 In Fig. 12.8 the following are the known co-ordinates of points P, Q and R:

Point		
P	247.6 E	171.3 N
Q	332.0 E	205.4 N
R	390.4 E	122.1 N

Calculate the radius of the curve that passes through all three points.

Solution

(a) Tan bearing PQ $= \dfrac{\Delta \text{ east}}{\Delta \text{ north}}$

$= \dfrac{332.0 - 247.6}{205.4 - 171.3}$

$= \dfrac{84.4}{34.1}$

Bearing PQ $=$ N 68° E
  $=$ 68° WCB

(b) Tan bearing PR $= \dfrac{\Delta \text{ east}}{\Delta \text{ north}}$

$= \dfrac{390.4 - 247.6}{122.1 - 171.3}$

$= \dfrac{142.8}{-49.2}$

Bearing PR $=$ S 71° E
  $=$ 109° WCB

(c) Distance QR $= \sqrt{\Delta \text{ east} + \Delta \text{ north}}$
  $= \sqrt{83.3^2 + 58.4^2}$
  $= 101.73$ m

(d) Angle QPR $= 109° - 68°$
  $= 41°$

(e) In triangle PQR,

$$\frac{p}{\sin P} = 2 \times \text{radius}$$

Therefore radius $= \dfrac{101.73}{2 \times \sin 41°}$

$= 77.53$ m

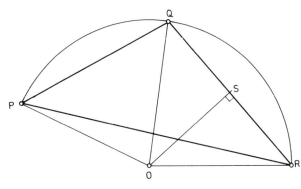

Figure 12.8

Exercise 12.5

1 Figure 12.9 shows three manholes A, B and C, which are part of a town centre development scheme. The centre is to be landscaped such that the manholes will lie within a circular grassed area. The edge of the grassed area is to be 1 metre clear of the manhole centres. The coordinates of the manholes are as follows:

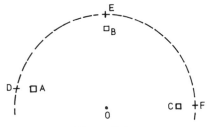

Figure 12.9

Manhole	Easting	Northing
A	10.0	10.0
B	15.0	12.5
C	19.0	8.0

Calculate:

(a) the radius of the circle of the required grass area,
(b) the coordinates of the centre of the circle.

5. Setting out curves (calculations)

The purpose of curve calculations is to enable the curve to be set in its predetermined position on the ground. The centre-line of the curve is positioned by a series of pegs, set at intervals chosen by the surveyor. (Setting out construction works is more fully dealt with in Chapter 13.)

Where curves are long and of large radius (over 100 m) a theodolite must be used to obtain the desired accuracy of setting out. Small radius curves can be set out quickly and accurately by using tapes only.

(a) Small radius curves

Method 1: Finding the centre
In Fig. 12.10 kerbs have to be laid at the roadway junction. Consider the right-hand curve. The deviation angle α is measured from the plan and the tangent lengths IT_1 and IT_2 ($= R \tan \alpha/2$) calculated. The procedure for setting the curve is then as follows:

1. From I, measure back along the straights the distances IT_1 and IT_2.
2. Hammer in pegs at these points and mark the exact positions of T_1 and T_2 by nails.
3. Hook a steel tape over each nail and mark the centre O at the point where the tapes intersect when reading R. Hammer in a peg and mark the centre exactly with a nail.
4. Any point on the curve is established by hooking the tape over the peg O and swinging the radius. This method is widely used where the radius of curvature is less than 30 m.

Method 2: Offsets from the tangent
When the deviation angle is small (less than 50°) the length of the curve short and the centre inaccessible, the curve can be set out by measuring offsets from the tangent. In the left-hand curve of Fig. 12.10, y is an offset from the tangent at a distance x metres from tangent point T_1. In the figure, the line AB is drawn parallel to the tangent until it cuts the radius. The length $AT_1 = y$ and the length $AO = (R - y)$.

In triangle OAB,

$$OA = \sqrt{OB^2 - AB^2} \text{ (by Pythagoras)}$$
$$\text{i.e. } (R - y) = \sqrt{R^2 - x^2}$$
$$\text{Therefore } y = R - \sqrt{R^2 - x^2}$$

Thus the offset y can be calculated for any distance x along the tangent and can be set by eye or by optical square.

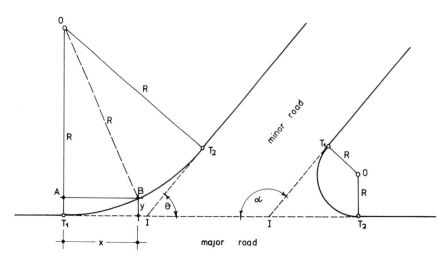

Figure 12.10

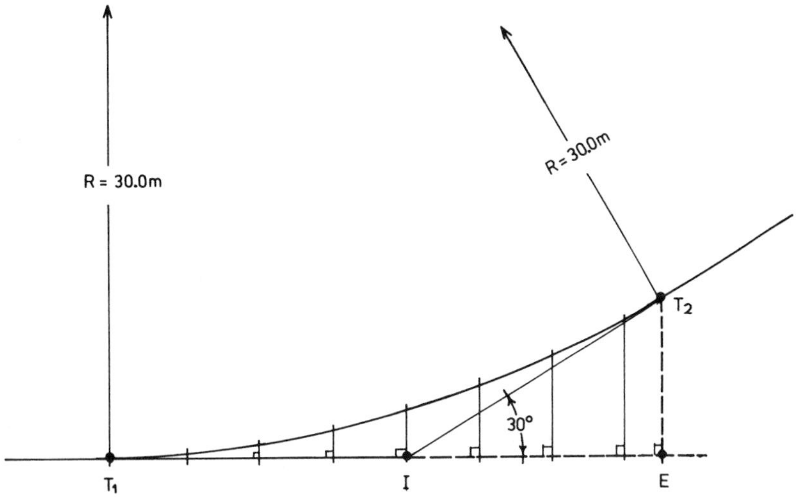

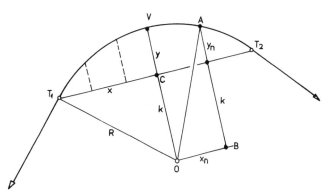

Figure 12.11

Example

6 Given that the deviation angle $\theta = 50°$ and the radius $R = 60$ m, calculate the offsets from the tangent at 5 metre intervals (Fig. 12.10).

(a) Tangent lengths $IT_1 = IT_2 = 60 \tan 25°$
$= 27.98$ m

(b) Offsets at
$$5 \text{ m} = 60 - \sqrt{60^2 - 5^2} \qquad = 0.210 \text{ m}$$
$$10 \text{ m} = 60 - \sqrt{60^2 - 10^2} \qquad = 0.840 \text{ m}$$
$$15 \text{ m} = 60 - \sqrt{60^2 - 15^2} \qquad = 1.905 \text{ m}$$
$$20 \text{ m} = 60 - \sqrt{60^2 - 20^2} \qquad = 3.430 \text{ m}$$
$$25 \text{ m} = 60 - \sqrt{60^2 - 25^2} \qquad = 5.456 \text{ m}$$
$$27.98 \text{ m} = 60 - \sqrt{60^2 - 27.98^2} = 6.923 \text{ m}$$

The procedure for setting out the curve is then as follows:

1. From I, measure back along the straights the distances IT_1 and IT_2 and drive in pegs to establish the exact positions of T_1 and T_2.
2. Establish pegs at 5 m intervals along the straights between I and the tangent points.
3. Using an optical square set out the appropriate offsets y at right angles to the tangents and drive in a peg at each point.

Exercise 12.6

1 In Fig. 12.11, the deviation angle between the minor road and major road is 30° and the radius of the curve is 30.00 m. Calculate:
(a) the tangent lengths IT_1 and IT_2,
(b) the length IE of the extension to the left-hand straight,

(c) the lengths of the offsets at 2 metre intervals, required to set out the curve, from tangent point T_1 to tangent point T_2.

Method 3: Offsets from the long chord
This method is suitable for curves of small radius. The curve is established by measuring offsets y at right angles to the long chord T_1T_2 at selected distances from the tangent points (Fig. 12.12).

VC is the major offset y at the mid-point C of the long chord and OC is constant, k.

$$\text{Major offset } y = (R - k)$$
$$\text{In triangle } OT_1C, k = \sqrt{R^2 - x^2}$$
$$\text{Therefore } y = R - \sqrt{R^2 - x^2}$$
$$\text{Any other offset } y_n = (AB - k)$$
$$\text{In triangle ABO, } AB = \sqrt{R^2 - x_n^2}$$
$$\text{Therefore } y_n = \sqrt{R^2 - x_n^2} - k$$

Any offset y can be calculated for any distance along the long chord, and can be set out by eye or prism square.

Figure 12.12

Example

7 A roadway kerb in Fig. 12.12 has a radius of curvature of 40 m. The length of the long chord is 60 m. Calculate the offsets from the chord at 10 m intervals.

Answer

$$\text{Major offset} = R - \sqrt{R_2 - x^2}$$
$$= 40 - \sqrt{1600 - 900}$$
$$= 40 - 26.46$$
$$= 13.54 \text{ m}$$
$$k = 26.46 \text{ m}$$

Offset $y_{10}(x = 20 \text{ m from C})$
$$y_{10} = \sqrt{R_2 - x^2} - k$$
$$= \sqrt{1600 - 400} - 26.46$$
$$= 34.64 - 26.46$$
$$= 8.18 \text{ m}$$

Offset $y_{20}(x = 10 \text{ m from C})$
$$y_{20} = \sqrt{R_2 - x^2} - k$$
$$= \sqrt{1600 - 100} - 26.46$$
$$= 38.73 - 26.46$$
$$= 12.27 \text{ m}$$

The offsets at 40 m and 50 m from T_1 are the same lengths as the offsets at 20 and 10 m respectively.

The procedure for setting out the curve is as follows:

1. Locate T_1 and T_2 and measure the distance between them. It should equal 60 metres.
2. At 10 m intervals along the long chord drive in pegs.
3. Set out the offsets at right angles to the long chord using a prism square and drive in pegs to mark the curve.

Exercise 12.7

1 A curved roadway kerb of radius 100 metres is to be set out using the method of offsets from the long chord. Given that the length of the long chord is 60 metres, calculate the lengths of the offsets at 10 metre intervals along the long chord.

Method 4: Offsets from chords produced

Length of chord In this and subsequent methods of setting out curves, chords have to be chosen such that the difference in length between the chord and arc is as small as possible.

The chord length should not be greater than one-tenth of the length of the radius. The error caused by assuming that the chord equals the arc is 1 in 2400, which is acceptable for much construction work. If greater accuracy is required the ratio of chord to radius must be reduced. The ratio of 1:20 gives errors of the order of 1 in 10 000.

Procedure for setting out the curve In Fig. 12.13(a), T is the tangent point set out as before by measuring back the distance IT ($= R \tan \theta/2$) from the intersection point I.

The procedure for setting out the curve is as follows:

1. Select a length, c, less than one-tenth the length of the radius and lay off the distance from the point T along the straight towards I. Mark the point B.
2. Calculate the offset y_1, and swing the tape from T through the arc y_1 to establish point C on the curve. Drive in a peg and mark C accurately with a nail.
3. Extend the chord TC for a further distance of c metres and mark point D.

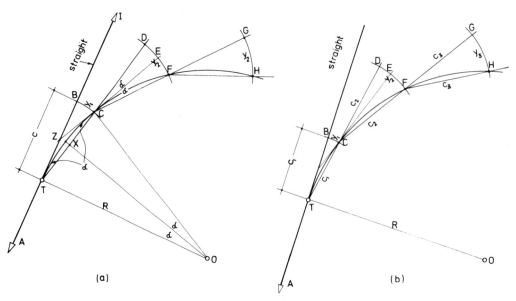

(a)   (b)

Figure 12.13

4. Calculate the second offset y_2 and swing the tape from C through the arc y_2 to establish point F on the curve.

5. Repeat operations (3) and (4) to establish further pegs H, etc., on the curve. The offset distance in each case is y_2.

Length of offsets In these calculations, the arc and chord lengths are assumed equal. In Fig. 12.13(a), C is joined to O to form angle TOC and O is joined to X, making OX the perpendicular bisector of TC.

Therefore angle TOX = angle XOC = α

Angle BTC is the angle between the tangent BT and the chord CT.

Therefore BTC = $\frac{1}{2}$ angle TOC
= angle TOX = α

Therefore sector BTC is similar to sector TOX.

$$\text{Now } \frac{BC}{TC} = \frac{TX}{TO}$$

$$\text{i.e. } \frac{y_1}{c} = \frac{c/2}{R}$$

$$\text{Therefore } y_1 = \frac{c^2}{2R} \text{ (first offset)}$$

If the tangent ZE is drawn through C:

Angle ZCT = α
Also, Angle ZCT = angle DCE
(vertically opposite angles)
Therefore angle DCE = α

Angle ECF is the angle between tangent EC and chord FC.

Therefore angle ECF = α
DCF = angle DCE + angle ECF
= 2α
= angle COF (angle at centre)

Therefore sector DCF is similar to sector COF.

$$\text{Now } \frac{DF}{CF} = \frac{CF}{CO}$$

$$\text{i.e. } \frac{y_2}{c} = \frac{c}{R}$$

$$\text{Therefore } y_2 = \frac{c^2}{R} \text{ (second and subsequent offsets)}$$

Similarly GH and other offsets are equal to c^2/R.

This method is fairly accurate and is not restricted to setting out curves of small radius (refer to Example 8).

(b) Curve composition

In setting out large radius curves, or in some cases small radius curves, pegs are set at regular intervals around the curve. The interval is commonly 10 or 20 metres and is measured as a running chainage, from the zero chainage point of the road system. Consequently, it would be very unlikely that either tangent point of the curve would coincide with a chainage which is at an exact tape length.

In Fig. 12.14, straights AI and IB deviate by 13° at intersection point I, where the chainage is 171.574 metres.

Tangent lengths IT_1 and IT_2 = 400 tan 6.5°
= 45.574 m
Therefore chainage T_1 = 171.574 − 45.574
= 126.000 m
Curve length = $2\pi R \times 13/360$
= 90.757 m
Therefore chainage T_2 = 126.000 + 90.757
= 216.757 m

The last peg on the straight, measured at 20 metre intervals from A, occurs at chainage 120 m; therefore the first peg on the curve, at chainage 140 m, lies at a distance of (140 − 126) m = 14 m from tangent point

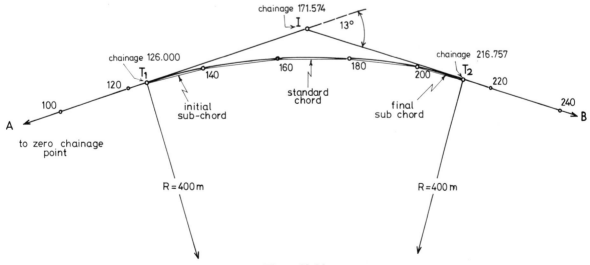

Figure 12.14

T_1. This short chord is called the initial sub-chord. Thereafter, pegs placed at standard chord intervals of 20 metres occur at chainages 160, 180 and 200 m. The final tangent point T_2 is reached at 216.75 m; therefore the final chord is $(216.757 - 200.000)$ m $= 16.757$ m. This short chord is called the final sub-chord.

Summarizing, the chord composition is derived as follows:

$$\text{Chainage } T_1 = 126.000 \text{ m}$$
$$\text{Chainage of first peg on curve} = 140.000 \text{ m}$$
$$\text{Therefore initial sub-chord} = (140.000 - 126.000)$$
$$= 14.000 \text{ m}$$
$$\text{Chainage of last peg on curve} = 200.000 \text{ m}$$
$$\text{Therefore number of standard chords}$$
$$= (200 - 140)/20 = 3$$
$$\text{Chainage } T_2 = 216.757 \text{ m}$$
$$\text{Therefore final sub-chord} = (216.757 - 200.000)$$
$$= 16.757 \text{ m}$$

In setting out large radius curves, the chords must be almost equal to the arcs that they subtend. An accuracy of about 1 part in 10 000 is obtainable, provided the chord length does not exceed one twentieth of the length of the radius, i.e. $c < R/20$.

(c) Setting out large radius curves

Method 1: Offsets from chords produced

The pegs are set on the curve at intervals of one tape length measured from the point of zero chainage A (Fig. 12.13(b)).

The first chord is less than one tape length. In the figure the initial chord is c_1 metres long and all others are the standard one-tape length of c_2 metres.

$$\text{As before, offset } y_1 = \frac{c_1^2}{2R} \text{ (initial sub-chord)} \quad (1)$$

Sector BTC is similar to sector DCE.

$$\text{Therefore } \frac{\text{DE}}{\text{BC}} = \frac{c_2}{c_1}$$

$$\text{and offset DE} = y_1 \times \frac{c_2}{c_1}$$

$$= \frac{c_1^2 \times c_2}{2R \times c_1}$$

$$= \frac{c_1 c_2}{2R}$$

$$\text{As before, EF} = \frac{c_2^2}{2R}$$

$$\text{Offset } y_2 = \text{DF} = (\text{DE} + \text{EF})$$

$$= \frac{c_1 c_2}{2R} + \frac{c_2^2}{2R}$$

$$= \frac{c_2(c_1 + c_2)}{2R} \text{ (first full chord)} \quad (2)$$

As before, $y_3 = \dfrac{c_3^2}{R}$ (second and subsequent full chords) $\quad (3)$

A general expression can be deduced from formula (2) which can be applied to find any offset, namely:

$$y_n = \frac{c_n(c_n + c_{(n-1)})}{2R} \quad (4)$$

Example

8 A 300 metre radius curve is to be set out by offsets from chords. The chainages of the first and second tangent points are 327.5 and 425.3 m respectively. Calculate the lengths of the offsets to set out pegs at regular chainages of 20 m.

Solution

Length of curve $= (425.3 - 327.5) = 97.8$ m

First peg on curve occurs at 340 m

$$\text{Therefore initial sub-chord} = (340.0 - 327.5)$$
$$= 12.5 \text{ m}$$

Curve is composed of:

(a) initial sub-chord	12.5 m
(b) four 20 m chords	80.0 m
(c) final sub-chord	5.3 m
	97.8 m

From formulae (1) to (4),

$$\text{First offset} = 12.5^2/600$$
$$= 0.260 \text{ m}$$
$$\text{Second offset} = 20(20 + 12.5)/600$$
$$= 1.083 \text{ m}$$
$$\text{Third, fourth and fifth offsets}$$
$$= 20^2/300$$
$$= 1.333 \text{ m}$$
$$\text{Final offset} = 5.3(5.3 + 20)/600$$
$$= 0.223 \text{ m}$$

The setting-out information is tabulated as in Table 12.1.

Chord number	Chord length (m)	Chord chainage (m)	Offset (m)
1	12.5	340.0	0.260
2	20.0	360.0	1.083
3	20.0	380.0	1.333
4	20.0	400.0	1.333
5	20.0	420.0	1.333
6	5.3	425.3	0.223
	97.8		

Table 12.1

Exercise 12.8

1 In Fig. 12.14, a 400 metre radius curve is to be set out by offsets from chords. The chainages of the first and last tangent points are 126.000 and 216.757 m respectively. Calculate the offsets required to set out pegs at 20 metre intervals of through chainages from the zero chainage point.

Method 2: Setting by tangential angles

Method   The method involves the use of tape and theodolite and is the common method of setting out large radius curves when accuracy is required. In Fig. 12.15, the tangent point T_1 at the beginning of the curve has been established as in previous methods. BC and CD are equal standard chords, c_2 and c_3, chosen such that their length is less then one-twentieth the length of the radius. TB is the initial sub-chord, c_1, which is shorter than the standard chords because the chainage of T_1 is irregular, and c_4 is the final sub-chord, which is also shorter than the standard chords.

Deflection angles   Angles α_1, α_2, α_3 and α_4 are the angles by which the curve deflects to the right or left. They are called *deflection* angles, though the terms chord angle and tangential angles are commonly used. In this book they are called deflection angles.

Their values must be calculated in order to set out the curve.

Procedure   Assuming for a moment that the deflection angles are known, the curve is set out as follows:

1. Set the theodolite at T and sight intersection point I on zero degrees.
2. Release the upper clamp and set the theodolite to read α_1 degrees.
3. Holding the end of a tape at T, line in the tape with the theodolite and drive in a peg B at a distance of c_1 metres from T.
4. Set the theodolite to read $(\alpha_1 + \alpha_2)$ degrees.
5. Hold the rear end of the tape at B and with the tape reading c_2 metres, i.e. a standard chord length, swing the forward end until it is intersected by the line of sight of the theodolite. This is the point C on the curve.
6. Set the theodolite to read $(\alpha_1 + \alpha_2 + \alpha_3)$ degrees. Repeat operation 5 to establish point D on the curve.

In most cases this operation will be repeated several more times to establish a number of pegs on the curve at one standard chord interval. In this example peg D is the last standard chord.

7. Set the theodolite to read $(\alpha_1 + \alpha_2 + \alpha_3 + \alpha_4) = \theta/2$ degrees.

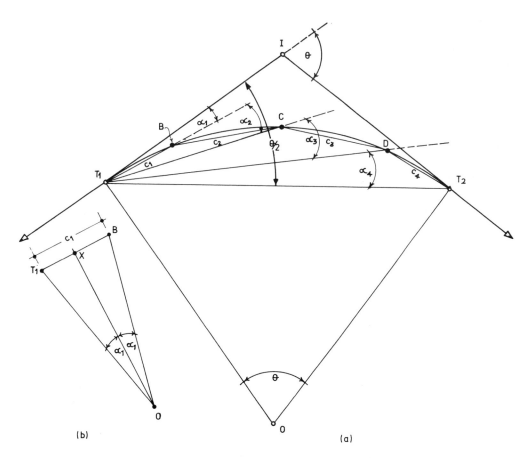

Figure 12.15

8. Holding the rear end of the tape at D swing the forward end until the reading of c_4 metres is intersected by the line of sight of the theodolite. This establishes tangent point T_2.

Calculation of deflection angles   Angle IT_1B is the angle between tangent T_1I and chord T_1B. Angle T_1OB is the angle at the centre subtended by chord T_1B.

$$\text{Therefore angle } IT_1B = \tfrac{1}{2}\text{ angle } T_1OB = \alpha$$

In Fig. 12.15(b), OX is the perpendicular bisector of chord T_1B. Therefore angle T_1OX = angle XOB = α.
  In triangle T_1OX,

$$\sin T_1OX = \frac{T_1X}{T_1O}$$
$$= \frac{c_1/2}{R}$$
$$= \frac{c_1}{2R}$$

The value of any deflection angle (α_1, α_2, α_3 and α_4) can similarly be found and the formula can be written in general terms:

$$\sin \alpha = \frac{c}{2R} \qquad (5)$$

When c is less then one-twentieth of R, an accurate value of α can be determined thus:

$$\sin \alpha = \alpha \text{ radians (since } \alpha \text{ is always small)}$$

$$\text{Therefore } \alpha \text{ radians} = \frac{c}{2R}$$

$$\text{Hence } \alpha \text{ degrees} = \frac{c}{2R} \times \frac{180}{\pi}$$

$$\text{and } \alpha \text{ minutes} = \frac{c}{2R} \times \frac{180}{\pi} \times 60$$

$$\text{i.e. } \alpha = \left(\frac{c}{R} \times 1718.9\right) \text{ minutes} \quad (6)$$

In Fig. 12.15(a),

$$\text{Standard-chord angle } \alpha_2 = \left(\frac{c_2}{R} \times 1718.9\right) \text{ minutes}$$

$$\text{or } \sin \alpha_2 = \frac{c_2}{2R}$$

$$\text{Initial sub-chord angle } \alpha_1 = \left(\frac{c_1}{R} \times 1718.9\right) \text{ minutes}$$

$$\text{or } \sin \alpha_1 = \frac{c_1}{2R}$$

From these calculations it can be seen that angles α_1 and α_2 are proportional to their chord lengths and the most convenient way to calculate deflection angles is firstly to calculate the standard-chord angle and then by proportion to calculate the initial and final sub-chord angles:

$$\text{Final sub-chord angle} = \alpha_2 \times \frac{c_4}{c_2}$$

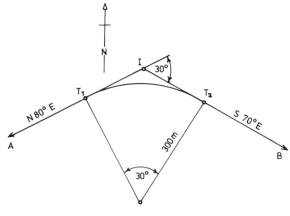

Figure 12.16

Tangential angles   The reading to which the theodolite is set is the tangential angle, so called because the angle being set out is measured from the tangent. Other titles include total angle, setting out angle, etc.
  Any tangential angle to a point on a curve is simply the summation of the deflection angles to that point. Thus, the tangential angle to point $C = \alpha_1 + \alpha_2$ and to point $D = \alpha_1 + \alpha_2 + \alpha_3$.

Example

9 Two straights AI and IB have bearings of 80° and 110° respectively. They are to be joined by a circular curve of 300 metres radius. The chainage of intersection point I is 872.485 m (Fig. 12.16). Calculate the data for setting out the curve by 20 m standard chords.

Answer
(a)   Tangent length = 300 tan 15°
                             = 80.385 m

(b)   Chainage of T_1 = (872.485 − 80.385) m
                             = 792.100 m

(c)   Curve length = 300 × θ radians
                             = 157.080 m

(d)   Chainage of T_2 = 949.180 m

(e) Number of chords
        Initial sub-chord = (800.00 − 792.10)
                             = 7.90 m
    7 full chords of 20 m = (7 × 20.0)
                             = 140.00 m
             Sub-total = 7.90 + 140.00
                             = 147.90 m
    Therefore final sub-chord
                             = (157.08 − 147.90)
                             = 9.18 m
             Total = 147.90 + 9.18
                             = 157.08 m
                             = curve length

(f) Tangential angle for standard 20 m chord

$$= \left(\frac{20}{300} \times 1718.9\right) \text{ minutes}$$
$$= 114.593 \text{ minutes}$$
$$= 01° \ 54' \ 36''$$

(g) Sub-chord angles

$$\text{Initial} = \frac{114.593 \times 7.90}{20.0} \text{ minutes}$$
$$= 45.264 \text{ minutes}$$
$$= 00° \ 45' \ 16''$$

$$\text{Final} = \left(\frac{114.593 \times 9.18}{20.0}\right) \text{ minutes}$$
$$= 52.598 \text{ minutes}$$
$$= 00° \ 52' \ 36''$$

For this and every other example, the setting-out information is presented in tabular fashion as in Table 12.2. Note that there is a discrepancy of 4″ in the final tangential angle in Table 12.2 due to the rounding off of the angles to whole seconds.

When curves are to turn to the left the tangential angles must be subtracted from 360°. For example, if

Chord number	Length (m)	Chainage (m)	Deflection angle	Tangential angle
(T₁)		792.10		
1	7.90	800.00	00° 45′ 16″	00° 45′ 16″
2	20.00	820.00	01° 54′ 36″	02° 39′ 52″
3	20.00	840.00	01° 54′ 36″	04° 34′ 28″
4	20.00	860.00	01° 54′ 36″	06° 29′ 04″
5	20.00	880.00	01° 54′ 36″	08° 23′ 40″
6	20.00	900.00	01° 54′ 36″	10° 18′ 16″
7	20.00	920.00	01° 54′ 36″	12° 12′ 52″
8	20.00	940.00	01° 54′ 36″	14° 07′ 28″
9 (T₂)	9.18	949.18	00° 52′ 36″	15° 00′ 00″
	157.08	157.08		15° 00′ 04″

Table 12.2

the straights in Example 9 had deviated to the left by 30° the final tangential reading would have been

$$(360° \ 00' \ 00'' - 15° \ 00' \ 00'') = 345° \ 00' \ 00''$$

Exercise 12.9

1 Two straight roadways, AB and BC, intersecting at B, have bearings of 5° 30′ and 353° 30′ respectively. They are to be joined by a circular curve of 500 metres radius. The curve is to be set out by the method of tangential angles. Pegs are to be set out at intervals of 20 metres of through chainage, from the point A. Given that the chainage of the point B is 842.75 m from A, calculate:
(a) the chainages of the tangent points of the curve,
(b) the chainages of the setting out pegs along the curve,
(c) the tangential angles required to set out the curve from the initial tangent point.

Method 3: Using two theodolites
In the methods of setting out discussed so far linear measurements were used. When the ground conditions are unsuitable for measuring, the curve may be set out by the use of two theodolites and linear measurements are dispensed with altogether.

In Fig. 12.17 the tangent points T_1 and T_2 have been located on the ground. The following procedure is used to set out the points on the curve.

1. Set theodolite 1 at T_1 and sight I on 00° 00′ 00″.
2. Calculate tangential angle α for the initial sub-chord T_1C and set the theodolite to this reading.
3. Set theodolite 2 at T_2 and sight T_1 on 00° 00′.
4. Set the theodolite to read α.
5. The intersection of the lines of sight of the two theodolites is a point on the curve, since $I\hat{T}_1C$ is the angle between tangent T_1I and chord T_1C which equals angle T_1T_2C, the angle at the circumference subtended by the chord T_1C. A peg is driven in at this point.

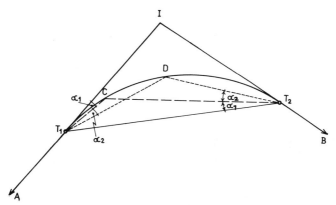

Figure 12.17

6. Further tangential angles are set on both theodolites, the intersection of the sight lines in all cases being a point on the curve. This method has the advantage over the others that each point is set out completely independently.

Method 4: Setting out by coordinates

With the increasing affordability and continued reliability of EDM equipment, setting out by coordinates is fast becoming normal practice in setting out construction works, the main advantage being that the equipment can set out horizontal distances.

When calculating setting-out distances, whether for curved roadways or otherwise, the plan distance is always obtained and has to be set out. However, it is often very difficult to measure distances horizontally; hence slope distances have frequently to be determined or inaccuracies in setting the pegs must occur. Furthermore, measuring by tape is slow, labour intensive and often uncomfortable.

Setting out by EDM methods requires that the coordinates of every proposed point be determined, usually by calculation. The coordinates are compared with those of the setting-out survey point and the horizontal distance and bearing between the two are computed.

On site, the EDM instrument is switched to the horizontal distance mode (tracking mode) and set on the correct bearing. The reflector is then aligned at the required horizontal distance and no slope correction is ever necessary.

Figure 12.18 shows a curve of 572.96 m radius, connecting two straights AI and IB that have bearings of 40° 00′ 00″ and 44° 30′ 00″ respectively. Point A (zero

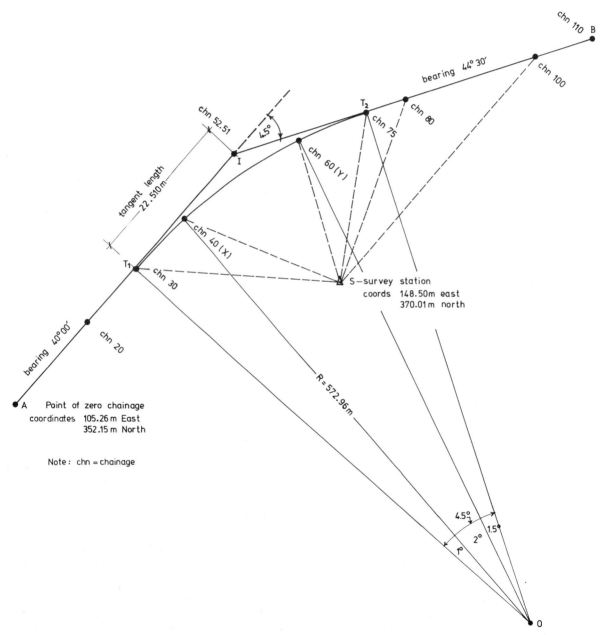

Figure 12.18

chainage) has known coordinates of 105.25 m east and 352.15 m north. Based on the theory of Sec. 2 (curve elements) and 5(c) (setting out large radius curves), the data required to set out the curve are as shown in Table 12.3.

---

1. Point A—point of zero chainage
   Coordinate 105.26 m east
                 352.15 m north
2. Radius of curve = 572.96 m
3. Deviation angle = 4.5°
4. Chainage of intersection point = 52.51 m
5. Tangent length = $572.96 \tan 2.25° = 22.51$ m
6. Chainage $T_1 = 52.51 - 22.51$ m = 30.0 m
7. Curve length = $2\pi R \times (4.5/360) = 45.0$ m
8. Chainage $T_2 = 30 + 45$ m = 75.0 m
9. Peg chainages
      0.0 m   Start
     20.0 m   On left-hand straight
     30.0 m   Tangent point T_1
     40.0 m   On curve (X)
     60.0 m   On curve (Y)
     75.0 m   Tangent point T_2
     80.0 m   On right-hand straight
     100 m   On right hand straight
     110 m   Point B
10. Curve composition
    Initial sub-chord = 10.0 m
    Standard chord = 20.0 m
    Final sub-chord = 15.0 m
11. Deflection angles
    Initial sub-chord = $\sin^{-1} 10/2R = 00° \, 30' \, 00''$
    Standard chord = $\sin^{-1} 20/2R = 01° \, 00' \, 00''$
    Final sub-chord = $\sin^{-1} 15/2R = 00° \, 45' \, 00''$
12. Centre angles (= $2 \times$ deflection angles)
      $T_1OX = 01° \, 00' \, 00''$
      $XOY = 02° \, 00' \, 00''$
      $YOT_2 = 01° \, 30' \, 00''$
      Total = $\overline{04° \, 30' \, 00''}$

Table 12.3

The curve is to be set out from a nearby survey traverse station S (with coordinates 148.50 m east, 370.01 m north) by the method of coordinates. Therefore the coordinates of all chainage points of the roadway scheme must first be calculated.

Several methods are available to compute those coordinates, of which the easiest to understand is the following:

1. Calculate the coordinates of points AT_1IT_2 and B as a traverse. The coordinates are computed in Table 12.4(a).
2. Calculate the coordinates of points T_1O and T_2 as a traverse.

$$\begin{aligned} \text{Forward bearing } AT_1 = &\quad 40° \, 00' \, 00'' \\ \text{Back bearing } T_1A = &\quad 220° \, 00' \, 00'' \\ - \text{ angle } AT_1O &\quad -90° \, 00' \, 00'' \\ \hline \text{Therefore bearing } T_1O = &\quad 130° \, 00' \, 00'' \end{aligned}$$

$$\begin{aligned} \text{Back bearing } OT_1 = &\quad 310° \, 00' \, 00'' \\ + \text{ angle } T_1OT_2 &\quad +4° \, 30' \, 00'' \\ \hline \text{Therefore forward bearing } OT_2 = &\quad 314° \, 30' \, 00'' \end{aligned}$$

The coordinates of points T_1, O and T_2 are computed in Table 12.4(b). The coordinates of point T_2 in Table 12.4(b) check with those in Table 12.4(a).

3. Calculate the coordinates of the remaining curve chainage points, namely X (chainage 40 m) and Y (chainage 60 m).

$$\begin{aligned} \text{Bearing } OT_1 = &\quad 310° \, 00' \, 00'' \\ + \text{ angle } T_1OX &\quad +1° \, 00' \, 00'' \\ \text{Forward bearing } OX = &\quad 311° \, 00' \, 00'' \\ + \text{ angle } XOY &\quad +2° \, 00' \, 00'' \\ \text{Forward bearing } OY = &\quad 313° \, 00' \, 00'' \\ + \text{ angle } YOT_2 &\quad +1° \, 30' \, 00'' \\ \text{Forward bearing } OT_2 = &\quad 314° \, 30' \, 00'' \end{aligned}$$

(Checks with previous result)

The coordinates of points X and Y are computed in Tables 12.4(c) and 12.4(d) respectively.

4. Calculate the bearings and distances that will be required to set out the various chainage points from station S. Table 12.5 shows the relevant calculation, based, of course, on the formulae:

Bearing between two points = $\tan^{-1}(\Delta \text{ east}/\Delta \text{ north})$

and

distance between two points = $\sqrt{\Delta \text{ east}^2 + \Delta \text{ north}^2}$

6. Setting out points by EDM methods

It must be pointed out, at this juncture, that the mechanics of setting an electromagnetic distance measuring instrument along a certain line on a specific bearing varies with the type of instrument. Only the general principle is described here.

1. The EDM instrument (total station) is set up, centred and levelled at survey station S, and the bearing to point A (247° 33′ 26″) is set on the instrument (Table 12.5).
2. A sight is taken to point A and the instrument clamped, reading 247° 33′ 26″. The instrument is set to tracking mode and the horizontal distance to point A is measured as a check. The distance should be 46.783 metres.
3. The bearing and horizontal distance to the first setting-out point, chainage 20 m, are keyed into the instrument. From Table 12.5 the relevant data are: bearing, 265° 13′ 24″ and distance, 30.490 m.
4. The survey assistant will, by this time, have gone to the approximate position of the point to be set out, and will hold the reflector vertically.

	Line	Length	Whole circle bearing	Δ east	Δ north	Easting	Northing	Station
(a)						105.260	352.150	A
	A–Ch20	20.00	40° 0′ 0″	12.856	15.321	118.116	367.471	Ch20
	Ch20–T1	10.00	40° 0′ 0″	6.428	7.660	124.544	375.131	T1
	T1–I	22.51	40° 0′ 0″	14.469	17.244	139.013	392.375	I
	I–T2	22.51	44° 30′ 0″	15.777	16.055	154.790	408.430	T2
	T2–Ch80	5.00	44° 30′ 0″	3.505	3.566	158.295	411.997	Ch80
	80–100	20.00	44° 30′ 0″	14.018	14.265	172.313	426.262	Ch100
	Ch100–B	10.00	44° 30′ 0″	7.009	7.133	179.322	433.395	Ch110(B)
(b)						124.544	375.131	T1
	T1–O	572.96	130° 0′ 0″	438.913	−368.292	563.457	6.839	O
	O–T2	572.96	314° 30′ 0″	−408.664	401.593	154.793	408.432	T2
(c)						563.457	6.839	O
	O–X	572.96	311° 0′ 0″	−432.418	375.896	131.039	382.735	X
(d)						563.457	6.839	O
	O–Y	572.96	313° 0′ 0″	−419.036	390.758	144.421	397.597	Y

Table 12.4

Ref Station	Easting	Northing				
S	148.500	370.010				

Station	Easting	Northing	Difference in eastings	Difference in northings	Set-out distance	Set-out bearing
A	105.260	352.150	−43.240	−17.860	46.783	247° 33′ 26.1″
Ch20	118.116	367.471	−30.384	−2.539	30.490	265° 13′ 23.7″
T1	124.544	375.131	−23.956	5.121	24.497	282° 3′ 58.8″
Ch40 (X)	131.039	382.735	−17.461	12.725	21.606	306° 5′ 00.0″
Ch60 (Y)	144.421	397.597	−4.079	27.587	27.887	351° 35′ 21.2″
T2	154.790	408.430	6.290	38.420	38.931	9° 17′ 52.1″
Ch80	158.295	411.997	9.795	41.987	43.114	13° 7′ 53.3″
Ch100	172.311	426.262	23.811	56.252	61.084	22° 56′ 33.1″
Ch110 (B)	179.320	433.395	30.820	63.385	70.481	25° 55′ 50.3″

Table 12.5

5. A sight is taken to the reflector and as soon as contact is made, the difference in bearing and distance between the setting-out point and the actual position of the reflector are displayed on the keypad of the instrument as dHA and dHD.

6. The instrument is then rotated and the reflector moved until dHA and dHD become zero.

7. The assistant marks the reflector position, inserts a peg, re-checks the complete operation and when satisfied that it is correct, moves to the next setting-out location, namely tangent point T_1.

8. The next bearing and distance are keyed in and the procedure repeated. All remaining pegs are set out in this manner.

Exercise 12.10

1 In Fig. 12.19, two straight roadways, PQ and QR, have bearings of 05° 00′ 00″ and 357° 20′ 00″ respectively and intersect at point Q (chainage 78.793 m).

They are to be joined by a curve of 429.718 metres radius.

(a) Using Table 12.6, calculate all of the curve elements.

(b) Calculate the coordinates of all setting-out pegs, from chainage 40 m on the left-hand straight to chainage 140 m (point R) on the right-hand straight, at through chainages of 20 metres.

(c) Given that all of the points are to be set out from tangent point T_1, calculate the bearings and distances required to set out the pegs in their correct locations.

7. Obstructions

Very often it is important to set out all the points on the curve from the tangent point because of some obstruction. In Fig. 12.20, the third point E on the curve has been set out by the tangential angle $(\alpha_1 + \alpha_2 + \alpha_2) = $ say 5°.

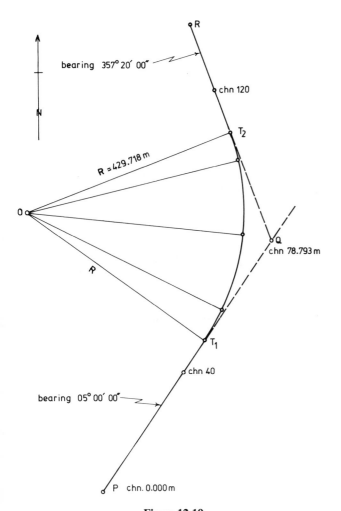

Figure 12.19

1. Point P—point of zero chainage
   Coordinates: 00.000 m east
                   00.000 m north
2. Radius of curve = 429.718 m
3. Deviation angle = $7°\ 40'\ 00''$
4. Chainage of intersection point = 78.793 m
5. Tangent length =
6. Chainage T_1 =
7. Curve length =
8. Chainage T_2 =
9. Peg chainages
   40.000 m on left-hand straight
        m tangent point T_1
        m on curve
        m on curve
        m on curve
        m tangent point T_2
   120.000 m on right-hand straight
   140.000 m on right hand straight
10. Curve composition
    Initial sub-chord =
    Two standard chords =
    Final sub-chord =
11. Deflection angles
    Initial sub-chord =
    Standard chord =
    Final sub-chord =
12. Centre angles (= 2 × deflection angles)
    T_1OA =
    AOB =
    BOC =
    COT_2 =

Table 12.6

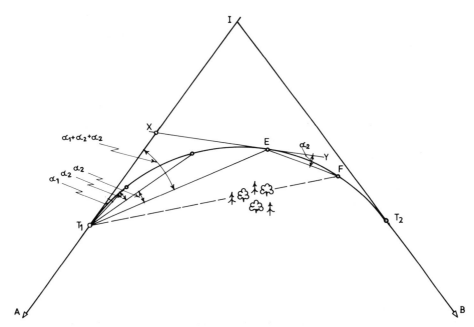

Figure 12.20

Owing to the trees, point F cannot be seen from T_1. In such a case the theodolite is removed to E and a sight taken back to the tangent point T_1 with the horizontal circle of the theodolite reading $180° - (\alpha_1 + \alpha_2 + \alpha_2) = 175°$ (say).

If a tangent is drawn to the circle at E, angle XT_1E equals angle T_1EX, i.e. angle $T_1EX = 5°$. The theodolite is then set to read $(175° + \text{angle } T_1EX) = (175° + 5°) = 180°$, in which case the line of sight is along tangent EX. The telescope is swung through a further 180° to point along the continuation of the tangent EY and the horizontal circle reads zero.

Point E is then treated as being a tangent point and F is set out by setting the circle to read α_2 degrees and measuring the standard chord length EF.

8. Vertical curves

(a) Summit and valley curves

Whenever roads or railways change gradient, a vertical curve is required to take traffic smoothly from one gradient to the other. When the two gradients form a hill, the curve is called a summit curve and when the gradients form a valley, a sag or valley curve is produced (Fig. 12.21).

(b) Percentage gradients

The gradients are expressed as percentages. A gradient of 1 in 50 is a 2 per cent gradient, i.e. the

gradient rises or falls by 2 units in 100 units. Similarly a gradient of 1 in 200 is a 0.5 per cent gradient.

In vertical curve calculations, the left-hand gradient is p per cent and the right-hand gradient is q per cent.

Example

10 A downhill gradient of 1 in 40 is to be connected to an uphill gradient of 1 in 18, by a vertical curve. Calculate the values of the gradients as percentages.

Answer

$$\text{Gradient 1 in 40 (uphill)} = +1/40$$
$$\text{Percentage grade} = (1/40) \times 100$$
$$= +2.5 \text{ per cent}$$
$$\text{Gradient 1 in 18 (downhill)} = 1/18$$
$$\text{Percentage grade} = (-1/18) \times 100$$
$$= -5.556 \text{ per cent}$$

Since the change of gradient from slope to curve is required to be smooth and gradual, parabolic curves are chosen. This form of curve is flat near the tangent point and calculations are reasonably simple. The form of the curve is $y = ax^2 + bx + c$, where

y = reduced level of any point on the curve

x = distance to that point measured from the start of the curve.

a = multiplying coefficient, which will be derived in Sec. 8(e).

b = value of the left-hand gradient

c = reduced level of the first point on the curve

(c) Properties of the parabola

1. The distance between the points T and T_1 as measured along (a) the curve TT_1, (b) the tangents TIT_1 and (c) the chord TT_1 are so close in length that they are considered equal (Fig. 12.22).

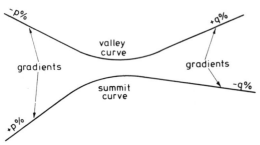

Figure 12.21

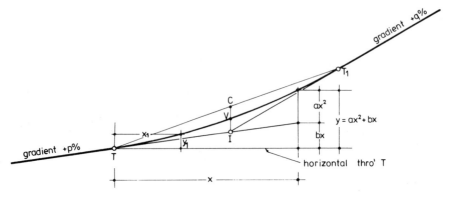

Figure 12.22

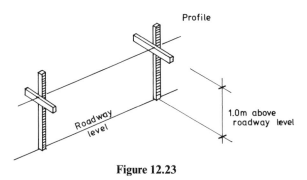

Figure 12.23

2. The intersection point I is treated as being midway between the points T and T_1; thus the lengths IT and IT_1 are equal. The curve is in fact often called an equal tangent parabola.

3. The height IV is called the correction in gradient and equals the height VC. In other words, the parabola bisects the length CI.

(d) Setting-out data

In setting out a vertical curve on the ground, the objective is to place large pegs at the required intervals along the line of the proposed roadway (Fig. 12.23) and to nail a cross-piece to each peg at a certain height (usually 1.0 m), above the proposed road level. These pegs are called profiles and the erection of these profiles is the standard method of setting out proposed levels on any construction site. A full explanation is given in Chapter 13.

The following information is required for any setting-out calculations:

1. The length of curve. The length of the curve is dependent upon: (a) the gradient of the straights; (b) the sight distance.

(a) Generally the steeper the approach gradients, the greater will be the centrifugal effect caused by the change of gradient from the slope to the curve. The curve length must be increased to reduce this effect when gradients are steep.

(b) The sight distance is the length required for a vehicle to stop from the moment a driver sees an obstruction over the brow of a summit curve. The sight distance includes thinking, braking, stopping, and safety margin distances.

The length of curve is taken from tables published by the Ministry of Transport.

2. The gradients of the slopes and the reduced level of one chainage point, preferably the intersection point.

(e) Calculation of data

In Fig. 12.24(a) a gradient of $+1$ per cent (i.e. 1 in 100) meets a gradient of $+4$ per cent (i.e. 1 in 25) at intersection point I, the chainage and reduced level of which are 500 and 261.30 m respectively. A 100 m long vertical curve is to be inserted between the straights.

The levels on the roadway curved surface at 25 metre intervals are required. These levels are called corrected grade elevations or curve levels.

Step 1 Calculate the reduced levels of the initial tangent point T, the final tangent point T_1 and the intersection point I. In Figure 12.24,

$$IT = IT_1 = 100/2 = 50 \text{ m}$$

Reduced level I = 261.30 m (given)

Reduced level T = reduced level I $-$ 1 per cent of 50
$$= 261.30 - (0.01 \times 50)$$
$$= 260.80 \text{ m}$$

Reduced level T_1 = Reduced level I $+$ 4 per cent of 50
$$= 261.30 + (0.04 \times 50)$$
$$= 263.30 \text{ m}$$

Step 2 Calculate the tangent levels, i.e. the levels that would obtain on the left-hand gradient if it were extended above or in this case below the right-hand gradient, towards the final tangent point. Mathematically any tangent level is

Tangent level = (reduced level T $+ bx$)
where b = left-hand gradient (p per cent)
x = distance from T

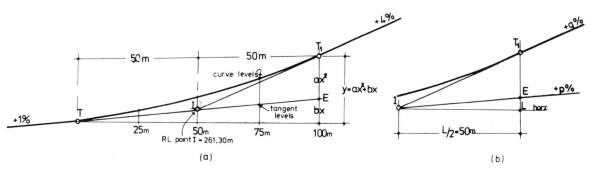

(a)

(b)

Figure 12.24

Tangent level at 25 m = reduced level T + bx
$$= 260.80 + 1 \text{ per cent of } 25 \text{ m}$$
$$= 261.05$$
Tangent level 50 m (I) = $260.80 + 1$ per cent of 50 m
$$= 261.30$$
Tangent level at 75 m = $260.80 + 1$ per cent of 75 m
$$= 261.55$$
Tangent level 100 m (E)
$$= 260.80 \text{ m}$$
$$+ 1 \text{ per cent of } 100 \text{ m}$$
$$= 261.80 \text{ m}$$

Step 3 Calculate the grade corrections at the required chainage points. The grade correction is the value of ax^2 which, when added to the various tangent levels in step 2, will give the level on the curve.

The term x is, of course, the distance of the chainage point from the initial tangent point. The value of a is unknown and has to be found in order to calculate these grade corrections.

In Fig. 12.24(b) the reduced levels of T_1 and E are 263.30 and 261.80 respectively; therefore reduced level T_1 − reduced level E = 263.30 − 261.80 = 1.50 m. This value of 1.50 m is really the grade correction at point E, i.e. the value that is applied to tangent level E to produce the curve level T_1. Therefore

Grade correction 1.50 m = ax^2 (where x = 100 m)
$$= a \times 100^2$$
Therefore $a = 1.50 \div 100^2$
$$= 1.50 \times 10^{-4}$$

Value of ax^2 at 25 m intervals from T:

Chainage	ax^2
25 m	$(1.50 \times 10^{-4}) \times 25^2 = 0.094$ m
50 m	$(1.50 \times 10^{-4}) \times 50^2 = 0.375$ m
75 m	$(1.50 \times 10^{-4}) \times 75^2 = 0.844$ m
100 m	$(1.50 \times 10^{-4}) \times 100^2 = 1.500$ m

The following is an alternative method of calculating the term a. In Fig. 12.24(b),

Reduced level T_1 = reduced level I + q per cent of $L/2$
$$= \text{RL I} + \frac{qL}{200}$$

Reduced level E = reduced level I + p per cent of $L/2$
$$= \text{RL I} + \frac{pL}{200}$$

Now ax^2 = reduced level T_1 − reduced level E
$$= \frac{qL}{200} - \frac{pL}{200}$$
$$= \frac{(q - p)L}{200}$$

Since $x = L$ at point T_1,
$$aL^2 = \frac{(q - p)L}{200}$$

Therefore $a = \dfrac{(q - p)}{200} \times \dfrac{L}{L^2}$
$$= \frac{q - p}{200L}$$

The formula applies in all situations where a is required. The proper sign convention for positive and negative gradients p or q must of course be used.

In this case, $q = +4$ per cent, $p = +1$ per cent and $L = 100$ m. Therefore

$$a = \frac{4 - 1}{200 \times 100} = \frac{3}{20\,000} = 1.5 \times 10^{-4}$$

Step 4 Calculate the curve level at the various chainage points. The curve level at any point is the algebraic addition of the tangent level (T + bx) and grade correction (ax^2):

i.e. curve level = tangent level + gradient correction

Therefore curve level 25 m = $261.05 + 0.094$
$$= 261.144 \text{ m}$$

Curve level 50 m = $261.30 + 0.375$
$$= 261.675 \text{ m}$$
$$75 \text{ m} = 261.55 + 0.844$$
$$= 262.394 \text{ m}$$
$$100 \text{ m} = 261.80 + 1.500$$
$$= 263.300 \text{ m}$$

In all examples the calculations are performed in tabular fashion as shown in Table 12.7.

Chainage (m)	Tangent level (m) T + bx	Grade correction (m) ax^2	Curve level (m) T + ax^2 + bx
0(T)	260.80	0	260.800
25	261.05	0.094	261.144
50	261.30	0.375	261.675
75	261.55	0.844	262.394
100(E)	261.80	1.500	263.300

Table 12.7

Example

11 A rising gradient of 1 in 40 is to be connected to a falling gradient of 1 in 75 by means of a vertical parabolic curve 400 m in length. The reduced level of the intersection point of the gradients is 26.850 m above Ordnance Datum. Calculate:
(a) the reduced levels of the tangent points,
(b) the reduced levels at 50 m intervals along the curve.

(SCOTEC-OND Building)

Solution

See Fig. 12.25 and Table 12.8.

$$IT = IT_1 = 400/2 = 200 \text{ m}$$

Reduced level I (given) = 26.850 m

$$\begin{aligned}
\text{Reduced level T} &= 26.850 - 200/40 \\
&= 21.850 \text{ m}
\end{aligned}$$

$$\begin{aligned}
\text{Reduced level T}_1 &= 26.850 - 200/75 \\
&= 24.183 \text{ m}
\end{aligned}$$

$$\begin{aligned}
\text{Reduced level E} &= 26.850 + 200/40 \\
&= 31.850 \text{ m}
\end{aligned}$$

$$\begin{aligned}
ET_1 = ax^2 \text{ at } 400 \text{ m} &= 24.183 - 31.850 \\
&= -7.667
\end{aligned}$$

$$\begin{aligned}
\text{Therefore } a &= -7.667/400^2 \\
&= -4.792 \times 10^{-5}
\end{aligned}$$

$$\text{or } a = \frac{q - p}{200L}$$

$$\begin{aligned}
p\% &= 1/40 \times 100 \\
&= +2.5 \text{ per cent}
\end{aligned}$$

$$\begin{aligned}
q\% &= -1/75 \times 100 \\
&= -1.333 \text{ per cent}
\end{aligned}$$

Therefore

$$a = \left(\frac{-1.333 - 2.5}{200 \times 400}\right)$$

$$= -4.792 \times 10^{-5},$$
as before

Chainage (m)	Tangent level (m) $T + bx$	Grade correction (m) ax^2	Curve level (m) $T + ax^2 + bx$
0(T)	21.850	0	21.850
50	23.100	−0.120	22.980
100	24.350	−0.479	23.871
150	25.600	−1.078	24.522
200	26.850	−1.917	24.933
250	28.100	−2.995	25.105
300	29.350	−4.313	25.037
350	30.600	−5.870	24.730
400(T$_1$)	31.850	−7.667	24.183

Table 12.8

Exercise 12.11

1 A downward gradient of 1 in 50 is to be connected to an upward gradient of 1 in 35 by means of a 60 metre long vertical parabolic curve. Given that the reduced level of the intersection point of the gradients is 26.000 m, calculate the reduced levels on the curve at 25 metre intervals.

(f) Highest (or lowest) point on any curve

The highest or lowest point on any curve is the turning point, i.e. the position where the gradient of the tangent is zero.

The gradient of the tangent is found by differentiating y with respect to x in the equation $y = ax^2 + bx$.

$$\frac{\mathrm{d}y}{\mathrm{d}x} = 2ax + b$$

$$\text{When } \mathrm{d}y/\mathrm{d}x = 0$$

$$x = -b/2a$$

In Example 11,

$$b = +2.5 \text{ per cent} = +2.5 \times 10^{-2}$$
$$a = -4.792 \times 10^{-5}$$

$$\begin{aligned}
\text{Therefore } x &= \frac{-2.5 \times 10^{-2}}{2 \times -4.792 \times 10^{-5}} \\
&= 10^3 \times 0.260\,85 \\
&= 260.851 \text{ m from T}
\end{aligned}$$

$$\begin{aligned}
\text{Level of highest point} &= 21.850 + ax^2 + bx \\
&= 21.850 - 3.261 + 6.521 \\
&= 25.110 \text{ m}
\end{aligned}$$

Exercise 12.12

1 Using the figures of Exercise 12.11, calculate the chainage and reduced level of the lowest point on the curve.

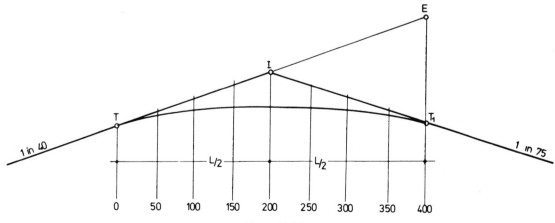

Figure 12.25

Exercise 12.13

1 A 40 m radius roadway kerb is to be set out at the junction of two roadways by the method of offsets from the long chord. Given that the length of the long chord is 30 m, calculate the lengths of the offsets from the chord at 5 m intervals.

2 Two straights AX and XB are to be connected by a 400 m radius circular curve. The bearings and lengths of the straights are as follows:

Straight	Whole circle bearing	Length (m)
AX	73° 10′	197.5
XB	81° 40′	—

Calculate:
(a) the intersection angle between the straights,
(b) the tangent lengths,
(c) the curve length,
(d) the chainages of the tangent points,
(e) the setting-out information to enable the curve to be set out at 20 m intervals of through chainage. (Present the information in the form of a setting out table.)

(ONC, Topographic Studies)

3 The following data refer to three straight sections of roadway:

Straight	Bearing	Length (m)
AB	N 75° E	610.00 m
BC	S 65° E	450.86 m
CD	N 45° E	343.10 m

The straights are to be connected by two curves of equal radius such that there is a straight portion of 100 m between them along the straight BC.
   Calculate:
(a) the radius of the curves,
(b) the chainages of the four tangent points.

4 Figure 12.26 shows two straights AB and CD which are to be joined by a curve of 330 m radius. The intersection point I is inaccessible and a traverse ABCD produced the results shown in the figure. Calculate:

Station	Northing	Easting
A	+75.38	+111.20
I	+154.60	+146.81
B	+128.70	+165.35

Table 12.9

(a) the tangent lengths,
(b) the chainages of the tangent points,
(c) the setting-out information to set out the curve at even chainages of 20 metres.

5 Two straight roadways AI and IB are to be joined by a circular curve of radius 40 metres. The total coordinates of A, I and B are shown in Table 12.9 Calculate:
(a) the distance from A and B to the tangent points of the curve,
(b) the *total* coordinates of each tangent point.

(OND, Building)

6 A vertical curve is to be used to connect a rising gradient of 1 in 60 with a falling gradient of 1 in 100 which intersect at a point having a reduced level 65.25 metres AOD. Given that the curve is to be 150 metres long calculate:
(a) the reduced levels of the tangent points,
(b) the levels at 30 metre intervals along the curve and the depths of cutting required,
(c) the distance from the tangent point on the 1 in 60 gradient to the highest point on the curve, and the reduced level of this point.

(OND, Building)

7 Two straights AB and BC having gradients of −1:40 and +1:50 respectively meet at point B at a level of 40.00 m AOD and chainage 1500.0 m in the direction AB. The gradients are to be joined by a vertical parabolic arc 240 m in length. Calculate the level and chainage of the lowest point on the curve.

(OND, Building)

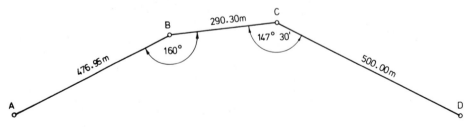

Figure 12.26

Figure 12.27

9. Answers

Exercise 12.1

1 (a) Figure 12.27

Point	Easting	Northing
A	0.000	0.000
B	859.230	151.505
$\Delta E_{AB} =$ 859.230	$\Delta N_{AB} =$ 151.505	
B	859.230	151.50
C	1423.046	−53.707
$\Delta E_{BC} =$ 563.816	$\Delta N_{BC} =$ −205.212	

$$\text{Length AB} = \sqrt{859.230^2 + 151.505^2}$$
$$= 872.485 \text{ m}$$

$$\text{Bearing AB} = \tan^{-1}(859.230/151.505)$$
$$= 80°\ 00'\ 00''$$

$$\text{Length BC} = \sqrt{563.816^2 + 205.212^2}$$
$$= 600.000 \text{ m}$$

$$\text{Bearing BC} = \tan^{-1}(563.816/-205.212)$$
$$= 110°\ 00'\ 00''$$

(b) Deviation angle θ = bearing BC − bearing AB
$$= 110°\ 00'\ 00'' - 80°\ 00'\ 00''$$
$$= 30°\ 00'\ 00''$$

(c) Tangent length BA = BC = $R \tan \theta$
$$= 300.000 \tan(30°\ 00'\ 00''/2)$$
$$= 80.385 \text{ m}$$

(d) Curve length = $2\pi R \times 30/360$
$$= 2\pi 300/12$$
$$= 157.080 \text{ m}$$

(e) Length AT_1 = AB − BT_1 = 872.485
$$-80.385$$
$$= \overline{792.100 \text{ m}}$$
Curve length T_1T_2 = + 157.080
Therefore chainage T_2 = $\overline{949.180 \text{ m}}$

Exercise 12.2

1 $T_1T_2 = 2R \sin \theta/2$
$$= 2R \sin (47°\ 09'\ 20'')/2$$
$$= 100 \sin (23°\ 34'\ 40'')$$
$$= 40.000 \text{ m}$$

Major offset CV = $R(1 - \cos \theta/2)$
$$= 50(1 - \cos 23°\ 34'\ 40'')$$
$$= 50(1 - 0.916\,518)$$
$$= 50 \times 0.083\,482$$
$$= 4.174 \text{ m}$$

Shortest distance VY = $R(\sec \theta/2 - 1)$
$$= 50 \times (1.090\,994 - 1)$$
$$= 4.550 \text{ m}$$

Tangent length YT_1 = YT_2 = $R \tan \theta/2$
$$= 50 \tan 23°\ 34'\ 40''$$
$$= 21.810 \text{ m}$$

Curve length = $2\pi R \times (47°\ 09'\ 20''/360°)$
$$= 100\pi \times 0.130\,988$$
$$= 41.151 \text{ m}$$

Exercise 12.3

1 Make a sketch from the given information (Fig. 12.28).

(a) Angle XPQ = 180° − QPA
$$= 28°\ 40'$$

Angle PQX = (180° − BQP)
$$= 36°\ 50'$$

Angle QXP = 180° − (28° 40' + 36° 50')
$$= 180° - 65°\ 30'$$
$$= 114°\ 30'$$

In triangle XQP,

$$XQ = \frac{\sin 28°\ 40' \times 55.0}{\sin 114°\ 30'}$$
$$= 28.995 \text{ m}$$

and $$XP = \frac{\sin 36°\ 50' \times 55.0}{\sin 114°\ 30'}$$
$$= 36.324 \text{ m}$$

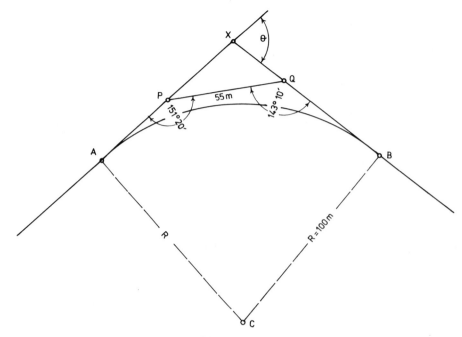

Figure 12.28

(b) Deviation angle $\theta = 180° - 114° 30'$
$\qquad\qquad = 65° 30'$
$\qquad$ Therefore $\theta/2 = 32° 45'$

$\qquad$ Tangent lengths XA and XB $= R \tan \dfrac{\theta}{2}$
$\qquad\qquad\qquad\qquad\qquad = 100 \tan 32° 45'$
$\qquad\qquad\qquad\qquad\qquad = 64.322$ m

(c) AP = AX − PX
$\qquad$ = 64.322 − 36.324
$\qquad$ = 27.998 m

$\qquad$ QB = BX − QX
$\qquad$ = 64.322 − 28.995
$\qquad$ = 35.327 m

(d) Curve $= 2\pi \times 100 \times (65.5/360)$
$\qquad\quad = 114.319$ m

Exercise 12.4

1 (a) Figure 12.29
$\qquad$ Bearing line AB = $37° 12'$
$\qquad$ Bearing line BC = $91° 02'$
$\qquad$ Therefore angle α = $53° 50'$
$\qquad\qquad$ and $\alpha/2$ = $26° 55'$

$\qquad$ Bearing line BC = $\quad91° 02'$
$\qquad$ Bearing line CD = $147° 14'$
$\qquad$ Therefore angle β = $\quad56° 12'$
$\qquad\qquad$ and $\beta/2$ = $\quad28° 06'$

$\qquad$ BY = BX = $R \tan \alpha/2 = R \tan 26° 55'$
$\qquad$ XC = $R \tan \beta/2 = R \tan 28° 06'$

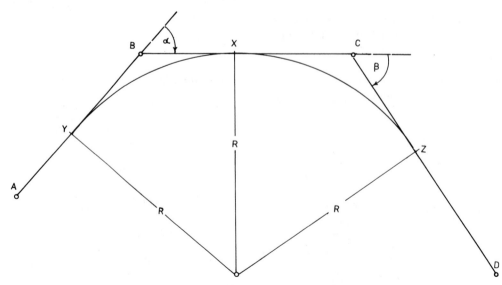

Figure 12.29

$$BC = (BX + XC)$$
$$= R \tan 26° 55' + R \tan 28° 06'$$
i.e. $312.7 = R (\tan 26° 55' + \tan 28° 06')$

Therefore $R = \dfrac{312.700}{\tan 26° 55' + \tan 28° 06'}$

$$= \dfrac{312.700}{0.507\,694\,8 + 0.533\,950\,3}$$

$$= 300.20 \text{ m}$$

(b) $BY = BX = R \tan 26° 55' = 152.41 \text{ m}$

The theodolite is set over peg B and made ready for sighting. A backsight is taken to station A and the instrument is locked along the line. The tape is laid out along this line and is kept in line by constant sights through the theodolite. When the required distance of 152.41 m is reached a peg is hammered securely into the ground and a nail is lined on to the peg at the required distance BY.

Exercise 12.5

1 Figure 12.30

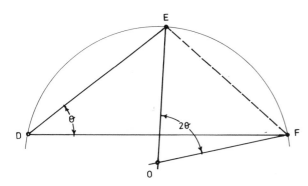

Figure 12.30

	Point	East (m)	North (m)
Existing coordinates	A	10.0	10.0
	B	15.0	12.5
	C	19.0	8.0
Proposed coordinates	D	9.0	10.0
	E	15.0	13.5
	F	20.0	8.0

Tan bearing $DE = \Delta$ easting/Δ northing
$$= (15 - 9)/(13.5 - 10)$$
$$= 6/3.5$$
Therefore bearing $DE = 59° 44' 37''$

Tan bearing $DF = \Delta$ easting/Δ northing
$$= (20 - 9)/(8 - 10)$$
$$= 11/-2$$
Therefore bearing $DF = 100° 18' 17''$

Distance $EF = \sqrt{\Delta \text{ east}^2 + \Delta \text{ north}^2}$
$$= \sqrt{(15 - 20)^2 + (13.5 - 8)^2}$$
$$= \sqrt{5^2 + 5.5^2}$$
$$= 7.43 \text{ m}$$

Angle $EDF(\theta) = 100° 18' 17'' - 59° 44' 37''$
$$= 40° 33' 40''$$

In Δ DEF,

$$\dfrac{EF}{\sin EDF} = 2R$$

Therefore $R = 7.43/(2 \sin 40° 33' 40'')$
$$= 5.71 \text{ m}$$

Angle $EOF = 2 \times$ angle EDF
$$= 2\theta$$
$$= 81° 07' 20''$$

Angles $(FEO + OFE) = 180° - 81° 07' 20''$
$$= 98° 52' 40''$$
Therefore angle $FEO = 49° 26' 20''$

Tan bearing $EF = \Delta$ easting/Δ northing
$$= (20 - 15)/(8 - 13.5)$$
$$= 5/-5.5$$
Therefore bearing $EF = 137° 43' 34''$

Bearing $EO =$ bearing $EF +$ angle FEO
$$= 137° 43' 34'' + 49° 26' 20''$$
$$= 187° 09' 54''$$

Easting $O =$ easting E
$$+ EO \sin 187° 09' 54''$$
$$= 15.0 + 5.71 \sin 187° 09' 54''$$
$$= 15.0 - 0.71$$
$$= 14.29 \text{ m}$$

Northing $O =$ northing E
$$+ EO \cos 187° 09' 54''$$
$$= 13.5$$
$$+ 5.71 \cos 187° 09' 54''$$
$$= 13.5 - 5.67$$
$$= 7.83 \text{ m}$$

Exercise 12.6

1 Tan length $IT_1 = IT_2 = R \tan \theta/2$
$$= 30 \tan 15°$$
$$= 8.038 \text{ m}$$

In Δ IET$_2$,

IE $= IT_1 \cos \theta$
$$= 8.038 \cos 30°$$
$$= 6.962 \text{ m}$$

$T_1E = IT_1 + IE$
$$= 8.038 + 6.962$$
$$= 15.000 \text{ m}$$

Offset at $2\,m = 30 - \sqrt{30^2 - 2^2}\ = 0.067\,m$
$4\,m = 30 - \sqrt{30^2 - 4^2}\ = 0.268\,m$
$6\,m = 30 - \sqrt{30^2 - 6^2}\ = 0.606\,m$
$8\,m = 30 - \sqrt{30^2 - 8^2}\ = 1.086\,m$
$10\,m = 30 - \sqrt{30^2 - 10^2} = 1.716\,m$
$12\,m = 30 - \sqrt{30^2 - 12^2} = 2.504\,m$
$14\,m = 30 - \sqrt{30^2 - 14^2} = 3.467\,m$
$15\,m = 30 - \sqrt{30^2 - 15^2} = 4.019\,m$

Exercise 12.7

1 Make a sketch from the given information (Fig. 12.31)

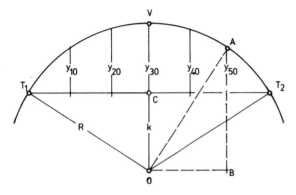

Figure 12.31

In $\triangle OTC$,

$$OT_1 = R = 100.0\,m$$
$$CT_1 = x = 30.0\,m$$

By Pythogoras' theorem

$$OC = k = \sqrt{R^2 - x^2}$$
$$= \sqrt{100^2 - 30^2}$$
$$= 95.394\,m$$

Major offset $VC = 100 - 94.394$
$$= 4.61\,m$$

In $\triangle ABO$,

$$OB = x_n = 20\,m$$
$$OA = R\ = 100\,m$$
$$AB = \sqrt{R^2 - x_n^2}$$

Offset y_{50} (and any other offset)

$$= \sqrt{R^2 - x_n^2} - k$$
$$= \sqrt{100^2 - 20^2} - 95.394$$
$$= 97.980 - 95.394$$
$$= 2.59\,m$$
Also $y_{10} = 2.59\,m$

Offset $y_{20} = y_{40} = \sqrt{R^2 - x_n^2} - k$
$$= \sqrt{100^2 - 10^2} - 95.394$$
$$= 99.499 - 95.394$$
$$= 4.11\,m$$

$y_0 = 0.00\,m$
$y_{10} = 2.59\,m$
$y_{20} = 4.11\,m$
$y_{30} = 4.61\,m$
$y_{40} = 4.11\,m$
$y_{50} = 2.59\,m$
$y_{60} = 0.00\,m$

Exercise 12.8

1 Length of curve $= 216.757 - 126.000$
$$= 90.757\,m$$

First peg on curve occurs at 140 m.
Therefore, initial sub-chord $= 140.000 - 126.000\,m$
$$= 14.000\,m$$

Last peg on curve occurs at 200 m.
Therefore number of standard chords $= 200 - 140$
$$= 3$$
Final sub-chord $= 216.757 - 200.000 = 16.757\,m$

From formula (4),

$$\text{First offset} = 14^2/800$$
$$= 0.245\,m$$
$$\text{Second offset} = 20(20 + 14)/800$$
$$= 0.850\,m$$
$$\text{Third, fourth offsets} = 20^2/400$$
$$= 1.000\,m$$
$$\text{Final offset} = 16.757(16.757 + 20.000)/800$$
$$= 0.770\,m$$

Exercise 12.9

1 Table 12.10

Point	Chainage	Tangential angle (left)	Theodolite reading (10")
T_1	790.20		
1	800.00	00° 33′ 41″	359° 26′ 20″
2	820.00	01° 42′ 27″	358° 17′ 30″
3	840.00	02° 51′ 13″	357° 08′ 50″
4	860.00	03° 59′ 59″	356° 00′ 00″
5	880.00	05° 08′ 45″	354° 51′ 10″
6(T_2)	894.92	06° 00′ 03″	354° 00′ 00″

Table 12.10

Exercise 12.10

1 (a) Table 12.11
(b) Table 12.12
(c) Table 12.13

1. Point P—point of zero chainage
   Coordinates: 00.000 m east
   00.000 m north
2. Radius of curve = 429.718 m
3. Deviation angle = 7° 40′ 00″
4. Chainage of intersection point = 78.793 m
5. Tangent length = 429.718 tan 3° 50′ 00″ = 28.793 m
6. Chainage T_1 = 78.793 − 28.793 = 50.000 m
7. Curve length = $2\pi R \times (7° 40′/360°)$ = 57.500 m
8. Chainage T_2 = 50.000 + 57.500 = 107.500 m

9. Peg chainages
   40.000 m on left-hand straight
   50.000 m tangent point T_1
   60.000 m on curve
   80.000 m on curve
   100.000 m on curve
   107.500 m tangent point T_2
   120.000 m on right-hand straight
   140.000 m on right hand straight

10. Curve composition
    Initial sub-chord = 60.0 − 50.0 m = 10.00 m
    Two standard chords = (20 × 2) m = 40.00 m
    Final sub-chord = 107.500 − 100.000 = 7.500 m

11. Deflection angles
    Initial sub-chord = $\sin^{-1}10/2R$ = 00° 40′ 00″
    Standard chord = $\sin^{-1}20/2R$ = 01° 20′ 00″
    Final sub-chord = $\sin^{-1}7.500/2R$ = 00° 30′ 00″

12. Centre angles (= 2 × deflection angles)
    T_1OA = 01° 20′ 00″
    AOB = 02° 40′ 00″
    BOC = 02° 40′ 00″
    COT_2 = 01° 00′ 00″
    Total = 07° 40′ 00″

Table 12.11

Line	Length	Whole circle bearing	Δ east	Δ north	Easting	Northing	Station
					0.000	0.000	P
P/40	40.00	5° 0′ 0″	3.486	39.848	3.486	39.848	40
40/T1	10.00	5° 0′ 0″	0.872	9.962	4.358	49.810	T1 (50)
T1/Q	28.79	5° 0′ 0″	2.509	28.683	6.867	78.493	Q
Q/T2	28.79	357° 20′ 0″	−1.340	28.762	5.528	107.255	T2 (107.50)
T2/120	12.50	357° 20′ 0″	−0.582	12.486	4.946	119.741	120
T20/140	20.00	357° 20′ 0″	−0.931	19.978	4.015	139.719	140 (R)
					0.000	0.000	P
P/T1	50.00	5° 0′ 0″	4.358	49.810	4.358	49.810	T1
T1/O	429.72	275° 0′ 0″	−428.083	37.452	−423.725	87.262	O
O/T2	429.72	87° 20′ 0″	429.253	19.993	5.528	107.255	T2
					−423.724	87.262	O
O/60	429.718	93° 40′ 0″	428.838	−27.481	5.114	59.781	60
					−423.724	87.262	O
O/80	429.718	91° 0′ 0″	429.653	−7.500	5.929	79.762	80
					−423.724	87.262	O
O/100	429.718	88° 20′ 0″	429.536	12.498	5.812	99.760	100
					−423.724	87.262	O
O/T2	429.718	87° 20′ 0″	429.253	19.993	5.529	107.255	T2

Table 12.12

Ref. Station	Easting ***	Northing ***
T1	4.358	49.810

Station	Easting	Northing	Difference in eastings	Difference in northings	Set-out distance	Set-out bearing
40	3.486	39.848	−0.872	−9.962	10.000	185° 00′ 00″
T1 (50)	4.358	49.810	0.000	0.000	0.000	– – –
60	5.114	59.781	0.756	9.971	10.000	4° 20′ 13″
80	5.929	79.762	1.571	29.952	29.993	3° 0′ 10″
100	5.812	99.760	1.454	49.950	49.971	1° 40′ 3″
T2 (107.50)	5.528	107.255	1.170	57.445	57.457	1° 10′ 1″
120	4.946	119.741	0.588	69.931	69.934	0° 28′ 55″
140	4.015	139.719	−0.343	89.909	89.910	359° 46′ 54″

Table 12.13

Exercise 12.11

1 Table 12.14

1. Left-hand grade-1 in (p)	−50.000	(p)% = −2.000
2. Right-hand grade 1 in (q)	35.000	(q)% = 2.857
3. Length of curve (1)	60.000	$a = 0.000\,404\,76$
4. Chainage initial tangent point	0.000	
5. Reduced level initial tangent point	26.600	

Chainage	Chainage interval	Tangent level	Grade correction	Curve level	Remarks
0.000		26.600		26.600	Start of curve
10.000	10.000	26.400	0.040	26.440	
20.000	10.000	26.200	0.162	26.362	
30.000	10.000	26.000	0.364	26.364	
40.000	10.000	25.800	0.648	26.448	
50.000	10.000	25.600	1.012	26.612	
60.000	10.000	25.400	1.457	26.857	End of curve

Table 12.14

Exercise 12.12

1 Chainage $x = \dfrac{-b}{2a}$

$$= \dfrac{+2.0 \times 10^{-2}}{2 \times 0.000\,404\,76}$$

$$= 24.710 \text{ m}$$

Level $y = ax^2 + bx + c$
$$= 0.000\,404\,76 \times 24.710^2 + ((-2/100)$$
$$\times 24.710) + 26.600$$
$$= 0.247 + (-0.494) + 26.600$$
$$= 26.353 \text{ m}$$

Exercise 12.13

1 (a) Offset at 5 m intervals = 1.65 m, 2.61 m,
    2.92 m, 2.61 m, 1.65 m

2 (a) 08° 30′   (b) 29.73 m   (c) 59.34 m
    (d) 167.77 m, 227.11 m

(e)

Chord number	Length	Chainage	Tangential angle
T_1		167.77	
1	12.23	180.00	00° 52′ 33″
2	20.00	200.00	02° 18′ 30″
3	20.00	220.00	03° 44′ 27″
4(T_2)	7.11	227.11	04° 15′ 00″

3 (a) 329.7 m
(b) 490.0 m, 720.2 m, 820.2 m, 1223.0 m

4 (a) 162.74
(b) 527.62, 830.00
(c) Angles, 1 at 1° 04′ 30″ initial sub-chord
              14 at 1° 44′ 10″
              1 at 0° 52′ 10″ final sub-chord

5 On AI tangent point is between A and I:

Distance A to TP = 17.29 m

On BI tangent point is outwith BI:

Distance B to TP = −37.71 m

Coordinates   TP1 118.29 mE,   91.15 mN
                TP2 197.94 mE,   109.73 mN

6 (a)   Reduced level left-hand TP = 64.000 m AOD
        Reduced level right-hand TP = 64.500 m AOD

(b)

Chainage	Curve level	Cut
0	64	0
30	64.42	0.08
60	64.68	0.32
90	64.78	0.32
120	64.72	0.08
150	64.50	0

(c) Highest point, chainage 93.757, reduced level 64.781 m in AOD

7 Chainage   1513.3 m
  Reduced level   41.34 m AOD

10. Project

Chapter 18 is a project covering all chapters of this textbook. It is intended that the project builds into a complete portfolio of the surveying work required in the survey and setting out of a building or engineering development.

If the reader wishes to continue work on the project or begin work at this stage, he or she should preferably study Chapter 13, Sections 4 and 5 (Setting out roadways), before attempting Sections 9 and 10 of the project on page 397.

Setting out construction works

Objective

After studying this chapter, the reader should be able to (a) make the necessary calculations to set out the horizontal and vertical positions of straight and/or curved construction works and (b) use a range of surveying instruments to set out the positions of buildings, roads, drains, sewers and associated works, on the ground.

1. Setting out—horizontal control

Introduction

Figure 6.1, here reproduced as Fig. 13.1, shows the development plan of the proposed construction works at the GCB Outdoor Centre site. The architectural development shows three houses and a private roadway serving them. Drainage is provided by a storm water drain and a foul sewer. The plan also shows the survey traverse stations A, B, C, D, E and F, which were used to survey the site (see Exercise 9.5). The coordinates of these stations are shown in Table 13.1.

The proposed positions of the roads, sewers and drains can be measured on the plan relative to the survey points using a scale rule and protractor, or more accurately by calculation from the coordinates. This information is taken into the field by the surveyor and these scaled or calculated dimensions are set out from the survey pegs to establish the ground positions of the proposed works.

On any construction site, it is general practice to construct firstly the roadways and sewers, in order to provide (a) access to the site and (b) main drainage to all buildings.

2. Setting out equipment

All roads, buildings, drains and sewers are set out using standard surveying equipment comprising:

1. *Steel tapes*
Steel tapes must always be used for setting out purposes as they are not subject to the same degree of stretching as are Fibron tapes. The accuracy of setting out work is largely dependent upon the condition of the tape and, or course, the expertise of the user.
2. *Levels*
Automatic levels are used on most sites. They compare favourably in price with optical levels and produce much more reliable results.
3. *Theodolites*
A wide variety of theodolites is now available but, as with levels, the more automation that can be provided, the greater will be the accuracy of the setting out. Thus, it is good practice to use theodolites with optical plumbing, automatic vertical circle indexing, electronic readout and electronic two axis levelling.
4. *Total stations*
Most setting out work, particularly roadways, can be readily and accurately accomplished by the method of coordinates. An EDM instrument is required to set out the distances and total stations can set out horizontal distances with ease.
5. *Autoplumb instruments*
These instruments are used to set out vertical lines in high-rise buildings. They are much more convenient to use for this purpose than theodolites. They save much time and greatly increase accuracy.
6. *Pegs*
Pegs are either wooden 50 mm × 50 mm × 500 mm stakes for use in soft ground or 25 mm × 25 mm × 300 mm angle irons for hard standing.

Pegs should be colour coded with paint and the code should be used throughout the duration of the contract to avoid confusion. Any code will suffice but

Station	Easting	Northing
A	00.00	00.00
B	−34.379	34.379
C	−51.503	13.013
D	−12.070	−15.102
E	5.348	−28.784
F	16.336	−14.428

Table 13.1

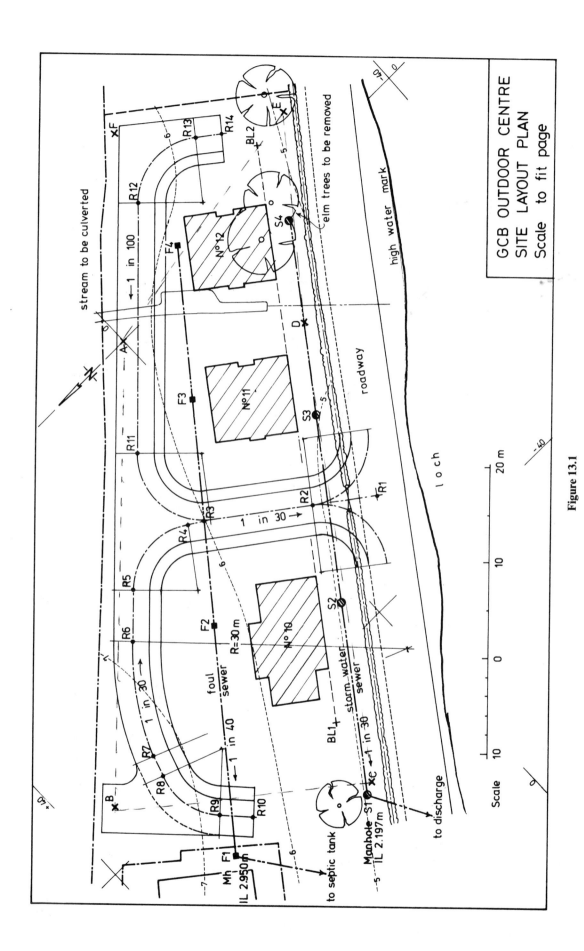

Figure 13.1

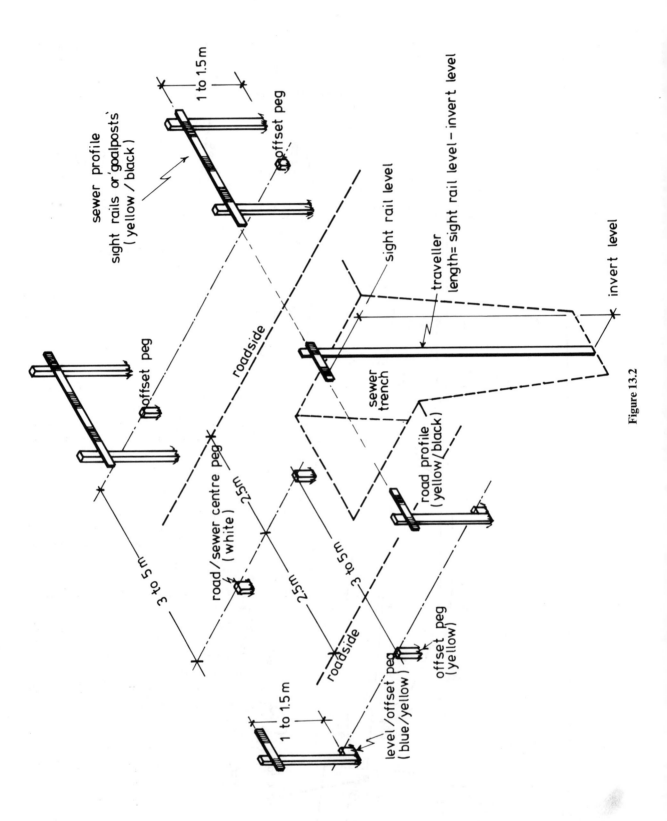

sewer profile
sight rails or 'goalposts'
(yellow/black)

offset peg

1 to 1.5 m

offset peg

roadside

sight rail level

traveller
length = sight rail level − invert level

invert level

sewer
trench

road profile
(yellow/black)

road/sewer centre peg
(white)

2·5m

2·5m

3 to 5 m

roadside

1 to 1.5 m

level/offset peg
(blue/yellow)

offset peg
(yellow)

3 to 5 m

Figure 13.2

in general centre line pegs are white (Fig. 13.2), offset pegs are yellow and level pegs are blue.

In setting out roads and sewers, the centre line pegs (white) are first established. Naturally they are removed during excavation, so offset pegs (yellow) are positioned 3–5 m to the right and left of centre lines (Fig. 13.2).

7. *Profiles*
A profile is a wooden stake to which a cross-piece, painted in contrasting coloured stripes, is nailed (Fig. 13.2). For sewer work, goalpost-type profiles, called sight rails, are preferable. The profiles are erected over the offset pegs in order to remain clear of the excavations.

8. *Travellers*
Travellers are really mobile profile boards used in conjunction with sight rails. The length of the traveller equals the sight rail level − sewer invert level. The length should be kept to multiples of 0.25 metres and travellers are usually about 2 metres long.

9. *Corner profiles*
During the construction of buildings, the pegs denoting the corners of the buildings are always removed during the construction work. The corner positions have, therefore, to be removed some distance back from the excavations on to corner profiles (see Fig. 13.7). These are constructed from stout wooden stakes 50 mm × 50 mm, on to which wooden boards 250 mm × 25 mm × 1.00 m long are securely nailed.

3. Setting out a peg on a specified distance and bearing

(a) Setting out on level ground

In order to set out the roads, buildings and sewers shown on the development plan (Fig. 13.1), a total of some forty to fifty pegs need to be accurately placed on the ground in their proposed positions. Fortunately, every peg is set out in exactly the same manner. It is not an easy task to physically set a peg in its exact proposed location and, in order to do so, the following sequence of operations is required.

In Fig. 13.3(a), a peg C is to be set out from a survey line AB. A surveyor and two assistants are usually required to complete the task.

Procedure
1. The theodolite is set over station B and correctly levelled and centred. On face left, a back sight is taken to station A with the theodolite reading zero degrees (the method varies with the type of theodolite).
2. The horizontal circle is set to read 65° 30′; thus the theodolite is pointing along the line BC.
3. The end of the tape is held against the nail in peg B and laid out approximately along the line BC by the assistants.

4. The 10.25 m reading on the tape is held against the *SIDE* of the proposed peg C (Fig. 13.3(b)), the tape is tightened and slowly swing in an arc, until the surveyor sees it clearly through the telescope of the theodolite.
5. The peg is carefully moved, on the observer's instructions, until the *bottom, front* edge of the peg is accurately bisected. The peg is then hammered home.
6. The tape is again held at peg B, by assistant 1, while assistant 2 tightens it and marks a pencil line across the peg C at distance 10.25 m.
7. A pencil is held vertically on this line by an assistant and is moved slowly along the line until the surveyor sees it bisected by the line of sight through the theodolite. The assistant marks this point on the peg.
8. The distance of 10.25 m is checked and the operation is repeated on face right. If all is well, the two positions of point C should coincide or differ by a very few millimetres. The mean is accepted and a nail hammered into the peg to denote point C.

It is not good practice to hook the end of the tape over the nail at peg B when setting out the distance, as excessive tension on the tape will move the nail head or even move the peg.

(b) Setting out on sloping ground

In all setting out operations, the horizontal distance is required, but frequently, because of ground undulations, it will be necessary to set out the slope distance. The setting-out procedure is very similar to that of Sec. 3(a).

1. The theodolite is set up at station B, backsighted to A reading zero and foresighted along the line BC reading 65° 30′. The instrument height is measured and noted. Let the height be 1.35 m.
2. The tape is stretched out in the proposed direction of line BC and a levelling staff held vertically at distance 10.25 m.
3. The instrument height (1.35 m) is read on the staff and the angle of inclination (vertical or zenith angle) is noted. Let the zenith angle be 84° 15′.
4. The distance to be set out on the slope will have to be increased to be the equivalent of 10.25 m of plan length, as follows (Fig. 13.3(c)):

Plan length P/slope length S
$$= \text{sin zenith angle}$$
Slope length S = plan length P/sin zenith angle
$$S = 10.250/\sin 84° \, 15′$$
$$= 10.302 \text{ m}$$

5. The procedure in setting out peg C then follows exactly the procedure detailed in Sec. 3(a), using the new setting out length of 10.302 m.

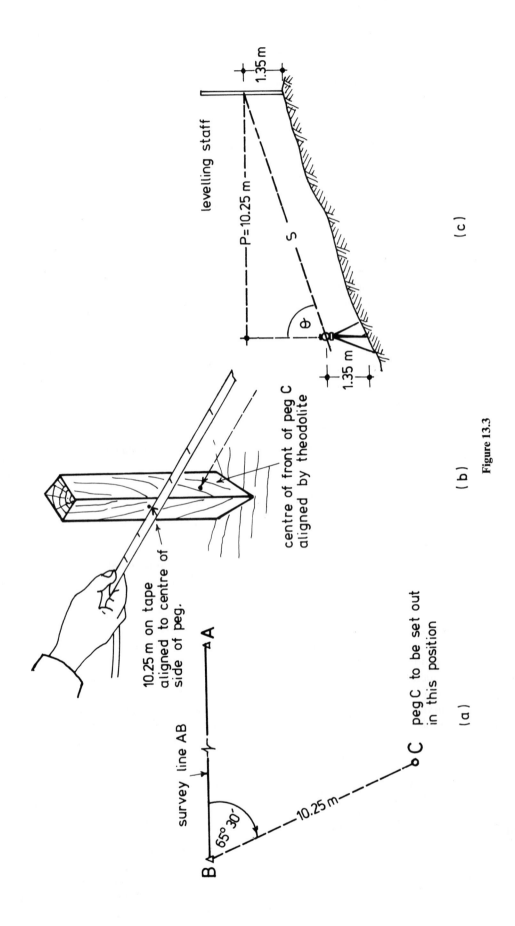

levelling staff

P= 10.25 m

1.35 m

1.35 m

S

θ

(c)

centre of front of peg C
aligned by theodolite

10.25 m on tape
aligned to centre of
side of peg.

survey line AB

A

65° 30'

10.25 m

B

C peg C to be set out
in this position

(a)

(b)

Figure 13.3

4. Setting out roadways—office work

Step 1 As stated in the introduction, the first step in setting out a roadway is the establishment of the centre line. In order to achieve this, the coordinates of all changes of direction, i.e. the intersection points of the roadway curves, must be scaled as accurately as possible from the development plan. On a 1:200 scale, coordinates can be measured with an accuracy of 50 mm. Figure 13.4 is a skeletal layout plan of the two roads R1–R3–R10 and R3–R14 taken from Fig. 13.1, drawn to a large scale and reduced to fit the page size. The coordinates of (a) the starting point R1(I1), (b) the finishing point R10 and (c) the intersection points I2, I3 and I4 of the western roadway, scaled from the plan, are shown in Table 13.2.

Point	Easting (m)	Northing (m)
R1(I1)	− 29.700	− 8.300
I2	− 14.600	+ 13.200
I3	− 27.400	+ 26.000
I4	− 39.800	+ 30.700
R10	− 45.600	+ 25.100

Table 13.2

Step 2 The bearings and distances of the roadway centre lines are next calculated from the coordinates.

Example

1 Using the figures of Table 13.2, calculate the bearing and distance between the points R1(I1) and I2 and between the points I2 and I3.

Answer

Point	Easting (m)	Northing (m)
R1(I1)	− 29.700	− 8.300
I2	− 14.600	+ 13.200

$$\Delta E(\text{RI–I2}) = + 15.100$$
$$\Delta N(\text{RI–I2}) = + 21.500$$

$$\text{Bearing R1–I2} = \tan^{-1}(\Delta \text{ east}/\Delta \text{ north})$$
$$= + 15.100/ + 21.500$$
$$= 35° 04' 53''$$
$$\text{Distance R1–I2} = \sqrt{\Delta \text{ east}^2/\text{north}^2}$$
$$= \sqrt{15.100^2 + 21.500^2}$$
$$= 26.273 \text{ m}$$
$$\text{Bearing I2–I3} = \tan^{-1}(\Delta \text{ east}/\Delta \text{ north})$$
$$\text{Length I2–I3} = \sqrt{\Delta \text{ east}^2/\Delta \text{ north}^2}$$
$$\text{I2–I3} = 315° 00' 00'': 18.102 \text{ m}$$

Exercise 13.1

1 Using the figures of Table 13.2, calculate the bearings and lengths of the lines I3–I4 and I4–R10.

Step 3 The various intersection points next have to be set out on the ground from some nearby survey station and, once set out, the last point must be checked on to some other station to prove the accuracy of the setting out.

In Fig. 13.4, the roadway starting point R1(I1) is to be set out from survey station D and the final roadway point R10 is to be checked on to survey station B. The bearings and lengths from station D to R1 and from R10 to station B are therefore required.

Example

2 The coordinates of survey stations D and B (taken from Table 13.1) and of roadway points R1 and R10 (taken from Table 13.2) are tabulated as in Table 13.3. Calculate the bearing and distance from survey station D to roadway point R1.

Point	Easting (m)	Northing (m)
D	− 12.070	− 15.102
R1	− 29.700	− 8.300
B	− 34.379	+ 34.379
R10	− 45.600	+ 25.100

Table 13.3

Answer
$$\text{Bearing} = \tan^{-1}(\Delta \text{ east}/\Delta \text{ north})$$
$$\text{Distance} = \sqrt{\Delta \text{ east}^2/\Delta \text{ north}^2}$$

Line D–R1
$$\Delta E = − 17.630$$
$$\Delta N = + 6.802$$

Therefore
$$\text{bearing D} \rightarrow \text{R1} = \tan^{-1}(−17.630/6.802)$$
$$= 291° 05' 51''$$
$$\text{Distance D} \rightarrow \text{R1} = \sqrt{17.630^2 + 6.802^2}$$
$$= 18.897 \text{ m}$$

Exercise 13.2

1 The coordinates of the final roadway point R10 and of the survey station B are shown in Table 13.3. Calculate the bearing and distance between the points, which will enable the setting out of the roadway to be checked.

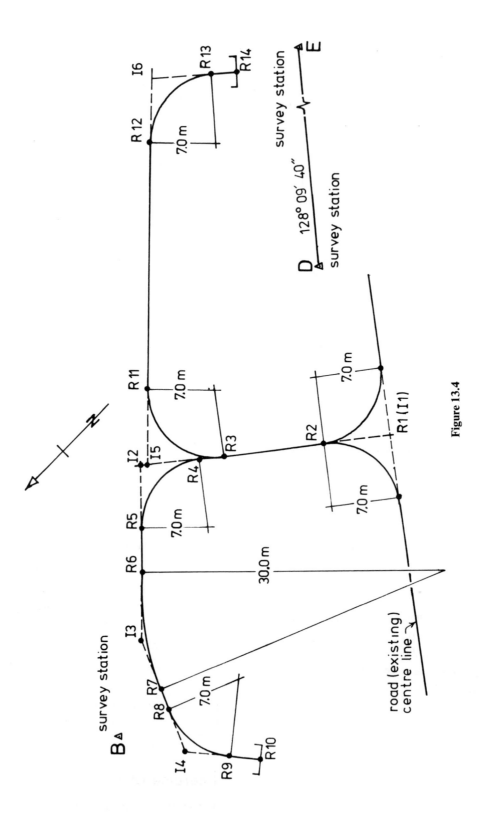

Figure 13.4

Station	Easting	Northing	Line	Length	Whole circle bearing
D	−12.070	−15.102	E–D		308° 9′ 40″
R1(I1)	−29.700	−8.300	D–R1	18.897	291° 5′ 51″
I2	−14.600	13.200	R1–I2	26.273	35° 4′ 53″
I3	−27.400	26.000	I2–I3	18.102	315° 0′ 0″
I4	−39.800	30.700	I3–I4	13.261	290° 45′ 29″
R10	−45.600	25.100	I4–R10	8.062	226° 0′ 18″
B	−34.379	34.379	R10–B	14.560	50° 24′ 42″

Table 13.4

Setting-out table (a)

Finally, when all bearings and distances have been computed a setting-out table showing all data from station D through the various roadway points to station B is compiled as in Table 13.4.

Step 4 The roadway may be set out perfectly adequately from the information shown in Table 13.4 but many surveyors prefer to set the theodolite to zero at each set-up point. Hence the clockwise angle between the two relevant lines at each station must be calculated by subtracting the bearing to the backsight point (the back bearing) from the bearing to the foresight point (the forward bearing).

Examples

3 In Fig. 13.4, the bearing from survey station D to survey station E is 128° 09′ 40″ and the bearing from station D to the first point R1 of the roadway is 291° 05′ 51″. Calculate the clockwise angle E–D–R1 to be set out at station D in order to position a peg at R1.

Answer

Back bearing D–E = 128° 09′ 40″
Forward bearing D–R1 = 291° 05′ 51″
Therefore angle E–D–R1 = 162° 56′ 11″

4 Similarly, given that the bearing from survey point D to roadway point I1(R1) is 291° 05′ 51″ and the bearing between roadway points I1 and I2 is 35° 04′ 53″ (from Table 13.4), calculate the angle D–I1–I2, to be set out at station I1, in order to place a peg at I2.

Answer
Back bearing I1–D = 291° 05′ 51″ − 180° 00′ 00″
= 111° 05′ 51″
Forward bearing I1–I2
= 35° 04′ 53″
Angle D–I1–I2 = 35° 04′ 53″ − 111° 05′ 51″
= 395° 04′ 53″ − 111° 05′ 51″
= 283° 59′ 02″

This is a difficult example but is typical of the work required in setting out; therefore much care is required in the calculations.

Exercise 13.3

1 In Table 13.4, the bearings of the centre lines of the roadway R1–R10 are as follows:

Line	Bearing
(R1)I1–I2	35° 04′ 53″
I2–I3	315° 00′ 00″
I3–I4	290° 45′ 29″
I4–R10	226° 0′ 18″

Calculate the values of the angles I1–I2–I3, I2–I3–I4 and I3–I4–R10.

Setting-out table (b)

A complete setting-out table showing all data from station D through the various roadway points to station B is compiled and distributed to the engineers and contracts managers and preparations are made to set the pegs physically into their ground locations. The setting out table is shown in Table 13.5.

Line	Length (m)	Forward whole circle bearing	Angle	Clockwise value
E–D		308° 9′ 40″		
D–R1	18.897	291° 5′ 51″	E–D–R1	162° 56′ 11″
R1–I2	26.273	35° 4′ 53″	D–R1–I2	283° 59′ 02″
I2–I3	18.102	315° 0′ 00″	R1–I2–I3	99° 55′ 07″
I3–I4	13.261	290° 45′ 29″	I2–I3–I4	155° 45′ 29″
I4–R10	8.062	226° 0′ 18″	I3–I4–R10	115° 14′ 48″
R10–B	14.560	50° 24′ 42″	I4–R10–B	04° 24′ 24″

Table 13.5

5. Setting out roadways—fieldwork

(a) Setting out the centre line

One surveyor and two assistants are required to set out the centre line of a roadway. Using the setting-out table, Table 13.5, the roadway shown in Fig. 13.4 is pegged out in the following manner:

1. The theodolite is set at station D and a backsight is taken to station E, with the theodolite reading zero degrees. (The method of setting to zero degrees varies with the type of theodolite.)
2. The horizontal circle is set to read $162° 56' 11''$, or as closely to that reading as is possible with the type of theodolite employed.
3. The assistants set out the horizontal distance D to R1 (18.897 m) and establish a peg at R1. The mechanics of actually setting out the peg are described in detail in Sec. 3(a). If the horizontal distance cannot be set out, the equivalent slope distance must be determined and the peg established in the manner described in Sec. 3(b).
4. The operation is repeated on face right and, assuming that no errors have occurred, the mean position of the peg R1 is established as the start of the roadway.
5. The theodolite is removed to station R1 and a zero backsight taken to the newly vacated station D. Angle D–R1–I2 ($283° 59' 02''$) is set on the theodolite and peg I2 is established at horizontal distance 26.273 m.
6. This procedure is repeated at points I2, I3 and I4, thus establishing the centre line from R1 to R10.
7. At peg R10, the horizontal angle and distance to survey peg B are measured and compared with the previously calculated values of $04° 24' 24''$ and 14.560 m, to check the accuracy of the setting out. Should there be an unacceptable discrepancy, the complete set of calculations and setting-out work must be repeated, until the error is discovered and eliminated.

(b) Setting out the tangent points of curves

1. Assuming that all intersection points have been correctly established and the work has been proven by checking on to survey point B, the next task is to set out the tangent points at the beginning and end of the four proposed curved sections of roadway. The curve elements, namely the tangent lengths, curve lengths and chainages of each curve, must first be computed.

Example

5 Figure 13.5 shows the skeletal layout of part of the proposed roadway scheme at the GCB Outdoor Centre site. Point R1 is the point of zero chainage and all curves have radii of 7.00 m except curve 3, which has a radius of 30.00 m.

The various lengths of the straights and angles have been extracted from Table 13.5. Calculate the chainages of points R2, R4 and R5.

Answer
Curve 1
Deviation angle $\theta_1 = 90° 00' 00''$
Radius of curve $= 7.000$ m
Therefore tangent length R1–R2
$= 7.000$ m
and chainage R2 $= 0.000 + 7.000$
$= 7.000$ m

Curve 2
Deviation angle $\theta_2 = 80° 04' 53''$
Radius of curve $= 7.000$ m
Tangent length $= R \tan \theta_2/2$
$= 5.881$ m
Therefore chainage R4 $=$ chainage R2
$+ (26.273 - 7.000$
$- 5.881)$
$= 7.000 + 13.392$
$= 20.392$ m
Curve length $= 2\pi R \times (80.0692°/360°)$
$= 9.782$ m
Therefore chainage R5 $= 20.392 + 9.782$
$= 30.174$ m

Exercise 13.4

1 Using the data shown in Fig. 13.5 and given that roadway tangent point R5 has a chainage of 30.174 m, calculate the chainages of roadway centre line points R6 to R10.

2. Each tangent point is established in the field by measuring either forward from or backward from the associated intersection point. Thus tangent point R4 is established by (a) setting a theodolite at intersection point I2 and sighting back to intersection point I1 and (b) measuring distance I2 to R4 = 5.881 metres back from I2 and aligning a peg on the line I2–I1 using the theodolite.

Similarly, the tangent point R5 is established by (a) aligning the theodolite, already set at intersection point I2, on to intersection point I3 and (b) measuring distance I2 to R5 = 5.881 metres forward from I2 and establishing peg R5 using the theodolite to align the peg.

Tangent points R2, R7, R8 and R9 are similarly positioned from their respective intersection points.

(c) Setting out the curve

1. The method of setting out the pegs on the curve really depends upon (a) the radius and length of the curve (b) whether any obstructions such as buildings, trees, etc. lie in the vicinity of the curve.

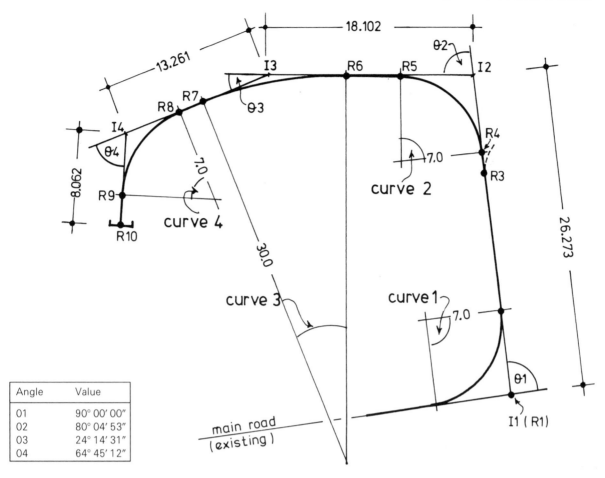

Angle	Value
01	90° 00′ 00″
02	80° 04′ 53″
03	24° 14′ 31″
04	64° 45′ 12″

Figure 13.5

Assuming that the area is clear of obstructions, the 7.0 m radius curves are set out by one of four methods all of which are fully detailed in Chapter 12, sec. 5(a).

These methods are (a) finding the centre of the curve and from that point, swinging an arc of 7.0 m radius; (b) measuring offsets from the straight tangents; (c) measuring offsets from the long chord; (d) setting out offsets from chords.

Example

6 In Fig. 13.5 curve 2 (points R4 to R5) is to be set out by offsets from the long chord. Calculate (to the nearest centimetre):
(a) the length of the long chord given that the radius of the curve = 7.00 m and the deviation angle = 80° 05′,
(b) the lengths of the offsets at 1.50 metre intervals along the chord.

Answer
(a) Long chord = $2R \sin \theta_2/2$
$$= 2 \times 7.00 \times \sin 40° 02′ 30″$$
$$= 9.00 \text{ m}$$

(b) At 4.5 m, i.e. the centre of the long chord,

$$\text{Major offset} = R - \sqrt{R^2 - x^2}$$
$$= 7 - \sqrt{7^2 - 4.5^2}$$
$$= 7 - 5.36$$
$$= 1.64$$
Therefore $k = 5.36$ m

$$\text{Offset } Y_{1.5} = Y_{7.5} = \sqrt{R^2 - x^2} - k$$
$$= \sqrt{7^2 - 3^2} - 5.36$$
$$= 6.32 - 5.36$$
$$= 0.96 \text{ m}$$

$$\text{Offset } Y_{3.0} = Y_{6.0} = \sqrt{R^2 - x^2} - k$$
$$= \sqrt{7^2 - 1.5^2} - 5.36$$
$$= 6.84 - 5.36$$
$$= 1.48 \text{ m}$$

2. The long chord points are set out using a steel tape at 1.5 m intervals from tangent point R4 and the calculated offsets, measured to the right at right angles, are set off from the long chord points using either tape geometry or a prism square.

Exercise 13.5

1 In Fig. 13.5, curve 4 is to be set out, using the method of offsets from the long chord R8–R9. The deviation angle $\theta_4 = 64°\ 45'$ and the radius of the curve is 7.00 m. Calculate (to the nearest centimetre):
(a) the length of the long chord R8–R9,
(b) the lengths of the offsets at 1.50 m intervals along the chord.

3. Curve 3 which has a radius of 30.0 metres is still a relatively small radius curve and could be set out by any of the four methods mentioned above. However, it would probably be set out by the method of tangential angles using a theodolite. As the curve length is short, the rule regarding the length of chords, namely chord length $= (R/20)$ metres, could be relaxed, since no appreciable error would be incurred over such a short length.

Example

7 Curve 3 in Fig. 13.5 is to be set out from tangent point R6 to tangent point R7 using the method of tangential angles. The deviation angle $\theta_3 = 24°\ 14'\ 30''$, the length of the radius $= 30$ m and the chord length $= (R/15)$ metres. Calculate:
(a) the length of the curve R6–R7,
(b) the curve composition when chord length $= R/15$ m,
(c) the chord deflection angles,
(d) the tangential angles required to set out the pegs around the curve.

Answer
(a) Curve length $= 2\pi R \times (24°\ 14'\ 30''/360°)$
 $= 12.693$ m

(b) Curve composition (chord $= R/15$ m)
 Standard chord $= 30/15 = 2.0$ m

Therefore there are 6 standard chords and one final sub-chord of 0.693 m.

(c), (d) Table 13.6

---

4. The curve is pegged out on the ground as follows:
(a) The theodolite is set over peg R6, backsighted to intersection point I_3 and set to read zero degrees.
(b) The upper plate is released and the circle set to the value of the first tangential angle, namely 358° 05' 23".
(c) The end of a tape is held at peg R6, the tape aligned with the theodolite and a peg 1 driven into the ground at distance 2.000 metres.
(d) The theodolite is set to read the value of the second tangential angle: 356° 10' 46".
(e) The tape end is held at peg 1 and the tape reading of 2.00 m is swung slowly in an arc, until the 2.00 m reading is intersected by the line of sight of the theodolite. Peg 2 is driven in at this point.
(f) The theodolite is set to read the third tangential angle and peg 3 on the curve is established at the point where the 2.00 m reading from peg 2 is intersected by the line of sight of the theodolite.
(g) This procedure is repeated until the roadway tangent point R7 is reached.

(d) Offset pegs

Once the centre line of the roadway, including curves, has been pegged out on the ground, offset pegs, usually coloured yellow, are set at right angles to the left and right of the centre line, using a prism square or site square, at distances such that they will not be disturbed by future excavation work.

6. Setting out small buildings

Dwelling houses are still largely traditionally built and small inaccuracies in the setting out can usually be tolerated. Large factory buildings, multi-storey buildings, schools, etc., are nowadays largely prefabricated and little, if any, inaccuracy can be tolerated in the setting out. Consequently the methods of setting out vary considerably.

The exact position that the building is to occupy on the ground is governed by the building line as defined by the Local Authority. In Figs 13.1 and 13.6,

Chord number	Length	Chainage (m) (from R6)	Deflection angle	Tangential angle	Theodolite reading
		0.000 (R6)			360° 00' 00"
1	2.00	2.000	01° 54' 37"	01° 54' 37"	358° 05' 23"
2	2.00	4.000	01° 54' 37"	03° 49' 14"	356° 10' 46"
3	2.00	6.000	01° 54' 37"	05° 43' 50"	354° 16' 10"
4	2.00	8.000	01° 54' 37"	07° 38' 27"	352° 21' 33"
5	2.00	10.000	01° 54' 37"	09° 33' 04"	350° 26' 56"
6	2.00	12.000	01° 54' 37"	11° 27' 41"	348° 32' 19"
7	0.693	12.693(R7)	00° 39' 42"	12° 07' 23"	347° 52' 37"

Table 13.6

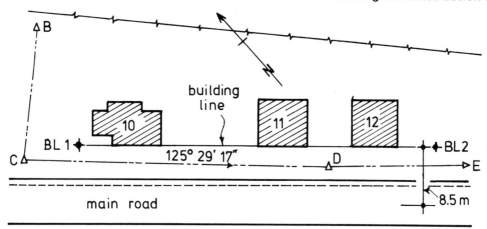

Figure 13.6

showing the development at the GCB Outdoor Centre, the building line is parallel to the main shore road at a distance of 8.5 metres from the centre line. The frontages of all three proposed buildings lie on this line. The building line must therefore be established first of all.

(a) Setting out the building line—office work

The building line may be set out by measuring two 8.5 m offsets to the north of the roadway centre line or by scaling the coordinates of two points on the line and setting them out from an existing survey station by theodolite.

Example

8 Figure 13.6, extracted from the development plan of Fig. 13.1, shows the positions of two points BL1 and BL2 on the building line and two traverse stations C and D. The coordinates of the points C, BL1 and BL2 are as follows:

Point	Easting	Northing
C	−51.503	+13.013
BL1	−45.000	+11.600
BL2	+5.000	−24.500

The bearing of line CD from traverse data is $125° 29' 17''$. Calculate the bearing and distance from:
(a) survey point C to building line point BL1,
(b) point BL1 to point BL2;
hence calculate:
(c) the angle required at survey point C to set out point BL1,
(d) the angle required at point BL1 to set out BL2.

Answer
(a) $\Delta E(\text{C–BL1}) = 6.503$ m
$\Delta N(\text{C–BL1}) = -1.413$
Tan bearing$(\text{C} \to \text{BL1}) = 6.503/-1.413$
$= 102° 15' 32''$
Distance$(\text{C–BL1}) = \sqrt{6.503^2 + 1.413^2}$
$= 6.655$ m

(b) $\Delta E(\text{BL1–BL2}) = 50.000$ m
$\Delta N(\text{BL1–BL2}) = -36.100$
Tan bearing$(\text{BL1} \to \text{BL2})$
$= 50.000/-36.100$
$= 125° 49' 45''$
Distance$(\text{BL1–BL2}) = \sqrt{50.000^2 + 36.100^2}$
$= 61.670$ m

(c) Back bearing$(\text{C} \to \text{D}) = 125° 29' 17''$
Forward bearing$(\text{C} \to \text{BL1}) = 102° 15' 32''$
Angle$(\text{D–C–BL1}) = 102° 15' 32''$
$- 125° 29' 17''$
$= 336° 46' 15''$

(d) Back bearing$(\text{BL1} \to \text{C}) = 282° 15' 32''$
Forward bearing$(\text{BL1} \to \text{BL2})$
$= 125° 49' 45''$
Therefore angle(C–BL1–BL2)
$= 125° 49' 45''$
$- 282° 15' 32''$
$= 203° 34' 13''$

(b) Setting out the building line—fieldwork

1. Set the theodolite over peg C and take a backsight reading to D with the horizontal circle set to zero degrees (face left).
2. Set the horizontal circle to read $336° 46' 14''$ and set out and peg BL1 at distance 6.655 m.
3. Repeat the operation on face right as a check.

4. Transfer the theodolite to BL1 and take a back sight reading to C, with the horizontal circle set to zero degrees (face left).

5. Set the theodolite to read 203° 34′ 13″ and set out peg BL2 at distance 61.670 m.

6. Repeat the operation on face right as a check.

(c) Setting out the building—fieldwork

In Figs 13.1 and 13.6, the three buildings fronting Shore Road have different shapes but, when setting out, each building is reduced to a basic rectangle, enabling checks to be easily applied.

Figure 13.7(a) shows the positions of house 10 Shore Road and the relevant building line. The building may be set out from the building line using either (a) a steel tape or (b) some form of surveying instrument, usually a site square or optical square.

Procedure

1. Using a scale rule, measure on the plan the distance between the building line starting point BL1 and the corner A of the house. The scaled dimension is 2.50 m.

2. Determine, from the plan, the dimensions of a basic rectangle to enclose the house. The scaled dimensions are 13.0 m by 8.0 m.

3. Using a steel tape, set out the distance 2.50 m along the building line from point BL1, to establish corner A of the house. Mark the point A by a nail driven into a wooden peg.

4. Measure the distance AB (13.0 m) along the building line and establish a peg B. Mark the point by a nail.

5. Using a basic 3:4:5 right angle, measure the lengths AD and BC (8.0 m) and establish pegs at C and D.

6. Check the lengths of the diagonals AC and BD (15.264 m). Both measurements should be equal, thus proving that the building is square.

Although the method of setting out a right angle using a 3:4:5 triangle is theoretically sound, in practice it tends to lead to inaccuracies in positioning. By calculating the length of the diagonal of the rectangle and using two tapes, the setting out can be accomplished much more accurately and speedily as follows:

1. As before, measure the length AB and mark the positions A and B by nails driven into the wooden stakes.

2. Calculate the diagonal size of the rectangle using the theorem of Pythagoras:

$$AC = \sqrt{13.0^2 + 8.0^2} = 15.264 \text{ m}$$

3. Hold the zero of tape 1 against point A; hold the zero of tape 2 against point B and stretch them out in the direction of the point C.

4. At the intersection of 15.264 m of the first tape and 8.000 m of the second tape, mark point C on a peg.

5. Repeat for point D by measuring AD = 8.00 m with tape 1 and BD = 15.264 m with tape 2.

6. Check that DC = 13.000 m.

7. In order to establish peg G, measure a distance of 3.00 m from D (Fig. 13.7(a)) along line DC and 3.00 m from A along line AB. Insert nails into pegs at these points. Measure 1.5 m from D along line DA and 1.5 m from C along line CB. Insert nails into pegs at these points. Stretch tapes or builder's lines between the two sets of nails. The intersection of the lines is point G.

8. Similarly establish pegs E and F.

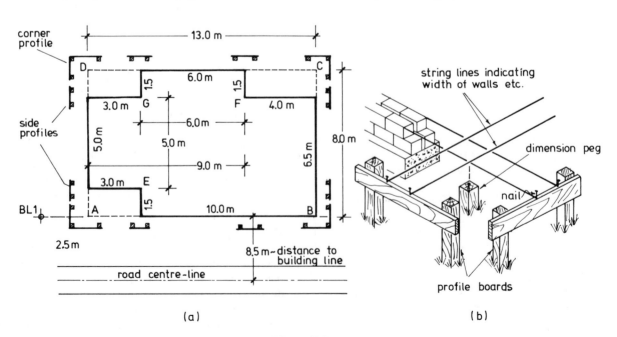

Figure 13.7

9. As a final check on the work, measure the dimensions EG, GF and FE. The distances should be EG = 5.00 m, GF = 6.00 m and EF = $\sqrt{5.0^2 + 6.0^2}$ = 7.810 metres.

Profile boards
During the excavation of the foundations, the pegs A, B, C and D will be destroyed and it is necessary to establish subsidiary marks on profile boards (Fig. 13.7(b)).

Profile boards are stout pieces of timber 150 mm by 25 mm cut to varying lengths. The boards are nailed to 50 mm square posts hammered into the ground, well clear of the foundations. Once the profiles have been established, builder's lines are strung between them and accurately plumbed above the pegs A, B, C and D, and nails are driven into the boards to hold the strings and mark the positions of the walls, foundations, etc.

Setting out on sloping ground
When setting out buildings on sloping ground, it must be remembered that the dimensions taken from the plan are horizontal lengths and consequently the tape must be held horizontally and the method of step taping used.

The diagonals must also be measured horizontally and in practice considerable difficulty is experienced in obtaining checks under such conditions. Besides, the method is laborious and time consuming.

It is possible to dispense with measuring the diagonals if a site square is used to set out the right angles at A and B (Fig. 13.8). The instrument is capable of setting out right angles with an accuracy of 1 in 2000. It consists of two small telescopes fixed rigidly at right angles on a small tripod. When the site square is set up at peg A the observer simply sights peg B through one telescope and lines in peg D through the other.

Similarly, peg C is set out from B and a check is provided on the work by erecting the instrument at D and checking that angle CDA is right angled.

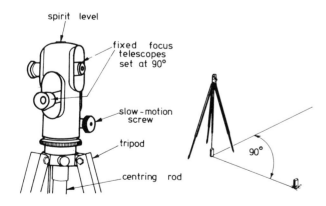

Figure 13.8

7. Setting out large buildings

In setting out large buildings which are mainly prefabricated, accuracy is absolutely essential. The factory-built components cannot be altered on site and even though some allowance is made for fitting on site, faulty setting out causes loss of time and money in corrective work.

In setting out such buildings, cognisance must be taken of the effects of calibration, temperature and tension on tape measurements, and the effects of instrumental errors in angular observations.

(a) Tape measurements

Fabric-based tapes should never be used for setting out. Good quality steel tapes are always employed and due allowance made for the following potential sources of error:

1. *Calibration*
Calibration errors can be ignored when good-quality tapes are used. After long use, however, the tape should be tested against a standard. If the tape has been broken and repaired it should not be used for accurate setting out.

2. *Temperature*
The length of a steel tape varies with temperature. The tape is the standard length at 20°C only. If left lying in direct sunlight the tape may reach an abnormally high temperature. In winter the temperature may well be at freezing point. In such cases, a correction has to be applied for differences in temperature but a fairly accurate estimation is obtained by allowing 1 mm per 10 metres per 10°C difference in temperature. For example, when using a 30 m steel tape on a winter day when the temperature is 0° an allowance of $-(1 \times 3 \times 2) = -6$ mm should be made for each tape length measured.

3. *Tension*
Steel tapes should be used on the flat with a tension of 4.5 kg. Without a spring balance few people can judge tensions and tests have shown that errors of 10 mm in 30 m can be caused by exerting excessive pressure on the tape.

When the tape is allowed to sag there is even more error. The natural tendency is to apply excessive tension to correct the sag, and in many cases the error due to stretching of the tape is greater than that due to sag.

A constant-tension handle has now been developed which applies a compromise tension. The tension applied is such that the effect of sag is compensated by the effect of stretching the tape.

4. *Slope*
It should be clear from error source 3 above that sagging of the tape should be avoided if possible, which suggests that measurements should be made

along the ground. This, however, introduces slope errors, and so measurements of slope must be made and corrections applied.

5. *Taping procedure*

For highly accurate measurements the following procedure should be carried out:

(a) Use a good quality tape which has never been broken.

(b) Lay the tape on wooden blocks if measuring on concrete, etc., to allow air to circulate around the tape.

(c) Whenever possible do not allow the tape to sag.

(d) Measure the temperature.

(e) Use a spring balance or constant-tension handle.

(f) Measure the ground slope.

(g) Compute the various corrections and apply the correction to the distance set out.

(b) Angular measurements

Angular measurements must always be made on both faces of the instrument because of the effect of instrumental maladjustments, which were fully explained in Chapter 7.

(c) Procedure for setting out

Figure 13.9 shows the plan of a large factory building and office block. The columns of the office block are of reinforced concrete, the cladding being prefabricated panels. The factory columns are of steel supporting steel latticed roof trusses. The column centres must be placed at exactly the correct distance apart and must be perfectly in line.

The following procedure is necessary to ensure that the requirements are met:

1. Establish a line AB from the site traverse stations at a predetermined distance x metres from the centre line of the left-hand columns. Measure the distance accurately.

2. Set out peg D by face left and face right observations from A, positioning D at some predetermined distance from the centre line of the right-hand columns.

3. Set out C by double face observations from D, making CD = AB.

4. Finally, check angle C and distance CB to ensure that ABCD is a perfect rectangle.

5. Set out the column centres along each line on stout profiles or preferably on pegs embedded in concrete. The centres must be set out by steel tape with due allowance being made for temperature, etc.

6. The column centres are defined by the intersections of wires strung between the appropriate reference marks or by setting the theodolite at a peg on one side of the building, sighting the appropriate peg on the other side and lining in the columns directly from the instrument.

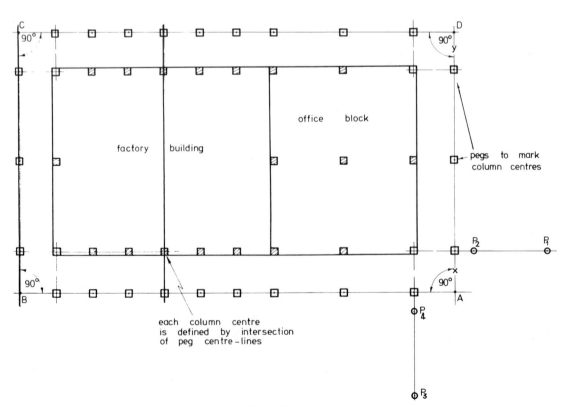

Figure 13.9

No hard and fast rules can be formulated for setting out since conditions vary greatly from site to site and some ingenuity is called for on the part of the engineer.

Exercise 13.6

1 Figure 13.10 shows the plan of a site where an access road and a building are to be set out

(a) Show on the plan suitable setting-out points to locate the building and centre lines of the roadway straights on the ground.
(b) Using a scale rule, determine the coordinates of those setting-out points.
(c) Calculate the angles and distances required to locate the setting-out points on the ground.

(Scotvec—Higher National Certificate in Civil Engineering)

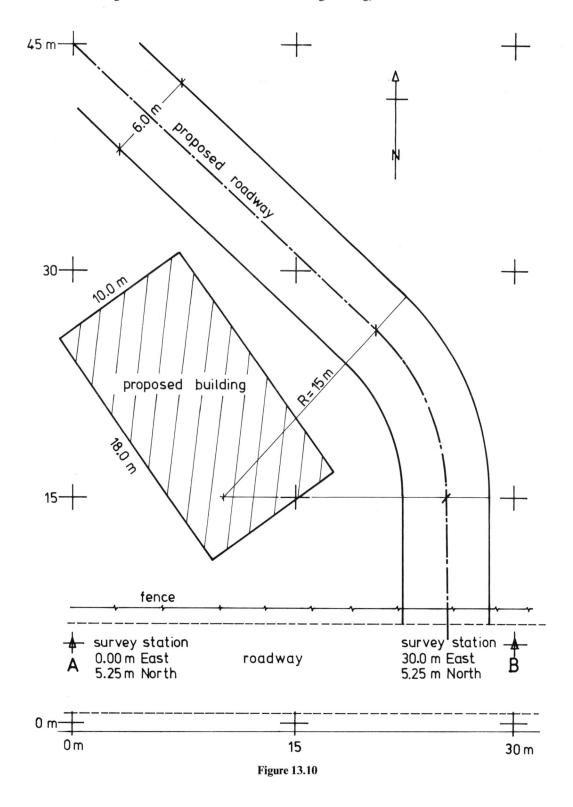

Figure 13.10

8. Checking verticality

In Fig. 13.11 the office block rises to a height of five storeys and the columns must be checked at every storey for verticality.

Several methods are available to the engineer using (a) theodolite, (b) plumb-bob, (c) special instruments.

(a) Theodolite methods

1. When setting out the framework for horizontal control, further marks are established at the points P_1 to P_4, i.e. on two lines at right angles to each other. Four pegs are required for each corner. In Figs 13.9 and 13.11 the pegs are shown for the bottom right-hand corner only.

The theodolite is set up on face left over the outer peg P_1 and sighted on to the inner peg P_2. The telescope is then raised to any level on the building and a mark made on the outside of the column. The procedure is repeated on face right. If the theodolite is in good adjustment the two marks will coincide. If not, the mean position is correct.

The instrument is removed to peg P_3 and the whole of the above procedure repeated to establish a second mark on the column. Thus, it is possible to determine the amount of deviation of the column from the vertical. Each corner is checked in this way.

2. The centre lines of the building are established and extended on all four sides of the building. Two permanent marks are placed on each line. The marks are shown by the numbers 1 to 8 in Fig. 13.11.

The theodolite is placed on each of the outer marks, the telescope is raised to the required floor level and a mark established on all four sides by double face observations. The intersection of lines strung between the appropriate marks locates the centre of the building. From the lines measurements may be made to corners, etc.

3. *Diagonal eyepiece method*

Most theodolites have an eyepiece which is interchangeable with the diagonal eyepiece shown in Fig. 13.12. The diagonal eyepiece allows the telescope to be placed vertically and direct plumbing is possible to any height with high accuracy.

The engineer must arrange for a hole 500 mm square to be left in each floor so the theodolite can sight vertically upwards. Sights are taken to a specially designed target shown in Fig. 13.12.

The following procedure is carried out when establishing a vertical line by diagonal eyepiece:

(a) The theodolite is set accurately over a predetermined mark on the ground floor and is centred and levelled in the usual manner. The horizontal circle is set to zero.

(b) The altitude spirit level is centralized and the vertical circle set to read zenith (the reading varies

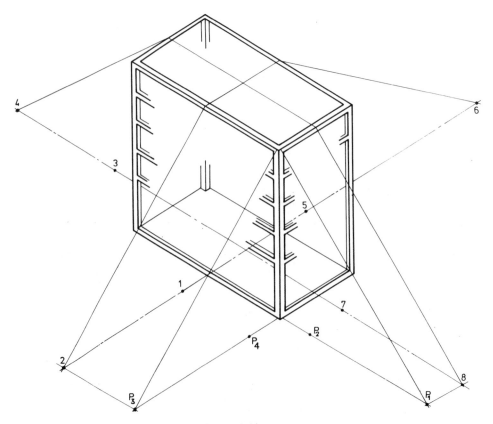

Figure 13.11

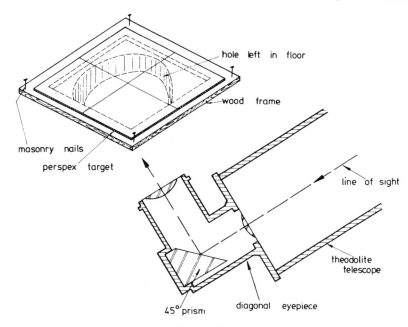

Figure 13.12

with the type of instrument). The telescope is therefore pointing vertically.

(c) A assistant on the upper floor moves to the Perspex target and frame over the hole until the horizontal cross-hair lies along the centre line of the target. The frame is nailed to the floor using masonry nails and the ends of the target centre line lightly marked on the frame with a pencil.

(d) The instrument is turned until the horizontal circle reads 180°. The altitude spirit level is recentralized and the vertical circle set to zenith.

(e) If the instrument is in good adjustment the horizontal cross-hair will lie along the centre line of the target. If not, the assistant is directed to move the target until it does.

(f) The centre line of the target is again pencilled lightly on the frame and the mean position clearly marked. This mean line is the true vertical plane.

(g) The operations are repeated with horizontal circle readings of 90° and 270°. Two more sets of marks are lightly pencilled on the frame and the mean position clearly marked, to establish a second vertical plane at right angles to the first. The intersection of the planes so marked is the true plumb point above the instrument.

(b) Plumb-bob method

Heavy plumb-bobs are suspended from adjustable reels on piano wire over marks on the floor level. The positions of the marks are generally on the centre line at a known distance x metres from the centre of the building.

The plumb-bobs have usually to be immersed in barrels of water to damp the oscillations of the wire set up by the wind currents. The barrels therefore sit

over the marks on the floor and the marks have to be referenced to some form of staging built around them.

The reels are adjusted on the upper floor until the wires are correctly positioned relative to the marks on the staging. The wires form a base line from which measurements may be made to the various columns on the upper floor.

(c) Automatic plumbing

Several instrument manufacturers produce instruments that set out vertical lines of sight automatically. The Sokkia PD3 precise optical plummet (Fig. 13.13) and the Topcon VS-A1 are typical examples. The

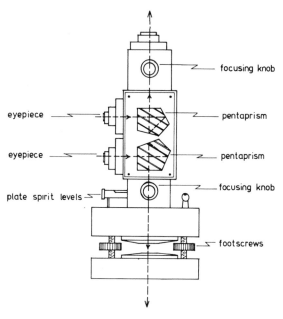

Figure 13.13

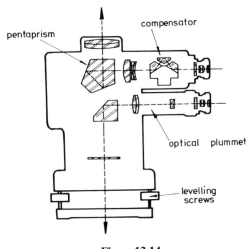

Figure 13.14

modern PD3 has two separate telescopes, which make possible simultaneous upward and downward plumbing, as does the arrangement shown in Fig. 13.14. The telescopes of these instruments are positioned such that the upward and downward lines of sight coincide and a perpendicular accuracy of 1 in 40 000 is claimed by the manufacturer.

Arrangements must be made for leaving a hole in each floor for upward sighting. Sights are taken to a target similar to that used with the diagonal eyepiece method. The procedure for setting vertical with the Autoplumb is identical to that of the theodolite diagonal eyepiece method.

9. Setting out—vertical control

Whenever any proposed level is to be set out, sight rails (profiles) must be erected either at the proposed level in the case of a floor level or at some convenient

height above the proposed level in cases of foundation levels, formation levels and invert levels. Suitable forms of sight rails or profiles are shown in Fig. 13.2. The rails should be set at right angles to the centre lines of drains, sewers, etc.

A traveller or boning rod is really a mobile profile which is used in conjunction with sight rails. The length of the traveller is equal to the difference in height between the rail level and the proposed excavation level. Figure 13.2 shows the traveller in use in a trench excavation.

10. Setting out a peg at a predetermined level

The basic principle of setting out a profile board at a predetermined level is shown in Fig. 13.15. Point A is a temporary bench mark (RL 8.55 m AD). Profile boards B and C are to be erected such that the level of board B is 9.000 m and that of board C is 8.500 m. These levels may represent floor levels of buildings or may represent a level of, say, 1.00 or 2.00 m above a drain invert level or a roadway formation level.

Setting up profile boards at different levels is the same operation and, once mastered, the methods may be used for any number of profiles on a site.

(a) Procedure—method 1

1. The observer sets up the levelling instrument at a height convenient for observing a site bench mark (RL 8.55 m AD) and takes a backsight staff reading (1.25 m). The height of collimation (HPC) is therefore

RL bench mark + BS reading

i.e. HPC = 8.55 m + 1.25 m = 9.80 m.

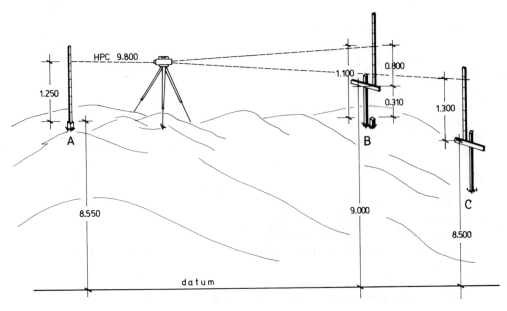

Figure 13.15

2. The assistant firmly hammers home a small peg, 300 mm long, beside the profile peg and a foresight reading (1.11 m) is taken to a staff held vertically upon it. The level of the top of the small peg is therefore

$$RL = HPC - FS \text{ reading}$$
$$= 9.80 - 1.11 = 8.69 \text{ m}$$

3. The difference in level between the top of the small peg (8.69 m) and the required profile level (9.00 m) is calculated:

$$\text{Difference} = 9.00 \text{ m} - 8.69 \text{ m} = 0.31 \text{ m}$$

Using a tape, the assistant measures this height against the profile peg and marks it in pencil.
4. A profile board is nailed securely to the profile peg, such that the upper edge of the board is against the pencil mark, and is thus at a level of 9.00 m.

This method is widely used on construction sites because of its simplicity. However, it has the disadvantage that the observer has to rely upon the assistant (often untrained) to correctly mark the final height on the profile peg. The disadvantage is overcome by using the following method.

(b) Procedure—method 2

1. The observer sets up the levelling instrument, takes a backsight to the bench mark and computes the height of collimation (HPC) as before:

$$HPC = RL \text{ bench mark} + BS \text{ reading}$$
$$= 8.55 \text{ m} + 1.25 \text{ m} = 9.80 \text{ m}$$

2. The staffman holds the staff against the profile peg and moves it slowly up or down until the base of the staff is at height 9.00 m, exactly. This will occur when the observer reads $9.80 \text{ m} - 9.00 \text{ m} = 0.80 \text{ m}$ on the staff, since

$$HPC = 9.80 \text{ m}$$
$$\text{Required profile level} = 9.00 \text{ m}$$
$$\text{Therefore staff reading} = 0.80 \text{ m}$$

3. The base of the staff is marked in pencil against the profile peg and the profile board is nailed securely to the peg, such that the upper edge of the board is against the pencil mark.
4. Profile board C is established in exactly the same manner, but since the board is to be erected at a different level, a new calculation is required:

$$HPC = 9.80 \text{ m}$$
$$\text{Required profile level} = 8.50 \text{ m}$$
$$\text{Therefore staff reading} = (9.80 - 8.50) \text{ m} = 1.30 \text{ m}$$

11. Setting out floor levels

Floor levels are set out on profile boards in exactly the manner described above in Sec. 10. Profile pegs are set

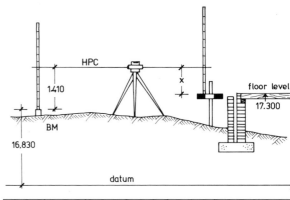

Figure 13.16

BS	IS	FS	HPC	RL	Remarks
1.410				16.830	Bench mark
		x		17.300	Floor level

around the perimeter of the house as required. A levelling instrument is set up and the height of collimation (HPC) of the level is determined. The proposed floor level is subtracted from the HPC and the resultant staff reading used to set out the profile boards.

Example

9 In Fig. 13.16 the floor level (17.30 m AD) of a building is to be set out from a nearby bench mark (16.830 m AD). Calculate the staff reading (x) required to set out a profile board at floor level.

Answer
$$HPC = 16.830 + 1.410$$
$$= 18.240$$
$$\text{Required reading } x = 18.240 - 17.300$$
$$= 0.940 \text{ m}$$

Exercise 13.7

1 The floor levels of a split level house are upper level 25.500 m AD and lower level 24.300 m AD. They are to be set out in relation to a nearby bench mark (23.870 m AD). Table 13.7 shows the relevant readings. Calculate the staff readings x and y required to set out profiles at both floor levels.

12. Setting out invert levels

(a) Setting out invert levels—office work

In setting out drains and sewers, it is not possible to set out the proposed levels of the drain, i.e. the invert

BS	IS	FS	HPC	Required level	Remarks
2.360				23.870	Bench mark
	x			25.500	Upper floor level
		y		24.300	Lower floor level

Table 13.7

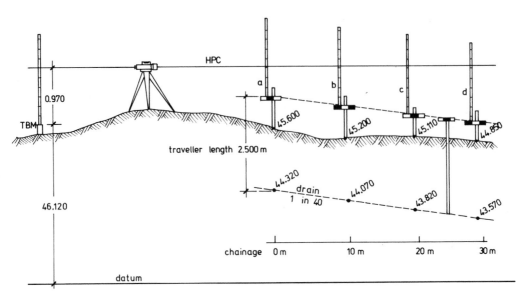

Figure 13.17

levels, since they are always below the ground. Hence, profiles must be set at some convenient height above the invert levels. This height is chosen by the surveyor but should always be a multiple of 250 mm. The height is known as the traveller length and is determined as follows:

1. In Fig. 13.17, a drain, 30 metres long, is to be excavated, at a gradient of 1 in 40. The invert level at the start of the drain, chainage 0.00 m, is 44.320 m above datum.
2. Ground levels are taken at regular intervals along the centre line of the proposed drain. The levels, taken at 10 metre intervals, are shown in Table 13.8, column 2.

1	2	3	4
Chainage (m)	Surface level	Invert level	Depth (m)
0	45.600	44.320	1.300
10	45.200	44.070	1.130
20	45.110	43.820	1.290
30	44.850	43.570	1.280

Table 13.8

3. The invert levels of the drain are next calculated. In this case, the drain beginning at invert level 44.320 m falls at a gradient of 1 in 40. The fall over 10 metres is one-fortieth of 10 m = 0.250 m, and the invert level at chainage 10 m is therefore 44.320 − 0.250 = 44.070 m. Since the chainage intervals are regular, the fall must also be regular, resulting in the various invert levels shown in column 3 of Table 13.8.
4. The depth from surface to invert level is found by subtracting the invert level from the surface level. Thus, at 0 m chainage, the depth is 45.620 m (surface) − 44.320 (invert) = 1.300 m. The depths at the various chainage points are shown in column 4 of Table 13.8.
5. From the table, the maximum depth is 1.300 m and since profiles should be about 1 metre above ground level the length of traveller should be 1.300 m + 1.00 m = 2.30 m. The traveller would probably be made 2.5 m long.

Having determined the length of the traveller, the levels of the profile boards to be erected along the line of the drain are next calculated. The profile level is 2.50 m greater than the invert level at any chainage. The profile levels are therefore as shown in Table 13.9, which becomes in effect the setting-out table.

Chainage	Invert level	Profile level
0	44.320	46.820
10	44.070	46.570
20	43.820	46.320
30	43.570	46.070

Table 13.9

(b) Setting out invert levels—field work

In Fig. 13.17, the levelling instrument has been set up to sight to the bench mark and to the various profile board positions. Table 13.10 shows the relevant data. The staff readings a, b, c and d are required to set the profile boards at their correct levels.

The relevant readings are calculated as follows:

1. Calculate HPC = 46.120 + 0.970 = 47.090
2. Calculate:
(a) = 47.090 − 46.820 = 0.270
(b) = 47.090 − 46.570 = 0.520
(c) = 47.090 − 46.320 = 0.770
(d) = 47.090 − 46.070 = 1.020

The profile boards are then set out by one of the methods described in Sec. 10.

Example

10 Figure 13.18 shows a sewer, 55 metres long, which is to be set out on a gradient of 1 in 50 falling from chainage 0 m to chainage 55 m. The following data have been obtained:

Invert level at 0 m chainage = 24.210 m
Ground level at 0 m chainage = 25.690 m
BS reading 0.665 m to bench mark
(RL 25.685 m)

Calculate the staff readings required to set out sight rails at 0, 30, and 55 m chainages.

Answer
(a) Invert level at 0 m chainage = 24.210 m
   Fall = 1/50th of distance
   Therefore fall
   (0 to 30 m) = 1/50 × 30 = −0.600 m
   Invert level at 30 m = 23.610 m
   Fall 30 to 55 m = 1/50 × 25 = −0.500 m
   Invert level at 55 m = 23.110 m

(b) Ground level at 0 m chainage
                              = 25.690 m
   Sight rail level
   (1 m above ground) = 26.690 m
   Therefore traveller = rail level
   − invert level = 26.690 − 24.210
                  = 2.480 m

BS	IS	FS	HPC	Required level	Remarks
0.970				46.120	Bench mark
	(a)			46.820	Profile 0 m
	(b)			46.570	Profile 10 m
	(c)			46.320	Profile 20 m
		(d)		46.070	Profile 30 m

Table 13.10

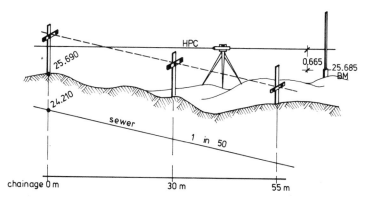

Figure 13.18

This length of traveller is unsuitable and would be rounded up to 2.500 m.

(c) Sight rail level = invert level + traveller
Therefore at 0 m chainage
Rail level = 24.210 + 2.500 = 26.710 m
At 30 m chainage
Rail level = 23.610 + 2.500 = 26.110 m
At 55 m chainage
Rail level = 23.110 + 2.500 = 25.610 m

(d) HPC = 25.685 + 0.665 = 26.350

(e) Staff readings required to set out rails =
HPC − rail level
At   0 m = 26.350 − 26.710 = −0.360
At 30 m = 26.350 − 26.110 = 0.240
At 55 m = 26.350 − 25.610 = 0.740

It should be noted that at chainage 0 m, the line of the sight through the telescope (HPC) is actually lower than the proposed sight rail level, and the resultant staff reading is therefore a negative value (−0.360 m). In order to establish the rail, an inverted staff reading of 0.360 m is required. In practice it usually proves impossible to turn the staff upside down since it would be resting on the ground, in which case the line of sight is marked against the profile peg in pencil and 0.360 m is measured upwards using a tape to establish the profile level.

Exercise 13.8

1 Figure 13.1 shows the proposed development at the Glasgow College of Building and Printing Outdoor Centre. Table 13.11 shows the data relating to pro-

posed foul manholes F1 to F4. The foul sewer is to fall from F4 to manhole F1 at a gradient of 1 in 40. Calculate:
(a) the proposed invert level at manholes F2, F3 and F4,
(b) the profile board levels at all four manholes, given that the length of the traveller is 4.00 m,
(c) the heights to be measured up from the ground level pegs to establish the profile boards at their correct levels.

13. Setting out roadway levels

Setting out roadway profiles is essentially the same operation as setting out drain and sewer profiles. The profile board must be set at some predetermined height above the roadway formation levels in the same way as sewer profiles are set some height above the invert level.

The problem therefore lies simply in calculating the various formation levels along roadway slopes and vertical curves and adding a suitable height, say 1.0 metre, to all of them to produce profile levels. The profiles are then set out in the manner previously described.

Exercise 13.9

1 In Sec. 8(e) in Chapter 12 (roadway curves) an example was used where a roadway on a rising gradient of 1 in 100 (1 per cent) changed to a rising gradient of 1 in 25 (4 per cent) by means of a vertical curve 100 metres long (Fig. 13.19). The resultant formation levels of the roadway at 25 m intervals were calculated in that section and are shown in Table 13.12.

Manhole number	Chainage (m)	Ground level (pegged)	Proposed invert level
F1	0.0000	6.66	2.950
F2	24.000	6.41	—
F3	48.000	5.87	—
F4	64.000	6.02	—

Table 13.11

Chainage (m)	Formation level	Profile level (formation level + 1.000 m)
0	260.800	261.800
25	261.144	262.144
50	261.675	262.675
75	262.394	263.394
100	263.300	264.300

Table 13.12

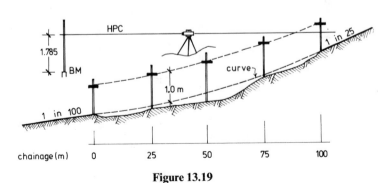

Figure 13.19

The levels are to be set out from a nearby bench mark (RL 262.065 m AD) using an automatic level. The backsight to the bench mark is 1.785 m. Calculate the staff readings required to set out the profile levels.

14. Large-scale excavations

Consider Fig. 13.20 in which a large area of high ground is to be reduced to formation level 50.00 m to form a sports arena. The following is the usual procedure adopted for setting out the sight rails:
1. The sight rail level is calculated. This is normally 1.00 m above formation level = 51.00 m.
2. A levelling is made from a nearby TBM and the 50 m contour is traced on the ground. Sight rail uprights are driven in around the contour at intervals. If large earth-moving plant is being used it is wise to move the uprights outside the area of excavation.

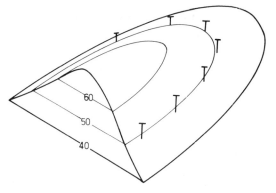

Figure 13.20

3. The cross pieces are nailed to the uprights at 51.00 m level using the collimation system of levelling.
4. The length of traveller in this case is 51.00 − 50.00 = 1.00 m. The length of the traveller is written on the back of the sight rail in paint or waterproof chalk.

15. Vertical control using laser instruments

In setting out the proposed levels of construction works, a levelling instrument is set up in order to project a plane of collimation over the site. Measurements are then made downwards from the plane of collimation.

Since the early eighties, laser instruments have been developed for general use on building projects. The instruments can either sweep out a plane of infra-red laser light over the sight or set out a beam of visible laser light on any predetermined gradient.

Figure 13.21 shows the LB 2 self-levelling laser level developed by Laser Alignment Inc. of Michigan, USA. There are several other similar types of level currently in use, namely Gradomat, Stolz Baulaser and Topcon RL20.

Rechargeable nickel–cadmium batteries provide the power to generate an invisible infra-red laser beam. This beam is rotated in a complete circle around the instrument.

The instrument is either set on a tripod or free standing and is levelled by two levelling screws and

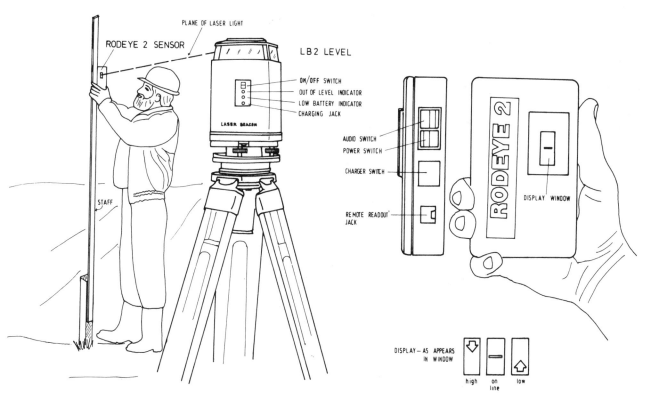

Figure 13.21

spirit level. This brings the instrument within its automatic self-levelling range. The instrument automatically shuts off if the self-levelling unit is jarred out of its position and has to be relevelled.

The unit is switched on and the beam is swept out across the site over a range of 100 metres. Being invisible, the beam has to be detected at any point of setting out. This is accomplished by the Rodeye 2 sensor unit (Fig. 13.21). It weighs less then 300 grammes, is charged by a 9 volt rechargeable battery and is easily carried in the pocket. The Rodeye is either hand held against a levelling staff or against some vertical object such as a column or wall. In use it is moved slowly vertically until an indication is received in the sensor window that the sensor is on the line of the beam. A mark is made against the column or levelling staff and the setting out level is measured up or down from this mark. The Rodeye incorporates an audio signalling device which can be used to indicate the vertical position of the sensor. The instrument has an accuracy of ± 1 mm.

The instrument may also be laid on its side and when properly levelled it will sweep out a vertical plane of laser light. In this position, it is ideal for setting columns and framework in their true vertical positions.

Figure 13.22 shows Spectra Physics Dialgrade laser alignment unit for use in setting out drains and sewers, etc. The instrument is connected to a 12 volt d.c. electrical supply which produces a 2.0 MW helium neon laser beam. The instrument can be used in a free-standing position or may be tripod mounted or attached to a bar clamped between the walls of a sewer pipe.

In operation, the unit is roughly levelled using the bullseye level. When switched on, the wide levelling range allows the instrument's levelling motors to adjust it automatically from there. Using the finger touch panels, the gradient of the sewer is dialled into the unit and the laser beam is projected along the trench or through a pipe at the correct gradient. Gradients of ± 7 per cent may be set out with the Dialgrade. The instrument has proved to be robust, waterproof and reliable in rugged site conditions.

Exercise 13.10

1 A drain is to be set out using an engineer's tilting level from the following information (see Fig. 13.23):

> Length of drain AB = 150 m
> Gradient AB, falling from A to B at 1 in 100
> Invert level A = 64.350 m
> Length of traveller = 2.000 m

A backsight of 1.200 m has been taken to a nearby bench mark, the reduced level of which is 67.650 m. Calculate the staff readings necessary to locate sight rails over A and B.

2 The readings in Table 13.13 were taken along the line of a drain:
(a) Copy the field book entries and complete the booking.

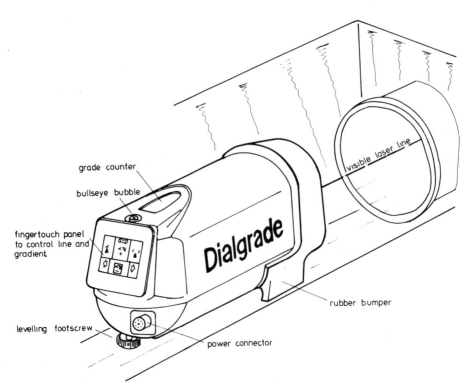

Figure 13.22

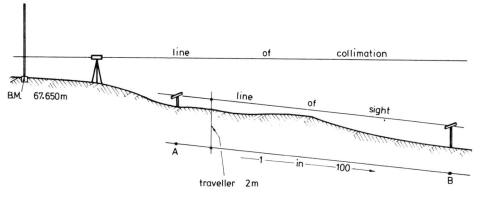

Figure 13.23

BS	IS	FS	Height of collimation	RL	Distance (m)	Remarks
0.860			97.240			TBM
2.240		0.550		0		MH 5
	2.590			48.000		MH 4
	1.690			64.000		MH 3
1.322		0.975		85.600		MH 2
		0.937		118.400		MH 1

Table 13.13

(b) The invert level of MH 1 is to be 96.811 m and the pipe runs are to slope down at the following gradients:

MH 1 to MH 2	1 in 125
MH 2 to MH 3	1 in 115
MH 3 to MH 4	1 in 100
MH 4 to MH 5	1 in 80

Calculate the invert levels of each manhole.
(c) How should the depth and gradient of the drainage trench be controlled?

(City & Guilds of London Institute—Const. Tech. Cert.)

16. Answers

Exercise 13.1

1 Bearing I3–I4 = 290° 45′ 29″: length = 13.261 m
Bearing I4–R10 = 226° 00′ 18″: length = 8.062 m

Exercise 13.2

1 Line R10 → B: $\Delta E = +11.221$ m, $\Delta N = +9.279$ m
Bearing R10 → B $= \tan^{-1}(11.221/9.279)$
$= 50° 24′ 42″$
Distance R10–B $= \sqrt{11.221^2 + 9.279^2} = 14.560$
m

Exercise 13.3

1 Angle I1–I2–I3 = 99° 55′ 07″
Angle I2–I3–I4 = 155° 45′ 29″
Angle I3–I4–R10 = 115° 14′ 48″

Exercise 13.4

1 Curve 3
Deviation angle $\theta_3 = 24° 14′ 31″$
Radius of curve = 30.000 m
Tangent length $= R \tan \theta_3/2 = 6.443$ m
Therefore chainage R6
$= 30.174$
$+ (18.102 - 5.881 - 6.443)$ m
$= 30.174 + 5.778$
$= 35.952$
Curve length $= 2\pi R \times (24.2419°/360°)$
$= 12.693$ m
Therefore chainage R7
$= 48.645$ m

Curve 4
Deviation angle $\theta_4 = 64° 45′ 12″$
Radius of curve = 7.000 m
Tangent length $= R \tan \theta_4/2 = 4.438$ m
Therefore chainage R8
$= 48.645$
$+ (13.261 - 6.443 - 4.438)$ m
$= 48.645 + 2.380$
$= 51.025$ m
Curve length $= 2\pi R \times (64.7533°/360°)$
$= 7.911$ m
Therefore chainage R9
$= 51.025 + 7.911$ m
$= 58.936$ m
Chainage R10 = R9(58.936)
$+ (8.062 - 4.438)$
$= 58.936 + 3.624$
$= 62.560$ m

Exercise 13.5

1 (a)   Long chord $= 2R \sin \theta/2 = 7.50$ m
At 3.75 m,
major offset $y = R - \sqrt{R^2 - x^2}$
$= 7.00 - 5.91$
$= 1.09$ m
$k = 5.91$ m

(b) For offsets $Y_{1.5} = Y_{6.0}$,
$x = 3.75 - 1.5 = 2.25$
$Y_{1.5} = \sqrt{R^2 - x^2} - k$
$= 6.63 - 5.91$
$= 0.72$ m
For offset $Y_3 = Y_{4.5}$,
$x = 0.75$ m
$Y_3 = \sqrt{R^2 - x^2} - k$
$= 6.96 - 5.91$
$= 1.05$ m

Exercise 13.6

1 Figure 13.24

Point	East (m)	North (m)
A	00.00	5.25
B	30.00	5.25
C	25.40	3.50
D	25.40	21.70
E	00.00	45.00
F	9.40	10.80
G	17.50	16.70

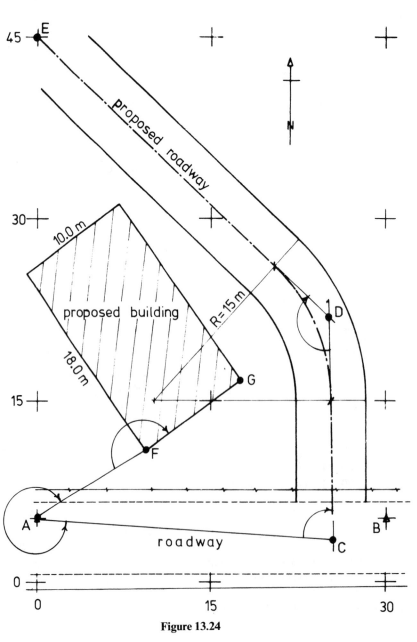

Figure 13.24

Line	Δ East	Δ North	Length (m)	Bearing
AB	30.00	00.00	30.00	90° 00′ 00″
AC	25.40	−1.75	25.46	93° 56′ 29″
CD	00.00	18.20	18.20	360° 00′ 00″
DE	−25.40	23.30	34.47	312° 31′ 51″
AF	9.40	5.55	10.92	59° 26′ 29″
FG	8.10	5.90	10.02	53° 55′ 50″

Angle	Value
BAC	3° 56′ 29″
ACD	86° 03′ 31″
CDE	132° 31′ 51″
CAF	325° 30′ 00″
AFG	174° 29′ 21″

Exercise 13.7

1 Table 13.14

BS	IS	FS	HPC	Reduced level	Remarks
2.360			26.230	23.870	Bench mark
	0.730		26.230	25.500	Upper floor level
		1.930	26.230	24.300	Lower floor level

Table 13.14

Exercise 13.8

1 Table 13.15

Manhole number	Chainage (m)	Ground level (pegged)	Proposed invert level	Profile level	Height from peg to profile
F1	0.00	6.660	2.950	6.95	0.29
F2	24.00	6.410	3.550	7.55	1.14
F3	48.00	5.870	4.150	8.15	2.28
F4	64.00	6.020	4.550	8.55	2.53

Table 13.15

Exercise 13.9

1 Table 13.16

BS	IS	FS	HPC	Reduced level	Profile level	Remarks
1.785			263.850	262.065		Bench mark
	2.050		263.850		261.800	Profile 0 m
	1.706		263.850		262.144	Profile 25 m
	1.175		263.850		262.675	Profile 50 m
	0.456		263.850		263.394	Profile 75 m
		−0.450	263.850		264.300	Profile 100 m

Note. The required staff reading to the profile peg (100 m) shows that an inverted staff reading 0.450 m is required.

Table 13.16

Exercise 13.10

1

$$\begin{array}{r} \text{Bench mark RL} = 67.650 \text{ m} \\ \text{Backsight} = \underline{1.200} \\ \text{Therefore height of collimation} = 68.850 \text{ m} \\ \text{Invert level A} = 64.350 \text{ m} \\ \text{Length of traveller} = \underline{2.000} \\ \text{Sight rail level} = 66.350 \text{ m} \\ \text{Staff reading to sight rail A} = 68.850 - 66.350 \\ = 2.500 \text{ m} \end{array}$$

Gradient AB = 1 in 100 falling
Length AB = 150 m
Therefore fall A to B = 150/100 = 1.500 m
Sight rail level B = 66.350 − 1.500
          = 64.850 m
Staff reading to sight rail B = 68.850 − 64.850
               = 4.000 m

2 (a)

RL	Remarks
97.240	TBM
97.550	MH5
97.200	MH4
98.100	MH3
98.815	MH2
99.200	MH1

(b)

IL	Remarks
95.600	MH5
96.200	MH4
96.366	MH3
96.550	MH2
96.811	MH1

(c) Sight rails and traveller

17. Project

Chapter 18 is a project covering all chapters of this textbook. It is intended that the project build into a complete portfolio of the surveying work required in the survey and the setting out of a building or engineering development.

If the reader wishes to continue work on the project or begin work at this stage, he or she should now turn to page 397 and attempt Section 9 then Section 10.

Objective
*After studying this chapter the reader
should be able to (a) assess the most
suitable method of calculating areas
(whether regular or irregular in shape) and
(b) make the necessary computations to
determine the areas of construction works
projects.*

CHAPTER 14

Mensuration— areas

On even the smallest site, calculations have to be made of a wide variety of areas and volumes, e.g. the area of the site itself, the volume of earthworks, cuttings, embankments, etc. Many of the figures encountered can be calculated by the direct application of the accepted mensuration formulae, but very often the figures are irregular in shape.

Regardless of the shape of the area, relevant data have to be obtained in some way in order to make the calculations:

1. The data are gathered in the field by some form of survey. The area is then calculated directly from these notes.

2. The data are converted into coordinates or a plotted plan from which the area is computed.
3. The data already exist in the form of a map or plan, e.g. Ordnance Survey plans.

1. Regular areas

Figure 14.1 shows the common regular figures and the formulae required to calculate their areas. In modern construction practice, many developments include shapes comprised of several of the common geometrical figures.

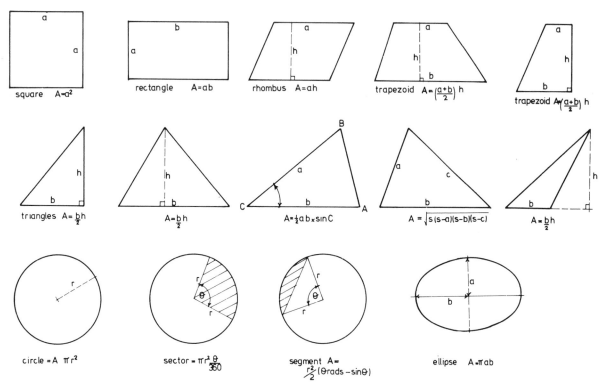

Figure 14.1

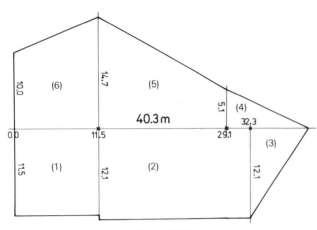

Figure 14.2

Answer
Area
(1) Rectangle
$$15.00 \times 10.00 \qquad\qquad = 150.00 \text{ m}^2$$
(2) Trapezoid
$$1/2\,(15.50 + 10.00) \times 9.50 = 121.13 \text{ m}^2$$
(3) Quarter circle
$$\pi/4 \times 15.00^2 \qquad\qquad = 176.71 \text{ m}^2$$
(4) Triangle

$$\text{Area} = \sqrt{s(s-a)(s-b)(s-c)}$$

$a = 13.44$	$s - a = 6.27$
$b = 10.98$	$s - b = 8.73$
$c = 15.00$	$s - c = 4.71$
$2s = 39.42$	Check $= 19.71$
$s = 19.71$	

$$\text{Area} = \sqrt{s(s-a)(s-b)(s-c)}$$
$$= \sqrt{5081.47} \qquad = \underline{71.28 \text{ m}^2}$$
$$\text{Total area of plot} = \overline{519.12 \text{ m}^2}$$

Examples

1 The central piazza of a town centre development is drawn to scale 1.500 in Fig. 14.2. The piazza has been split into six figures on the plan in order to calculate its area. Using the relevant surveyed dimensions, calculate the total area of the piazza in m².

Answer
Area
(1) Square
$$11.5 \times 11.5 \qquad\qquad = 132.25$$
(2) Rectangle
$$12.1 \times 20.8 \qquad\qquad = 251.68$$
(3) Triangle
$$0.5 \times 12.1 \times 8.0 \qquad = 48.40$$
(4) Triangle
$$0.5 \times 5.1 \times 11.2 \qquad = 28.56$$
(5) Trapezoid
$$0.5 \times (5.1 + 14.7) \times 17.6 \; = 174.24$$
(6) Trapezoid
$$0.5 \times (14.7 + 10.0) \times 11.5 = \underline{142.03}$$
$$\text{Total} = \overline{777.16 \text{ m}^2}$$

2 Figure 14.3 shows the dimensions of a grassed area at the Glasgow College of Building. Calculate the total area of the plot in square metres.

Exercise 14.1

1 Figure 14.4 shows the surveyed dimensions of the concourse of a shopping centre which is to be floored with terrazzo tiles. The central area is to be a water feature with a fountain. Calculate the total area to be tiled.

2. Irregular areas

A plot of ground having at least one curved side is considered to be irregular in shape, unless, of course, the curved side forms part of a circle. The curved side or sides preclude the use of the regular geometric formulae and use has to be made of two rules, namely (a) the trapezoidal rule and (b) Simpson's rule, to calculate the area of the irregularly shaped plot.

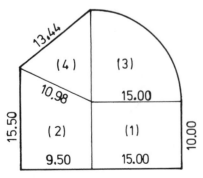

Figure 14.3

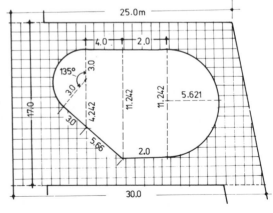

Figure 14.4

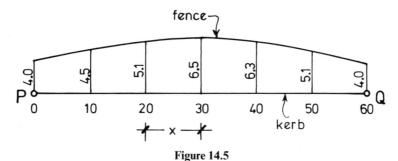

Figure 14.5

(a) Trapezoidal rule

Figure 14.5 shows the field notes of the linear survey of a plot of ground, lying between a straight kerb PQ and a curved fence. The length of the kerb is 60 metres and the offsets are taken at regular intervals of 10 metres.

Calling each offset y, the area between *any* two offsets is calculated thus:

$$\text{Area between chainage 20 m and chainage 30 m} = \tfrac{1}{2}(y_{20} + y_{30}) \times 10$$

Therefore total area

$$= \tfrac{1}{2}(y_0 + y_{10}) \times 10 + \tfrac{1}{2}(y_{10} + y_{20}) \times 10$$
$$+ \tfrac{1}{2}(y_{20} + y_{30}) \times 10 + \cdots + \tfrac{1}{2}(y_{50} + y_{60}) \times 10$$
$$= \tfrac{1}{2} \times 10(y_0 + y_{10} + y_{10} + y_{20} + y_{20} + y_{30}$$
$$+ \cdots + y_{50} + y_{60})$$
$$= \tfrac{1}{2} \times 10(y_0 + y_{60} + 2y_{10} + 2y_{20} + 2y_{30}$$
$$+ 2y_{40} + 2y_{50})$$
$$= 10\left(\frac{y_0 + y_{60}}{2} + y_{10} + y_{20} + y_{40} + y_{50}\right)$$

This is the *trapezoidal rule* and is usually expressed thus:

Area = strip width × (average of first and last offsets + sum of others)

In Fig. 14.5 the area is as follows:

$$\text{Area} = 10\left(\frac{4 \times 4}{2} + 4.5 + 5.1 + 6.5 + 6.3 + 5.1\right)$$
$$= 315.0 \text{ m}^2$$

(b) Simpson's rule

The area can be found slightly more accurately by Simpson's rule. A knowledge of the integral calculus is required to prove the rule but it can be shown to be

Area = $\tfrac{1}{3}$ strip width (first + last offsets + twice sum of odd offsets + four times sum of even offsets).

Note. (a) There must be an *odd* number of offsets.
(b) The offsets must be at regular intervals.

Using Simpson's rule the area between the line PQ and the road is as follows:

$$\text{Area} = 10/30[y_0 + y_{60} + 2(y_{20} + y_{40})$$
$$+ 4(y_{10} + y_{30} + y_{50})]$$
$$= 10/3[4 + 4 + 2(5.1 + 6.3) + 4(4.5 + 6.5 + 5.1)]$$
$$= 10/3[8 + 2(11.4) + 4(16.1)]$$
$$= 317.3 \text{ m}^2$$

Example 3 illustrates the method of dealing with irregularly shaped areas. In general terms, ordinates, i.e. the offsets, are measured at right angles to a base line and Simpson's or trapezoidal rules applied.

Example

3 Figure 14.6 shows an irregularly shaped pond, drawn to scale 1:500.
(a) Divide the figure into a suitable number of strips of the same width, using a 1:500 scale rule.
(b) Draw lines at right angles (offsets) at each strip width interval and measure the width of the pond along each offset.
(c) Using the trapezoidal rule and Simpson's rule, calculate the area of the pond.

Answer
(a) Since Simpson's rule is to be used, the pond must be divided into an even number of strips which produces an *odd* number of offsets, namely five. The strip width is therefore 8.0 metres.

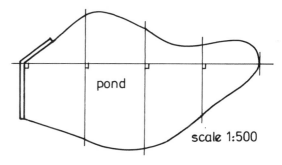

Figure 14.6

(b) The lengths of the offsets, measured by scale rule, are 7.60, 15.9, 16.3, 9.5 and 0.0 m.

(c) Using the trapezoidal rule,

$$\text{Area} = 8.0\left[\left(\frac{7.6 + 0.0}{2}\right) + 15.9 + 16.3 + 9.5\right]$$
$$= 8.0 + 45.5$$
$$= 364 \text{ m}^2$$

Using Simpson's rule,

$$\text{Area} = 8.0/3[7.6 + 0.0 + 2(16.3)$$
$$\qquad + 4(15.9 + 9.5)]$$
$$= 8.0/3(141.80)$$
$$= 378.1 \text{ m}^2$$

3. Areas from field notes

(a) Linear surveys

In a linear survey the area is divided into triangles, and the lengths of the three sides of each triangle are measured. The area contained within any one triangle ABC is found from the formula.

$$\text{Area} = \sqrt{s(s-a)(s-b)(s-c)}$$

where s is the semi-perimeter.

The boundaries of the linear survey are established by measuring offsets from the main lines. In Fig. 14.7 the area between the survey line and stream is composed of a series of trapezoids and triangles. The area of each figure must be computed separately.

The area between survey line PQ and the road is again composed of a series of separate figures. It must be noticed, however, that the offsets are at regular intervals of 15 metres in this instance and there is an

odd number of offsets. Simpson's rule is therefore used to calculate the area.

Finally, along the line RP, there is an even number of offsets between R and P at regular 10 metre spacings. The area is calculated by the trapezoidal rule.

Exercise 14.2

1 In Fig. 14.7, PQR is the area of a proposed factory development taken from an OS map. PR is a new boundary fence denoting the northern limit of the site. A linear survey was made to determine the area of the site. The lengths of the sides are

$$PQ = \quad 60.0 \text{ m}$$
$$QR = 104.6 \text{ m}$$
$$RP = \quad 70.0 \text{ m}$$

All offset dimensions are shown on the figure. Calculate the area of the site.

(b) Levelling

> **Example**
>
> **4** In Fig. 14.8 the survey of a proposed cutting shows that the depths at 20 m intervals are 0.0, 0.9, 1.5, 3.2 and 3.3 m. Given that the roadway is to be 5 m wide and that the cutting has 45° side slopes, calculate:
>
> (a) the plan surface area of the excavation, ABCD,
>
> (b) the actual area of the side slopes, ABE and CDF.

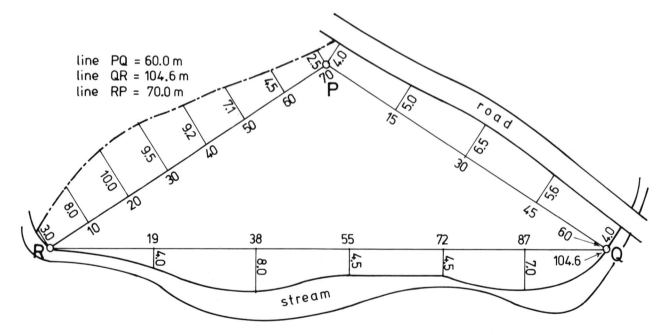

line PQ = 60.0 m
line QR = 104.6 m
line RP = 70.0 m

Figure 14.7

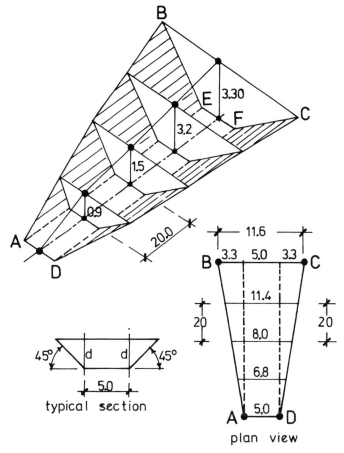

Figure 14.8

Answer

(a) Area ABCD

Chainage	Depth (d)	Top width [= (5.0 m + 2d)]
0	0.0	5.0
20	0.9	6.8
40	1.5	8.0
60	3.2	11.4
80	3.3	11.6

Using Simpson's rule:

$$\text{Area} = \frac{20}{3}[5.0 + 11.6 + 2(8.0)$$
$$+ 4(6.8 + 11.4)]\,\text{m}^2$$
$$= \frac{20}{3} \times 105.4$$
$$= 702.7\,\text{m}^2$$

(b) Plan areas ABE and CDF

Area ABE + area CDF
$$= \text{area ABCD} - \text{area ADEF}$$
$$= 702.7 - (5 \times 80)$$
$$= 302.7\,\text{m}^2$$

Therefore area ABE = area CDF
$$= 151.35\,\text{m}^2$$
Actual side area = plan area ÷ cos 45°
$$= 151.35 \div 0.7071$$
$$= 214.04\,\text{m}^2$$

Alternatively, the side slope areas may be computed independently. At each chainage point the side width = depth since the slopes have 45° gradients.

Using Simpson's rule:

$$\text{Side area} = \frac{20}{3}[0.0 + 3.3 + 2(1.5)$$
$$+ 4(0.9 + 3.2)]$$
$$= \frac{20}{3} \times 22.7$$
$$= 151.33\,\text{m}^2\,(\text{plan})$$

Using the trapezoidal rule:

$$\text{Side area} = 20\left[\frac{0.0 + 3.3}{2} + (0.9 + 1.5 + 3.2)\right]$$
$$= 145.0\,\text{m}^2\,\text{plan}$$

Actual area = 151.33 ÷ cos 45° = 214.02 m² or
145.0 ÷ cos 45° = 205.1 m²

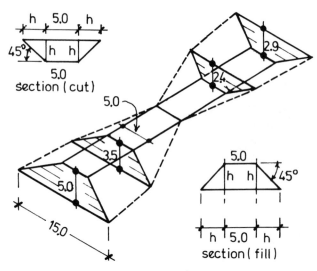

Figure 14.9

Exercise 14.3

1 In Fig. 14.9, the survey of a proposed roadway earthwork shows the heights, at 30 metre intervals, to be 5.0 m (fill), 3.5 m (fill), 0.0 m, 2.4 m (cut) and 2.9 m (cut). The roadway is to be 5.0 m wide and the embankment sides are to slope at 45°. Calculate:
(a) the plan view surface area of the embankment,
(b) the actual area of the side slopes that have to be grassed.

(c) Areas from coordinates

The calculation of areas is not made directly from the actual field notes but from the coordinate calculations made from the notes.

Method of double longitudes
The meanings of the terms, difference in easting (ΔE) difference in northings (ΔN), easting (E) and northing (N), have already been made clear in Chapter 9 and should be revised at this point.

One further definition is required, namely the longitude of any line is the easting of the mid-point of the line. In other words, the longitude of any line is the mean of the eastings of the stations at the ends of the line. In Fig. 14.10.

Longitude of line AB = $\frac{1}{2}$(easting A + easting B)

Therefore,

Double longitude AB = (easting A + easting B)
= 30.0 + 90.0
= 120.0 m

Figure 14.10 is an example of a coordinated figure where the coordinates of the stations are:

Station	Easting (m)	Northing (m)
A	30.0	60.0
B	90.0	100.0
C	120.0	20.0

Area of triangle ABC = area of trapezium 3BC1
— area of trapezium 3BA2
— area of trapezium 2AC1

i.e. triangle ABC = $\dfrac{E_3 - E_1}{2} \times (N_3 - N_1)$

$- \dfrac{E_3 + E_2}{2} \times (N_3 - N_2)$

$- \dfrac{E_2 + E_1}{2} \times (N_2 - N_1)$

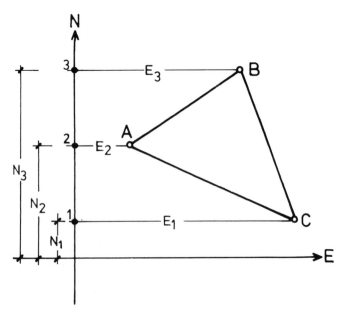

Figure 14.10

Station	Total coordinates		Line	Double longitude	Differences in northings	Double areas (m^2)	
	East (m)	North (m)				+	−
C	+120.0	+20.0					
B	+90.0	+100.0	CB	+210.0	+80.0	16 800	
A	+30.0	+60.0	BA	+120.0	−40.0		4 800
C	+120.0	+20.0	AC	+150.0	−40.0		6 000
						16 800	10 800
						−10 800	
					÷ 2 =	$\dfrac{6\,000}{3\,000}$ m²	

Table 14.1

Therefore 2 × triangle ABC
$$= (E_3 + E_1) \times (N_3 - N_1) - (E_3 + E_2) \times (N_3 - N_2)$$
$$- (E_2 + E_1) \times (N_2 - N_1)$$
$$= (210 \times 80) - (120 \times 40) - (150 \times 40)$$
or $= (+210 \times +80) + (+120 \times -40)$
$$+ (+150 \times -40)$$

The values of $+210$, $+120$ and $+150$ are the double longitudes of lines CB, BA and AC respectively, while the values $+80$, -40 and -40 are the partial latitudes of the same lines when moving around the figure in an anticlockwise direction.

Therefore 2 × triangle ABC = 16 800 − 4800 − 6000
$$= 6000 \text{ m}^2$$
$$\text{Triangle ABC} = 3000 \text{ m}^2$$

In order to find the area of any polygon, the following is the sequence of operations:

1. Find the double longitude and difference in northings (ΔN) of each line.
2. Multiply double longitude by Δnorth.
3. Add these products algebraically.
4. Halve the sum.

The calculations are more neatly set out in tabular form as in Table 14.1 where the area of triangle ABC is calculated. It should be noted that a negative result is perfectly possible. For example, if the coordinates in Table 14.1 were written in a clockwise direction the area would have been − 3000 m². In such cases the negative sign is ignored.

Example

5 The coordinates shown below refer to a closed theodolite traverse ABCDEA:

Station	Easting (m)	Northing (m)
A	+51.0	−150.2
B	+300.1	−24.6
C	+220.1	+151.3
D	−50.0	+175.0
E	−125.2	−51.1

Calculate the area in hectares enclosed by the stations.

Answer (Table 14.2)

Station	Total coordinates		Line	Double longitude	Differences in northings	Double areas (m^2)	
	East (m)	North (m)				+	−
A	+51.0	−150.2					
B	+300.1	−24.6	AB	+351.1	+125.6	44 098.16	
C	+220.1	+151.3	BC	+520.2	+175.9	91 503.18	
D	−50.0	+175.0	CD	+170.1	+23.7	4 031.37	
E	−125.2	−51.1	DE	−175.2	−226.1	39 612.72	
A	+51.0	−150.2	EA	−74.2	−99.1	7 353.22	
					Area =	$\dfrac{186\,598.65}{93\,299.325}$ m²	
					=	9.3299 hectares	

Table 14.2

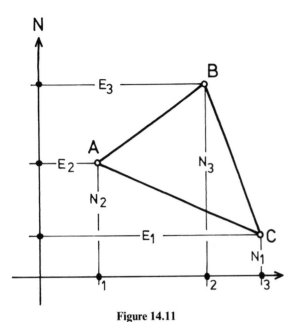

Figure 14.11

= [sum (easting of station
$\times$ northing of preceding station)
$-$ sum (easting of station
$\times$ northing of following station)]

The final result is easily remembered by writing the array in the following manner:

$$2 \times \text{area of figure} = \begin{array}{cc} E_1 & N_1 \\ E_2 & N_2 \\ E_3 & N_3 \\ E_1 & N_1 \end{array}$$

negative *positive*

The following is the sequence of operations in the calculation of the area of any polygon:

1. Write the array of eastings and northings as shown above. Note that E_1 and N_1 are repeated at the bottom of the columns.
2. Multiply the departure of each station by the latitude of the preceding station and find the sum.
3. Multiply the departure of each station by the latitude of the following station and find the sum.
4. Find the algebraic difference between operations 2 and 3 above.
5. Halve this figure to give the area of the polygon. Again the results are always set out in tabular form as in Table 14.3.

Method of total coordinate products

Figure 14.11 shows the same triangular area as Fig. 14.10, where the coordinates of the stations are:

Station	Easting (m)	Northing (m)
A	30.0	60.0
B	90.0	100.0
C	120.0	20.0

Area of triangle ABC = area of trapezium 1AB2
+ area of trapezium 2BC3
$-$ area of trapezium 1AC3

i.e.

$$\text{area of triangle ABC} = \left(\frac{N_2 + N_3}{2}\right) \times (E_3 - E_2)$$
$$+ \left(\frac{N_3 + N_1}{2}\right) \times (E_1 - E_3)$$
$$- \left(\frac{N_2 + N_1}{2}\right) \times (E_1 - E_2)$$

Therefore (2 $\times$ area) of triangle ABC
$= (N_2 + N_3) \times (E_3 - E_2) + (N_3 + N_1) \times (E_1 - E_3)$
$- (N_2 + N_1) \times (E_1 - E_2)$
$= N_2E_3 - N_2E_2 + N_3E_3 - N_3E_2 + N_3E_1 - N_3E_3$
$+ N_1E_1 - N_1E_3 - N_2E_1 + N_2E_2 - N_1E_1 + N_1E_2$

Six of these twelve terms cancel. Therefore, when the remaining six terms are rearranged, the double area of the triangle ABC is

$$2\,ABC = N_1E_2 + N_2E_3 + N_3E_1 - N_1E_3$$
$$- N_2E_1 - N_3E_2$$
$$= (E_1N_3 + E_2N_1 + E_3N_2)$$
$$- (E_1N_2 + E_2N_3 + E_3N_1)$$

Example

6 The coordinates listed below refer to a closed traverse PQRS:

Station	Easting (m)	Northing (m)
P	+ 35.2	+ 46.1
Q	+ 162.9	+ 151.0
R	+ 14.9	+ 218.6
S	$-$ 69.2	$-$ 25.2

Calculate the area in hectares enclosed by the stations.

Solution (Table 14.4)

	Table coordinates		Areas	
Station	East (m) (E)	North (m) (N)	E_2N_1 +	E_1N_2, etc. $-$
A	+ 30.0	+ 60.0		
B	+ 90.0	+ 100.0	+ 5 400	+ 3 000
C	+ 120.0	+ 20.0	+ 12 000	+ 1 800
A	+ 30.0	+ 60.0	+ 600	+ 7 200
			+ 18 000	+ 12 000
			6 000	
		$\div 2 =$	3 000 m^2	

Table 14.3

			Area	
Station	Easting (E)	Northing (N)	$E_2N_1 (+ve)$	$E_1N_2 (-ve)$
P	+35.2	+46.1		
Q	+162.9	+151.0	7 509.69	5 315.20
R	+14.9	+218.6	2 249.90	35 609.94
S	−69.2	−25.2	−15 127.12	−375.48
P	+35.2	+46.1	−887.04	−3 190.12
			−6 254.57	37 359.54
			−37 359.54	
			−43 614.11	
		÷ 2 =	21 807.06 m²	

Table 14.4

Exercise 14.4

1 The coordinates listed below refer to the closed traverse at the GCB Outdoor Centre, which was the subject of Exercise 9.5:

Station	Easting (m)	Northing (m)
A	0.00	0.00
B	−34.39	34.39
C	−51.50	13.01
D	−12.07	−15.10
E	5.35	−28.78
F	16.34	−14.43

Calculate the areas enclosed by the stations, using
(a) the method of double longitudes,
(b) the method of products.

4. Measuring areas from plans

Several methods are available for calculating the area of a figure from a survey plot.

(a) Graphically

A piece of transparent graph paper is laid over the area, the squares are counted and the area is calculated by multiplying the area of a square by the number of squares.

(b) By Simpson's or trapezoidal rules

The area is divided into a series of equidistant strips. The ordinates are measured and the rules applied as in previous examples.

(c) Mechanically

The area is measured, using a mechanical device known as a planimeter.

Example

7 Figure 14.12 shows an irregular area drawn on a plan to a scale of 1:500. Calculate the area of the top of the embankment by the following methods:
(a) counting squares,
(b) Simpson's and trapezoidal rules.

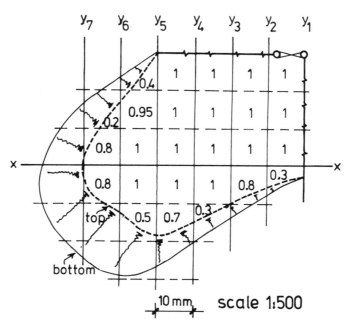

Figure 14.12

Solution

(a) The graph paper shown superimposed on the area has squares of 10 mm side; therefore, for each square,

$$\text{Ground area} = (10 \times 500 \times 10 \times 500)\ \text{mm}^2$$
$$= 100 \times 0.25\ \text{m}^2$$
$$= 25\ \text{m}^2$$

$$\text{Area} = (25 \times \text{number of squares})\ \text{m}^2$$
$$= (25 \times 21.75)\ \text{m}^2$$
$$= 543.8\ \text{m}^2$$

(b) Consider the line marked xx as a base line and every line of the graph paper as an ordinate y, thereby producing seven in total (y_1 to y_7). The lengths of the respective ordinates are, by scaling, 16, 18.3, 20, 22.5, 23.8, 15.3 and 0 m, and the spacing of the ordinates is 5 m along the base line.

By Simpson's rule:

$$\text{Area} = \tfrac{5}{3}[16 + 0 + 2(20 + 23.8)$$
$$+ 4(18.3 + 22.5 + 15.3)]$$
$$= 546.67\ \text{m}^2$$

By trapezoidal rule:

Area
$$= 5\left[\frac{(16 + 0)}{2} + 18.3 + 20 + 22.5 + 23.8 + 15.3\right]$$
$$= 539.50\ \text{m}^2$$

(d) Planimeter

The area of any irregular figure may be found from a plan, by using a mechanical device for measuring areas known as a planimeter.

Two kinds of planimeter are available:

(a) a fixed index model,
(b) a sliding bar model.

Construction

The construction of both models is essentially the same (Fig. 14.13), consisting of:

1. An arm of fixed length, known as the polar arm. The polar arm rests within the pole block P, which, in turn, rests upon the plan in a stationary position.
2. A tracer arm carrying a tracer point, T, which can be moved in any direction across the plan.
3. Attached to both of these arms is the measuring unit, M, which is, in effect, a rolling wheel. As the tracer point moves, the rolling wheel rotates. The wheel is divided into ten units each of which is subdivided into ten parts. The drum therefore reads directly to hundredths of a revolution and a vernier reading against the drum allows thousandths of a revolution to be measured. A horizontal counting wheel is directly geared to the rolling wheel and records the

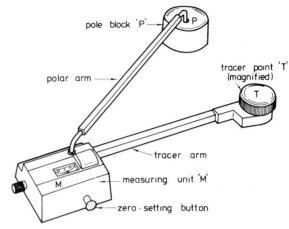

Figure 14.13

number of complete revolutions. The counting wheel can be made to read zero by simply pressing the zero-setting button.

Use of planimeter

1. *Fixed index model*

If only a small area is to be measured the following procedure is carried out:

(a) Position the pole block outside the area.
(b) Set the tracer point over a well-defined mark such as the intersection of two fences.
(c) Press the zero-setting button. The instrument of course reads 0.000 revolutions.
(d) Move the tracer point carefully around the boundary of the area being measured and return to the starting point.
(e) Note the reading; let it be 3.250 revolutions.
(f) Repeat all of the operations twice more and obtain a mean value of the number of revolutions of the wheel.

This particular model of the Stanley Allbrit planimeter reads the area in square centimetres. Each revolution of the measuring wheel is equivalent to 100 cm^2 of area.

In the example the area is therefore

$$(100 \times 3.250)\ \text{cm}^2 = 325.0\ \text{cm}^2$$

If the scale of the plan is full size, the actual area measured is 325 cm^2. If, as is likely, the scale is much smaller, for example 1:500, the actual area must be obtained by calculation.

$$\text{On 1:500 scale, 1 cm} = 500\ \text{cm}$$
$$\text{Therefore 1 cm}^2 = (500 \times 500)\ \text{cm}^2$$
$$= \left(\frac{500 \times 500}{100 \times 100}\right)\ \text{m}^2$$
$$= 25\ \text{m}^2$$
$$\text{Therefore 325 cm}^2 = (25 \times 325)\ \text{m}^2$$
$$= 8125\ \text{m}^2$$

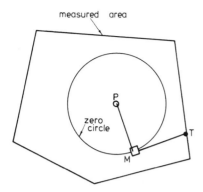

measured area

Figure 14.14

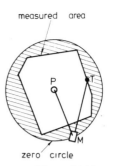

measured area

Figure 14.15

If a large area is to be measured, account must be taken of the instrument's 'zero circle'. Every planimeter has a zero circle. When the polar arm and tracer arm form an angle of, say, 90° and the angle is maintained as the tracer point is moved round in a circle, the drum will not revolve and the area of the circle swept out on the plan by the polar arm will be zero. The manufacturer supplies the actual area of the zero circle in the form of a constant which is added to the number of revolutions counted on the drum.

If the enclosure in Fig. 14.14 is on a scale of 1:2500 and its area is required, the pole block is placed within the enclosure and the tracer point is moved around the boundary as before. The average number of revolutions after following the boundary three times is perhaps 5.290.

$$\begin{aligned}
\text{Number of revs} &= 5.290 \\
\text{Add zero circle constant} &= 22.300 \\
\text{Total revs} &= 27.590 \\
\text{Total cm}^2 &= 27.59 \times 100 \\
&= 2759 \text{ cm}^2
\end{aligned}$$

On 1:2500 scale, $1 \text{ cm}^2 = \left(\dfrac{2500 \times 2500}{100 \times 100}\right) \text{m}^2$

$$= 625 \text{ m}^2$$

Therefore total area of

$$\begin{aligned}
\text{enclosure} &= 2759 \times 625 \\
&= 1\,724\,375 \text{ m}^2 \\
&= 172.438 \text{ hectares}
\end{aligned}$$

Great care must be taken when using the planimeter with the pole block inside the area. It is perfectly possible that the zero circle is larger in area than is the parcel of land being measured (Fig. 14.15). In such a case the rolling drum actually moves backwards and the second reading is subtracted from the first to obtain the area in square centimetres. If the first reading is called 10.000 instead of 0.000 the subtraction is simple; for example:

$$\begin{aligned}
\text{First reading} &= 10.000 \\
\text{Second reading} &= 7.535 \\
\text{Number of revs} &= 2.465
\end{aligned}$$

This area is, however, the area of the shaded portion of Fig. 14.15 and the true area is found by subtracting the number of revolutions from the zero circle constant:

$$\begin{aligned}
\text{Number of revs} &= 22.300 - 2.465 \\
&= 19.835 \\
&= 1983.5 \text{ cm}^2
\end{aligned}$$

If the scale of the plan is 1:100:

$$\text{True area} = 1983.5 \text{ m}^2$$

2. *Sliding bar model*
On the Allbrit sliding bar planimeter the tracer arm is able to slide through the measuring unit (Fig. 14.16) and can be clamped at any position. The arm is graduated and any setting can be made against the index mark, which is attached to the measuring unit. A table on the base of the instrument gives the graduation reading for various scales in common use.

The number of revolutions is obtained by the methods previously described and the true area is calculated by multiplying by the appropriate conversion factor supplied by the manufacturer for the scale being used. For example, when the graduation setting for a scale 1:2500 is 9.92, one revolution of the wheel corresponds to a true area of 4 hectares.

If the pole block was positioned inside the enclosure, the mean readings for four measurements were:

First reading: 0.000   0.000   0.000   0.000
Second reading: 4.320   4.320   4.321   4.323

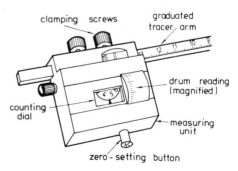

Figure 14.16

The true area is found as follows:

Mean of four measurements = 4.321
Constant for zero circle = 25.730

$$\overline{\hspace{2cm}30.051}$$

True area = 30.051 × 4 hectares
= 120.204

Example

8 In Example 7, the area of the top of an embankment was calculated by (a) counting squares and (b) using trapezoidal and Simpson's rules. Check the area of the embankment in Fig. 14.12, which is drawn to scale 1:500, using a planimeter.

Answer

By planimeter (fixed index):

Number of revolutions = 0.2158
At 1:500, 1 rev = 2500 m^2
Therefore area = 539.5 m^2

Exercise 14.5

1 Figure 14.17 shows an irregular area of ground lying between a straight kerb and a curved fence drawn to scale 1.500. Calculate the area by:
(a) Simpson's and trapezoidal rules,
(b) counting squares,
(c) planimeter.

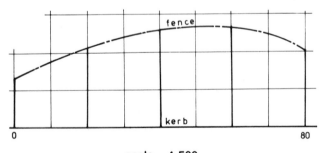

scale   1.500

Figure 14.17

Exercise 14.6

1 Figure 14.18 is the sketch of a building site in the shape of a quadrilateral, the lengths of the sides being as follows:

AB 325 m, AD 195 m, DB 410 m,
DC 392 m, CB 260 m

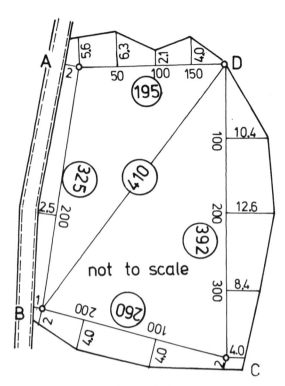

not to scale

Figure 14.18

Calculate the area enclosed by the survey lines, in square metres.

2 The site in Fig. 14.18 is bounded by fences along the sides AD, DC and CB and by a roadway along the side AB. The offsets to the fences and roadway are:

Line chainage (m)	Offset (m)	Line chainage (m)	Offset (m)
AB 0 (A)	2.0 right	AD 0 (A)	5.6 left
200	2.5 right	50	6.3 left
325 (B)	1.0 right	100	2.1 left
		150	4.0 left
		195 (D)	0.0
DC 0 (D)	0.0 left	CB 0 (C)	2.0 left
100	10.4 left	100	4.0 left
200	12.6 left	200	4.0 left
300	8.4 left	260 (B)	2.0 left
392 (C)	4.0 left		

Calculate the area of the site, in square metres.

3 Stations M, N, O, P and Q form a closed traverse. The following coordinates refer to the stations.

Station	Total latitude	Total departure
M	+2000	+2000
N	+3327	+1242
O	+4093	+2048
P	+3141	+3035
Q	+1192	+3572

Calculate the area in hectares enclosed by the stations.

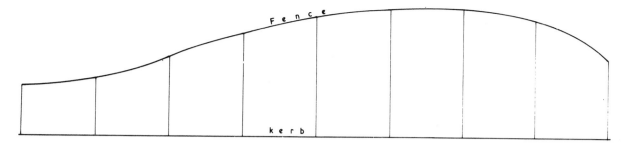

scale 1.500

Figure 14.19

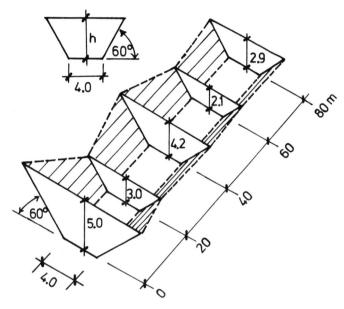

Figure 14.20

4 Figure 14.19 shows an irregular parcel of ground bounded by a kerb on the south side and a fence on the north (scale 1.500). Calculate the area by:
(a) counting squares,
(b) Simpson's rule,
(c) trapezoidal rule,
(d) planimeter.

5 Figure 14.20 shows a roadway cutting with cross sections drawn at 20 metre intervals. Calculate:
(a) the top width of each cross section,
(b) the total plan area occupied by the cutting.
(c) the cross-sectional area of the trapezoidal sections at chainages 0.0 and 40.0 m.

5. Answers

Exercise 14.1

1 Overall area (trapezoid)

$$\text{Area} = 0.5 (25.0 + 30.0) \times 17.0 = 27.5 \times 17$$
$$= 467.50^2$$

Water feature
1. Semi-circle
$$= \pi \times 5.621^2 \times 0.5 \qquad = 49.63$$
2. Rectangle
$$= 11.242 \times 2.0 \qquad = 22.48$$
3. Trapezoid
$$= 0.5 (11.242 + 7.242) \times 4.0 = 36.97$$
4. Triangle
$$= 0.5 (3 \times 3) \qquad = 4.50$$
5. Sector
$$= \pi \times 3^2 \times (135 \div 360) \qquad = 10.60$$
$$\text{Total} \qquad = 124.18 \text{ m}^2$$
Area to be tiled $= 467.50 - 124.18$
$$= 343.32 \text{ m}^2$$

Exercise 14.2

1 In triangle PQ $= r = 60.0 \text{ m}$ $s - r = 57.3$
PQR   QR $= p = 104.6 \text{ m}$ $s - p = 12.7$
   RP $= q = 70.0 \text{ m}$ $s - q = 47.3$
Perimeter of PQR $= 234.6 \text{ m}$ Check $= 117.3 = s$

Therefore semi-perimeter $s = 117.3$ m

Area of triangle PQR
$$= \sqrt{s(s-r)(s-p)(s-q)}$$
$$= \sqrt{117.3 \times 57.3 \times 12.7 \times 47.3}$$
$$= 2009.3 \text{ m}^2$$

Area between line RQ and stream is as follows:

$$\text{Area of triangle (1)} = \tfrac{1}{2} \times 19 \times 4 = 38.0$$
$$\text{Trapezoid (2)} = \tfrac{1}{2}(4+8) \times (38-19) = 114.0$$
$$\text{Trapezoid (3)} = \tfrac{1}{2}(8+4.5) \times (55-38) = 106.25$$
$$\text{Rectangle (4)} = 4.5 \times (72-55) = 76.5$$
$$\text{Trapezoid (5)} = \tfrac{1}{2}(4.5+7) \times (87-72) = 86.25$$
$$\text{Triangle (6)} = \tfrac{1}{2}(104.6-87) \times 7 = 61.6$$
$$\text{Total} = 482.6 \text{ m}^2$$

Area between line RP and fence is as follows:

$$\text{Area} = 10\left(\frac{3+2.5}{2} + 8 + 10 + 9.5 + 9.2 + 7.1 + 4.5\right)$$
$$= 510.5 + 5.0 \text{ (5.0 is the area of the small}$$
$$\text{triangular parcel at P)}$$
$$= 515.5 \text{ m}^2$$

Area between PQ and road is as follows:

$$\text{Area} = \frac{15}{3}[4 + 4 + 2(6.5) + 4(5.0 + 5.6)]$$
$$= 317.0 \text{ m}^2$$

Total area of linear survey
$$= 2009.3 + 482.6 + 317.0 + 515.5$$
$$= 3324.4 \text{ m}^2$$

Exercise 14.3

1

Chainage (m)	Height h (m)	Bottom/top width (m) (5.0 + 2h)
0	5.0	15.0
30	3.5	12.0
60	0.0	5.0
90	2.4	9.8
120	2.9	10.8

Using Simpson's rule,

$$\text{Plan area} = (30/3)[15.0 + 10.8 + 2(5.0)$$
$$+ 4(12.0 + 9.8)]$$
$$= 1230 \text{ m}^2$$

$$\text{Road area} = 120 \times 5.0$$
$$= 600 \text{ m}^2$$

Therefore plan area of side slopes
$$= (1230 - 600) \text{ m}^2$$
$$= 630 \text{ m}^2$$

$$\text{Actual area} = 630/\cos 45°$$
$$= 891 \text{ m}^2$$

Exercise 14.4

1 (a) Table 14.5
(b) Table 14.6

Exercise 14.5

1 (a) Using Simpson's rule:
$$\text{Area} = \frac{10}{3}[6.5 + 10.0 + 2(12.8)$$
$$+ 4(10.5 + 13.2)] \text{ m}^2$$
$$= \frac{10}{3}[136.9]$$
$$= 456.3 \text{ m}^2$$

Using the trapezoidal rule:

$$\text{Area} = 10\left[\frac{6.5 + 10.0}{2} + (10.5 + 12.8 + 13.2)\right]$$
$$= 447.5 \text{ m}^2$$

(b) By counting squares (Fig. 14.17):
$$\text{Ground area} = 18 \times 25 \text{ m}^2 = 450 \text{ m}^2$$

(c) By planimeter:
$$\text{Area} = 25 \times 18.1$$
$$= 452.5 \text{ m}^2$$

Station	Easting	Northing	Line	Double longitude	Difference in northings	Double area
A	0.00	0.00	—	—	—	—
B	−34.39	34.39	AB	−34.39	−34.39	1182.67
C	−51.50	13.01	BC	−85.89	21.38	−1836.33
D	−12.07	−15.10	CD	−63.57	28.11	−1786.95
E	5.35	−28.78	DE	−6.72	13.68	−91.93
F	16.34	−14.43	EF	21.69	−14.35	−311.25
A	0.00	0.00	FA	16.34	−14.43	−235.79
						−3079.58
					Area =	−1539.79 m²

Table 14.5

Station	Easting	Northing	Line	Areas	
				E_2N_1	E_1N_2
A	0.00	0.00		—	—
B	−34.39	34.39	AB	0.00	0.00
C	−51.50	13.01	BC	−1771.09	−447.41
D	−12.07	−15.10	CD	−157.03	777.65
E	5.35	−28.78	DE	−80.79	347.37
F	16.34	−14.43	EF	−470.27	−77.20
A	0.00	0.00	FA	0.00	0.00

−2479.166 − 600.410 =	−2479.166 −3079.576	600.410

Area =	1539.79 m^2

Table 14.6

Exercise 14.6

1 80 300 m^2

2 85 910.4 m^2

3 319.5805 hectares

4 (a) 42.5 squares × 25 m^2 = 1062.5 m^2
   (b) Simpson's rule = 1062 m^2
   (c) Trapezoidal rule = 1047 m^2
   (d) 4.247 revs × 100 = 42.47 × 25 m^2
       = 1061.8 m^2

5 (a)

Chainage (m)	Top width (m)
0	9.77
20	7.46
40	8.85
60	6.42
80	7.34

(b) Plan area occupied by cutting = 602 m^2

(c)

Chainage (m)	Cross-sectional area (m^2)
0	34.43
40	26.99

6. Project

Chapter 18 is a project covering all chapters of this textbook. It is intended that the project build into a complete portfolio of the surveying work required in the survey and the setting out of a building or engineering development.

If the reader wishes to continue work on the project or begin work at this stage, he or she should now turn to page 399 and attempt Section 11 (part).

CHAPTER 15

Mensuration— volumes

Objective
After studying this chapter, the reader should be able to (a) assess the most suitable method of calculating volumes (whether regular or irregular in shape) and (b) make the necessary calculations to determine the volumes of construction work projects.

On almost every construction site, some form of cutting or embankment is necessary to accommodate roads, buildings, etc. In general the earthworks fall into one of two categories:

(a) long narrow earthworks of varying depths— roadway cutting and embankments,
(b) wide flat earthworks—reservoirs, sports pitches, car parks, etc.

1. Cuttings (with vertical sides)

In Chapter 6, Fig. 6.4, partly reproduced below to a smaller scale as Fig. 15.1, the longitudinal and cross-sectional areas of a proposed sewer are drawn to scale. The reader should revise the particular section in Chapter 6 in order to appreciate the sources of data used below.

In Fig. 15.1 the sewer track has vertical sides. The depth varies along the length of the trench, the width

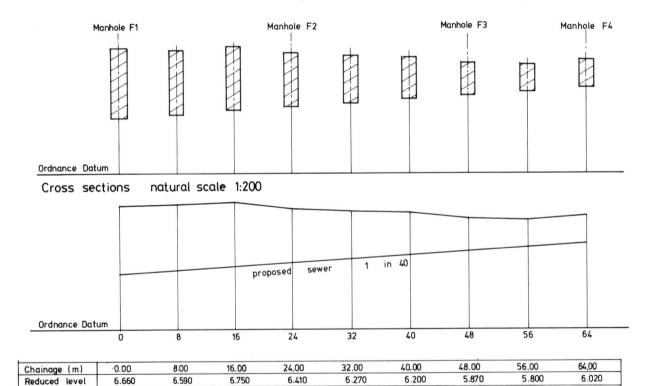

Chainage (m)	-0.00	8.00	16.00	24.00	32.00	40.00	48.00	56.00	64.00
Reduced level	6.660	6.590	6.750	6.410	6.270	6.200	5.870	5.800	6.020
Invert level	2.950	3.150	3.350	3.550	3.750	3.950	4.150	4.350	4.550
Cut(+) Fill(-)	3.710	3.440	3.400	2.850	2.520	2.250	1.720	1.450	1.470

Longitudinal section along sewer F1- F4   Scale horz :-1:500   vert. :- 1:200

Figure 15.1

is constant at 0.8 m and there is no ground slope across the section.

In this case, and *only* in this case, the volume may be calculated by *either* of the following methods.

(a) Method 1: side area

Computing the side area of the trench ABCD by Simpson's rule of the trapezoidal rule and multiplying the area by width 0.8 m.

(b) Method 2: cross sections

Computing the cross-sectional area of the trench at each chainage point and entering the values into Simpson's rule to produce the volume directly. The rule is the same as the rule for area except that cross-sectional areas are substituted for ordinates in the formula. Therefore, in an earthwork having five cross sections, A_1 to A_5, at a chainage interval of d metres, the formula is

$$\text{Volume} = \frac{d}{3}[A_1 + A_5 + 2 \times A_3 + 4 \times (A_2 + A_4)] \text{ m}^2$$

Example

1 In Fig. 15.1, the accompanying longitudinal and cross sections show that the depth of the trench, at 8 metre intervals, is as follows:

Chainage (m)	Depth (m)
0	3.71
8	3.44
16	3.40
24	2.85
32	2.52
40	2.25
48	1.72
56	1.45
64	1.47

Given that the trench is to be 0.8 metre wide and has vertical sides, calculate the volume of material to be removed to form the excavation.

Answer

Method 1

(a) Using Simpson's rule and calling the chainage interval D, the area A of the side of the trench is found from the formula:

$A = (D/3)[\text{first} + \text{last} + 2(\text{odds}) + 4(\text{evens})]$ m
$= (8/3)[(3.71 + 1.47) + 2(3.40 + 2.52 + 1.72)$
$\qquad + 4(3.44 + 2.85 + 2.25 + 1.45)]$
$= (8/3)[5.18 + 2(7.64) + 4(9.99)]$
$= (8/3)[60.42]$
$= 161.12$ m

$$\text{Volume} = (161.12 \times 0.8) \text{ m}$$
$$= 128.9 \text{ m}^3$$

(b) Using the trapezoidal rule.

$A = D[\text{average of first and last depths}$
$\qquad + \text{sum of others}]$
$= 8[(3.71 + 1.47)/2 + 3.44 + 3.40 + 2.85$
$\qquad + 2.52 + 2.25 + 1.72 + 1.45]$
$= 161.76 \text{ m}^2$

$$\text{Volume} = (161.76 \times 0.8)$$
$$= 129.4 \text{ m}^3$$

Method 2
The cross-sectional area at each chainage point is calculated and the values entered directly into Simpson's rule to produce the volume.

Chainage (m)	Depth (m)	Area ($D \times 0.8$) m^2
0	3.71	2.97
8	3.44	2.75
16	3.40	2.72
24	2.85	2.28
32	2.52	2.02
40	2.25	1.80
48	1.72	1.38
56	1.45	1.16
64	1.47	1.18

$\text{Volume} = (8/3)[(2.97 + 1.18)$
$\qquad + 2(2.72 + 2.02 + 1.38)$
$\qquad + 4(2.75 + 2.28 + 1.80 + 1.16)] \text{ m}^3$
$= (8/3)[4.15 + 2(6.12) + 4(7.99)]$
$= (8/3)[48.35]$
$= 128.9 \text{ m}^3$

Exercise 15.1

1 Figure 6.10, reproduced here as Fig. 15.2, shows the longitudinal and cross sections of the proposed storm drain at the GCB Outdoor Centre site. Using the chainages and depths shown in the data table and given that the drain is 0.75 metre wide, calculate the volume of material to be removed in excavating the drain track.

2. General rule for calculating volume

Example 1 and Exercise 15.1 clearly demonstrate that two methods are available for calculating volumes of earthworks.

It must be emphasized, however, the method 1, i.e. calculating the area of the side of the trench and multiplying that area by the width, may be used *only* when the trench has vertical sides. In effect, the vertical side of the trench is the base of a prism, the shape of which remains constant over the width of the

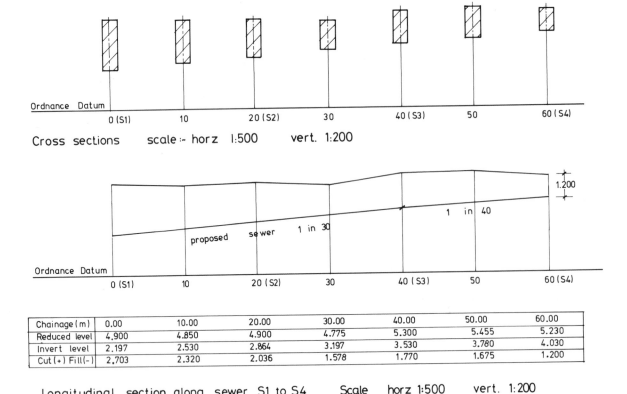

Cross sections   scale :- horz 1:500   vert. 1:200

Chainage (m)	0.00	10.00	20.00	30.00	40.00	50.00	60.00
Reduced level	4.900	4.850	4.900	4.775	5.300	5.455	5.230
Invert level	2.197	2.530	2.864	3.197	3.530	3.780	4.030
Cut (+) Fill(-)	2.703	2.320	2.036	1.578	1.770	1.675	1.200

Longitudinal section along sewer S1 to S4   Scale   horz 1:500   vert. 1:200

Note : Gradient of sewer from mh. S3 to mh. S4 = 0.5m rise in 20m = 1 in 40

Figure 15.2

trench. This area may therefore be multiplied by any width to produce the volume of the trench.

In every other case, i.e. where trench *sides are not vertical*, method 2 must be used. Using this method, the cross-sectional area of the earthwork at regular intervals must be calculated first of all. These areas are then used directly in Simpson's rule to produce the volume.

The use of Simpson's rule presents a problem where there is an even number of cross sections, since the rule only works when there is an *odd* number of sections. In such cases, Simpson's rule is applied to the maximum odd number of sections, leaving an end portion that is easily computed using the prismoidal formula.

The trapezoidal rule, in any form, should not be used, since major errors occur in cases where cross-sectional areas differ substantially.

In general, therefore, the steps in calculating the volume of trenches, cuttings and embankments are:

1. Calculate the cross-sectional area of the earthwork at regular intervals.
2. In cases where there is an odd number of sections, use those sections in Simpson's rule.
3. In a case where there is an even number of sections, use the maximum odd number of cross-sectional areas in Simpson's rule and use the prismoidal rule to

calculate the volume of the remaining prismoid. The prismoidal rule is the subject of Sec. 5(b) (following).

3. Cuttings and embankments with sloping sides

From the foregoing, it is clear that a major part of any volume calculation is the calculation of the cross-sectional areas of the earthworks. There are three distinct types of cross section:

1. One level section
This type of section is developed from the longitudinal section of any earthworks. The one level, which is known, is the centre line ground level at any particular chainage. In Fig. 15.3(a), showing a cutting, and in Fig. 15.3(b), showing an embankment, the ground level (g) has been obtained by levelling. On the cross section, the ground surface is treated as being horizontal through the centre line level. The outline of the earthwork is added from a knowledge of the formation level (f), formation width (w) and side slope gradients (s). The drawing of cross sections is treated fully in Sec. 3(b) in Chapter 6.

2. Three level section
This type of section is usually developed from contour plans or from field work levelling, where levels are taken on either side of the centre line. The three

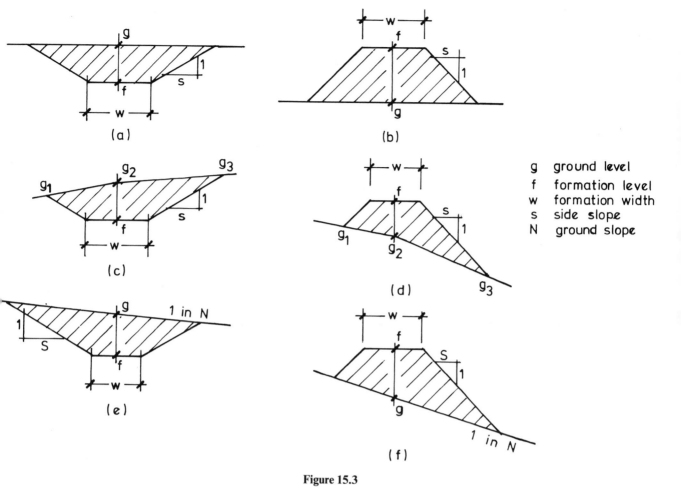

g   ground level
f   formation level
w   formation width
s   side slope
N   ground slope

Figure 15.3

known levels (g) are shown in Fig. 15.3(c) and (d), where they are denoted as g_1, g_2 and g_3. Again the outline of the cutting or embankment is added from a knowledge of the formation level (f), the formation width (w) and side slope gradients (s). The drawing of this type of section is treated fully in Exercises 6.4 and 6.5.

3. Cross fall section

This type of section is developed from the longitudinal section of the earthworks. The centre line level at any particular chainage is observed and the gradient N of the slope across the line of the proposed earthwork is measured using a clinometer. Figure 15.3(e) shows a cutting and Fig. 15.3(f) an embankment, drawn from a knowledge of the gradient 1 in N. This method of obtaining cross sections is not normal practice and is seldom used.

4. Calculation of cross-sectional areas

(a) One level section

In Fig. 15.4, reproduced from Fig. 6.3, the longitudinal and cross sections of roadway R1–R10 at the GCB Outdoor Centre are shown at chainage intervals of 10 metres.

At each chainage point, only the centre line level was known. Each cross section was produced by assuming that the ground across the section was horizontal. By adding the roadway formation level and side slopes, a trapezoidal cross section was produced.

In Fig. 15.5 the cross section is trapezium shaped. Dimensions w and c are known, as explained above, and D is the only unknown. Since the sides slope at 1 to s:

$$D = cs + w + cs$$
$$= 2cs + w$$

$$\text{Area of trapezium} = \left(\frac{D+w}{2}\right) \times c$$

$$= \left(\frac{2cs + w + w}{2}\right) \times c$$

$$= (cs + w) \times c$$

Alternatively, the side slope gradient may be expressed as an angle of θ degrees. In Fig. 15.5, dimension $cs = c/\tan\theta$ and the area of the trapezium is as follows:

$$\text{Trapezium } (cs + w)c = (c/\tan\theta + w)c$$
$$= (c \cot\theta + w)c$$

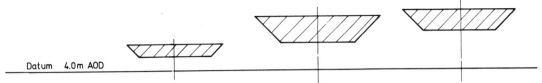

Cross sections: natural scale 1:200

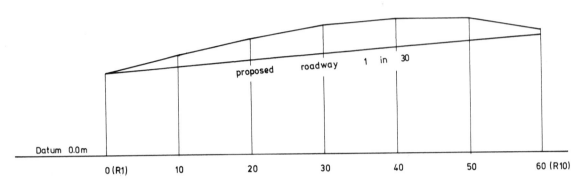

Chainage (m)	0.00	10.00	20.00	30.00	40.00	50.00	60.00
Reduced level	4.400	5.420	6.265	6.895	7.250	7.320	6.585
Formation level	4.400	4.733	5.067	5.400	5.733	6.067	6.400
Cut (+) fill (−)	0.000	0.687	1.198	1.495	1.517	1.253	0.185

Longitudinal section along road R1– R10   Scales:- horz...1:500   vert....1:200

Figure 15.4

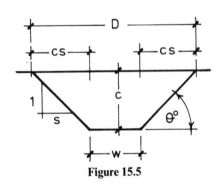

Figure 15.5

Examples

2 From the table of data shown in Fig. 15.4 (the GCB Outdoor Centre), calculate:

(a) the cross-sectional area of the roadway R1–R10 at each chainage point, given that the road is 4.0 metres wide and that the cutting sides slope 1 to 1, i.e. 45°,

(b) the volume of material excavated to form the earthworks, using Simpson's rule.

Answer

(a) Since the side slope gradient is 1 to 1, the cross-sectional area $A = c(cs + w) \, \text{m}^2$

At chainage 0.0 m,

$$A = 0.00 \, \text{m}$$

and at chainage 10.0 m,

$$A = 0.687[(0.687 \times 1) + 4.000]$$
$$= 0.687 \times 4.687$$
$$= 3.22 \, \text{m}^2$$

When several cross sections are to be calculated, a tabular solution is best:

Chainage	c	cs	(cs + w)	c(cs + w)
0	0.000	0.000	4.000	0.00
10	0.687	0.687	4.687	3.22
20	1.198	1.198	5.198	6.23
30	1.495	1.495	5.495	8.22
40	1.517	1.517	5.517	8.37
50	1.253	1.253	5.253	6.58
60	0.185	0.185	4.185	0.77

(b) Volume (V) by Simpson's rule:

$$V = (D/3)[\text{first} + \text{last} + 2(\text{odds}) + 4(\text{evens})]$$
$$= (10/3)[0.00 + 0.77 + 2(6.23 + 8.37)$$
$$+ 4(3.22 + 8.22 + 6.58)] \, \text{m}^3$$
$$= (10/3)[0.77 + 29.2 + 72.08] \, \text{m}^3$$
$$= 340.2 \, \text{m}^3$$

3 The reduced ground level and formation level of an embankment at 0, 30 and 60 m chainages are as follows:

Chainage (m)	0	30	60
RL (m)	35.10	36.20	35.80
FL (m)	38.20	38.40	38.60

Given that the formation width of the top of the embankment is 6.00 m, that the transverse ground slope is horizontal and that the embankment sides slope at 1 unit vertically to 2 units horizontally, calculate the cross-sectional areas at the various chainages.

Answer (Fig. 15.6)

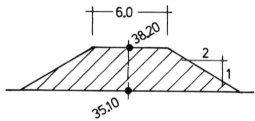

Figure 15.6

Chainage 0 m

$$c = 38.20 - 35.10$$
$$= 3.10 \text{ m}$$
$$w = 6.00 \text{ m (given)}$$
$$D = 2cs + w$$
$$= (2 \times 3.10 \times 2) + 6.00$$
$$= 12.40 + 6.00$$
$$= 18.40 \text{ m}$$

$$\text{Area} = \left(\frac{D+w}{2}\right) \times c$$
$$= \left(\frac{18.40 + 6.00}{2}\right) \times 3.10$$
$$= 12.2 \times 3.10$$
$$= 37.82 \text{ m}^2$$

Alternatively,

$$\text{Area} = (cs + w) \times c$$
$$= (3.10 \times 2 + 6.00) \times 3.10$$
$$= 12.20 \times 3.10$$
$$= 37.82 \text{ m}^2$$

Tabular solution

Chainage	RL	FL	c	cs	(cs + w)	(cs + w)c
0	35.10	38.20	3.10	6.20	12.20	37.82 m²
30	36.20	38.40	2.20	4.40	10.40	22.88 m²
60	35.80	38.60	2.80	5.60	11.60	32.48 m²

4 In Fig. 15.7, the depths of excavation required to form a cutting at 20 m intervals are 0.00, 0.90, 1.50, 3.20 and 3.30 m. Given that the roadway is 5.0 m wide and the side slopes at gradient 30°, calculate the cross-sectional areas of the cutting and hence the volume of excavated material.

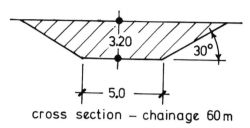

cross section – chainage 60 m

Figure 15.7

Answer
Cross-sectional areas

Chainage (m)	Depth c (m)	(c/tan 30°) (m)	(c/tan 30° + w) (m)	c(c/tan 30° + w) (m²)
0	0.000	0.000	5.000	0.000
20	0.900	1.559	6.559	5.900
40	1.500	2.598	7.598	11.400
60	3.200	5.543	10.543	33.740
80	3.300	5.716	10.716	35.360

Volume (V) by Simpson's rule:

$$V = (20/3)[0.00 + 35.36 + 2(11.40)$$
$$+ 4(5.90 + 33.74)] \text{ m}^3$$
$$= (20/3)[35.36 + 22.80 + 158.56) \text{ m}^3$$
$$= 1444.8 \text{ m}^3$$

In practice there is usually a large number of cross sections in any earthworks. The volume is most easily calculated by using a simple computer program or a spreadsheet.

In Chapter 9, programs were used to calculate co-ordinates and the basic concept of looping was introduced. A loop is a repetitive calculation and since every cross section of an earthwork is a repeat of any other, the computer is ideal for solving earthworks problems.

The simplest program involving looping, for any number of cross sections, is given with an explanation in Sec. 6 in Chapter 17. The program is used to compute Example 3.

Exercise 15.2

1 Figure 15.8 shows part of a roadway, 5.0 m wide, which is partly in cutting and partly on embankment. The relevant earthworks data are given below:

Chainage (m)	Depth of cutting	Height of filling	Side slope gradient
0	1.1	—	1 to 2
10	2.5	—	1 to 2
20	0.0	0.0	—
30	—	1.0	1 to 1
40	—	2.3	1 to 1

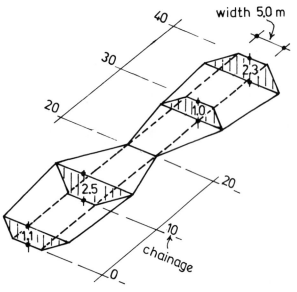

width 5.0 m

Figure 15.8

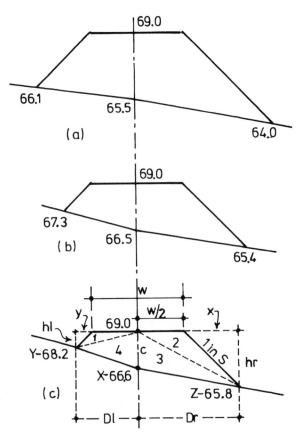

69.0

66.1

65.5

(a)

64.0

69.0

67.3

66.5

(b)

65.4

w

w/2

x

y

hl

69.0

Y-68.2

1

4   c

2

3

1 in s

hr

X-66.6

(c)

Z-65.8

Dl

Dr

Figure 15.9

Calculate:
(a) the cross-sectional area at each chainage point,
(b) the volume of material excavated from the cutting,
(c) The volume of material infilled to form the embankment.

(b) Three level section

Chapter 5 dealt with the subject of contouring and intersection of surfaces. Figure 5.24 shows a contoured plan of a sloping area of ground across which an embankment is to be built. The reader should revise, in Chapter 5, Exercise 5.4, Question 1 (page 118), the method of drawing the outline of the embankment and the cross sections at points A, B and C.

Figure 15.9 is a reproduction of these cross sections and shows clearly (a) the formation level of the embankment, (b) the ground level on the centre line of the embankment and (c) the two ground levels at the base of the side slopes of the embankment. (The figure is reproduced from Fig. 5.28.)

Since the figure is not a regular geometrical figure, it has to be split into triangles where the base and altitude can be readily calculated. The differences between the formation level and reduced levels X, Y and Z produce the dimensions c, hl and hr. Since the sides slope at 1 to s,

$$x = (hr \times s) \quad \text{and} \quad y = (hl \times s)$$

and

$$Dr = x + w/2 = (hr \times s) + w/2$$
$$Dl = y + w/2 = (hl \times s) + w/2$$

In all four triangles, area $= \frac{1}{2}$ base $\times$ altitude

Area triangle 1 $= \frac{1}{2}(w/2 \times hl) = (w/4 \times hl)$
Area triangle 2 $= \frac{1}{2}(w/2 \times hr) = (w/4 \times hr)$

Area triangle 3 $= \frac{1}{2}(c \times Dr) = c/2 \times Dr$
Area triangle 4 $= \frac{1}{2}(c \times Dl) = c/2 \times Dl$

Therefore

$$\text{Total area} = (w/4 \times hl) + (w/4 \times hr) + (c/2 \times Dr) + (c/2 \times Dl)$$
$$= (w/4)(hl + hr) + (c/2)(Dr + Dl)$$

Examples

5 In Fig. 15.9(c), the width of the roadway is 5.0 m and the sides slope at gradient 1 to 1. Calculate the cross-sectional area of the embankment.

Solution
Cross section 15.9(c)

At Y, $hl = 69.0 - 68.2 = 0.80$ m
At X, $c = 69.0 - 66.6 = 2.40$ m
At Z, $hr = 69.0 - 65.8 = 3.20$ m

Since the sides slope at 1 to 1 and $w/2 = 2.50$ m,

$$Dl = (0.80 + 2.50) = 3.30 \text{ m}$$
$$Dr = (3.20 + 2.50) = 5.70 \text{ m}$$

Area of triangle 1
$$= (w/4 \times hl) = 1.25 \times 0.80 = \quad 1.000$$
Area of triangle 2
$$= (w/4 \times hr) = 1.25 \times 3.20 = \quad 4.000$$
Area of triangle 3
$$= \tfrac{1}{2}(c \times Dr) = 1.20 \times 3.30 = \quad 3.960$$
Area of triangle 4
$$= \tfrac{1}{2}(c \times Dl) = 1.20 \times 5.70 = \quad 6.840$$
$$\text{Total area} = \underline{15.800 \text{ m}^2}$$

Alternatively,

$$\text{Total area} = (w/4)(hl + hr) + (c/2)(Dr + Dl)$$
$$= 1.25(0.80 + 3.20)$$
$$+ 1.20(3.30 + 5.70) \text{ m}^2$$
$$= 15.80 \text{ m}^2$$

6 In Fig. 15.9, calculate the areas of the (a) and (b) cross sections, given that the roadway is 5.0 m wide and the sides slope at gradient 1 to 1.

Answer
Cross section (a) (top)

$hl = 2.9$	$Dl = 5.4$	Area $= 32.45$ m^2
$c = 3.5$	$Dr = 7.5$	
$hr = 5.0$		

Cross section (b) (middle)

$hl = 1.7$	$Dl = 4.2$	Area $= 19.50$ m^2
$c = 2.5$	$Dr = 6.1$	
$hr = 3.6$		

Exercise 15.3

1 Figure 5.29 is the solution to Exercise 5.4, question 2 (page 123). In this exercise, a roadway cutting is to be constructed across an area of undulating ground and cross sections are to be drawn at points X, Y and Z. Figure 15.10 is a reproduction of the solution and shows the three cross sections with their respective ground levels and formation levels.

Calculate:
(a) the cross-sectional areas at X, Y and Z,
(b) the volume of material excavated in constructing the cutting.

(c) Cross fall section

In Fig. 15.11 a roadway cutting ABCD is to be formed in ground that has a regular gradient across the longitudinal line of the roadway. From the field work, the reduced ground level F on the centre line is known, together with the formation level E, side slope values 1 in *S*, and transverse ground slope 1 in *N*.

Since the figure is not a geometrically regular figure, it must be split into figures whose areas are easily calculated. Perpendicular lines are drawn through B and C to split the figure into two triangles

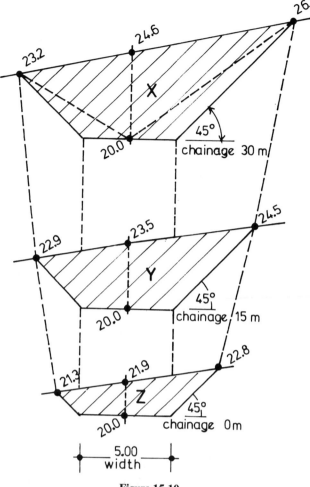

Figure 15.10

AHB and GDC and one trapezium BCGH. Horizontal lines are also drawn through A and D to meet the perpendiculars at J and K respectively.

The central height *c* is the difference between the reduced level F and formation level E. Thus *w* and *c* are known and the area of the trapezium BCGH $= (c \times w)$ m^2.

The areas of the triangles AHB and GDC are of course $\tfrac{1}{2}$(base $\times$ altitude).

$$\text{Therefore area } \Delta AHB = \tfrac{1}{2}(cl \times wl)$$
$$\text{and area } \Delta GDC = \tfrac{1}{2}(cr \times wr)$$

In Fig. 15.11 the following information is known:

Formation width $= 5$ m
Central height $c = 4$ m
Side slopes are 1 to 2
Transverse ground slope is 1 to 5

Since the ground slopes at 1 to 5, $cr = 1/5$ of $w/2$ greater than c, while cl is the same value less than c. Therefore,

$cr = c + (1/5 \times 2.5)$	and	$cl = c - (1/5 \times 2.5)$
$= 4.0 + 0.5$		$= 4.0 - 0.5$
$= 4.5$ m		$= 3.5$ m

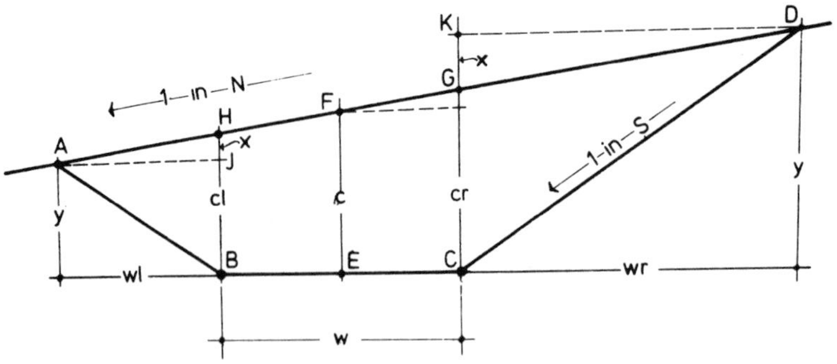

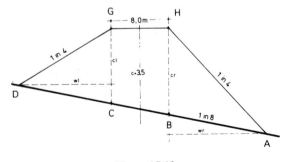

Figure 15.11

In $\triangle ABJ$, $y/wl = \frac{1}{2}$ and therefore $y = wl/2$ and in $\triangle AHJ$, $x/wl = \frac{1}{5}$ and therefore $x = wl/5$. Now

$$cl = (y + x) = \frac{wl}{2} + \frac{wl}{5}$$

$$= \frac{5wl + 2wl}{5 \times 2}$$

$$= \frac{wl(5 + 2)}{5 \times 2}$$

Substitution of the general S for 2 and N for 5 shows that

$$cl = wl\frac{N + S}{NS}$$

$$\text{and } wl = cl\frac{NS}{N + S}$$

Therefore $wl = 3.5 \times \frac{(5 \times 2)}{(5 + 2)} = 5.00$ m

Similarly, in $\triangle KDC$, $y/wr = \frac{1}{2}$ and therefore $y = wr/2$ and in $\triangle GDK$, $x/wr = \frac{1}{5}$ and therefore $x = wr/5$.

$$\text{Now } cr = (y - x) = \frac{wr}{2} - \frac{wr}{5}$$

$$= \frac{5wr - 2wr}{5 \times 2}$$

$$= \frac{wr(5 - 2)}{5 \times 2}$$

Again substitution of S for 2 and N for 5 shows that

$$cr = \frac{wr \times (N - S)}{N \times S}$$

$$\text{and } wr = cr \times \frac{NS}{N - S}$$

Therefore $wr = 4.5 \times \frac{(5 \times 2)}{5 - 2} = 15.0$ m

A general formula can be derived showing that

$$\text{Horizontal length} = \text{vertical length} \times \frac{NS}{N \pm S}$$

The positive sign applies when the ground slope and side slope run in opposite directions, i.e. one up and one down, while the negative sign applies when both gradients are in the same direction, i.e. either both up or both down.

The areas of the three component figures in Fig. 15.11 are:

$$\text{Trapezium HGCB area} = (c \times w)$$
$$= 4 \times 5$$
$$= 20 \text{ m}^2$$

$$\text{Triangle GDC} = cr/2 \times wr$$
$$= \frac{1}{2}(4.5 \times 15.0)$$
$$= 33.75 \text{ m}^2$$

$$\text{Triangle ABH area} = cl/2 \times wl$$
$$= \frac{1}{2}(3.5 \times 5)$$
$$= 8.75 \text{ m}^2$$

$$\text{Total cross-sectional area} = 62.5 \text{ m}^2$$

Examples

7 In Fig. 15.12, a roadway is to be built on ground having a transverse ground slope of 1 in 8. The road is 8.0 m wide, has a central height of 3.5 m and 1 to 4 side slopes. Calculate the cross sectional area of the embankment.

Figure 15.12

Solution

$c = 3.5$, $w = 8.0$ m (given); therefore $w/2 = 4.0$ m, $cr = c + (\frac{1}{8} \times 4.0) = 4.0$ m and $cl = c - (\frac{1}{8} \times 4.0) = 3.0$ m.

$$wr = \frac{4.0 \times NS}{N - S} \qquad wl = \frac{3.0 \times NS}{N + S}$$

$$= \frac{4.0 \times 32}{4} \qquad = \frac{3.0 \times 32}{12}$$

$$= 32.0 \text{ m} \qquad = 8.0 \text{ m}$$

Area HGCB $= c \times w$ Area ABH $= \dfrac{cr}{2} \times wr$

$$= 3.5 \times 8.0$$
$$= 28 \text{ m}^2 \qquad = 2 \times 32.0$$
$$= 64.0 \text{ m}$$

Area CGD $= \dfrac{cl}{2} \times wl$

$$= 1.5 \times 8.0$$
$$= 12.0 \text{ m}^2$$

Total area $= 104.0$ m^2

8 Figure 15.13 shows a side hill section where the ground slope gradient is known. Given the following information, calculate the area of cutting and area of filling.

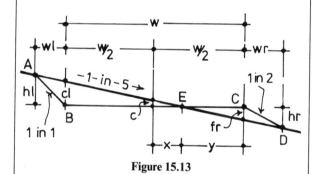

Figure 15.13

Answer

Formation width $w = 15$ m
Central height $c = 0.5$ m
Side slopes
   Cutting $= 1$ in 1
   Embankment $= 1$ in 2
Ground slope 1 in $N = 1$ in 5

   Cutting $cl = c + (\frac{1}{5}$ of $w/2)$ m as before
$$= 0.5 + (\frac{1}{5} \times 7.5) \text{ m}$$
$$= 2.0 \text{ m}$$

Point of no cutting or filling occurs at x metres from the centre line of the formation:

$$x = 5 \times c$$
$$= 2.5 \text{ m}$$
Therefore $y = 5.0$ m

   Filling $fr = \frac{1}{5}$ of y
$$= 1.0 \text{ m}$$

By principle of converging gradients:

$$wl = \frac{cl \times (5 \times 1)}{5 - 1} \qquad wr = \frac{fr \times (5 \times 2)}{5 - 2}$$

$$= \frac{2.0 \times 5}{4} \quad \text{and} \quad = \frac{1.0 \times 10}{3}$$

$$= 2.5 \text{ m} \qquad = 3.33 \text{ m}$$

Slope of cutting $= 1$ in 1
Therefore $hl = 2.5$ m

Slope of embankment $= 1$ in 2
Therefore $hr = 1.67$ m

Area of cutting $=$ triangle ABE
$$= \frac{1}{2}(w/2 + x) \times hl$$
$$= \frac{1}{2} \times 10.0 \times 2.5$$
$$= 12.5 \text{ m}^2$$

Area of filling $=$ triangle ECD
$$= \frac{1}{2}y \times hr$$
$$= \frac{1}{2} \times 5.0 \times 1.67$$
$$= 4.17 \text{ m}^2$$

Exercise 15.4

1 In Fig. 15.14, a roadway is to be built on ground that has a transverse ground slope of 1 in 10. The road is to be 5.00 metres wide, with a central height of 2.00 metres and 1 to 2 side slopes. Calculate the cross-sectional area of the embankment.

5. Calculation of volume

(a) Calculation of volume by Simpson's rule

Once the various cross-sections have been calculated the volume of material contained in the embankment is calculated by Simpson's volume rule. The rule has already been used in Examples 1, 2 and 3, where there were nine, seven and five cross sections respectively.

The rule applies only when there is an *odd* number of cross sections. Should there be an *even* number. Simpson's rule is used to calculate the maximum odd

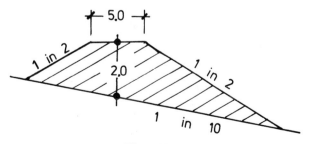

Figure 15.14

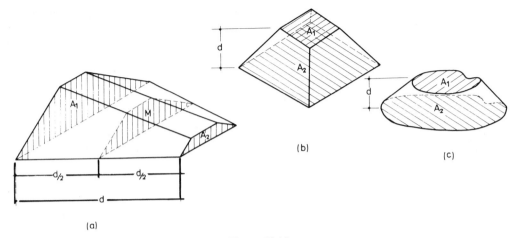

Figure 15.15

number of sections and the prismoidal rule applied to the remainder of the earthworks.

(b) Calculation of volume by the prismoidal rule

A prismoid is defined as any solid having two plane parallel faces, regular or irregular in shape, which can be joined by surfaces either plane or curved on which a straight line may be drawn from one of the parallel ends to the other. Examples of prismoids are shown in Fig. 15.15.

Consider Fig. 15.15(a). In order to determine the volume by Simpson's rule, it is necessary to split the figure such that there is an odd number of equidistant cross sections. Three is the minimum number which fulfills this condition.

Calling the mid-section M, the volume by Simpson's rule is

$$\text{Volume} = (\tfrac{1}{3} \times d/2)\,[A_1 + A_2 + 2(\text{zero}) + 4M]$$
$$= (d/6)\,[A_1 + A_2 + 4M]$$

This is Simpson's prismoidal rule which can be used to find the volume of any prismoid, provided the area M of the central section is determined.

Note. The area of M is *not* the mean of areas A_1 and A_2. The area must be computed from its own dimensions, which are the means of the heights and widths of the two end sections.

Examples

9 Figure 15.16 shows a proposed cutting where the following information is known:

$$\begin{aligned}
\text{Length of cutting} &= 30\text{ m}\\
\text{Formation width} &= 8\text{ m}\\
\text{Depth at commencement} &= 8\text{ m}\\
\text{Depth at end} &= 5\text{ m}\\
\text{Side slopes} &= 1\text{ in }1
\end{aligned}$$

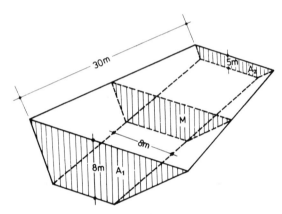

Figure 15.16

Using the prismoidal formula, calculate the volume of material to be removed.

Solution
Section A_1

 Formation width $= 8$ m
   Top width $= (8 + 2c)$ m
 Central depth $c = 8$ m
Therefore top width $= (8 + 16)$ m
       $= 24$ m

Section A_2

 Formation width $= 8$ m
   Top width $= (8 + 2c)$ m
 Central depth $c = 5$ m
Therefore top width $= (8 + 10)$ m
       $= 18$ m

Section M

 Formation width $= 8$ m
   Top width $= (8 + 2c)$ m
 Central depth $c = $ average of depths of A_1
       and A_2
     $= \tfrac{1}{2}(8 + 5)$ m
     $= 6.5$ m

Therefore top width = 8 + 13 m
$$= 21 \text{ m}$$
$$= \text{average of widths of } A_1 \text{ and } A_2$$

Cross-sectional areas (trapezia)

$$A_1 = \tfrac{1}{2}(8 + 24) \times 8 \quad = 128 \text{ m}^2$$
$$A_2 = \tfrac{1}{2}(8 + 18) \times 5 \quad = 65 \text{ m}^2$$
$$M = \tfrac{1}{2}(8 + 21) \times 6.5 = 94.25 \text{ m}^2$$

$$\text{Volume} = (30/6)[128 + 65 + (4 \times 94.25)] \text{ m}^3$$
$$= 2850 \text{ m}^3$$

10 Figure 15.15(b) shows a mound of excavated material stored on site in the form of a truncated pyramid (prismoid). Top area A_1 is 4.00 metres square, base area A_2 is 10.000 metres square, while height d is 4.50 metres. Calculate the volume of material contained in the pyramid using the prismoidal rule.

Answer

$$\text{Top area} = 4.00 \times 4.00 = 16.00 \text{ m}^2$$
$$\text{Base area} = 10.0 \times 10.0 = 100.00 \text{ m}$$

Mid dimensions M
$$= \tfrac{1}{2}(4 + 10) \text{ m square} = 7.00 \text{ m}$$
$$\text{Mid area } M = (7.00 \times 7.00) \text{ m}^2$$
$$= 49.00 \text{ m}^2$$

Volume
$$= (d/6)[A_1 + A_2 + 4M] \text{ m}^3$$
$$= (4.5/6)[16.00 + 100.00 + (4 \times 49.00)] \text{ m}^3$$
$$= 234 \text{ m}^3$$

Note. It is fairly common practice on site to use the end areas formula to calculate the volume of a prismoid. Using this method, the mean of the two end areas is multiplied by the height to produce a volume.

In Example 10, the volume, as computed by this method, is $4.5[(100 + 16)/2] = 261 \text{ m}^3$, which produces an error of 12 per cent. As the top area tends towards zero, the error becomes progressively larger and the use of the end areas formula is not recommended.

Exercise 15.5

1 A proposed service roadway, 5.5 m wide, is to be built along a centre line XY. The embankment is to have side slopes of 1 to 2. Given the following data, calculate the volume of material required to construct the embankment:

Chainage (m)	40	60	80	100	120	140
Ground level (m)	10.00	9.60	9.50	10.40	7.30	10.40
Formation level (m)	11.00	11.10	11.20	11.30	11.40	11.50

6. Volumes of large-scale earthworks

Whenever the volumes of large-scale earthworks have to be determined, e.g. the formation of sports fields, reservoirs, large factory buildings, the field work consists of covering the area by a network of squares and obtaining the reduced levels. The volume is then determined either from the grid levels themselves or from the contours plotted therefrom.

(a) Volumes from spot levels

Figure 15.17 shows a small section of a grid. The total area is to be excavated to a formation level of 90.00 m to form a car park. The sides of the excavation are to be vertical.

The solid formed by each grid square is a vertical truncated prism, i.e. a prism where the end faces are not parallel.

Volume of each prism = mean height × area of base

Mean height of each truncated prism above 90.00 m level is

$$\text{Prism } 1 = (1.0 + 3.0 + 2.0 + 2.0) \div 4 = 2.0 \text{ m}$$
$$2 = (3.0 + 4.0 + 3.0 + 2.0) \div 4 = 3.0 \text{ m}$$
$$3 = (2.0 + 3.0 + 2.0 + 1.0) \div 4 = 2.0 \text{ m}$$
$$4 = (2.0 + 2.0 + 1.0 + 3.0) \div 4 = 2.0 \text{ m}$$

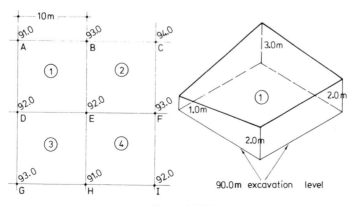

Figure 15.17

Area of base of each truncated prism = 10 × 10
$$= 100 \text{ m}^2$$

Therefore

$$\text{Volume of } 1 = 100 \times 2.0 = 200 \text{ m}^3$$
$$2 = 100 \times 3.0 = 300 \text{ m}^3$$
$$3 = 100 \times 2.0 = 200 \text{ m}^3$$
$$4 = 100 \times 2.0 = 200 \text{ m}^3$$
$$\text{Total volume of excavation} = \underline{900 \text{ m}^3}$$

Alternatively, the volume can be found thus:

Volume = mean height of excavation × total area

The mean height of the excavation is the mean of the mean heights of the truncated prisms. It is *not* the mean of the spot levels.

$$\text{Mean height excavation} = (2.0 + 3.0 + 2.0 + 2.0) \div 4$$
$$= 2.25 \text{ m}$$
$$\text{Total area of site} = 20 \times 20 = 400 \text{ m}^2$$
$$\text{Therefore total volume} = 2.25 \times 400 = 900 \text{ m}^3$$

When examined closely, it is seen that spot level A is used only once in obtaining the mean height of the excavation, spot level B twice, while spot level E is used four times in all. This mean height, and hence the volume, can be readily found by tabular solution as in Table 15.1.

Grid station	Height above formation level	Number of times used	Product
A	1.0	1	1.0
B	3.0	2	6.0
C	4.0	1	4.0
D	2.0	2	4.0
E	2.0	4	8.0
F	3.0	2	6.0
G	3.0	1	3.0
H	1.0	2	2.0
I	2.0	1	2.0
		Sum 16	Sum 36.0

Mean height of excavation = 36.0/16 m
= 2.25 m as before

Table 15.1

The various spot heights are tabulated in column 2 and the number of times they are used, in column 3. Column 4 is the product of columns 2 and 3. The mean height of the excavation is found by dividing the sum of column 4 by the sum of column 3.

Example

11 Figure 15.18 shows spot levels at 20 metre intervals over a site which is to be excavated to 47.00 m to accommodate three tennis courts. Calculate the volume of material to be removed assuming that the excavation has vertical sides.

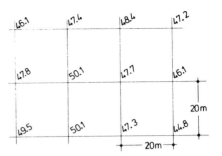

Figure 15.18

Answer

Height above formation level	Number of times used	Product
−0.9	1	−0.9
0.4	2	0.8
1.4	2	2.8
0.2	1	0.2
0.8	2	1.6
3.1	4	12.4
0.7	4	2.8
−0.9	2	−1.8
2.5	1	2.5
3.1	2	6.2
0.3	2	0.6
−2.2	1	−2.2
	24	25.0

$$\text{Mean height of excavation} = \left(\frac{25}{24}\right) \text{ m}$$
$$= 1.042 \text{ m}$$

$$\text{Total area of site} = 60 \times 40$$
$$= 2400 \text{ m}^2$$

$$\text{Therefore volume of excavation} = 1.042 \times 2400$$
$$= 2500 \text{ m}^3$$

Exercise 15.6

1 Figure 15.19 shows spot levels at 10 metre intervals over a small site, which is to be made level at 5.70 m AOD. Calculate the volume of material to be excavated, assuming the excavation has vertical sides.

(b) Volumes from contours

Figure 15.20 shows a mound that has been contoured. If the mound is to be removed the volume of material can be calculated by considering the solid to be split along the contours into a series of prismoids. The volume can then be calculated by successive applications of the prismoidal rule or, where circumstances are favourable, by direct application of Simpson's rule.

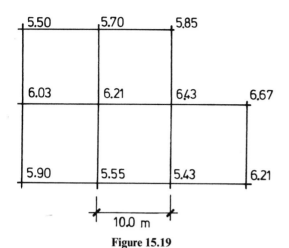

Figure 15.19

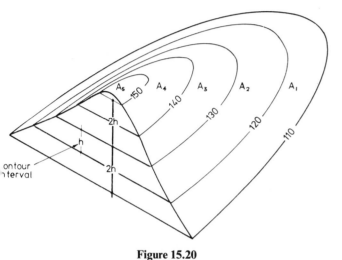

Figure 15.20

When using the prismoidal rule, three contours are taken at a time and the central one is used as the mid-area. The accuracy of the volume depends basically on the contour vertical interval. Generally, the closer the contour interval the more accurate is the volume.

Taking the prismoid formed by contours 110 m and 130 m, the areas enclosed by the contours are determined from the plan by planimeter. The mid-area enclosed by the 120 m contour is likewise determined and the volume of the prismoid is therefore

$$\text{Volume} = \frac{2h}{6}[A_1 + 4A_2 + A_3]$$

Similarly, the volume between contours 130 and 150 m is

$$\text{Volume} = \frac{2h}{6}[A_3 + 4A_4 + A_5]$$

Adding these results gives the volume between the 110 and 150 m contours:

$$\text{Volume} = \frac{2h}{6}[A_1 + 4A_2 + A_3] + \frac{2h}{6}[A_3 + 4A_4 + A_5]$$

$$= \frac{h}{3}[A_1 + A_5 + 2A_3 + 4(A_2 + A_4)]$$

which is the volume by Simpson's rule.

The part of the solid lying above the 150 m contour is not included in the above calculations. It must be approximated to the nearest geometrical solid and calculated separately. In general, the nearest regular solid is a cone or pyramid where the volume = $\frac{1}{3}$ base area × height.

Example

12 Figure 15.21 shows ground contours at 1 metre vertical intervals. ABCD is a proposed factory building where the floor level is to be 32.00 m. The volume of material to be excavated is required. The side slopes of any earthworks are 1 in 2.

(a) The earthwork contours are drawn at 1-metre vertical intervals, i.e. 2 metres horizontally apart.
(b) The surface intersections are found and the outline of the cutting drawn (broken line).
(c) The area enclosed by each contour is obtained by planimeter. The 32 m contour is bounded by A1CD, the 33 m contour by all points numbered 2, the 34 m contour by points numbered 3, the 35 m contour by points numbered 4, while the 36 m contour (point 5) has no area.
(d) The respective areas are:

Contour	32	33	34	35	36
Area (m²)	315.0	294.5	125.0	30.0	0.0

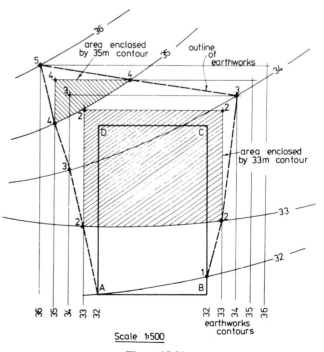

Scale 1:500

Figure 15.21

(e) Volume by Simpson's rule:

$$V = \tfrac{1}{3}[315.0 + 0.0 + (2 \times 125.0)$$
$$+ 4(294.5 + 30.0)] \, m^3$$
$$= 621.0 \, m^3$$

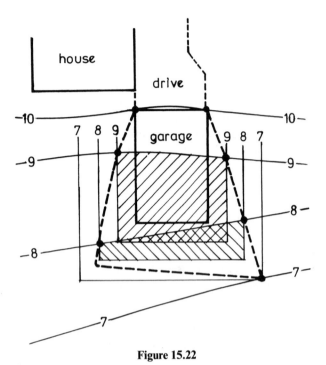

Figure 15.22

Exercise 15.7

1 Figure 15.22 shows ground contours at 1 metre vertical intervals. A lockup garage, measuring 6.0 m × 4.0 m, is to be built to the rear of an existing house, with a floor level of 10.00 m. The garage base requires to be built up on an embankment, the sides of which are to slope at 45°. Calculate:
(a) the areas of the contour planes 10, 9 and 8 m.
(b) the volume of material required to construct the embankment.

Exercise 15.8

1 A sewer 0.75 m wide is to be excavated along a line AB. The sides are to be vertical. Given the following data, calculate the volume of material to be excavated to form the sewer track.

Chainage (m)	0	20	40	60	80	100	120	140
Depth (m)		1.20	1.70	0.95	2.21	2.27	2.21	0.95	1.82

2 The reduced ground level and formation level of an embankment at 0, 30, 60 and 90 m chainages are shown below:

Chainage (m)	0	30	60	90
RL (m)	55.30	56.40	56.00	58.00
FL (m)	58.40	58.60	58.80	59.00

Given that the formation width of the top of the embankment is 6.00 m, that the transverse ground slope is horizontal and that the embankment sides slope at 1 unit vertically to 2 units horizontally, calculate:
(a) the cross-sectional areas at the various chainages,
(b) the volume of material contained in the embankment.

3 Calculate the volume of earth required to form the embankment shown in Fig. 15.23.

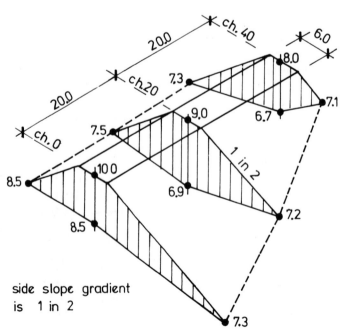

Figure 15.23

4 The central heights of an embankment at two points A and B 90 m apart are 4.0 and 6.5 m respectively. The embankment is built on ground where the maximum slope is 1 in 10 at right angles to the line of the embankment.

Given that the formation width of the embankment is 6 m and the side slope at 1 in 2, calculate the volume of material in cubic metres contained between A and B.

5 Figure 15.24 shows a rectangular grid with levels at 10 m intervals. The whole area is to be covered with waste material to form a car park formation level 86.5 m. Calculate the volume of material to be deposited.

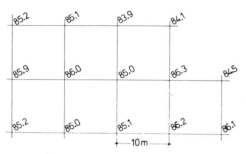

Figure 15.24

6 Figure 15.25 shows contours over an area where it is proposed to erect a small building with a formation level of 23.00 m AOD. Draw on the plan the outline of any earthworks required and thereafter calculate the volume of material required to be cut and filled to accommodate the building. All earthworks have side slopes of 1 in 1.

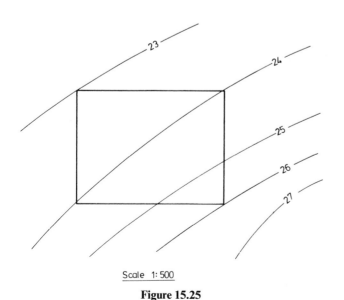

Scale 1:500

Figure 15.25

7. Answers

Exercise 15.1

1

Chainage (m)	Depth d (m)	Area (d × 0.75) (m²)
0.00	2.703	2.027
10.00	2.320	1.740
20.00	2.036	1.527
30.00	1.578	1.184
40.00	1.770	1.328
50.00	1.675	1.256
60.00	1.200	0.900

Volume (Simpson's rule)

$$V = (10/3)\,[(2.027 + 0.900) + 2(1.527 + 1.328) \\ + 4(1.740 + 1.184 + 1.256)]\,\mathrm{m}^3 \\ = 84.5\,\mathrm{m}^3$$

Exercise 15.2

1 (a)

Chainage (m)	0	10	20	30	40
Cross-sectional area (m²)	7.92	25.00	0.00	6.00	16.79

(b) Volume of cutting = 359.7 m³
(c) Volume of embankment = 136.0 m³

Exercise 15.3

1 Cross-sectional area X = 45.23 m³
Cross-sectional area Y = 30.95 m³
Cross-sectional area Z = 13.77 m³

Volume = 914 m³

Exercise 15.4

1 Cross-sectional area = 18.9 m²

Exercise 15.5

1

Chainage (m)	Depth c (m)	Width w (m)	(w + 2c)	Area (w + 2c)c (m²)
5 sections (40 m–120 m)				
40	1.00	5.50	7.50	7.50
60	1.50	5.50	8.50	12.75
80	1.70	5.50	8.90	15.13
100	0.90	5.50	7.30	6.57
120	4.10	5.50	13.70	56.17
2 sections (120 m–140 m)				
120	4.10	5.50	13.70	56.17
130 (mid)	2.60	5.50	10.70	27.82
140	1.10	5.50	7.70	8.47

Volume

Sections (40–120 m) by Simpson's rule:

$$\text{Volume} = \frac{20}{3}[7.50 + 56.17 + 2(15.13)$$
$$+ 4(12.75 + 6.57)]$$
$$= \frac{20}{3}[63.67 + 30.26 + 77.28]$$
$$= 1141.40 \text{ m}^3$$

Sections (120–140 m) by prismoidal rule:

$$\text{Volume} = \frac{20}{6}[56.17 + 4(27.82) + 8.47]$$
$$= 586.4 \text{ m}^3$$

$$\text{Total volume} = 1727.8 \text{ m}^3$$

Exercise 15.6

1

Station	Height above or below FL	Weighting	Product
1	−0.2	1	−0.2
2	0.0	2	0.0
3	0.15	1	0.15
4	0.33	2	0.66
5	0.51	4	2.04
6	0.73	3	2.19
7	0.97	1	0.97
8	0.20	1	0.20
9	−0.15	2	−0.30
10	−0.27	2	−0.54
11	0.51	1	0.51
		20	5.68

$$\text{Mean height of excavation} = 5.68/20 = 0.284 \text{ m}$$
$$\text{Area} = 5 \times 100 = 500 \text{ m}^2$$
$$\text{Volume} = 500 \times 0.284 \text{ m}$$
$$= 142 \text{ m}^3$$

Exercise 15.7

Area of 10 m contour plane = 24.0 m²
Area of 9 m contour plane = 34.8 m²
Area of 8 m contour plane = 14.4 m²

$$\text{Volume (10–8 m)} = (2/6)[24.0 + 14.4 + 4(34.8)] \text{ m}^3$$
$$= 59.2 \text{ m}^3$$

$$\text{Volume (8–7 m)} = 1/3 \times 1.0 \times 14.4$$
$$= 4.8 \text{ m}^3$$

$$\text{Total volume} = 64 \text{ m}^3$$

Exercise 15.8

1 Area of side of trench (trapezoidal rule) = 236 m²

$$\text{Volume} = 236 \times 0.75$$
$$= 177 \text{ m}^3$$

2

Chainage	RL	FL	c	cs	(cs + w)	(cs + w)c
0	55.30	58.40	3.10	6.20	12.20	37.82 m²
30	56.40	58.60	2.20	4.40	10.40	22.88 m²
60	56.00	58.80	2.80	5.60	11.60	32.48 m²
90	58.00	59.00	1.00	2.00	8.00	8.00 m²
75	57.00	58.90	1.90	3.80	9.80	18.62 m²

Volume 0–60 m (Simpson's rule) = 1618.20 m³
60–90 m (prismoidal rule) = 574.80 m³
Total volume = 2193 m³

3 Area   0 m = 17.10 m²
Area 20 m = 18.18 m²
Area 40 m = 8.38 m²
Volume = 654.7 m³

4 8236 m³

5 740.0 m³

6 430 m³ (see Fig. 15.26)

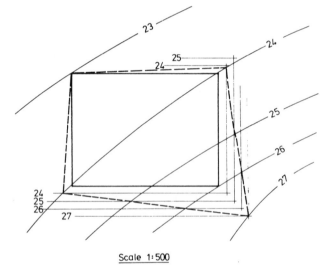

Scale 1:500

Figure 15.26

8. Project

Chapter 18 is a project covering all chapters of this textbook. It is intended that the project builds into a complete portfolio of the surveying work required in the survey and the setting out of a building or engineering department.

If the reader wishes to continue work on the project or begin work at this stage, he or she should now turn to page 399 and attempt Section 11.

CHAPTER 16

Surveys of existing buildings

Objective

After studying this chapter, the reader should be able to make a plan and elevation survey of a simple building and draw the relevant plans to scale.

The building technician and, in particular, the building surveyor will at some time in their careers be concerned with the extension, repair, alteration or demolition of existing buildings. In all of these cases, planning departments and building control departments of local authorities require accurate plans of the existing and proposed buildings, and it is the surveyor's task to take sufficient measurements to enable these to be made. The survey of even a small building will necessitate a large number of measurements being taken.

Besides being capable of conducting the survey, the surveyor must also be aware of (a) the Building Regulations and (b) current building construction practice.

(a) *Building Regulations*
Before any building works can proceed, the local authorities must be satisfied that the Building Regulations have been observed.

The surveyor does not require a detailed knowledge of these regulations but should certainly be aware of their implications.

(b) *Building construction*
The surveyor must have a sound knowledge of building construction and must understand thoroughly the construction of foundations, solid and cavity walls, roof, floors and windows in order to be able to draw a building convincingly.

1. Classification of drawings

In all construction schemes, several classes of drawings are required, the classification depending upon the particular information that is to be disseminated to users.

In British Standards Institution BS 1192, *Recommendations for Building Drawing Practice*, the following classifications are recommended.

(a) Design stage

Sketch drawings show the designer's general intentions.

(b) Production stage

1. *Location drawings*
   (a) Block plans to identify the site and locate outlines of buildings in relation to town plan wherever possible. It is recommended that this plan should be made from the appropriate OS 1:1250, although most authorities accept a scale of 1:2500.
   (b) Site plans to locate the position of buildings in relation to setting-out points, means of access, general layout of site. The plans should also contain information on services, drainage, etc. The recommended scale of these drawings is 1:200, but again local authorities accept 1:250 or 1:500.
   (c) General location drawings to show the position occupied by the various spaces in building, the general construction, the overall dimensions of new extensions, alterations, etc. The recommended scale of these plans is 1:50 or 1:100.
2. *Component drawings*
This classification includes ranges of components, details of components and assembly drawings, which are not really the concern of this chapter.

2. Principles of measurement

In general, the principles involved in measuring a building are those used in the measurement of areas of land. In particular, the principles of linear surveying are applied most often, since buildings can usually be measured completely by taping.

Occasionally a building is of such a complex nature that a theodolite is required.

357

When floor levels are to be related to outside ground and drainage levels, a level and staff are required.

3. Conducting the survey

Figure 16.1 shows the location of a holiday cottage situated in pleasant rural surroundings on the bank of a loch (or lake). The cottage is old and requires renovation and extension. No drawings exist and a complete survey of the premises is to be made.

The following sections describe and illustrate some of the survey work required.

At the conclusion of the chapter the reader should attempt to plot the survey.

(a) Preparation

Before surveying any properties, it is good practice to study the OS sheet and any old drawings that may exist. From these plans, a knowlege of the north direction can usually be gained and a list of adjoining addresses compiled. The proprietors or tenants of these adjoining properties may have to be contacted before permission to build any extension will be given by local authorities.

From the study, it may also be possible to gain some knowledge of difficulties that may arise during the subsequent survey.

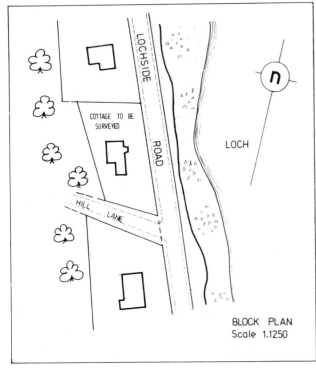

BLOCK PLAN
Scale 1.1250

Figure 16.1

(b) Reconnaissance

As in all surveys, time spent in reconnaissance is well spent. During the 'recce', attention is paid to the shape of the building, the number of floors, type of roof, position of doors and windows.

Squared paper is a necessity, and all sketches made during the reconnaissance should be roughly to scale. The scale depends upon the complexity of the building, but in general 1:100 scale proves adequate.

In measuring buildings, the relationship between rooms is all important, and a plan view of each floor is preferable to a room-by-room sketch.

Elevations will also be required, and here again the whole side of a building should be sketched in preference to a floor-by-floor elevation. The rule of working from the whole to the part is thereby adhered to.

(c) Equipment

In most cases the method of taping will be used, in which case the following equipment will be required:

(a) 20 m steel tape with locating hook,
(b) 5 m steel tape,
(c) 2 m folding rule,
(d) plumb-bob and string, chalk, light hammer and short nails,
(e) measurement book or loose-leaf pad containing a supply of squared paper,
(f) soft pencils, hard pencils and eraser.

The ideal number of surveyors is two, one of whom should be fairly experienced.

(d) Procedure

1. *Site survey*
The site survey uses the principles of linear surveying, namely trilateration and offsetting. On most small sites the details can be surveyed by trilateration alone.

In many cases it is possible to use the building sides as base lines and extend them to the boundaries, supplementing the dimensions with additional diagonal checks (Fig. 16.2).

Whenever possible, running dimensions should be taken, as this procedure is normally physically easier and leads to fewer errors.

The drainage arrangements may be shown on this plan if the system is simple; otherwise a separate sketch is drawn.
2. *Building location*
Plan During the reconnaissance survey, a plan view detailed sketch of the outside of the building is drawn on squared paper (Fig. 16.3). The measuring procedure is arranged in a systematic orderly manner, so that every feature of the building is recorded. The measurements are taken in a clockwise direction and,

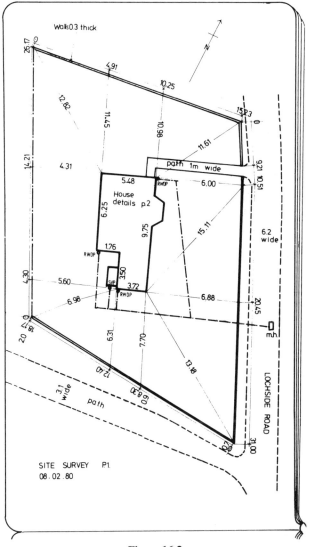

Figure 16.2

wherever possible, running sizes are taken along a complete side of the building.

The tape must be fitted with a locating hook; otherwise one surveyor will have to be employed simply to hold the tape at the zero chainage point of each side. When hooked, the tape is run out along the side of the building; each feature is noted in turn and booked as in Fig. 16.3.

Bay windows present a problem, since the running chainage must be terminated against the window and individual dimensions taken round the window. In Fig. 16.3 these measurements are clearly shown. A check dimension is made by holding the tape in line with the front of the building and measuring the running dimensions to the window and then to the back of the building. The projection of the window must also be obtained.

Elevation Elevations prove to be more difficult than plans, because of the inaccessibility of eaves, high windows, etc. Windows are usually surveyed first

of all. The overall width and height of each window are taken, and the width checked against the building location measurements.

The vertical dimensions from ground level to sill, sill to lintel, and lintel to soffit are measured next as running sizes when possible. Generally this is not possible without the aid of a ladder, and the dimensions are measured individually using the 2 m folding rule.

In the next section it will be seen that measurements must be made internally from the sill to floor level, and from lintel to ceiling level, thereby establishing relationships between floor, ceiling and ground level.

Heights are then measured to the eaves, either from ground level or from a convenient window lintel. A ladder is usually necessary to reach the eaves but, if it is dangerous, a 4 or 5 m levelling staff is usually long enough to enable the measurement to be made.

From the paragraphs above it will be obvious that vertical measurement is difficult, sometimes dangerous. However, a simple convenient, accurate method of determining heights exists and yet is seldom used. A level is set up in a convenient position and, in a very few minutes, staff readings can be made to ground levels and window sills. The staff is then inverted and reading taken to lintels, soffits and eaves. If the level has been judiciously placed, a reading can be taken through an open window to floor level, and an inverted reading taken to ceiling level.

The disadvantage of this method is, of course, the fact that time must be spent in reducing the levels. However, the ease with which the levels are obtained (and their accuracy) is a great advantage and the method should be used whenever possible.

Table 16.1 and Fig. 16.4 show the levels obtained along the western gable and extension of the house.

3. *Internal survey*

Plan Once again a neat accurate sketch of the interior of the building should be prepared on squared paper. In the case of a simple building, the sketch is combined with the building location sketch (Fig. 16.3).

The measuring procedure is arranged to start at the entrance hall and proceed in a clockwise direction around the hall. Each room is then taken in a clockwise direction and similarly measured.

Whenever possible, running dimensions are taken, but individual measurements of window openings, cupboards, etc., may have to be made in case paintings, pictures or curtains are damaged by stretching a long tape along a wall. Furthermore, it may prove impossible to hook a tape handle to a point that is convenient for running dimensions.

Particular attention must be paid at door and window openings when obtaining the thickness of internal and external walls. A sound knowledge of building construction practice is very useful in this context.

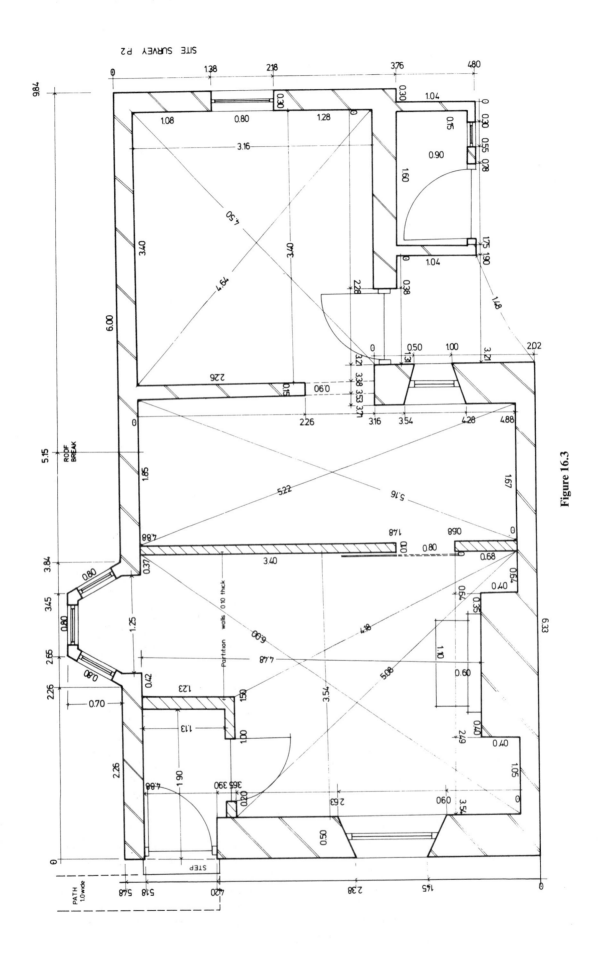

Figure 16.3

360

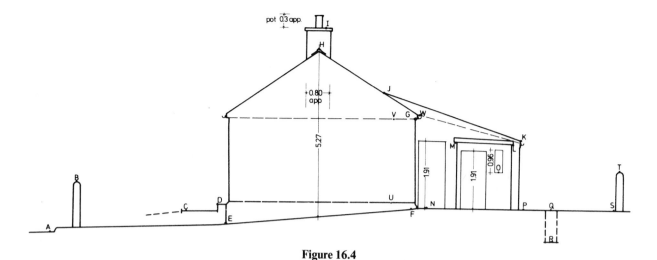

Figure 16.4

BS	IS	FS	Rise	Fall	Reduced level	Remarks
1.51					10.00	A Ground level
	0.62		0.89		10.89	B Top of wall
	1.29			0.67	10.22	C Path
	1.14		0.15		10.37	D Step
	1.41			0.27	10.10	E Ground level
	1.15		0.26		10.36	F Ground level
	−1.85		3.00		13.36	G Soffit
	−3.99		2.14		15.50	H Ridge
	−4.70		0.71		16.21	I Chimney stack
−2.65		−2.65		2.05	14.16	J Roof change
	−1.11			1.54	12.62	K Roof of extension
	−0.96			0.15	12.47	L Roof of WC
	−1.01		0.05		12.52	M Roof of WC
	1.15			2.16	10.36	N Floor level extension
	0.20		0.95		11.31	O Window WC
	1.21			1.01	10.30	P Ground level
	1.20		0.01		10.30	Q Manhole cover
	2.40			1.20	9.11	R Invert level
	1.31		1.09		10.20	S Ground level
	0.31		1.00		11.20	T Top of wall
	0.99			0.68	10.52	U Floor level living room
	−1.88		2.87		13.39	V Ceiling level living room
		−2.10	0.22		13.61	W Ceiling level extension
−1.14	−4.75	−4.75	13.34	9.73	13.61	
−(−4.75)			−9.73		−10.00	
+3.61			+3.61		+3.61	

Table 16.1

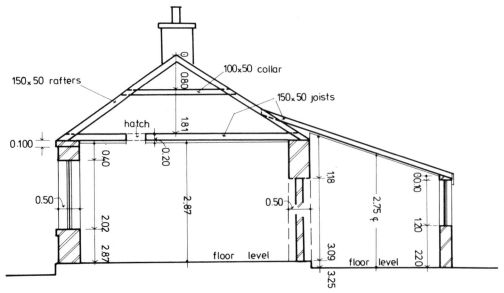

Figure 16.5

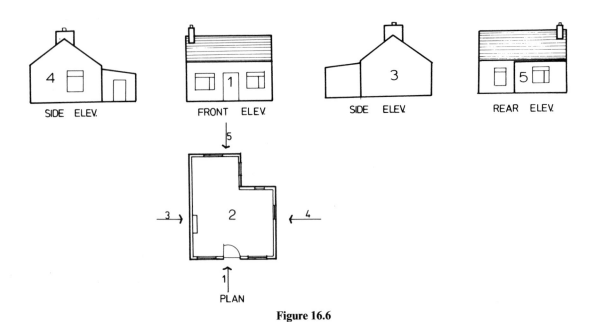

Figure 16.6

○ RWP	RAINWATER PIPE
○ VP	VENTILATION PIPE
○	MANHOLE (surface water)
▢	MANHOLE (soil)
—·—·—	DRAIN
▬▬▬	BUILDING
———	GENERAL DETAILS
———	DIMENSION LINES
⊙	EXISTING TREES
▨▨	BRICKWORK

Figure 16.7

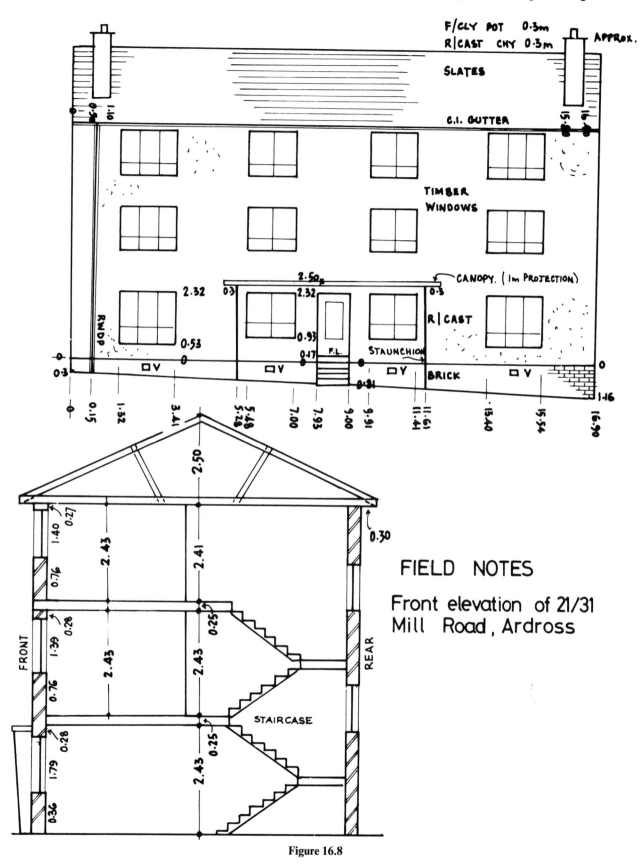

Figure 16.8

The booking of the results is most important, and time must be taken to ensure that cross-checks, etc., are properly dimensioned. Figure 16.3 shows the booking required in this example.

Internal heights Heights must be taken for the purposes of drawing cross sections and for determining ceiling heights in possible dormer extensions, etc. A floor-to-ceiling height is taken in each room, in the middle of the room and at each window. The dimensions should be measured from the ceiling downwards to internal lintels and sills of windows (Fig. 16.5).

A spirit level is a useful item of equipment when measuring heights to sills, since sills are not horizontal. The spirit level is laid across the bottom rail of the window and appropriate dimensions measured to the sill internally and externally.

All door heights are also measured.

When a building consists of more than one floor, the floor-to-floor heights are usually easily measured at the stair well. If difficulty is experienced, a levelling may have to be made up the stairs. Checks may be made externally by hanging a tape from a window on one floor down to a window on the next.

Measurements must also be taken from the floor of the uppermost room to the ceiling and into the roof space. These measurements are made through the ceiling hatch to the top of the ceiling joist, and from there to the apex of the roof.

4. Sections, services, etc.
Depending upon the purpose of the survey, sections may be required through the roof space, ground floor window heads and door thresholds.

The construction details are accurately sketched on squared paper and the relevant sizes obtained using a short 5 m tape. Figure 16.5 shows the relevant survey details for drawing a section through the roof space.

Services are traced individually and separate sketches made to show the run of electrical cables and conduits, and all waste and soil pipes.

4. Plotting the survey

In drawing buildings, several views from different angles are required. These include a plan and elevations of all sides of the building. It is essential that these views be presented in a systematic manner. BS 1192, *Building Drawing Practice*, recommends that only orthographic first-angle projection be used for building drawings.

Plotting of the survey information begins with the plan view, and from it all other views are constructed. In deciding where the plan view is to be placed on the sheet, the surveyor must decide which elevation is to be the principal view. Generally this is the front elevation. If it is possible to place all of the views on one sheet, the principal view is placed left of centre (position 1) in the top half of the sheet (Fig. 16.6).

The plan view is then positioned immediately below the front view (position 2). The side views from left and right are placed on the right (position 3) and left (position 4) respectively of the front view. The view from the rear is usually placed on the extreme right (position 5), although, if more convenient, it may be placed on the extreme left.

Signs and symbols BS 1192, *Building Drawing Practice*, recommends that the principal symbols used on building drawings be graphical. For a complete list of the symbols and abbreviations, readers are referred to this publication. However, in order in enable the reader to plot the exercise following, a few of these symbols are shown in Fig. 16.7.

Exercise 16.1

1 Using the survey information of Figs 16.1 to 16.5 and Table 16.1, draw:
(a) a site plan to a scale of 1:250,
(b) a general location drawing of the building to a scale of 1:50, showing the internal arrangements,
(c) an elevation of the western side of the building to a scale of 1:50,
(d) a north–south section through the building to a scale of 1:50.

2 Figure 16.8 shows the front elevation of a three-storey local authority flatted dwelling. The front elevation is to be renovated.

The various floor levels were located by measuring up the stairwell. The windows were observed to be vertically aligned and chimney dimensions were estimated. Plot the front elevation to a scale of 1:50.

5. Answers

The answer to the first exercise is contained in Figs 16.1 to 16.5 and the answer to the second exercise is Fig. 16.8. These figures are actually scaled versions of the answers and have been changed in size to fit the page format of this textbook.

CHAPTER 17

Computer programs and spreadsheets

Objective

After studying this chapter, the reader should be able to (a) write a simple program in basic code and (b) use a spreadsheet, to solve surveying problems.

1. Introduction

Throughout this textbook the use of computer programs has been advocated for the solution of repetitive surveying calculations. It must be emphasized at the outset that this textbook does not presume to teach computing; after all a computing textbook does not teach surveying.

The alternative to writing simple programs is to purchase them. Once the reader has experimented with the programs in this chapter and realized the saving in time and effort that accrue, he or she may well wish to purchase some of the excellent advanced programs currently available. On the other hand, the use of the following programs will certainly add interest to any surveying course and perhaps spur the reader to greater things.

The chapter assumes that the reader has virtually no knowledge of computing. The programs are presented as tools to enable the reader to solve basic surveying problems, simply by typing them into any microcomputer.

Each microcomputer has its own idiosyncracies but there is a large measure of commonality with all of them. The following programs were used on an AMSTRAD PCW 8512 computer and printer. Other microcomputers may require different printer instructions, in which case the reader should amend any LPRINT statement to enable the particular printer to be used.

2. BASIC programming

The reader will find greater reward and enjoyment if he or she understands the rudiments of programming and how the computer acts upon receipt of the user's instructions.

The following programs are written in a code called BASIC, which stands for Beginners All-purpose Symbolic Instruction Code. There are many versions of

this code and all computer manufacturers have created their own version by amending the original in some way.

The code is understood by the user who writes instructions to the computer in this BASIC code. Inside the computer is an in-built machine code which interprets the basic code and converts it into a series of electrical signals, which the computer, being an electronic device, readily comprehends and acts upon.

In practice, the user presents the computer with a problem, by giving it certain data, called the INPUT. The computer is given instructions on how to handle this data, usually in the form of formulae. The problem is solved by the computer and stored within the machine. The user then instructs the computer to present the solution, either on the screen or in the form of a printed statement.

The following extremely simple example is used to give the reader the opportunity to write a short program, use a computer and obtain a result speedily.

Example

1 The sum (S) and difference (D) of two numbers, A and B, are required. In a BASIC program, the user has to

(a) INPUT A then INPUT B.

(b) Instruct the computer that the sum (S) of A and B is $S = A + B$ and that the difference (D) of A and B is $D = A - B$.

(c) Instruct the computer to print the answers on the screen by telling it to PRINT S and PRINT D.

The computer needs to know the order of handling the data, so each instruction is numbered and the computer then handles each numbered instruction in sequence. The numbers are usually incremented in tens.

The computer is loaded with its own particular version of BASIC language and at the prompt, usually (OK), the program is typed exactly as follows:

```
10   INPUT A
20   INPUT B
30   S=A+B
40   D=A-B
50   PRINT S
60   PRINT D
```

In order to work on the program, the computer is instructed to operate, by typing the word RUN. The computer asks for the data by showing a question mark on the screen. The user responds by typing, say, 100. A second question mark appears and the user types, say, 20. The computer calculates the sum and difference of 100 and 20 and prints S (120) and D (80) on the screen.

Usually, the surveyor requires a hard copy of the results. The computer is instructed to print the results on to paper via a printer. Lines 50 and 60 are therefore replaced by 50 LPRINT S and 60 LPRINT D.

In order to compute a further two numbers, the instruction RUN is again required. However, the program may be made to repeat itself, by the incorporation of a GOTO statement, which tells the computer to go to some line of the program, whereupon the program resumes at that line. The required instruction, in this case is 70 GOTO 10.

Exercise 17.1

1 (a) Write a program which will compute the product of three numbers.
(b) Use the program to find the product of the numbers (i) 2, 3 and 5 and (ii) 100, 56 and 123.75.

Surveying programs use the same programming instructions as those of Example 1. In a traverse survey, the plan lengths and bearings of the lines are known. The partial coordinates are required as a first step in the plotting of the survey. The plan length (p) and the bearing (b) are input on the lines 10 and 20. The difference in eastings (x) and the difference in northings (y) are computed from the appropriate formulae, on lines 30 and 40, and printed on lines 50 and 60. The program is made to repeat itself by the GOTO statement on line 70.

In BASIC programming, angles may be entered in sexagesimal or decimal units but must be converted to radians before calculations are made. The conversion is made by multiplying the angular units by $\pi/180$.

3. Calculation of coordinates

Example

2 Calculate the coordinates of an open traverse ABCDE, given the following data:

Line	AB	BC	CD	DE
Bearing	30	110	225	295
Length	50.0	70.0	82.0	31.2

Note. This traverse is calculated manually in Chapter 9.

Program 1

```
10   INPUT p
20   INPUT b
30   x=p*SIN(b*3.141593/180)
40   y=p*COS(b*3.141593/180)
50   PRINT x
60   PRINT y
70   GOTO 10
```

Explanation
10   p is plan length of line
20   b is bearing of line
30   Computer calculation; b is changed to radians
40   Computer calculation; b is changed to radians
50   Answer: x is part departure (i.e. difference in eastings)
60   Answer: y is part latitude (i.e. difference in northings)
70   Computer goes back to line 10 to receive instructions for the next line of the survey

User instructions
1. Type: RUN
2. Computer response: ? type 50.0
3. Computer response: ? type 30
4. Computer prints x, y on screen like this: 25.0, 43.0
Repeat for next line

This simple Program 1 may be enhanced, refined and enlarged. Program 2 adds the station designation and prints the results on one line.

Program 2

```
10   INPUTS n$
20   INPUT p
30   INPUT b
40   x=p*SIN(b*3.141593/180)
50   y=p*COS(b*3.141593/180)
60   PRINT n$,x,y
70   GOTO 10
```

Explanation
10 n$ is the station reference of any station
20 *p* is plan length of line
30 *b* is bearing of line
40 Computer calculation
50 Computer calculation
60 Results on one line on screen
70 Returns to start

User instructions
1. Type: RUN
2. Computer response: ? type B
3. Computer response: ? type 50.0
4. Computer response: ? type 30
5. Computer prints N$, x, y on screen like this:
    B   25.0,   43.4
    Repeat for next line

Program 3 shows the input information entered on one line. A prompt has been added to tell the operator what information is required. The results have headings and are printed on a line printer using the command **LPRINT** (the command varies with the type of computer).

Program 3

```
10   PRINT "enter stn.ref.,length,brg.of line"
20   INPUT n$,p,b
30   x=p*SIN(b*3.141593/180)
40   y=p*COS(b*3.141593/180)
50   LPRINT "STN.REF", "DIFF EAST", "DIFF NORTH"
60   LPRINT n$,x,y
70   GOTO 10
```

Explanation
10 Prompt
20 Station reference, length, bearing of line
30 Computer calculation
40 Computer calculation
50 Headings printed on line printer
60 Results printed below headings
70 Return to start

User instructions
1. Type: RUN
2. Computer response:
   Enter station reference, length and bearing of line—type B, 50, 30
3. Computer prints headings and results

Solution

STN REF	DIFF EAST	DIFF NORTH
B	25.000	43.301
STN REF	DIFF EAST	DIFF NORTH
C	65.778	−23.941

In Program 3 the results heading is repeated for each line. Program 4 shows how this is amended to give only one heading.

Program 4

```
10  LPRINT "STN REF", "DIFF EAST", "DIFF NORTH"
20  PRINT "enter stn.ref,length,bearing"
30  INPUT n$,p,b
40  x=p*SIN(b*3.14159/180)
50  y=p*COS(b*3.14159/180)
60  LPRINT n$,x,y
70  GOTO 20
```

Explanation

10   Headings for results
20   Prompt
30   Station reference, length, bearing of line
40   Computer calculation
50   Computer calculation
60   Results printed below headings
70   Return to line 20

User instructions

1. Type: RUN
2. Computer response:
   Prints headings on printer, prints prompt on screen—type B, 50, 30
3. Computer prints results on printer below headings
   Repeat for next line

Solution

STN REF	DIFF EAST	DIFF NORTH
B	25.000	43.301
C	65.778	-23.941
D	-57.983	-57.983
E	-28.278	13.186

In each of these four programs, only the partial coordinates were calculated. When the total coordinates are to be computed, a deeper knowledge of BASIC programming is required, particularly the technique of looping. A loop is a repetitive part of a program. In any traverse the loop is as follows:

'Start on total coordinates of a point X; calculate the partial coordinates of line XY; add those to the coordinates of X to produce the coordinates of Y.'
Repeat for next point

In Program 5, arrangements are made to enter and list the total coordinates of the starting point on lines 10–40. Lines 80–140 form the loop (k) which is to be repeated from 1 to m times. Lines 50–70 prompt the operator to enter the number of times (n) that the loop is to be repeated.

Program 5

```
10   PRINT "Enter Stn.ref.,East, North of starting point"
20   INPUT n$, E,N
30   LPRINT "STN.REF","EASTING","NORTHING"
40   LPRINT n$,E,N
50   PRINT "Enter no. of traverse lines"
60   INPUT m
70   FOR k=1 TO m
80   PRINT "Enter stn.ref.,length,brg.of line"
90   INPUT m$,p,b
100  x=p*SIN(b*3.141593/180)
110  y=p*COS(b*3.141593/180)
120  E=(E+x)
130  N=(N+y)
140  LPRINT m$,E,N
150  NEXT k
```

Explanation

10   Prompt
20   Station reference, departure, latitude, of starting point
30   Headings for results
40   Coordinates of starting point
50   Prompt
60   m is number of lines
70   Loop—repeat calculation of n survey lines
80   Prompt
90   Station reference, length, bearing
100 Computer calculation
110 Computer calculation
120 E is easting of station B (= east A + difference in east line AB)
130 N is northing of station B (= north A + difference in north line AB)
140 Results of station B on printer
150 Return to line 80

User instruction

1. Type: RUN
2. Computer response:
   Enter station reference, etc.—Type A, 0.0, 0.0
3. Computer response:

   ```
   Prints    STN REF    EAST    NORTH
   Prints       A        0.00    0.00
   ```

   Prints 'enter number of lines' on screen—type 4
4. Computer response:
   Enter station reference, length and bearing—type B, 50, 30
5. Computer response:
   Prints total coordinates of station B on printer below those of A thus: B, 25.0, 43.3

Solution

STN REF	EAST	NORTH
A	0.00	0.00
B	25.00	43.30
C	90.78	19.36
D	32.80	-38.62
E	4.53	-25.44

Exercise 17.2

1 Calculate the coordinates of an open traverse ABCDE using a BASIC program, from the following data:

Line	AB	BC	CD	DE
WCB	35°	94°	58°	17°
Length (m)	79.0	59.0	52.0	44.0

(This is Example 10 of Chapter 9 (page 207), where the traverse was calculated manually.)

4. Calculation of bearings

Example

3 In Chapter 9, the bearings of the lines of a theodolite traverse were calculated, using the conventional rules of bearings computation. The traverse data are as follows:

Traverse ABCDEF Starting bearing (forward)
line AB = 65° 34′ 20″

Measured angles   ABC   110° 05′ 20″
BCD   219° 50′ 40″
CDE   134° 52′ 50″
DEF   250° 58′ 30″

The bearings of the traverse lines are computed in the solution to Program 6.

Program 6

```
10   REM BEARINGS CALCULATION
20   PRINT "enter ref. of start line"
30   INPUT m$
40   PRINT "enter FORWARD bearing of start line(d,m,s)"
50   INPUT d1,m1,s1
60   LPRINT "LINE","FORWARD WCB."
70   LPRINT m$,d1;m1;s1
80   b=d1+m1/60+s1/3600
90   PRINT "enter no. of lines to be calculated"
100  INPUT n
110  FOR k=1 TO n
120  PRINT "enter ref. of next line"
130  INPUT n$
140  PRINT "enter measured or adjusted angle(d,m,s)"
150  INPUT d2,m2,s2
160  a=d2+m2/60+s2/3600
```

```
170 b=b+a
180 IF b>540 THEN 210
190 IF b>180 THEN 230
200 IF b<180 THEN 250
210 b=b-540
220 GOTO 260
230 b=b-180
240 GOTO 260
250 b=b+180
260 b1=(b-INT(b))*60
270 b2=(b1-INT(b1))*60
280 LPRINT n$,INT(b);INT(b1);INT(b2)
290 NEXT k
```

Explanation

10	Remark
20	Prompt
30	m$ is reference of line
40	Prompt
50	d,m.s—degrees, minutes, seconds
60	Final headings on printer
70	Printout of starting bearing details
80	Starting bearing decimalized
90	Prompt
100	n is number of lines to be calculated
110	k is the loop
120	Prompt
130	n$ is reference line
140	Prompt
150	Angle in degrees, minutes, seconds
160	Angle decimalized
170	b is bearing of next line
180	If b exceeds 540° go to line 210
190	If b exceeds 180° go to line 230
200	If b is less then 180° go to line 250
210	Following from line 18—subtract 540° from b
220	Instruction to computer—omit lines 230–250
230	Following from line 190—subtract 180° from b
240	Instruction to computer—omit line 250
250	Following from line 200—add 180° to b
260–270	b converted to deg min s
280	Print next line reference and bearing on printer
290	Repeat calculation from line 110

User instructions

1. Type RUN
2. Computer response:
   Enter starting line reference—type AB
3. Computer response:
   Enter starting bearing—type 65,34,20
4. Computer response:
   (a) Printer:

LINE	WCB		
AB	65	34	20

   (b) Screen: enter reference next line—type BC
5. Computer response:
   Enter measured angle—type 110,05,20
6. Computer response:

BC	355	39	40

Solution

LINE	FORWARD WCB		
AB	65	34	20
BC	355	39	40
CD	35	30	20
DE	350	23	10
EF	61	21	40

Exercise 17.3

1 Exercise 9.2 (page 194) shows an open traverse ABCD, in which the following angles were measured:

Angle	Accepted value
ABC	93° 15′ 50″
BCD	274° 31′ 10″

Given that the forward bearing of the line AB is 65° 00′ 00″, use BASIC program 6 to calculate the forward bearings of lines BC and CD of the traverse.

5. Calculation of tacheometric data

Stadia tacheometry

Example

4 Table 17.1 shows tacheometric observations to three points B, C and D. (The observations are those of Chapter 10, Table 10.8 (page 226), where they were calculated manually.)

Calculate (a) the horizontal distances from station A to stations B, C and D and (b) the reduced levels of B, C and D.

Instrument station	A				
Height of instrument	1.390				
Reduced level	116.210				

Target station	Horizontal circle	Zenith angle	Stadia Top bottom	S	Mid
RO	00° 00′				
			2.040		
B	12° 30′	84° 20′ 00″	1.600	0.440	1.820
			1.670		
C	34° 15′	87° 34′ 00″	0.820	0.560	1.390
			1.380		
D	63° 26′	98° 10′ 00″	1.000	0.380	1.190

Table 17.1

Program 7

```
10   LPRINT "STADIA TACHMTRY (VERT STAFF)"
20   LPRINT " "
30   LPRINT " "
40   PRINT "enter stn.ref.and red.lev.of starting point"
50   INPUT n$,L
60   PRINT "enter inst.height"
70   INPUT ih
80   LPRINT "INST STN:-";n$;"RED LEV:-";L
90   LPRINT "***************************"
100  LPRINT " "
110  LPRINT "TARGET STN","HORZ.DIST.","RED.LEV."
120  PRINT "enter no. of target points"
130  INPUT n
140  FOR k=1 TO n
150  PRINT "enter stn.ref."
160  INPUT m$
170  PRINT "enter stadia readings-high,mid,low"
180  INPUT r1,r2,r3
190  PRINT "Enter Zenith angle as (d,m,s)"
200  INPUT d,m,s
210  a=(d+m/60+s/3600)*3.141593/180
220  S=r1-r3
230  h=100*S*SIN(a)*SIN(a)
240  v=h/TAN(a)
250  L1=ih+v-r2+L
260  LPRINT m$,h,L1
270  NEXT k
```

Explanation

10	Title
20, 30	Blank lines on printout
40	Prompt
50	n$ is instrument station: L is its reduced level
60	Prompt
70	ih is instrument height
80	Print on printer—4 items
90	Underline these items with a series of asterisks
100	Blank line on printout
110	Print on printer—3 items
120	Prompt
130	n is number of points observed
140	k is a loop (a repeat calculation) from 150 to 270
150	Prompt
160	m$ is any station name
170	Prompt
180	r1 is highest staff reading, r2 is mid reading, r3 is lowest reading
190	Prompt
200	d, m s is the vertical angle
210	Vertical angle decimalized and changed to radians
220	S is the stadia intercept (high–low) staff reading
230	h is horizontal distance

240   *v* is vertical distance from theodolite centre to
mid staff reading

250   L1 is reduced level of observed point

260   Print on printer—name of point, horizontal
distance and reduced level

270   Repeat calculation from line 150

User instructions

1. Type   : RUN
   Screen response : Enter station reference and reduced level of starting point

2. Type   : A, 116.21
   Screen response : Enter instrument height

3. Type: 1.39
   Printer response : INST STN:--- A RED LEV:--- 116.21
                 : *******************************
   Screen response : Enter number of targets

4. Type   : 3
   Screen response : Enter station reference

5. Type   : B
   Screen response : Enter stadia readings—highest, mid, lowest

6. Type   : 2.040, 1.820, 1.600
   Screen response : Enter zenith angle (d,m,s)

7. Type   : 84, 20, 00
   Printer response : B   43.57   120.10
   Screen response : Enter station reference

8. Type   : C, etc., and repeat

Solution

```
STADIA TACHMTRY (VERT.STAFF)

INST ST:- A      RED.LEV.:- 116.21
*********************************
TARGET STN.  HORZ. DIST.  RED. LEV
    B            43.57       120.10
    C            55.90       118.59
    D            37.23       111.07
```

Exercise 17.4

1 The following tacheometric observations were
made, using a theodolite, where the staff was held vertically.

Instrument station	Instrument height	Staff station	Horizontal circle reading	Zenith angle	Staff readings Top	Mid	Bottom
A	1.35 m	B	350° 10′	87° 20′	4.200	3.000	1.800
		C	40° 10′	86° 00′	3.760	2.500	1.240
		D	78° 10′	84° 40′	3.050	2.000	0.950

Calculate:

(a) the horizontal distances AB, AC and AD,

(b) the reduced levels of B, C and D, relative to
station A (25.31 m AOD).

(c) Compare the answers with those of Exercise 10.3,
question 3 (page 228).

6. Calculation of cross-sectional areas

In Chapter 15, cross-sectional areas of an embankment were calculated in Example 3 (data reproduced below). Program 8 computes that example. Program 9 extends the example and shows how the volume of material in the embankment is computed.

Example

5 The reduced ground level and formation level of an embankment at 0, 30 and 60 m chainages are shown below:

Chainage (m)	0	30	60
RL (m)	35.10	36.20	35.80
FL (m)	38.20	38.40	38.60

Given that the formation width of the top of the embankment is 6.00 m, that the transverse ground slope is horizontal and that the embankment sides slope at 1 unit vertically to 2 units horizontally, calculate the cross-sectional areas at the various chainages.

Program 8

```
10  REM CROSS-SECTIONAL AREAS
20  PRINT "Enter formation width"
30  INPUT w
40  PRINT "Side slope gradient is 1 in s: enter s"
50  INPUT s
60  LPRINT "Chainage (m)", "Area (sq. m)"
70  LPRINT " ","cut(+) fill(-)"
80  PRINT "enter no. of cross sections, n"
90  INPUT n
100 FOR k = 1 to n
110 PRINT "Enter chainage of section"
120 INPUT d
130 PRINT "For chn.";d;"enter red.lev.,form.lev"
140 INPUT r,f
150 c=(r-f)
160 b=ABS(r-f)
170 a=[w+(s*b)]*c
180 LPRINT INT(d*100+0.5)/100, INT(a*100+0.5)/100
190 NEXT k
```

Explanation

10	Remark—title
20–70	This is the part of the program that is constant for each section
20	Prompt
30	*w* is the formation width
40	Prompt
50	*s* is side slope gradient
60–70	Print on printer the headings 'chainage' and 'area', etc.

80–100	Loop details
80	Prompt
90	n is the number of times the loop is repeated
100	k is the loop (lines 100–160) repeated n times
110	Prompt
120	d is the chainage of the section
130	Prompt
140	r is the reduced level, f is the formation level of chainage d
150	c is the central cut (positive) or fill (negative)
160	b is the positive value of c, called the absolute value
170	a is the cross-sectional area
180	Print on printer, below chainage and area, the respective values
190	Repeat lines 110–150 for next cross-section

User instructions

Type:	RUN	
Screen response:	Enter formation width	Type: 6.0
Screen response:	Enter gradient s	Type: 1
Printer response:	Chainage (m)	Area (m²)
Screen response:	Enter number of cross-sections	Type: 3
Screen response:	Enter chainage	Type: 0
Screen response:	Enter reduced level	Type: 35.10
Screen response:	Enter formation level	Type: 38.20
Printer response:	-37.82	
Screen response:	Enter chainage	Type: 30, etc.

Solution

Chainage (m)	Area (m²) Cut (+) Fill (−)
0	− 37.82
30	− 22.88
60	− 32.48

Exercise 17.5

1 The reduced ground level and formation level of an embankment at 0, 30, 60 and 90 m chainages are shown below:

Chainage (m)	0	30	60	90
RL (m)	55.30	56.40	56.00	58.00
FR (m)	58.40	58.60	58.80	59.00

Given that the formation width of the top of the embankment is 6.00 m, that the transverse ground slope is horizontal and that the embankment sides slope at 1 unit vertically to 2 units horizontally, calculate the cross-sectional areas at the various chainages. Compare the answers with those of Exercise 15.8, question 2 (page 356).

7. Calculation of volume

Program 9 is an extension of Program 8 and calculates earthwork volumes from a knowledge of cross-sectional areas.

Example

6 Continue Example 5 to find the volume of material in the embankment.

Program 9

```
10   REM SIMPSON'S RULE-VOLUME FROM X.SECTS
20   PRINT "enter distance x between sections"
30   INPUT x
40   LET t=0
50   LPRINT "CHN.","AREA","MULT FACT.","AREA*M"
60   PRINT "enter form.width,w"
70   INPUT w
80   PRINT "side slope grade is 1 in s,enter  s"
90   INPUT s
100  PRINT "enter no. of x.sects,n"
110  INPUT n
120  FOR k=1 TO n
130  PRINT "Enter chn.,red.lev.,form.lev. (d,r,f)"
140  INPUT d,r,f
150  PRINT "enter mult.factor for SIMPSONS rule i.e. 1,4 or 2"
160  INPUT m
170  c=(r - f)
180  b=ABS(c)
190  a=(w+(s*b))*c
200  y=a*m
210  t=t+y
220  LPRINT d, INT(a*100+0.5)/100,m,INT(y*100+0.5)/100
230  NEXT k
240  v=t*(x/3)
250  LPRINT
260  LPRINT "*********************************"
270  LPRINT "VOLUME=";INT(v*100+0.5)/100;"cub.m."
```

Explanation

All variables in Program 9 have the meanings assigned to them in Program 8. Additionally:

line 30 x is the distance between the cross-sections
line 40 t is the cumulative total of the cross-sectional areas beginning with t = zero
line 160 m is Simpson's rule multiplying factor, namely 1 for first and last, 2 for odd and 4 for even sections
line 220   Prints the areas, a, to two decimal places
line 250   Prints a blank line
line 270   Prints the volume, v, to two decimal places

After the program has been entered, type RUN. The reader should now be capable of following the prompts presented by the program to obtain the volume of the embankment.

Solution

```
CHN.      AREA      MULT FACT.    AREA*M
  0      -37.82         1         -37.82
 30      -22.88         4         -91.52
 60      -32.48         1         -32.48
*********************************
VOLUME= -1618.2   cub.m.
```

Exercise 17.6

1 A proposed service roadway, 5.5 m wide, is to be built along a centre line XY. The embankment is to have side slopes of 1 to 2. Given the following data, calculate the volume of material required to construct the embankment.

Chainage (m)	40	60	80	100	120	140
Ground level (m)	10.00	9.60	9.50	10.40	7.30	10.40
Formation level (m)	11.00	11.10	11.20	11.30	11.40	11.50

Compare the answers with those of Exercise 15.5, question 1 (page 355).

8. Calculation of tri-dimensional coordinates

In Chapter 11, Example 1, three points of a radial positioning survey were calculated manually. The same exercise is shown, computed by a BASIC program, in the following example.

Example

7 Table 17.2 shows the survey data relating to three points, B, 1 and 2, observed from station A. Using BASIC program 10, compute the easting, northing and reduced level of the points B, 1 and 2.

Instrument station	A
Easting	100.000 m
Northing	200.000 m
Reduced level	35.210 m AD
Instrument height	1.350 m

Target point	Target height	Horizontal circle reading	Zenith angle	Slope distance	Remarks
RO		00° 00′ 00″	—	—	Reference object
B	1.350	36° 30′ 00″	71° 30′ 10″	18.325	Survey station
1	1.350	147° 29′ 10″	92° 04′ 00″	21.110	Fence
2	0.850	231° 15′ 30″	87° 13′ 40″	14.676	Fence

Table 17.2

Program 10

```
10   REM TRI-DIMENSIONAL COORDINATES
20   PRINT "Enter Inst.stn, Inst.height (eg. A, 1.30)"
30   INPUT N$, i
40   PRINT "Enter Easting, Northing, Red.level of Inst stn.(eg. 100.00,
     200.00,25.55)"
50   INPUT E1,N1,R1
60   LPRINT "Station","Easting","Northing","Red.level"
70   LPRINT
80   LPRINT "----------------------------------------------------------"
90   LPRINT N$, E1, N1, R1
100  PRINT "Enter the number of Target stns"
110  INPUT M
120  FOR K = 1 TO M
130  PRINT "Enter Target stn, Target height (eg. B, 1.45)"
140  INPUT T$,T
150  PRINT "Enter WCB to target in deg, min, sec, (eg. 20,25,43)"
160  INPUT D1,M1,S1
170  B=(D1+M1/60+S1/3600)*3.1415927/180
180  PRINT "Enter ZENITH angle to target in d,m,s(eg. 87,30,25)"
190  INPUT D2,M2,S2
200  Z=(D2+M2/60+S2/3600)*3.1415927/180
210  PRINT "Enter slope length to target (eg. 123,45)"
220  INPUT S
230  H=S*SIN(Z)
240  V=S*COS(Z)
250  X=H*SIN(B)
260  Y=H*COS(B)
270  E2=E1+X
280  N2=N1+Y
290  R2=R1+i+V-T
300  LPRINT T$, INT(E2*100+0.5)/100,INT(N2*100+0.5)/100,INT(R2*100+0.5)/100
310  NEXT K
```

Explanation

10   Program title
20   Prompt
30   N$ is the station reference, i is the instrument
     height
40   Prompt
50   E1, N1 and R1 are the x, y and z coordinates of
     the instrument station
60   Print on printer the final headings
70   Underline the headings
80   Leave a blank line on printout
90   Print on printer the reference and x, y and z
     coordinates of station
100 Prompt
110 M is the number of target stations
120 K is the loop calculation from lines 130 to 320
     to be repeated M times
130 Prompt
140 T$ is the reference of the target
150 Prompt
160 D1, M1, S1 is the bearing to the first target
170 B is the bearing in radians
180 Prompt

190 D2, M2, S2 is the zenith angle
200 *Z* is the zenith angle in radians
210 Prompt
220 *S* is the slope length to the target
230 *H* is the horizontal length to the target
240 *V* is the difference in elevation from the theodolite axis to the target reflector
250 X is the difference in east of the line
260 Y is the difference in north of the line
270 E2 is the easting of the target
280 N2 is the northing of the target
290 Prompt
300 Print on printer station reference, easting, northing, reduced level of target to 2 decimal places
310 Repeat calculation 130–300 for the next target station

Solution

Station	Easting	Northing	Reduced level
A	100.00	200.00	35.21
B	110.34	231.97	41.02
C	111.34	182.21	34.45
D	88.57	190.83	36.42

Exercise 17.7

1 Table 17.3 shows the data of part of a radial positioning survey at the GCB Outdoor Centre. (The table is actually a reproduction of Table 11.5.) Calculate the three-dimensional coordinates of stations 1 to 6 and compare the answers with those given in Table 11.5 (page 248).

9. Calculation of levels

It is generally agreed that a computer program of levels reduction is one of the more difficult programs. Consequently, it has been deliberately left to the end of this section.

Example

8 Table 17.4 shows levels observed as part of a proposed roadway scheme. (The example is that of Exercise 4.4, page 83.) Using Program 11, calculate the reduced level of the various chainage points, given that the reduced levels of point A is 13.273 m AOD.

Target station	Target height	Slope length	Bearing	Zenith angle
1	1.35	3.02	72° 07′ 30″	92° 06′ 35″
2	1.35	11.10	324° 09′ 45″	89° 21′ 00″
3	2.35	30.02	318° 55′ 20″	87° 08′ 35″
4	1.35	29.36	306° 49′ 50″	89° 21′ 50″
5	1.35	19.34	307° 13′ 30″	90° 04′ 16″
6	0.35	9.35	310° 12′ 00″	96° 13′ 30″

Table 17.3

BS	IS	FS	HPC	Reduced level	Remarks
3.105			16.378	13.273	BM A chainage 1200 m
	1.456			14.922	Chainage 1230 m
	0.350			16.028	Chainage 1260 m
	0.296			16.082	Chainage 1290 m
0.573		1.727	15.224	14.651	Change point
	3.393			11.831	Chainage 1320 m
	2.960			12.264	Chainage 1350 m
		2.342		12.882	Chainage 1380 m

Table 17.4

Program 11

```
10   REM LEVELS REDUCTION
20   PRINT "enter bench mark ref."
30   INPUT a$
40   PRINT "enter red.lev.of bench mark"
50   INPUT L
55   LPRINT "------------------------------------------------------------"
60   LPRINT TAB(3);"BS";TAB(9);"IS";TAB(16);"FS";TAB(23);"HPC";TAB(30);
     "RED.LEV";TAB(40);"REMARKS"
61   LPRINT "------------------------------------------------------------"
70   LPRINT
80   PRINT "enter B.S.staff reading"
90   INPUT r1
100  h=L+r1
101  LPRINT INT(r1*1000+0.5)/1000;TAB(21);h;TAB(30);L;TAB(40);a$
110  PRINT "enter status of next reading:(i for I.S.):(f for F.S.):(lf for last
     F.S.)"
120  INPUT b$
130  IF b$="i" THEN 160
140  IF b$="f" THEN 230
150  IF b$="lf" THEN 300
160  PRINT "enter stn. ref."
170  INPUT c$
180  PRINT "enter staff reading"
190  INPUT r2
200  L1=h-r2
210  LPRINT TAB(7);r2;TAB(30);L1;TAB(40);c$
220  GOTO 110
230  PRINT "enter stn.ref."
240  INPUT c$
250  PRINT "enter staff reading"
260  INPUT r2
270  L2=h-r2
280  PRINT "Enter next back sight reading"
290  INPUT r3
292  h=L2+r3
293  LPRINT r3;TAB(14);r2;TAB(21);h;TAB(30);L2;TAB(40);c$
294  GOTO 110
300  PRINT "enter stn.ref."
310  INPUT c$
320  PRINT "enter staff reading"
330  INPUT r4
340  L3=h-r4
350  LPRINT TAB(14);r4;TAB(30);L3;TAB(40); $
360  LPRINT "------------------------------------------------------------"
370  END
```

Exercise 17.8

1 Table 17.5 shows the readings observed during the levelling survey of a tunnel, constructed as part of a civil engineering project. Calculate the levels of the points B to G.

Note. Compare the answer with that of Exercise 4.5 (page 95).

BS	IS	FS	HPC	Reduced level	Remarks
1.10				10.00	TBM A
	−2.05				B Roof
−2.68		−1.45			C Roof
	1.05				D Floor
	−1.30				E Roof
	−0.46				F Roof
		2.78			G Floor

Table 17.5

10. Spreadsheets

Introduction

A spreadsheet is an electronic worksheet consisting of a set number of columns and rows. Any surveying calculation which is repetitive and can be tabulated can be neatly and speedily computed on a spreadsheet.

A spreadsheet is a commercial software package. Many spreadsheets are currently available including SuperCalc, VisiCalc and Microsoft's Multiplan, all of which are basically similar in layout and execution.

The following reasonably short explanation and instructions will enable the reader to use the Multiplan version and promote an understanding of the other spreadsheets. It must be emphasized that there are differences, none of them fundamental, and the reader should become acquainted with the appropriate manual before using any other spreadsheet.

11. Use of spreadsheets

In order to operate the system, the computer is switched on, the spreadsheet disk is loaded into drive A and the user's disk into drive B. Using Multiplan, the user types MP followed by RETURN.

The screen displays the spreadsheet (Fig. 17.1) or, rather, a small section of the spreadsheet. There are in fact 63 columns and 255 lines in this Multiplan version.

(a) Commands

The command line appears on the bottom of the screen. The commands that are of most use to the beginner are:

1. Alpha
Type 'a', then type any message in the illuminated cell and press return. The message is printed on the screen in the selected cell.
2. Blank
Type 'b' and press return. The contents of the illuminated cell are immediately blanked. Groups of cells may be blanked.

3. Copy
Type 'c'. The contents of the illuminated cell may be copied either down, right or left through any specified number of rows or columns.
4. Value
Type 'v', then type any numeric value or formula in the illuminated cell and press return. The value is printed on the screen in the selected cell.
5. Print
Type 'p'. The screen contents may then be printed in hard copy on the printer.
6. Transfer
Type 't', then type 's' for 'save' the program; type 'l' for 'load', an already saved program; type 'c' for 'clear' any program.

(b) Using the spreadsheet

The simplest way to learn about a spreadsheet is to use it. The principles, once mastered, operate for all types of surveying programs. Only the formulae are different.

Surveying problems require the input of data, the manipulation of these data via formulae and the output of the results. The secret of success in the use of spreadsheets for surveying lies in the organization and entering of the formulae.

A simple example will illustrate the fundamentals.

> **Example**
>
> **9** Calculate the sum and mean value of three pairs of numbers (x,y), (p,q), (a,b). The sum $(s) = x + y$ and the mean value $(m) = s/2$.
> (*Data.* $x = 2$, $y = 4$, $p = 10$; $q = -5$, $a = 1526.3$, $b = -3575.5$.)

1. Load Multiplan.
2. Select Alpha command.
3. Type all column headings on the spreadsheet in rows 1–5, columns 1–4 as shown in Table 17.6.
4. Enter the value of x (i.e. 2) in R6C1 and the value of y (i.e. 4) in R6C2.
5. Cell R6C3 requires the formula $s = x + y$.

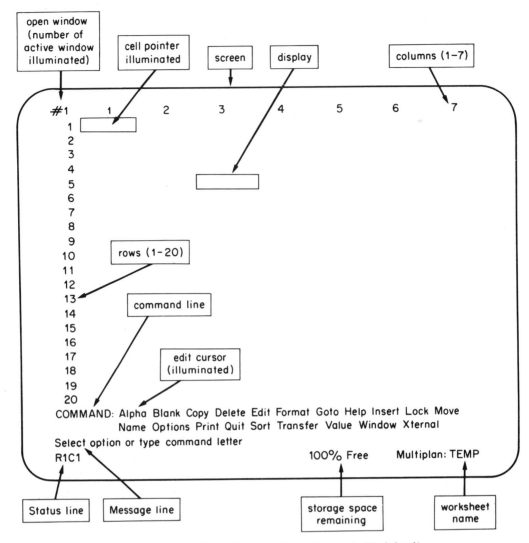

(Copyright © Microsoft Corporation – 'Electronic Worksheet'.)

Figure 17.1

	Column 1	Column 2	Column 3	Column 4
R1	Title:	Numbers		
R2	********************			
R3	First	Second	Sum of	Mean of
R4	Number	Number	Numbers	Numbers
R5	**************************************			
R6				

Table 17.6

(a) Place cursor in the cell R6C3.
(b) Press command 'value' or press = key.
(c) Screen response: Enter a formula.
(d) Place the cursor over *x* in R6C1.
(e) Press key +.

(f) Place the cursor over *x* in R6C2.
(g) Press return.
(h) The formula [RC(−2) + RC(−1)] has been entered, meaning 'the sum equals the cell contents in the same row two columns back added to the cell contents in the same row one cell back'.

6. Cell R6C4 requires the formula $m = s/2$. Place the cursor in the cell and enter the formula [RC(−1)/2] as above.

7. Copy down these formulae for any required number of rows—in this case three.

(a) Select COPY command.
(b) Screen response: RIGHT, or LEFT, or DOWN.
(c) Use the TAB key to move the command cursor over DOWN; type D.

1	2	3	4
"Title"	"NUMBERS"		
"*******"	"*******"	"***********"	"*******"
"First"	"Second"	"Sum of"	"Mean of"
"Number"	"Number"	"numbers"	"numbers"
"*******"	"*******"	"***********"	"*******"
2	4	RC(-2)+RC(-1)	RC(-1)/2
10	-5	RC(-2)+RC(-1)	RC(-1)/2
1526.3	-3575.5	RC(-2)+RC(-1)	RC(-1)/2

Table 17.7

```
Title         NUMBERS
*********************************************
First         Second      Sum of     Mean of
Number        Number      numbers    numbers
*********************************************
2             4           6          3
10            -5          5          2.5
1526.3        -3575.5     -2049.2    -1024.6
```

Table 17.8

(d) Screen response: copy DOWN number of cells: ?

(e) Type 2 RETURN

8. Move the cursor to R6C1; select V then type the data: 2.

Move the cursor to R6C2; type the data: 4 RETURN.

9. The results 6 and 3 will appear automatically in R6C3 and R6C4 respectively.

10. Repeat for data (10, −5) and (1526.3, −3575.5). The spreadsheet, as acted upon by the computer (but not presented on screen to the user), is shown, complete with formulae, in Table 17.7.

The spreadsheet that appears on screen is shown in Table 17.8.

Examples

10 Calculate the coordinates of an open traverse ABCDE, given the following data:

Line	AB	BC	CD	DE
WCB	30°	110°	225°	295°
Length (m)	50.0	70.0	82.0	31.2

Compare the answer with that shown in Table 9.7 (page 204).

Solution

1. Refer to Table 17.9. Select Alpha command and type rows 1 to 6 as they appear in the table.

2. Enter the formulae in the manner of Example 9 (parts 5 to 7) in row 8. Row 7 should be left clear of the formulae to enable the total coordinates of origin A to be entered.

3. Enter the survey data in rows 8 to 11 and columns 1 to 3. The results will appear automatically (Table 17.10).

11 The reduced ground levels and specification for a proposed roadway are shown below. Calculate the volume of cut or fill required.

Chainage (m)	20.0	40.0	60.0	80.0
Reduced level (m)	1.90	3.50	6.20	7.30

Roadway specification—

gradient	1 in 20 rising
width	5.00 m
side slope	gradient 1 in 1

Solution

1. Refer to Table 17.11. Select Alpha command and type rows 1 to 14 as they appear in the table.

2. Enter the formulae in the manner of Example 9 (parts 5 to 7) in rows 15 and 16.

3. Enter the survey data in rows 15 to 19. The results will appear automatically (Table 17.12).

Table 17.9

line	Length	Whole circle bearing	partial departure	partial latitude	total departure	total latitude	station
1	Spreadsheet Rect. Co-ords			Amstrad PCW8512		Surv. for Const.	
2							
3	***"	******	Whole	partial	partial	total	total	
4	line	Length	circle	departure	latitude	departure	latitude	station
5	***"	******	bearing	*********	********	*********	********	*******
6								
7						0	0	A
8	AB	50	30	RC(-2)**SIN(RC(-1)*3.14159/180)	RC(-3)*COS(RC(-2)*3.14159/180)	R(-1)C+RC(-2)	R(-1)C+RC(-2)	B
9	BC	70	110	RC(-2)*SIN(RC(-1)*3.14159/180)	RC(-3)*COS(RC(-2)*3.14159/180)	R(-1)C+RC(-2)	R(-1)C+RC(-2)	C

Table 17.10

Spreadsheet		Rect. Co-ords.		Amstrad PCW8512		Surv. for Constr.	
line	Length	Whole circle bearing	partial departure	partial latitude	total departure	total latitude	station
**********	******	******	*********	********	*********	********	*******
**		***	***	**	0.00	0.00	A
AB	50.00	30.00000	25.00	43.30	25.00	43.30	B
BC	70.00	110.00000	65.78	-23.94	90.78	19.36	C
CD	82.00	225.00000	-57.98	-57.98	32.80	-38.62	D
DE	31.20	295.00000	-28.28	13.19	4.52	-25.44	E

```
Line
No.     1          2          3          4                      5                 6             7              8                   9        10

1
2                  Cross sectional areas.   This program will calculate areas
3                  of trapezoidal shaped sections
4
5                  Enter in COL7, the following
6
7                  1. Formation width.(w)..............    Enter w,..............   ********************
8                  2. Gradient of formation, 1 in (n)..... Enter n,..............   *                 *
9                  3. Side slope gradient 1 in (s)........ Enter s,..............   *                 *
10                 4. Chainage interval (d)..............  Enter d,..............   *                 *
                                                                                    ********************
```

```
11   ****    ********   ******   ********************   **************   ************   ****************   ****************   ****   ****************
12   Sect.   Chainage   Reduced  Formation              Central          ABS.           Area               Cut or             Mult   Product
13   No.                Level    Level                  Height           (c)            sq. m              Fill               fact   Col19*Col10
14   ****    ******     ******   ********************   **************   ************   ****************   ****************   ****   ****************
```

Line No.	1 Sect. No.	2 Chainage	3 Reduced Level	4 Formation Level	5 Central Height	6 ABS. (c)	7 Area sq. m	8 Cut or Fill	9 Mult fact	10 Product Col19*Col10
15	1	0	0	0	RC(-1)-RC(-2)	ABS(RC(-1))	(R7C7+R9C7*RC(-1))*RC(-2)	IF(RC(-3)<0,"cut","fill")	1	RC(-3)*RC(-1)
16	2	20	1.9	R(-1)C+(RC(-2)-R(-1)C(-2))/R8C7	RC(-1)-RC(-2)	ABS(RC(-1))	(R7C7+R9C7*RC(-1))*RC(-2)	IF(RC(-3)<0,"cut","fill")	4	RC(-3)*RC(-1)
17	3	40	3.5	R(-1)C+(RC(-2)-R(-1)C(-2))/R8C7	RC(-1)-RC(-2)	ABS(RC(-1))	(R7C7+R9C7*RC(-1))*RC(-2)	IF(RC(-3)<0,"cut","fill")	2	RC(-3)*RC(-1)
18	4	60	6.2	R(-1)C+(RC(-2)-R(-1)C(-2))/R8C7	RC(-1)-RC(-2)	ABS(RC(-1))	(R7C7+R9C7*RC(-1))*RC(-2)	IF(RC(-3)<0,"cut","fill")	4	RC(-3)*RC(-1)
19	5	80	7.3	4	RC(-1)-RC(-2)	ABS(RC(-1))	(R7C7+R9C7*RC(-1))*RC(-2)	IF(RC(-3)<0,"cut","fill")	1	RC(-3)*RC(-1)

```
21   _____
22                                                                                            Total =   Sum(R(-3)C:R(-7)C)
23   Answer: Volume=(R10C7/3)*(R22C11) cub m.
```

Table 17.11

```
Cross sectional areas.      This program will calculate areas
of trapezoidal shaped sections

  Enter in COL 7, the following
                                              **********
  1. Formation width,(w)............... Enter w,.....*   5.00 *
  2. Gradient of formation, 1 in (n).. Enter n,.....*  20.00 *
  3. Side slope gradient 1 in (s)..... Enter s,.....*   1.00 *
  4. Chainage interval (d)............ Enter d,.....*  20.00 *
******************************************************************************
Sect.  Chainage  Reduced   Formation  Central  ABS.   Area    Cut or  Mult  Product
No.      ***      Level     Level     Height   (c)    sq.m    Fill    fact  Col9*Col10
******************************************************************************
                                       0.00    0.00    0.00   fill    1      0.00
1       0.00      0.00      0.00      -0.90    0.90   -5.31    cut     4    -21.24
2      20.00      1.90      1.00      -1.50    1.50   -9.75    cut     2    -19.50
3      40.00      3.50      2.00      -3.20    3.20  -26.24    cut     4   -104.96
4      60.00      6.20      3.00      -3.30    3.30  -27.39    cut     1    -27.39
5      80.00      7.30      4.00

                                                      Total    =       -173.09

        Answer Volume=-1153.93 cub m.
```

Table 17.12

12. Answers

Exercise 17.1

1 (a) 10   INPUT A
       20   INPUT B
       30   INPUT C
       40   P=A*B*C
       50   LPRINT P
       60   GOTO 10

   (b) (i) 30, (ii) 693 000

Exercise 17.2

1

STN REF	EAST	NORTH
A	00.00	00.00
B	45.31	64.71
C	104.17	60.60
D	148.27	88.15
E	161.13	130.23

Exercise 17.3

1

Line	Forward bearing
AB	65° 00′ 00″
BC	338° 15′ 50″
CD	72° 47′ 00″

Exercise 17.4

1 STADIA TACHMTRY (VERT.STAFF)
   INST STN:- A   RED.LEV.:- 25.31

TARGET	HORZ. DIST.	RED. LEVEL
B	239.48	34.81
C	250.77	41.70
D	208.19	44.09

Exercise 17.5

1

Chainage (m)	Area (m²) cut(+) fill(−)
0	37.82
30	22.88
60	32.48
90	8.00

Exercise 17.6

1

Chainage	Area	Multiplication factor (m)	Area × m
40	−7.5	1	−7.5
60	−12.75	4	−51.00
80	−15.13	2	−30.26
100	−6.57	4	−26.28
120	−56.17	1	−56.17

Volume = −1141.4 m³

120	−56.17	1	−56.17
130	−27.82	4	−111.28
140	−8.47	1	−8.47

Volume = −586.4 m³

Exercise 17.7

1

Station	Easting	Northing	Reduced level
A	0.00	0.00	6.48
1	2.87	0.93	6.37
2	−6.50	9.00	6.61
3	−19.70	22.60	6.98
4	−23.50	17.60	6.81
5	−15.40	11.70	6.46
6	−7.10	6.00	6.47

Exercise 17.8

1

BS	IS	FS	HPC	Reduced level	Remarks
1.10			11.10	10.00	TBM A
	−2.05			13.15	B Roof
−2.68		−1.45	9.87	12.55	C Roof
	1.05			8.82	D Floor
	−1.30			11.17	E Roof
	−0.46			10.33	F Roof
		2.78		7.09	G Floor

CHAPTER 18

Project

Objective
The objective of this chapter is to present the reader with the opportunity to calculate and plot a complete survey and to solve the various area, volume and setting-out computations, which are a necessary part of site development work.

The various sections of the project are arranged such that they may also be used (a) as a revision tool by students, in preparation for examinations or assessments, and (b) as examination questions by university or college lecturers.

1. Introduction

This chapter is a practical project, which the reader should complete, either in part, after studying the relevant chapters, or as a whole, near the end of a course of study.

There is no substitute for practical experience and, whenever possible, the reader should obtain surveying instruments and carry out practical surveying exercises. However, on university and college courses of study, timetabling constraints and vagaries of weather usually curtail practical exercises to a greater or lesser extent. This project is an attempt to replace, or at least supplement, these practical sessions.

This project is based on a practical survey at the Glasgow College of Building Outdoor Centre. The project simulates the work required in developing a small construction site and is designed here, in this chapter, as a series of exercises, in which the reader may join at any stage. Guidance is given throughout on how to join the project at intermediate stages.

2. Objective of project

The Outdoor Centre is a residential complex, catering for a variety of sports and field studies. An area of the grounds is to be further developed to provide chalet accommodation for twelve people, a jogging track, boat storage and a recreation area.

An accurately surveyed plan is required for the location of the facilities and associated services. The student is required to use the surveyed data (following) to

(a) construct a plan drawn to scale 1:250 showing contour lines at 1 metre vertical intervals

(b) to extract from the plan and field notes, sufficient data to compute areas and volumes of the construction works and

(c) to use the plan as the basis for setting out these works.

3. Section 1: General site location

(a) Study topics:
The reader should study Chapters 1 and 2 before attempting this section.

(b) Project task:
Figure 18.1 shows the location of the proposed development site on a 1:1250 scale map of the area.

1. Describe the location of the Centre grounds in relation to the surrounding topography.
2. From the grid north lines, determine the approximate orientation of the site.
3. State the features that have the following 10 metre grid references:
   (a) 0912,
   (b) 1203,
   (c) 1308,
   (d) 0503,
   (e) 0606,
   (f) 0913.
4. State the 10 metre grid reference of
   (a) the centre of the site (shown hatched),
   (b) the centre of the main building of the GCB Centre.
5. Describe the parish boundary shown on the plan.
6. State the parcel number and area in hectares and acres of the complete GCB Centre complex.
7. Using the grid coordinates determine the plan reference number.

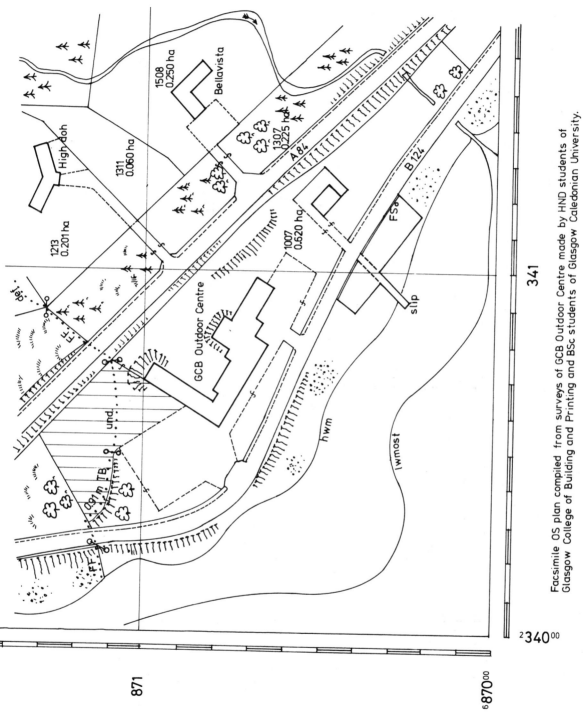

Facsimile OS plan compiled from surveys of GCB Outdoor Centre made by HND students of
Glasgow College of Building and Printing and BSc students of Glasgow Caledonian University.

Figure 18.1

389

4. Section 2: Linear survey

(a) Study topics:
The reader should study Chapter 3 before attempting this section.

(b) Project task:
Figures 18.2 and 18.3 show the field notes of a linear survey of the proposed development site.

1. Calculate the plan lengths of the lines and then plot the field notes to scale 1:250.

5. Section 3: Levelling

(a) Study topics:
The reader should study Chapter 4 (Levelling) before attempting this section.

(b) Project task:
Table 18.1 shows the field notes of a flying levelling made around the pegs of the linear survey in order to establish temporary bench marks.

1. Calculate the reduced levels of the tops of the pegs.
2. Add the reduced levels to the site plan.

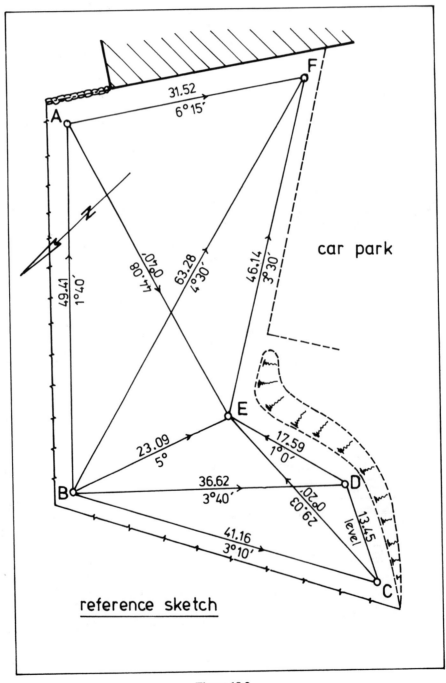

Figure 18.2

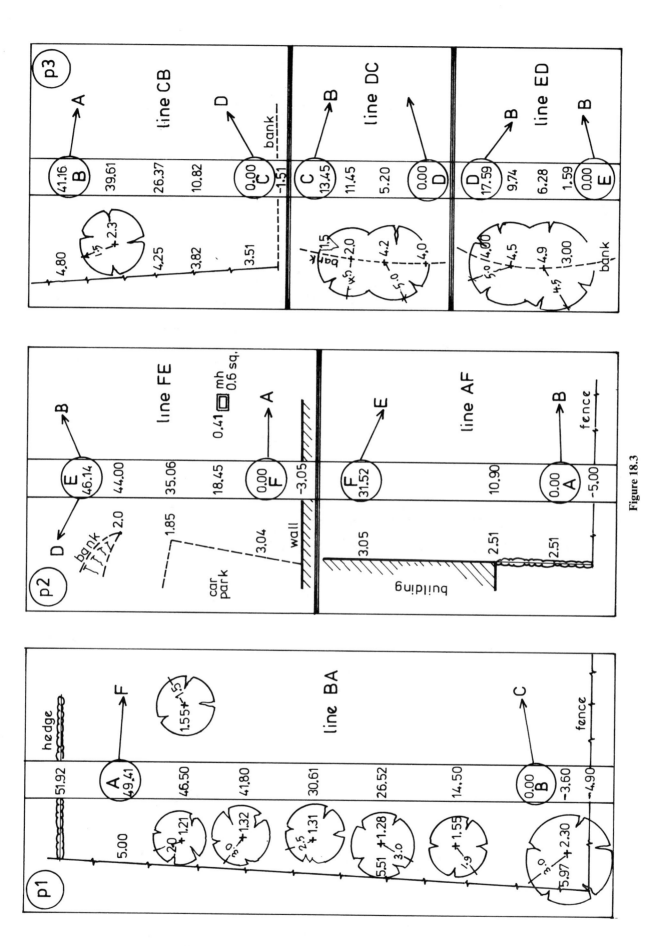

Figure 18.3

391

BS	IS	FS	Rise or fall	Reduced level	Remarks
3.612				5.210	BM
2.463		1.420			Change point
3.130		0.584			F
2.073		0.225			E
2.252		2.330			D
2.319		2.255			C
0.042		0.065			B
0.048		1.513			A
1.560		2.882			Change point
0.668		2.152			F
1.410		2.545			Change point
		3.597			BM

Table 18.1

Figure 18.4 shows the location of a grid of points levelled during a contour survey of the site and Table 18.2 shows the actual readings observed during the levelling.

3. Calculate the reduced levels of the points of the grid.
4. Add the grid and the reduced levels to the site plan.

6. Section 4: Contouring

(a) Study topics:
The reader should study Chapter 5 before proceeding further with this section.

(b) Joining the project:
The reader may join the project here without having completed the previous sections provided that Fig. 18.9 is studied to gain familiarity with the grid of levels covering the site. The figure should be enlarged on a photocopier to bring the plan to scale 1:250 and should be used for the remainder of this Section 4.

(c) Project task:
Using the site plan showing the grid of levels, interpolate the positions of the ground contour lines at vertical intervals of 1 metre and add the contour lines to the plan.

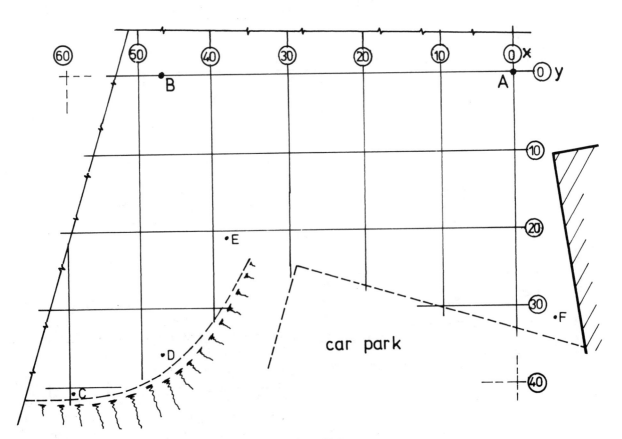

Figure 18.4

BS	IS	FS	Rise or fall	Reduced level	Distance	Remarks
3.182				9.281		Station F
	2.800					0, 30
	2.300					0, 20
	2.273					0, 10
	1.297					10, 10
	2.847					0, 33
	1.925					10, 20
	2.277					10, 30
	0.685					20, 10
	1.472					20, 20
	1.688					20, 27
	0.172					30, 10
	0.889					30, 20
	1.409					24, 26
		1.409				30, 25
2.573				9.281		Station F
	1.832					−4, 10
		2.704				−8, 30
0.417				14.180		Station B
	0.536					50, 0
	0.810					40, 0
	1.115					30, 0
	1.472					20, 0
	1.825					10, 0
	2.018					0, 0
	0.530					53, 0
	1.526					50, 10
	1.872					40, 10
		1.388				56, 10
2.246				11.926		Station C
	2.305					60, 40
	1.598					60, 30
	1.168					59, 20
	1.875					50, 30
	1.363					50, 20
	2.562					50, 37
	2.285					40, 20
	2.672					40, 30
	1.040					50, −6
	1.200					40, −6
	1.370					30, −6
	1.535					20, −5
	1.905					10, −5
	2.215					0, −5
		4.432				−7, 20

Table 18.2

7. Section 5: Sectioning and intersection of surfaces

(a) Study topics:
The reader should study Chapters 4, 5 and 6 before proceeding further with this section.

(b) Joining the project:
The reader may join the project at this point without having completed the previous sections. Figure 18.10 should be studied to gain familiarity with the site layout and general topography. The figure should be enlarged on a photocopier to bring it to scale 1:250 and used throughout the remainder of this section.

(c) Development specification:
Figure 18.5 shows the proposed development of the site. The positions of the residential chalet, access road, sewer, jogging track, boat storage area and recreational area are clearly shown. The specification details are as follows:

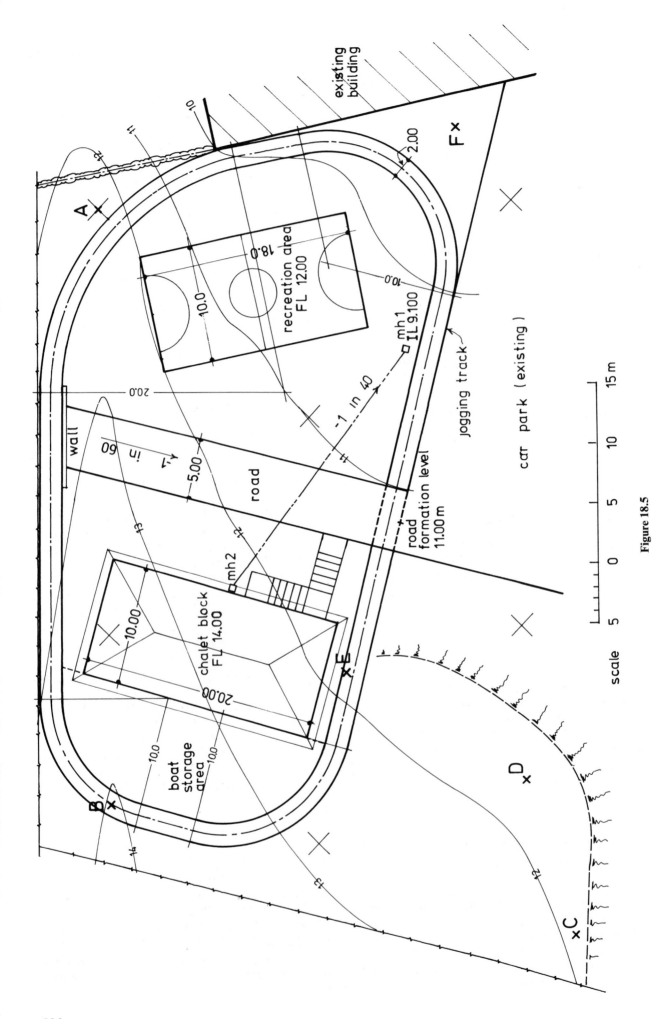

Figure 18.5

1. Chalet:
Frontage: 20.00 m
Gable: 10.00 m
Floor level: 14.00 m AOD
Access: Steps to roadway
2. Access road:
Width: 5.00 m
Length: 30.0 m
Formation level (chainage 0.00 m): 11.00 m AOD
Gradient: 1 in 60 rising
Earthworks: Side slopes 1 to 3
3. Drainage:
Combined foul and storm water system from manhole 2 (proposed building) to manhole 1 (existing manhole)
Width of sewer track: 0.80 m
Invert level (existing manhole):   9.100 m AOD
Gradient MH1 to MH2:   1 in 40 rising
Earthworks: Trench side slopes 1 to 1
4. Jogging track:
Width   2.00 m
Curve radii: Figure 18.5
Intersection angles, lengths of straights:
                  see Figure 18.7
5. Boat storage area:
Tarmac
6. Recreational area:
Length: 18.00 m
Width: 10.00 m
Formation level: 12.00 m AOD
Earthworks: Side slopes 1 to 1.5
Markings: Netball, volleyball, football

Project task:
The reader should transfer the proposed development from Fig. 18.5 to his own site plan and use it for the remainder of this section.
1. Table 18.3 shows the results of a levelling made along the centre line of the proposed sewerage track. Using the specification data in (3) above:
  (a) Calculate the reduced levels.
  (b) Draw a longitudinal section along the line of the track showing clearly the ground surface and bottom of the sewerage track (horizontal scale 1:250, vertical scale 1:100).

  (c) Calculate the depth of cut at 5 metre intervals along the track.
  (d) Draw natural cross sections of the cutting at chainages 10 m and 25 m.
2. Draw on the site plan the outline of the earthworks of
  (a) the proposed recreational area and
  (b) the access roadway.
3. Draw a longitudinal section along the line of the proposed road to show the existing ground surface, the proposed roadway and the proposed retaining wall to horizontal scale 1:250 and vertical scale 1:100.
4. Draw cross sections of the proposed roadway earthworks at 0, 15 and 30 m chainages to a natural scale of 1:250 and obtain, from either the plan view or cross sections of the earthworks, the ground levels at the centre of the proposed road and at the points where the side slopes meet the ground surface.

8. Section 6: Traversing

(a) Study topics:
The reader should study Chapters 7, 8 and 9 before attempting this section.
(b) Joining the project:
The reader may join the project here without having completed the previous sections, provided that Fig. 18.8 is studied to gain familiarity with the survey point peg layout A to F, which is used as the framework for the traverse survey (following).
(c) Project task:
The construction works are to be set out accurately by theodolite. Hence an accurately coordinated framework is required around the site. Table 18.4 shows the field notes of a closed traverse of the site, using the same survey points as the linear survey detailed in Section 2. The forward bearing of the line AB, measured from the Ordnance Survey map, is 315° 00′ 00″ and station A is the origin of the survey (100.000 m east, 100.000 m north).

1. Determine, from Table 18.4, the measured horizontal angles, zenith angles and slope lengths.
2. Adjust the horizontal angles to effect closure.

BS	IS	FS	Rise or fall	Reduced level	Distance	Remarks
0.534				12.186		BM E
	0.400				0.000	MH 2
	0.910				5.000	Centre line
	1.270				10.000	Centre line
	1.780				15.000	Centre line
	2.080				20.000	Centre line
		2.350			25.000	MH 1

Table 18.3

Instrument station	Station observed	Face left	Face right	Mean observed angle	Line	Face	Zenith angle	Measured length		Length	Remarks
								Rear	Forward		
A	F	00° 00′ 00″	180° 00′ 00″					0.100	49.513		
	B	100° 52′ 40″	280° 52′ 20″		AB	L	88° 28′ 00″	0.150	49.563		
B	A	00° 00′ 00″	180° 00′ 20″					0.150	41.300		
	C	106° 29′ 00″	286° 30′ 00″		BC	L	93° 09′ 00″	0.200	41.344		
C	B	00° 00′ 00″	179° 59′ 40″					0.370	13.820		
	D	56° 42′ 20″	236° 42′ 20″		CD	L	89° 59′ 00″	0.100	13.550		
D	C	00° 00′ 00″	180° 00′ 00″					0.230	17.820		
	E	138° 12′ 00″	318° 12′ 00″		DE	L	89° 06′ 00″	0.115	17.705		
E	D	00° 00′ 00″	180° 00′20″					0.200	46.340		
	F	251° 25′ 40″	71° 25′ 40″		EF	L	93° 34′ 00″	0.150	46.290		
F	E	00° 00′ 00″	180° 00′ 00″					0.370	31.904		
	A	66° 17′ 20″	246° 17′ 00″		FA	L	83° 45′ 00″	0.050	31.584		

Table 18.4

3. Calculate the plan lengths and bearings of the lines of the traverse.
4. Calculate the coordinates, and hence the accuracy, of the traverse.
5. Using Bowditch's rule, adjust the coordinates to close.
6. Plot the coordinates to scale 1:250 and compare the plotting with the linear survey plot (Fig. 18.8).

9. Section 7: Tacheometry

(a) Study topics:
The reader should study Chapter 10 before attempting this section.

(b) Joining the project:
The reader may join the project here without having completed the previous sections, provided that Fig. 18.5 is studied to gain familiarity with the proposed development of the site.

(c) Project task:
Table 18.5 shows the field notes of a short tacheometric survey, carried out to check the levels of three points in connection with the setting out of the road and sewer. The theodolite was set over peg F (Fig. 18.14) and a zero reading taken to station E as the back sight.

Instrument station	F
Instrument height	1.3
Reduced level	9.28

Observation station	Horizontal reading	Zenith angle	Stadia readings		
			Top	Mid	Bottom
E	0° 0′ 0″				
T1	355° 31′ 0″	86° 41′ 20″	1.449	1.300	1.151
T2	356° 29′ 0″	87° 11′ 0″	1.476	1.300	1.124
T3	2° 1′ 0″	86° 39′ 0″	1.393	1.300	1.207

Table 18.5

Calculate and plot on the site plan:
 (i) the horizontal distances F–T1, F–T2 and F–T3,
 (ii) the reduced levels of points T1, T2 and T3.

10. Section 8: Radial positioning

(a) Study topics:
The reader should study Chapter 11 before attempting this section.
(b) Joining the project:
The reader may join the project here without having completed the previous sections, provided that Fig. 18.14 is studied to gain familiarity with the survey point peg layout A to F.
(c) Project task:
Table 18.6 shows the field notes of a radial positioning survey of the pegs A, C, D, E and F surveyed from peg B where the instrument height was 1.35 m.

Calculate the easting, northing and reduced level of points C, D, E and F.

11. Section 9: Setting out

(a) Study topics:
The reader should study Chapter 13 before attempting this section.
(b) Joining the project:
The reader may join the project here without having completed the previous sections, provided that Fig.

18.5 is studied to gain familiarity with the development work required to be set out.
(c) Project task:
The construction work (Fig. 18.6) is to be set out, using a tape and theodolite as follows:

1. Chalet block: points H1 and H2, from station B, starting line B–A
2. Recreation area: points R1 and R2, from station A, starting line A–B
3. Jogging track intersection points I1 to I4, from station E, starting line E–F.

Figure 18.6 shows the surveyed coordinates of the relevant traverse stations together with the coordinates of the construction works points detailed above, as measured from the development plan (Fig. 18.5). Determine, from the coordinates, the setting-out data (bearings and distances) required to locate the various works on the ground.

12. Section 10: Curves

(a) Study topics:
The reader should study Chapter 12 before attempting this section.
(b) Joining the project:
The reader may join the project here, but should preferably have completed Section 9.

Instrument station	B (65.07 m east, 134.91 m north, 14.18 m AOD)
Back station	A (100.00 m east, 100.00 m north, 12.71 m AOD)
Back bearing B–A	135° 00′ 00″

Target station	Target height (m)	Slope length (m)	Whole circle bearing	Zenith angle
C	1.35	41.16	241° 28′ 45″	93° 08′ 20″
D	1.35	35.62	223° 01′ 50″	93° 37′ 25″
E	1.35	23.08	198° 08′ 40″	94° 57′ 29″
F	1.35	63.46	164° 05′ 10″	94° 25′ 40″

Table 18.6

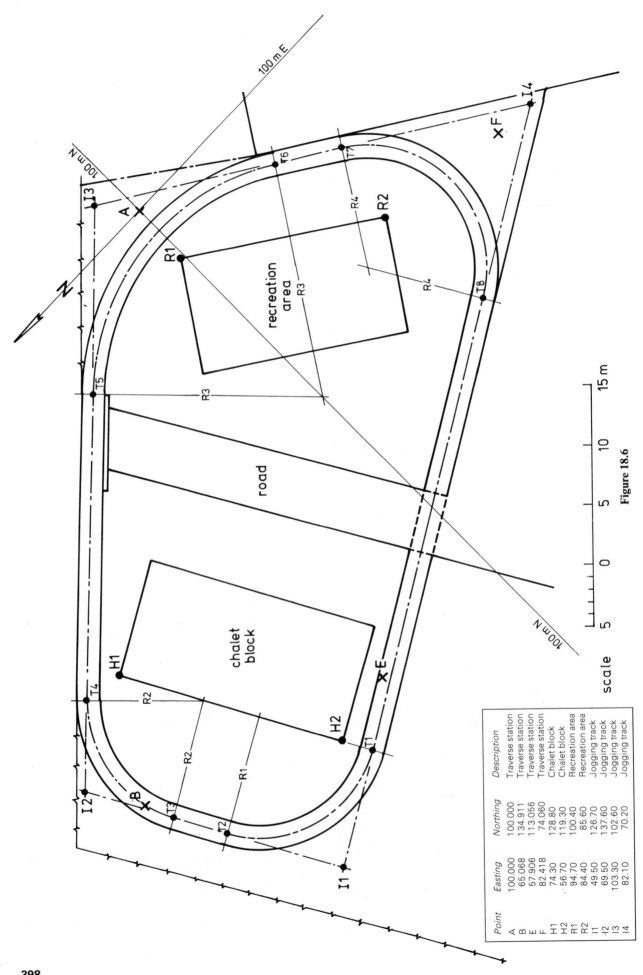

Point	Easting	Northing	Description
A	100.000	100.000	Traverse station
B	65.068	134.911	Traverse station
E	57.906	113.056	Traverse station
F	82.418	74.060	Traverse station
H1	74.30	128.80	Chalet block
H2	56.70	119.30	Chalet block
R1	94.70	100.40	Recreation area
R2	84.40	85.60	Recreation area
I1	49.50	126.70	Jogging track
I2	69.50	137.60	Jogging track
I3	103.30	102.60	Jogging track
I4	82.10	70.20	Jogging track

recreation area

road

chalet block

100 m E

100 m N

100 m N

N

15 m

10

5

0

5

scale

Figure 18.6

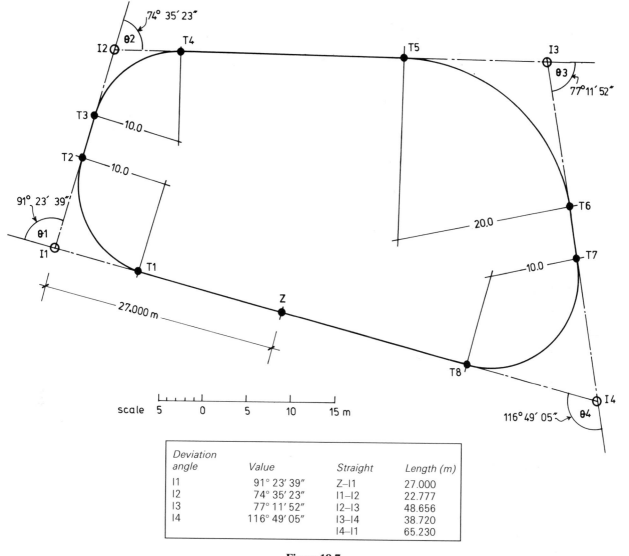

Deviation angle	Value	Straight	Length (m)
I1	91° 23′ 39″	Z–I1	27.000
I2	74° 35′ 23″	I1–I2	22.777
I3	77° 11′ 52″	I2–I3	48.656
I4	116° 49′ 05″	I3–I4	38.720
		I4–I1	65.230

Figure 18.7

(c) Project task:
Figure 18.7 shows the jogging track layout with deviation angles and lengths of straights calculated from Section 9 data.

1. Calculate for each curve:
  (a) the tangent length and length of curve,
  (b) the chainage of each tangent point from the zero chainage point Z, shown on the figure,
  (c) the total length of the jogging track.
2. The three 10 metre radius curves are to be set out by the method of offsets from the long chord. Calculate for each curve:
  (a) the length of the long chord,
  (b) the interval between offsets given that five offsets are to be set out at regular intervals,
  (c) the lengths of the required offsets.
3. The 20 metre curve is to be set out by the method of tangential angles (using theodolite and tape), from tangent point T5. Since the length of the curve is short, chords of 3 metres are to be used. Construct a setting-out table to enable the curve to be located on the ground.

13. Section 11: Areas and volumes

(a) Study topics:
The reader should study Chapters 14 and 15 before attempting.
(b) Joining the project:
The reader may join the project here without having completed the previous sections.
(c) Project task:
1. Figure 18.5 shows the proposed boat storage compound of the new development which is to be tarred. Calculate the area of the compound.
2. Table 18.13 shows the coordinates of the closed traverse of the site. Calculate the area within the coordinated framework.

3. Figure 18.11 shows the longitudinal and cross sections of the proposed drain. Calculate the volume of material to be removed in excavating the trench.

4. Figure 18.12 shows the outline of the proposed roadway cutting. Calculate:
   (a) the surface area occupied by the cutting,
   (b) the tarmacadamed area of the road,
   (c) the actual area of the side slopes of the cutting which require to be seeded (side slope gradient 1 in 3).

5. Figure 18.13 shows the longitudinal and cross sections of the proposed roadway. Calculate the volume of material to be removed in forming the cutting (side slope gradient 1 in 3).

6. Figure 18.12 shows the earthworks of the proposed recreation area. Calculate the area of each contour plane and using Simpson's rule calculate the volume of earth required to construct the earthwork (side slope gradient 1 in 1.5).

14. Answers

Section 1

1. The GCB Outdoor Centre lies between the main A84 road and the secondary B124 road. It faces south-westwards towards the loch and is separated from it by the coastal road and narrow shingle beach.

2. 135 degrees

3. 0912—trees, 1203—flagstaff, 1308—brace (area), 0503—low-water mark Ordinary Spring tide, 0606—embankment, 0913—boundary mereing

4. (a) 0511, (b) 0707

5. From west to east: face of fence; 0.91 m from top of bank; undefined boundary; face of fence; defaced

6. 1007—0.620 hectares, 1.55 acres

7. NS 3487 SW

Section 2

Figure 18.8

Section 3

Tables 18.7 and 18.8 and Fig. 18.9

Section 4

Figure 18.10

Section 5

1. (a) Table 18.9
1. (b), (c), (d) Figure 18.11
2. (a), (b) Figure 18.12
3. Figure 18.13
4. Figure 18.13

Section 6

1. Table 18.10
2, 3. Tables 18.11 and 18.12
4, 5. Table 18.13
6. Figure 18.14

Section 7

Table 18.14

Section 8

Table 18.15

BS	IS	FS	Rise or fall	Reduced level	Remarks
3.612				5.210	BM
2.463		1.420	2.192	7.402	Change point
3.130		0.584	1.879	9.281	F
2.073		0.225	2.905	12.186	E
2.252		2.330	−0.257	11.929	D
2.319		2.255	−0.003	11.926	C
0.042		0.065	2.254	14.180	B
0.048		1.513	−1.471	12.709	A
1.560		2.882	−2.834	9.875	Change point
0.668		2.152	−0.592	9.283	F
1.410		2.545	−1.877	7.406	Change point
		3.597	−2.187	5.129	BM

Table 18.7

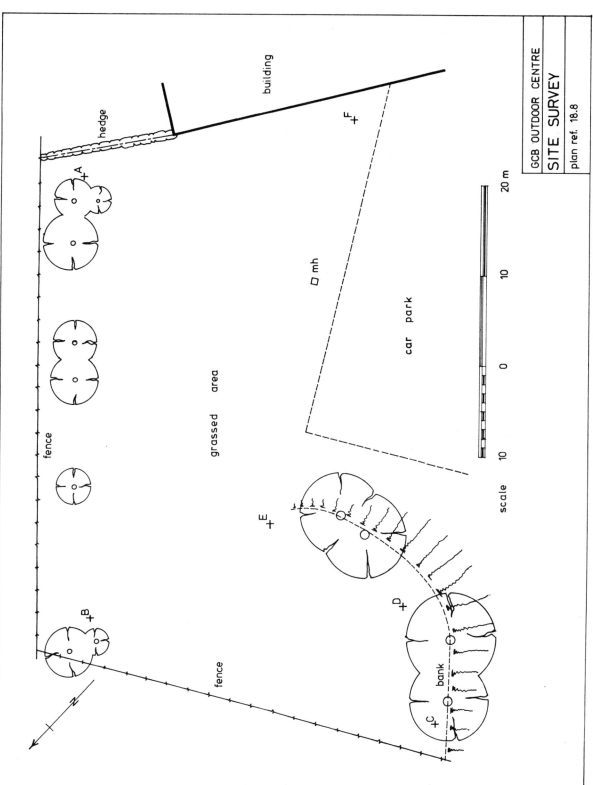

GCB OUTDOOR CENTRE
SITE SURVEY
plan ref. 18.8

Figure 18.8

BS	IS	FS	Rise or fall	Reduced level	Distance	Remarks
3.182				9.281		Station F
	2.800		0.382	9.663		0, 30
	2.300		0.500	10.163		0, 20
	2.273		0.027	10.190		0, 10
	1.297		0.976	11.166		10 10
	2.847		−1.550	9.616		0, 33
	1.925		0.922	10.538		10, 20
	2.277		−0.352	10.186		10, 30
	0.685		1.592	11.778		20, 10
	1.472		−0.787	10.991		20, 20
	1.688		−0.216	10.775		20, 27
	0.172		1.516	12.291		30, 10
	0.889		−0.717	11.574		30, 20
	1.409		−0.520	11.054		24, 26
		1.409	0.000	11.054		30, 25
2.573				9.281		Station F
	1.832		0.741	10.022		−4, 10
		2.704	−0.872	9.150		−8, 30
0.417				14.180		Station B
	0.536		−0.119	14.061		50, 0
	0.810		−0.274	13.787		40, 0
	1.115		−0.305	13.482		30, 0
	1.472		−0.357	13.125		20, 0
	1.825		−0.353	12.772		10, 0
	2.018		−0.193	12.579		0, 0
	0.530		1.488	14.067		53, 0
	1.526		−0.996	13.071		50, 10
	1.872		−0.346	12.725		40, 10
		1.388	0.484	13.209		56, 10
2.246				11.926		Station C
	2.305		−0.059	11.867		60, 40
	1.598		0.707	12.574		60, 30
	1.168		0.430	13.004		59, 20
	1.875		−0.707	12.297		50, 30
	1.363		0.512	12.809		50, 20
	2.562		−1.199	11.610		50, 37
	2.285		0.277	11.887		40, 20
	2.672		−0.387	11.500		40, 30
	1.040		1.632	13.132		50, −6
	1.200		−0.160	12.972		40, −6
	1.370		−0.170	12.802		30, −6
	1.535		−0.165	12.637		20, −5
	1.905		−0.370	12.267		10, −5
	2.215		−0.310	11.957		0, −5
		4.432	−2.217	9.740		−7, 20

Table 18.8

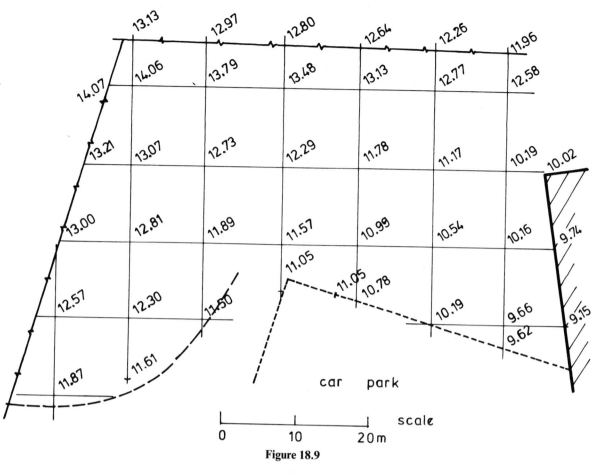

Figure 18.9

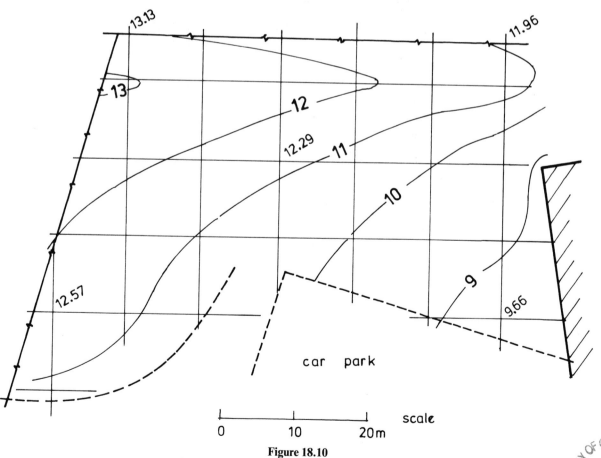

Figure 18.10

BS	IS	FS	Rise or fall	Reduced level	Distance	Remarks
0.534				12.186		BM E
	0.400		0.134	12.320	0.000	MH 2
	0.910		−0.510	11.810	5.000	Centre line
	1.270		−0.360	11.450	10.000	Centre line
	1.780		−0.510	10.940	15.000	Centre line
	2.080		−0.300	10.640	20.000	Centre line
		2.350	−0.270	10.370	25.000	MH 1

Table 18.9

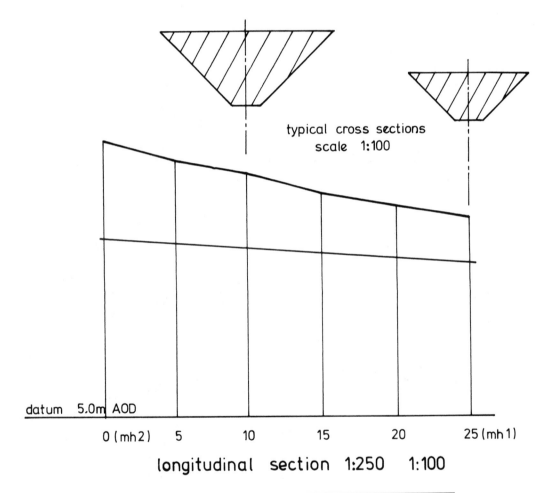

typical cross sections
scale 1:100

datum 5.0m AOD

0 (mh2)   5   10   15   20   25 (mh1)

longitudinal section 1:250   1:100

Chainage	0	5	10	15	20	25
Ground level	12.320	11.810	11.450	10.940	10.640	10.370
Invert level	9.725	9.600	9.475	9.350	9.225	9.100
Cut	2.595	2.210	1.975	1.590	1.415	1.270

Figure 18.11

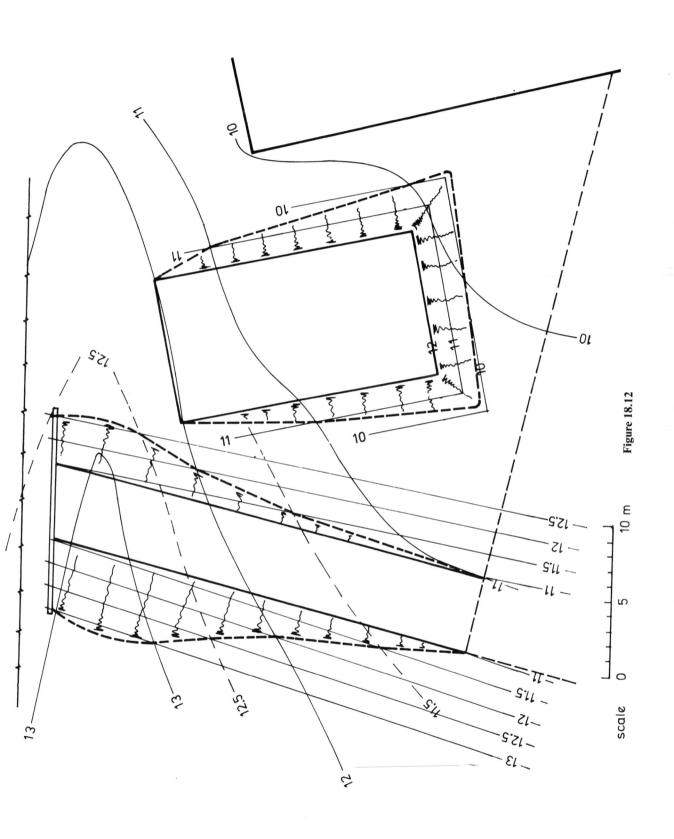

Figure 18.12

scale

10 m

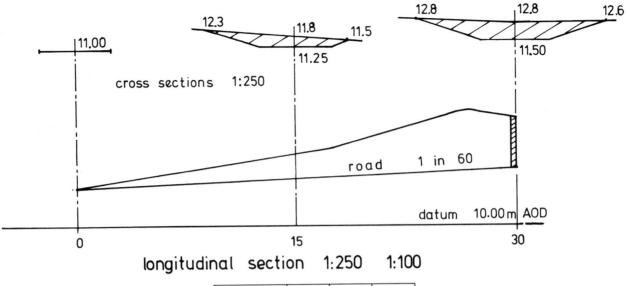

cross sections   1:250

road   1 in 60

datum   10.00 m AOD

longitudinal section   1:250   1:100

Chainage	0.00	15.00	30.00
Ground level	11.00	11.25	11.50
Road level	11.00	11.80	12.80
Cut (centre)	0.00	0.55	1.30

Figure 18.13

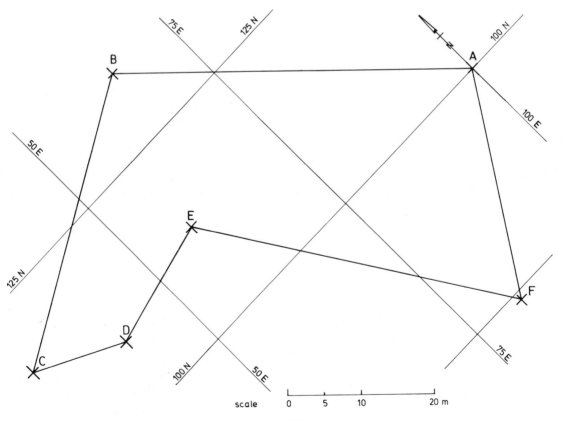

Figure 18.14

(1)

Instrument station	Station observed	Face left	Face right	Mean observed angle	Line	Face	Zenith angle	Measured length			Remarks
								Rear	Forward	Length	
A	F	00° 00′ 00″	180° 00′ 00″					0.100	49.513	49.413	
	B	100° 52′ 40″	280° 52′ 20″	100° 52′ 30″	AB	L	88° 28′ 00″	0.150	49.563	49.413	
										49.413	
B	A	00° 00′ 00″	180° 00′ 20″					0.150	41.300	41.150	
	C	106° 29′ 00″	286° 30′ 00″	106° 29′ 20″	BC	L	93° 09′ 00″	0.200	41.344	41.144	
										41.147	
C	B	00° 00′ 00″	179° 59′ 40″					. 0.370	13.820	13.450	
	D	56° 42′ 20″	236° 42′ 20″	56° 42′ 30″	CD	L	89° 59′ 00″	0.100	13.550	13.450	
										13.450	
D	C	00° 00′ 00″	180° 00′ 00″					0.230	17.820	17.590	
	E	138° 12′ 00″	318° 12′ 00″	138° 12′ 00″	DE	L	89° 06′ 00″	0.115	17.705	17.590	
										17.590	
E	D	00° 00′ 00″	180° 00′ 20″					0.200	46.340	46.140	
	F	251° 25′ 40″	71° 25′ 40″	251° 25′ 30″	EF	L	93° 34′ 00″	0.150	46.290	46.140	
										46.140	
F	E	00° 00′ 00″	180° 00′ 00″					0.370	31.904	31.354	
	A	66° 17′ 20″	246° 17′ 00″	66° 17′ 10″	FA	L	83° 45′ 00″	0.050	31.584	31.354	
										31.354	
			Sum =	719° 59′ 00″							

Table 18.10

(2) (3)

BS	IS	FS	Horizontal angle	Line	Forward bearing
—	—	—	— — —	AB	315° 0′ 0″
A	B	C	106° 29′ 30″	BC	241° 29′ 30″
B	C	D	56° 42′ 40″	CD	118° 12′ 10″
C	D	E	138° 12′ 10″	DE	76° 24′ 20″
D	E	F	251° 25′ 40″	EF	147° 50′ 0″
E	F	A	66° 17′ 20″	FA	34° 7′ 20″
F	A	B	100° 52′ 40″	AB	315° 0′ 0″

Table 18.11

(3)

Line	Slope length	Zenith angle	Plan length
AB	49.413	88° 28′ 0″	49.395
BC	41.147	93° 9′ 0″	41.085
CD	13.450	89° 59′ 0″	13.450
DE	17.590	89° 6′ 0″	17.588
EF	46.140	93° 34′ 0″	46.050
FA	31.534	83° 45′ 0″	31.347

Table 18.12

(4)   (5)

Line	Length	Whole circle bearing	Difference in eastings	Difference in northings	Correction to difference in east	Correction to difference in north	Corrected difference in east	Corrected difference in north	Easting	Northing	Station
									100.000	100.000	A
AB	49.395	315° 0′ 0″	−34.928	34.928	−0.005	−0.016	−34.932	34.911	65.068	134.911	B
BC	41.085	241° 29′ 30″	−36.103	−19.609	−0.004	−0.013	−36.107	−19.623	28.961	115.289	C
CD	13.450	118° 12′ 10″	11.853	−6.356	−0.001	−0.004	11.852	−6.361	40.813	108.928	D
DE	17.588	76° 24′ 20″	17.095	4.134	−0.002	−0.006	17.094	4.128	57.906	113.056	E
EF	46.050	147° 50′ 0″	24.516	−38.981	−0.004	−0.015	24.512	−38.996	82.418	74.060	F
FA	31.347	34° 7′ 20″	17.584	25.950	−0.003	−0.010	17.582	25.940	100.000	100.000	A
	198.915	Errors	0.018	0.065	−0.018	−0.065	0.000	0.000			
		Corrections	−0.018	−0.065							

Correction $(k_1) = -9.19 \times 10^{-5}$
Correction $(k_2) = -3.257 \times 10^{-4}$
Accuracy of survey is 1 in 2955

Table 18.13

Instrument station   F
Instrument height   1.3
Reduced level   9.28

Observation station	Horizontal reading	Zenith angle	Stadia readings			S	Plan length	Vertical height	Difference in height	Reduced level
---------------------	--------------------	--------------	Top	Mid	Bottom	---	-------------	-----------------	----------------------	---------------
			Top	Mid	Bottom					
E	0° 0′ 0″									
T1	355° 31′ 0″	86° 41′ 20″	1.449	1.300	1.151	0.298	29.70	1.72	1.72	11.00
T2	356° 29′ 0″	87° 11′ 0″	1.476	1.300	1.124	0.352	35.12	1.73	1.73	11.01
T3	2° 1′ 0″	86° 39′ 0″	1.393	1.300	1.207	0.186	18.54	1.09	1.09	10.37

Table 18.14

Tri-dimensional coordinates

	Instrument station	B		Backsight station	A
	Instrument height	1.35			
	Easting	65.07			
	Northing	134.91			
	Reduced level	14.18			

Target station	Target height (m)	Slope length (m)	Whole circle bearing	Zenith angle	Easting (m)	Northing (m)	Reduced level (m)	Target station
C	1.35	41.16	241° 28′ 45″	93° 8′ 20″	28.96	115.29	11.93	C
D	1.35	35.62	223° 1′ 50″	93° 37′ 25″	40.81	108.93	11.93	D
E	1.35	23.08	198° 8′ 40″	94° 57′ 29″	57.91	113.05	12.19	E
F	1.35	63.46	164° 5′ 10″	94° 25′ 40″	82.42	74.06	9.29	F

Table 18.15

Section 9

1. Chalet block

Line	Length	Bearing	Angle	Clockwise value
A–B	49.385	315° 00′ 00″		
B–H1	11.069	123° 30′ 12″	A–B–H1	348° 31′ 12″
H1–H2	20.000	241° 38′ 28″	B–H1–H2	298° 08′ 16″

2. Recreation area

B–A	49.385	135° 00′ 00″		
A–R1	5.315	274° 18′ 58″	B–A–R1	319° 18′ 58″
R1–R2	18.000	214° 50′ 10″	A–R1–R2	120° 31′ 12″

3. Jogging track

F–E	46.050	327° 50′ 51″		
E–I1	16.024	328° 20′ 36″	F–E–I1	180° 29′ 45″
I1–I2	22.777	61° 24′ 35″	E–I1–I2	273° 03′ 59″
I2–I3	48.656	135° 59′ 58″	I1–I2–I3	254° 35′ 23″
I3–I4	38.720	213° 11′ 51″	I2–I3–I4	257° 11′ 53″
I4–F	3.860	04° 42′ 35″	I3–I4–F	331° 30′ 43″
I4–I1	65.230	330° 0′ 56″		

Section 10

$$\text{Length Z–I1} = 27.000 \qquad 27.000 \text{ m}$$
$$\text{Tangent length I1–T1} = R \tan[(91° 23′ 39″)/2] = \underline{10.246}$$
$$\text{Chainage T1} = (27.000 - 10.246) \qquad = 16.754 \text{ m}$$
$$\text{Curve T1–T2} = 2\pi R \times \theta/360 \qquad = \underline{15.951}$$
$$\text{Chainage T2} = \qquad \underline{32.705 \text{ m}}$$

Similarly for remaining curves and straights resulting in:

Point	Z	T1	T2	T3	T4	T5	T6	T7	T8	Z
Chainage	0.000	16.754	32.705	37.619	50.637	75.711	102.6588	109.1522	129.5411	151.510

Length of jogging track = final chainage Z = 151.510 m

Long chord	Length(L) = (2R sin θ/2)	Offset interval = L/6	Offsets 1	2	3(mid)	4	5
T1–T2	14.313	2.386	1.805	2.728	3.015	2.728	1.805
T3–T4	12.118	2.020	1.193	1.839	2.045	1.839	1.193
T7–T8	17.036	2.839	2.993	4.350	4.759	4.350	2.993

Chord	Length (m)	Chainage (m)	Deflection angle	Tangential angle
T5	—	0.000		00° 00′ 00″
1	3.000	3.000	4° 18′ 4″	4° 18′ 4″
2	3.000	6.000	4° 18′ 4″	8° 36′ 8″
3	3.000	9.000	4° 18′ 4″	12° 54′ 12″
4	3.000	12.000	4° 18′ 4″	17° 12′ 16″
5	3.000	15.000	4° 18′ 4″	21° 30′ 20″
6	3.000	18.000	4° 18′ 4″	25° 48′ 24″
7	3.000	21.000	4° 18′ 4″	30° 06′ 28″
8	3.000	24.000	4° 18′ 4″	34° 24′ 32″
9 T6	2.947	26.947	4° 13′ 30″	38° 38′ 02″
	26.947	Total	38° 38′ 02″	

Note. Final angle should be 38° 35′ 57″
Angular error = 00° 02′ 06″
Displacement error of peg T6 = 0.016 m

Section 11

1. Boat storage area = 178 m^2
2. Area enclosed by traverse coordinates = 1570.8 m^2
3. Volume of material removed from drain trench = 122.2 m^3
4. Surface area of roadway cutting = 275.9 m^2
   Tarmac area of roadway = 150 m^2
   Actual area of side slopes = 132.7 m^2
5. Volume of material excavated from roadway cutting = 136.1 m^3
6. Recreation area
   Area of 12 m contour plane = 180 m^2
   Area of 11 m contour plane = 178 m^2
   Area of 10 m contour plane = 9 m^2
   Volume of material in earthworks = 301 m^3

Index